Welding and Other Joining Processes

Welding and Other Joining Processes

Roy A. Lindberg
University of Wisconsin

Norman R. Braton
University of Wisconsin

Allyn and Bacon, Inc.   Boston · London · Sydney

Library of Congress Cataloging in Publication Data

Lindberg, Roy A.
　　Welding and other joining processes.

　　Includes bibliographies and index.
　　1.　Welding.　　2.　Metal bonding.　　3.　Fasteners.
I.　Braton, Norman R., joint author.　II.　Title.
TS227.L58　　　　671.5　　　　75-30745
ISBN 0-205-05000-X

Contents

Preface

Specialization versus integration are ever present pedagogical problems. This book is written with a view toward integration of knowledge and at the level of the beginning engineer. Welding has traditionally been considered a separate entity with only scant attention given other materials joining methods. This is understandable since welding has fulfilled a more universal need and has an organized body of information. Now that a rapid technical and scientific advancement has occured in other areas such as adhesives and mechanical fasteners, they can no longer be legitimately ignored.

In this book the student is challenged to make process comparisons between whatever joining methods may be applicable. Advantages and disadvantages of each method are opened for examination and are also considered for economic aspects. The concept of materials joining is approached from the standpoint of energy required to produce a bond and then proceeds into each of the various processes.

Traditionally, books of this nature have dealt largely with the "how" rather than the "why" of bonding. The problems given at the end of each chapter require the student to make decisions based on his understanding of why one process may be chosen over another and thereby draw from his knowledge of the basic theory involved.

The authors wish to acknowledge the cooperation of many companies and professional societies for supplying information and making illustrations available.

Particular thanks is due Dr. Richard Moll of the University of Wisconsin Engineering Extension Department for his review of the chapter, "Metallurgy of Welding"; to Omer W. Blodgett, Design Consultant, The Lincoln Electric Company, Cleveland, Ohio, for his review of the chapter, "Weld Design"; and the staff at the American Welding Society headquarters in Miami, Florida, for review and permission to use illustrations.

R. A. Lindberg
N. R. Braton

Welding and Other Joining Processes

1

Introduction to Materials Joining

Today's emphasis on reliability, speed, and economy in assembly has led to the development of a wide variety of joining processes. Less than fifty years ago, aluminum parts joining, other than with mechanical fasteners, was considered difficult. There are now more than a score of welding processes alone that can be used.

If the design or manufacturing engineer is to make best use of the current processes, he must be thoroughly familiar with the joining options available. He must be able to assess which processes are applicable and weigh the advantages and disadvantages of each. As an example, many design applications specify the joining of dissimilar metals such as titanium to stainless steel, aluminum to stainless steel, beryllium to inconel, titanium to carbon steel, and so forth. Of course, mechanical fasteners or adhesives can be used, but if welding is specified, metallurgical interactions are likely to be encountered.

Shown in Fig. 1–1 are a variety of common joints for which each of the five main joining processes shown in Table 1.1 may be considered. In this chapter each of the joining processes will be discussed briefly to gain an overall view of its use and interrelationships.

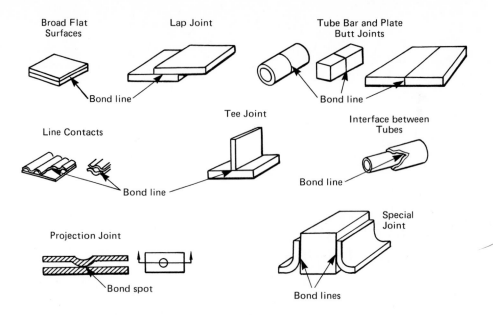

FIGURE 1–1. *A variety of joining problems, most of which will have several solutions; e.g., dependent upon permanence, stresses, equipment availability, environmental conditions to be encountered, and many other factors.*

WELDING

Although there are more than forty welding processes used in industry today, only a handful are of real industrial significance. As shown in Table 1.1, the three main types are gas welding, arc welding, and resistance welding. These processes are based on the fusion joining of metal, or what is essentially a small scale

TABLE 1.1. *Basic Joining Processes*

Welding	Brazing	Soldering	Adhesives	Mechanical Fasteners
Gas	Torch	Torch	Thermosetting	Screws
Oxyacetylene	Furnace	Dip	Thermoplastic	Bolts
Arc	Resistance	Resistance		Studs
Shielded metal arc	Induction	Furnace		Nuts
Submerged arc				Threaded inserts
Gas tungsten arc				Washers
Gas metal arc				Pins
Flux cored arc				Rivets
Plasma				Retaining rings
Resistance				Welded fasteners
Spot				Special purpose fasteners
Seam				

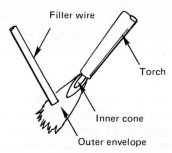

FIGURE 1–2. *The heat of fusion for gas welding is provided by a combustible gas such as acetylene. Oxygen is added to increase the rate of combustion, producing a flame of about 6000 °F (3316 °C).*

casting operation. The workpieces are melted along the common edges or surfaces so that the molten metal, usually combined with filler metal, forms a *weld melt* or puddle. The pieces are fused or welded together when the weld melt solidifies.

Gas Welding. The heat of fusion is supplied by a combination of a combustible gas such as acetylene with oxygen, Fig. 1–2. Welding skills for this process are fairly easy to master and the equipment is relatively inexpensive, but the process is slow compared to other modern welding methods. Gas welding is normally confined to repair and maintenance work rather than mass-production techniques. Gas welding equipment may be used for a variety of operations such as soldering, brazing, oxygen cutting, stress relieving, and surfacing.

Arc Welding. The most widely used form of welding today is arc welding. The heat is generated by an electric arc that is maintained in most cases between the electrode (rod) and the workpiece. The arc furnishes sufficient heat to melt the base metal in the immediate vicinity of the arc and usually the electrode. The subsequent cooling of the fused metal is very rapid.

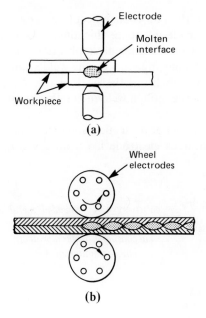

FIGURE 1–3. *Resistance welding. The interface between the two surfaces of the workpiece offers the greatest resistance to the current flow and is the point of the greatest current density and heat concentration, as shown in spot welding (a). Another variation of resistance welding is seam welding (b).*

3

Resistance Welding. In resistance welding, heat is generated by the resistance to the flow of a high amperage current across the interface of two mating surfaces, see Fig. 1–3. The heat generation is directly proportional to the resistance and the greatest current density offered by any point in the circuit. Since the interface of the joining surfaces is the point of greatest resistance in the circuit, it is usually the point of greatest heat generation. Fluxes or shielding are not required because the molten metal is not exposed to the atmosphere. The process is widely used in industry, mainly for mass-production work. A variation of resistance welding is seam welding. Wheel electrodes are used that can produce overlapping spots for a leak-proof seam.

ARC WELDING VARIATIONS

Originally, arc welding was done by means of two carbon electrodes. The welds tended to be brittle because of carbon contamination from the electrodes. The next technical advance, using one bare wire and maintaining an arc between the base metal and the electrode, eliminated the carbon problem, but gases were absorbed from the atmosphere. Finding methods to protect the molten metal from the atmosphere, particularly oxygen and nitrogen, led to several distinctly different variations of arc welding.

Shielded Metal Arc. One approach to shielding is the use of a chemical coating or flux applied to the electrode. The coating vaporizes from the heat of the arc and forms a cloud of protective gas around the molten metal, Fig. 1–4. This form of shielded metal arc welding is very common in industry today. Since the shielded metal arc electrode is a sticklike rod, about 14 in. long, it is often referred to as "stick electrode welding."

Submerged Arc Welding. Although the flux-coated electrode made arc welding a major industrial tool, it did not provide adequate shielding, appropriate welding metallurgy, or rapid enough metal deposition for all applications. The submerged arc process provides a mound of granular minerals on the weld seam ahead of the electrode, Fig. 1–5. The granular flux melts in the heat of the arc and forms a pool of molten slag floating on top of the molten weld metal, thus protecting it from atmospheric gases.

In addition to the advantages of better protection and higher deposition rates, the submerged process has a continuous electrode fed from a storage reel. Also,

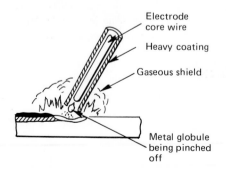

Electrode core wire

Heavy coating

Gaseous shield

Metal globule being pinched off

FIGURE 1–4. *Shielded metal arc or "stick electrode" welding. Vaporization of the chemical coating provides a protective gaseous shield around the molten metal.*

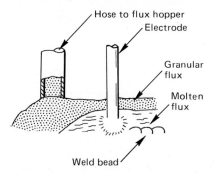

Hose to flux hopper
Electrode
Granular flux
Molten flux
Weld bead

FIGURE 1–5. *The submerged arc welding process provides a mound of granular mineral on the weld seam ahead of the electrodes. It is melted by the arc, forming a slag that floats over the weld deposit.*

the process can be made either semiautomatic or fully automatic. Although the wire electrode is fed automatically, the gun may be made for hand manipulation, much like an ordinary stick electrode, for semiautomatic welding. In other designs the gun is mounted on a carriage to guide the head automatically.

The high current density makes it possible to obtain good penetration at relatively high speeds. On heavy plate, where considerable weld deposit is required, tandem electrodes may be used. The process is limited to the common ferrous metals. Because of the large amount of molten metal, most of the welding is limited to the flat position.

Gas Tungsten Arc (GTA) Welding. Although the submerged arc system works well on ferrous metals, the shielding is unsatisfactory for more reactive metals such as aluminum and magnesium. By the late 1930s, engineers found inert gases that provide more adequate shielding. Eventually the switch was made to a nonconsumable tungsten electrode with helium for a shielding gas. Filler metal is supplied by a separate wire that does not carry electric current, Fig. 1–6. This process became known as the tungsten inert gas process or TIG welding. This term has now given way to the gas tungsten arc or the GTA process. Several different gas mixtures can be used and will be discussed in detail in Chapter 8.

The GTA welding process has the advantage of practically no postweld cleaning. The arc and the weld pool are always clearly visible to give the weldor maximum control. Since no metal is transferred through the arc, the process can be used in all positions. It does, however, have one serious drawback; it cannot be readily used to join materials that are both thick and highly conductive, such as thick plates of aluminum or magnesium. These materials form large heat sinks.

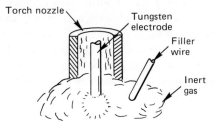

Torch nozzle
Tungsten electrode
Filler wire
Inert gas

FIGURE 1–6. *Gas tungsten arc welding (GTA).*

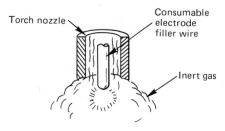

Torch nozzle

Consumable
electrode
filler wire

Inert gas

FIGURE 1-7. *Gas metal arc welding (GMA).*

Gas Metal Arc (GMA) Welding. By the late 1940s, technical problems had been overcome sufficiently to allow substitution of a consumable electrode for the nonconsumable tungsten electrode, Fig. 1-7. Since the center of melting is directly at the tip of the electrode, thermal conduction is less of a problem. The new process was first termed metal inert gas welding, or MIG, but it is now usually referred to as gas metal arc or GMA welding.

Initially the GMA process was based largely on argon gas mixtures, but the search for less costly gases led to carbon dioxide and mixtures of argon and carbon dioxide. Processes based primarily on carbon dioxide shielding are often referred to as CO_2 welding. GMA welding is now considered a simple, fast alternative to nearly all other commerical processes rather than merely being a means to weld thick thermally conductive plate. GMA provides faster, lower cost welding; while GTA provides better metallurgical properties in more critical applications, particularly in the nonferrous materials and the nickel base alloys.

Flux-Cored Electrodes. Shielding gases are somewhat expensive for simpler forms of welding, and submerged arc is limited to flat or nearly flat work. These problems stimulated the search for a process that would be simple, tidy, and low-cost enough for widespread use on common grades of steel. A solution to the problem was reached in the late 1950s when a chemical flux similar to that used on the outside of stick electrodes was placed inside a hollow continuous wire electrode, Fig. 1-8. In some versions of the flux-cored arc welding process, the electrode alone provides all the shielding, termed "self shielding," and in other versions supplementary CO_2 gas is used. In one sense the flux-cored process can be thought of as an automated version of stick electrode welding. If suplementary shielding gas is used, it can be considered a variation of the GMA process.

Plasma Arc Welding. Plasma arc welding basically resembles GTA welding, except some of the shielding gas, or sometimes a special *plasma gas*, is forced directly through the arc (instead of *just bathing* it) to form an extremely hot plasma. Plasma refers to a gas that becomes ionized. This occurs during any arc welding operation; however, when the term *plasma arc* is used, it has become associated with the processes employing a constricted arc. Arc constriction is brought about by forcing the gas to pass through a small nozzle or opening from the electrode to the workpiece, Fig. 1-9. The arc and plasma is thereby forced to take a columnar shape that is characteristic of plasma arcs.

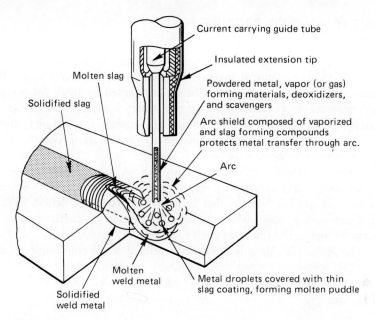

FIGURE 1–8. *Flux-cored arc welding.* (*Courtesy of Welding Design and Fabrication.*)

The plasma process is superior to ordinary GTA welding for extremely thin to moderately thick materials. Costs are often less due to the higher travel speeds that can be used with plasma welding. As an example, $\frac{1}{8}$ in. (3.2 mm) thick stainless steel tubing may be welded with the GTA process at 22 ipm (55.88 cm), whereas the plasma process can be used at 36 ipm (91.44/cm/min), or an increase in speed of 64%. (Metric conversion factors are provided in Table 1.A at the end of the chapter.)

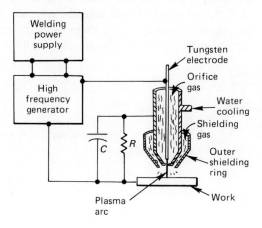

FIGURE 1–9. *Basic plasma welding. The high frequency current is superimposed on the regular welding current to lend greater stability to the arc.*

CHOOSING THE WELDING PROCESS

Manual welding, or stick electrode welding, is one of the most widely used processes simply because it involves very little capital investment. Other benefits include using in any position and being well suited to small lot production, such as in construction and one-of-a-kind machinery building that occupy so much of industry. Large jobs such as shipbuilding, pipeline fabrications, and the use of heavy plate tend to favor semiautomatic and automatic processes wherever labor costs become an important factor.

The gas-shielded processes are normally required for aluminum, magnesium, and titanium, or joints where high quality or extremely close control over metallurgical qualities are needed in the weld. To some extent, GTA is thought of as a premium process, where higher than normal properties are provided at the expense of higher than normal costs.

Resistance welding is normally confined to assembly-line type operations, since the tooling and equipment required are more easily justified under these conditions. The process is widely used on production lines for building automobiles and appliances.

BRAZING

Brazing includes a group of processes for joining metal parts by heating them to above 800 °F (427 °C), using a filler metal whose melting temperature is below that of the base metal. The filler metal flows into the joint by capillary action.

Torch Brazing. Torch brazing is done with a fuel gas flame to which oxygen is added. The flame is adjusted to have a slight excess of fuel gas, usually acetylene, and is not pointed directly at the joint opening, Fig. 1–10. Basically, all joints are simple lap or butt joints although many variations are used to satisfy specific requirements. Torch brazing is used on relatively small assemblies made from materials that do not oxidize at the brazing temperature or that can be protected from oxidation with flux. Filler metals include the aluminum–silicon alloys and

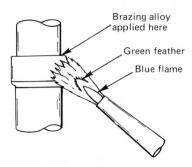

Brazing alloy applied here

Green feather

Blue flame

FIGURE 1–10. *Torch brazing used to join two sections of pipe.*

most silver base and copper zinc alloys; a flux is required with these alloys. Self-fluxing copper phosphorus alloys are also used for torch brazing.

Furnace Brazing.   Larger assemblies, and those that cannot be self-jigged, may be furnace brazed at high production rates. Furnace brazing is usually done in a protective atmosphere, although an ordinary furnace can be used if flux is applied to the joints. The type of atmosphere required depends upon the materials being brazed and the filler metals being used. Heat-resistant, high strength alloys that contain appreciable amounts of aluminum or titanium are brazed in a vacuum to prevent the formation of oxides, which inhibit wetting and flow of the filler metal.

Resistance brazing, induction brazing, quartz-lamp brazing and infrared brazing are other process variations used to a lesser extent than furnace brazing.

Brazing Advantages.   The advantages of brazing may be stated briefly as follows:

1. Materials of different thicknesses can be joined easily.
2. Brazed joints require little or no finishing.
3. There is less danger of changing the metallurgical structure of the joint than with fusion welding.
4. Complex assemblies can be brazed in several steps by using nonferrous filler metals with progressively lower melting points.
5. Dissimilar metals can be joined.
6. Assemblies can be brazed with a stress-free condition.

However, care must be exercised in selecting metals that are compatible. Reactions may occur during the brazing cycle that will produce undesirable intermetallic compounds, as may occur with aluminum and steel or titanium and steel.

Brazing is extensively used as a fabrication and repair technique for a wide variety of metals. Since a lower temperature is used, materials can be joined in less time and with less metallurgical affect than in welding.

SOLDERING

Soldering is a process that uses nonferrous, low melting-point [under 800 °F (427 °C)] alloys to join metal components. The molten solder alloys fill the space between the surfaces to be joined, adhere to these surfaces, and solidify. Solders adhere to the metal surfaces with a strength comparable to that of cohesive forces in the solder itself. The metal surface (of solderable metals) and the tin in the solder react to form a layer of intermetallic compound that permanently wets the surface.

Tin–lead alloys comprise the largest single group and are the most widely used of the soldering alloys. They are compatible with all types of base metal cleaners, fluxes, and heating methods. Most metals can be joined with these alloys. Since the solder is drawn in by capillary action, a joint clearance of

0.003 (0.08 mm) ±0.0005 in. (±0.013 mm) is recommended. Many different methods may be used to transmit heat to the metal surface or joint. The most common method is with a soldering iron, but other methods include the use of a torch, electrical resistance, hot plate, oven, and induction heating.

ADHESIVE BONDING

Adhesives such as rubber cement, household glues, Scotch™ tape, and epoxy repair kits are familiar to everyone. Some adhesives have been around a long time. Clays, rich in aluminum oxide, and silica mixed with water to make a paste, then air dried to form a solid bond, were used as early as 5400 B.C. Today the bulk of industrial adhesives are synthetics made of neoprene and butyl rubbers, petroleum resins, epoxies, phenolics, etc.

Adhesives are usually classified as *thermoplastic* or *thermoset*. A thermoplastic material will repeatedly soften when heated and harden again when cooled. The thermoset material undergoes a chemical reaction when subjected to heat, catalysts, ultraviolet light, and so forth, leading to a relatively infusible state.

Structural Adhesives. Engineers are mainly concerned with *structural adhesives*, adhesives capable of joining structural materials and maintaining the same resistance to specific stresses as the material itself. This definition is more the "ideal" than what is presently available. As shown in Table 1.2, lap shear strengths range from 2,000 psi (13.79 MPa) to 6,000 psi (41.37 MPa) at room temperature. Structural adhesives are usually thermosetting, although thermoplastics can be used when low strengths are satisfactory and high temperatures are not a factor.

Joint Design. Adhesive bonding should be tailored for the particular application; adapting joint designs from welding or riveting will often not be satisfactory. As much as possible, the principle stress in the joint should be placed in shear (or tension), Fig. 1–11. Most adhesives are weak in peel and cleavage, Fig. 1-12. When a joint is placed in shear, the total effective area is contributing to resist stress. To design a joint so that all of the load will be taken in pure tension stresses is quite difficult. Usually some peel and cleavage forces will be encountered.

Simple lap joints are often adequate for many applications. A beveled lap joint and a beveled double-butt lap joint are both excellent in minimizing all types of stresses, Fig. 1–13. The bevel lap joint is, of course, more expensive to produce.

Surface Preparation. The surface preparation prior to application of the adhesive may range from a simple solvent wipe, complete in seconds, to a multistage cleaning and chemical treatment requiring thirty minutes or more. The purpose is to produce sound surface layers on the parts to be joined, sound in the sense

---

™ Trademark of the 3M Company.

TABLE 1.2 *Classification of Structural Adhesives**

Chemical Type	Cure Temperature °F	(C°)	Service Temperature °F	(°C)	Lap Shear Strength psi @ °F	(MP @ °C)	Peel Strength at Room Temp. lb/in.	(kg)
Epoxy, formulated for room temp. cure	60 to 90	(15.6–32.2)	−67 to 180	(19.4–82)	2500 @ RT **(17.23) / 1500 @ 180	(10.34 @ 82)	5	(2.27)
Epoxy, formulated for elevated temp. cure	200 to 350	(93–177)	−67 to 350	(19.4–177)	2500 @ RT (17.23) / 1500 @ 350	(10.34 @ 177)	5	(2.27)
Epoxy–nylon	250 to 350	(121–177)	−423 to 180	(218–82)	6000 @ RT (41.37) / 2000 @ 180	(13.79 @ 82)	75	(34.02)
Epoxy–phenolic	250 to 350	(121–177)	−423 to 500	(218–260)	2500 @ RT (17.23) / 1500 @ 500	(10.34 @ 260)	10	(4.53)
Butyral–phenolic	275 to 350	(134–177)	−67 to 180	(19.4–82)	2500 @ RT (17.23) / 1000 @ 180	(6.89 @ 82)	10	(4.53)
Neoprene–phenolic	275 to 350	(134–177)	−67 to 180	(19.4–82)	2000 @ RT (13.79) / 1000 @ 180	(6.89 @ 82)	15	(6.80)
Nitrile–phenolic	275 to 350	(134–177)	−67 to 250	(19.4–121)	4000 @ RT (27.37) / 2000 @ 250	(13.79 @ 121)	60	(27.22)
Urethane	75 to 250	(23.9–121)	−423 to 180	(218–82)	2500 @ RT (17.23) / 1000 @ 180	(6.89 @ 82)	50	(22.68)
Polyimide	550 to 650	(288–343)	−423 to 1000	(218–538)	2500 @ RT (17.23) / 1000 @ 1000	(6.89 @ 538)	3	(1.37)

* Courtesy of *Machine Design*.

** RT means "reference temperature."

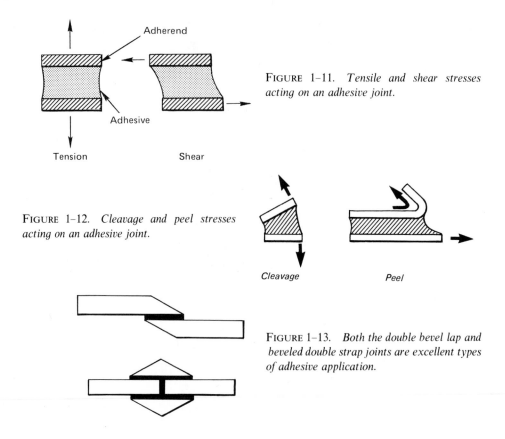

FIGURE 1–11. *Tensile and shear stresses acting on an adhesive joint.*

FIGURE 1–12. *Cleavage and peel stresses acting on an adhesive joint.*

FIGURE 1–13. *Both the double bevel lap and beveled double strap joints are excellent types of adhesive application.*

that the surface layers will not be the weak portions of the joint. Solvent wipes or even acid etches do not produce surfaces free of contamination, but will remove grease and oil and reduce the oxide layer.

ADVANTAGES AND DISADVANTAGES OF ADHESIVES

Advantages. They can be used to join any combination of similar or dissimilar materials. Thick and thin sections of any shape may be joined. Adhesives provide for a uniform distribution of stress on the joint with a larger stress-bearing area. Electrochemical corrosion is eliminated or minimized between dissimilar metals. Adhesive joints resist fatigue. The joints form a seal, which also insulates (heat and electricity) and helps damp out vibration. The setting heat is usually too low to affect the strength of the metal parts. Post assembly clean-up of parts is not difficult. Adhesive bonding is frequently faster and cheaper than conventional fastening.

Disadvantages. Careful preparation is required. Relatively long periods of time are sometimes required for setting the adhesive. The upper service temperature is about

350 °F (177 °C) for most adhesives, but some high temperature types such as the alumina-base materials have service temperatures in the range of 3,000 °F (1649 °C). Heat and pressure may be required to set the adhesive. Natural or vegetable origin materials are subject to attack by bacteria, mold, rodents, and vermin. These and other adhesives may also be attacked by moisture, solutions, and organic solvents.

OTHER ADHESIVE TYPES

Anerobic. Some adhesives have unique cure mechanisms. One such type, anerobic, remains fluid in the presence of oxygen but cures to a solid in the absence of it. Curing occurs in several hours at room temperature, but in a few minutes at slightly elevated temperatures. Anerobics are especially useful in securing and sealing threaded fasteners. Disassembly is possible by overcoming the torque strength of the adhesive or by heating to 450 °F (232 °C) and disassembling while hot.

Cyanoacrylates. Cyanoacrylates have an extremely short cure time (10 to 15 sec) if confined to a thin film between close fitting parts. The mechanism is one of an anionic polymerization induced by the presence of a base. A thin film of moisture usually present on surfaces exposed to the normal atmosphere is sufficient to harden the adhesive when squeezed to a thin film.

ADHESIVE SELECTION FACTORS

Some factors used in selecting an adhesive are:

1. Time and temperature required for cure.
2. Strength developed.
3. Environmental conditions subjected to:
   Temperature changes in the extremes of its range
   Vibration
   Loading stresses
4. Ease and form of application.

MECHANICAL FASTENING

Fasteners are produced in an almost infinite variety. One fastening innovation begets another until it becomes difficult to keep informed of all the types available for a given joining operation. The purpose of the following brief discussion is not to uncover all the various types available, but rather to provide some means of classification and reference to the bibliography where more detailed information can be found.

Classification. Fasteners may be classified into five broad categories as follows:

1. Threaded fasteners.
2. Rivets.

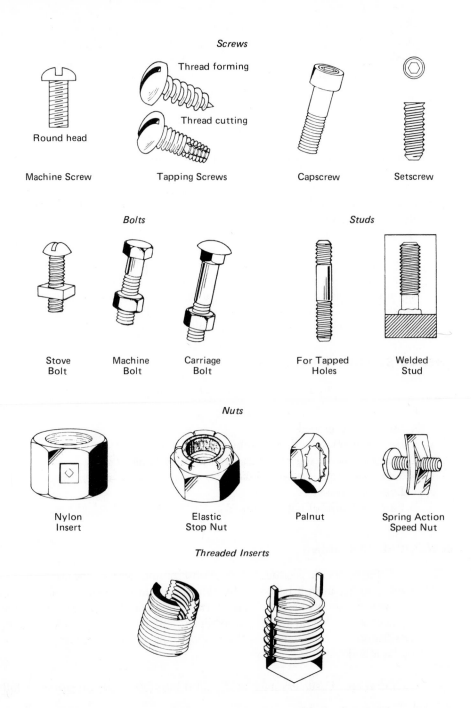

FIGURE 1-14. *Examples of thread fasteners. (Courtesy of American Machinist.)*

3. Pin fasteners.
4. Retaining rings.
5. Quick operating fasteners.

Threaded Fasteners.   Threaded fasteners include screws, bolts, studs, nuts, and threaded inserts, Fig. 1–14. Screws are generally regarded (in metal working) as fastening devices for use in tapped holes, but they also may be used with nuts, as in the case of machine screws and stove bolts. Screws, in general, are considered to be smaller than bolts; however, they may be as large as one inch in diameter.

Many varieties of nuts have been developed in recent years, most of which provide some type of locking or friction holding different from the thread itself. Other designs allow the nut to spin on but grip tightly when seated on the surface by a pinching action on the threads.

Standard studs are threaded on both ends and provide a blank space in the center to grip and turn into a tapped hole. Industry frequently uses resistance welds to attach studs to base plates.

Threaded inserts are especially useful in providing high strength threads in soft materials. Inserts are made to be driven, pressed, or screwed into the base material. Most of the inserts are self-locking and are made of steel or aluminum alloy.

Rivets.   Permanent fastening is often done with rivets. Sizes are designated by the body diameter and range from those for use in bridges to those for small toys and watches. Many different head styles are available to suit special conditions. Heading is done with a die in a powered hammer, the rivet being held in place with a backup bar or anvil.

Blind rivets derive their name from the characteristic that they can be headed from only one side of an assembly. Most are made with a central shank that is pushed or pulled to upset the head using some type of power tool, Fig. 1–15. Small explosive charges are also used to head the blind side.

Pin Fasteners.   Pin fasteners are often used in place of rivets or bolts as shown in Fig. 1–16. The most common types are groove, tapered, roll, and cotter.

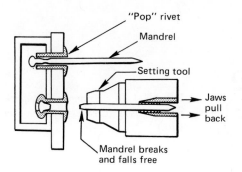

FIGURE 1–15.   *A sketch to show the installation of a blind rivet. (Courtesy of American Machinist.)*

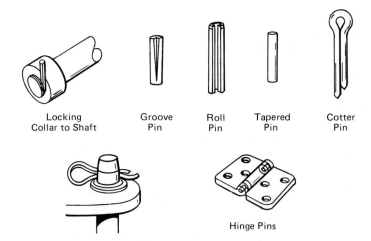

Locking Collar to Shaft	Groove Pin	Roll Pin	Tapered Pin	Cotter Pin

Hinge Pins

FIGURE 1–16. *Examples and applications of pin type fasteners.*

Retaining Rings. Retaining rings are inexpensive stampings or wire-formed circular metal discs made to fit securely into grooves, thereby acting as a stop or artificial shoulder, Fig. 1-17. They are usually applied as axial locators on shafts.

Quick Operating Fasteners. Most quick operating or "quick-release" fasteners are specialized items made to operate on or against spring pressure, Fig. 1–18. They are many, ingenious and varied. The most frequent application is for sheet metal assemblies.

FIGURE 1–17. *An example of one type of retaining ring used to hold a shaft in place.*

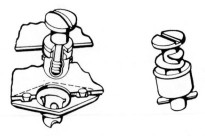

FIGURE 1–18. *Examples of quick-release type fasteners. (Courtesy of American Machinist.)*

PROCESS SELECTION SUMMARY

Selecting a joining process is based on a wide variety of factors, however, some of the main considerations are:

Strength required.   Appearance.
Type and direction of loading.   Reliability.
Ease of repair.   Ease of visual inspection.
In-place cost.

Required strength. A joint need not be stronger than the base material; in some cases it need not be as strong, as in fillet welds used on tee and lap joints welded on both sides, Fig. 1–19. Also, if unequal thicknesses are joined, the strength is expected to be that of the thinner one.

Type and Direction of Loading. The direction and type of loading is also important. This subject is discussed in more detail in Chapter 11, but some basic joints and the related stresses are shown in Fig. 1–20. Welding is recommended for the joints shown and the accompanying stresses. Adhesive bonding, brazing, soldering, and mechanical fasteners are recommended for the shear and compression joints.

Properly made butt welded joints are considered to be the same strength as the base metal. Fillet welds are required to produce a cross-section and strength equal to that of the smallest member. Properly made welds are as strong as their cross-section when at right angles to the direction in which they are loaded. This applies whether they are loaded in tension, shear, compression, torsion, or any combination.

Ease of Repair. As shown on the summary chart for joining small parts, Fig. 1–21, mechanical fasteners have the highest rating for ease of repair. (How easily defective parts can be unfastened and replaced with an effective part.) Most other joining methods rank low in repairability when one considers that the joint must be destroyed or removed before a repair can be effected. Figure 1–22 shows how a load is centered for fastening; Fig. 1–23 details threaded fastener nomenclature.

Square Tee Joint

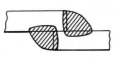
Double Fillet Lap Joint

FIGURE 1–19. *Examples of fillet welds as used on tee and lap joints.*

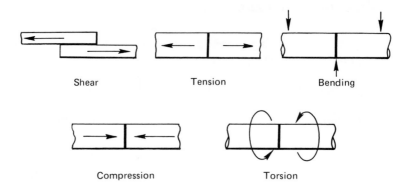

FIGURE 1–20. *Various types of stresses to which joints may be subjected.*

Joining Methods	Major Joining Qualities					
	Strength	In-place Cost	Appearance	Reliability	Visual Inspectability	Ease of Field Repair
Screws and Nuts	Best	Better	Better	Best	Best	Best
Resistance Welding	Best	Best	Good	Good	Good	Good
Arc Welding	Best	Good	Good	Best	Good	Good
Brazing	Good	Good	Best	Best	Good	Good
Riveting	Good	Good	Good	Good	Best	Good
Staking	Good	Best	Better	Good	Best	Good
Forming	Better	Best	Best	Best	Good	Good
Adhesive Bonding	Good	Good	Best	Best	Good	Good
Special Fasteners	Better	Good	Good	Best	Best	Better

Key: Good [▒] Better [ ] Best [■]

FIGURE 1–21. *A summary chart of methods used to join small assemblies and their corresponding qualities. (Courtesy of Machine Design.)*

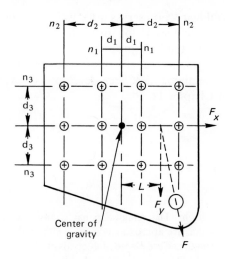

FIGURE 1–22. *A load is applied off-center to a group of equally sized fasteners.*

In-Place Cost. The cost of assembly is often regarded as a significant factor of the total product cost. Hence the axiom, "It's the in-place cost that counts, not the fastener cost." "In place" includes joint preparation, assembly machinery, labor, and overhead.

Appearance. The attractiveness of an assembly is often concerned with whether or not the joining medium is visible, and if visible, how well it blends in. Adhesive bonds are almost always invisible. Some fasteners are accentuated and made to be part of the eye appeal. Acceptable appearance levels vary greatly with a range of products and methods.

Reliability. The main concern in judging the reliability of the joining process is the likelihood of joint failure during the normal operating life of the assembly. The other concern of reliability is the repeatability or consistency of joint quality from one assembly to another.

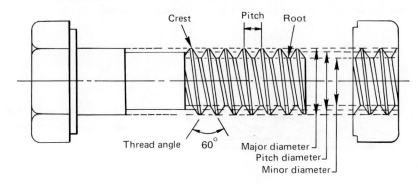

FIGURE 1–23. *Basic thread nomenclature.*

Ease of Visual Inspection. Visual inspection serves quite well for mechanical type fasteners, once they have been adequately tested for their intended application. Inspection is primarily a matter of checks that the proper installation procedure is carried out. Adhesive and welded joints are the most difficult to check visually; however, rapid nondestructive methods have been developed to overcome this objection and will be discussed in Chapter 11.

EXERCISES

Problems and Questions

1–1. (a) Shown in Fig. 1–1 are several joints, one of which is called special. Assume the center block is made out of two pieces of inch-square mild steel and the sheets bent up against the block are $\frac{1}{8}$ in. (3.17 mm) thick mild steel. The joint will be used in room temperature conditions with longitudinal shear stresses up to 5,000 psi of the joint. What three alternative methods can be used in making this joint?

(b) Which do you prefer and why?

1–2. Two pieces of aluminum tubing are to be joined. The joint will be a lap type, 2 in. (50.8 cm) long. The outside diameter (o.d.) of the large tube is 1–7/8 in. It has a wall thickness of 0.045 in. (1.14 mm). The next size smaller standard tube has an o.d. of 1–3/4 in. (44.45 mm). Each tubing joint will be stressed up to 10,000 lb in tension. There are 500 to be made. What method of joining do you recommend and why?

1–3. (a) Shown in Fig. P1–3 is a reinforcing member attached to the inside of an aluminum boat bottom. The reinforcing hat sections, also aluminum, are approximately the width of the boat in length and are placed at intervals of 12 in. (30.48 cm) apart. They must resist vibration and yet provide some flexibility. The joint must be able to resist gasoline, weathering, and oil. What methods can be used to join the strips to the inside of the boat hull? (b) What method would you prefer and why?

1–4. (a) Shown in Fig. P1–4 is a prepared butt weld for heavy [4 in. (10.16 cm) thick] plate. What method should be used to join the steel plates together if they are ten feet long and three feet wide? The plates can be turned over and all welding can be done in the flat position. (b) What welding method would be used for the same plates in the vertical position? (c) Same as in (b), but the material is aluminum?

1–5. Shown in Fig. P1–5 is a sectional view of a fan on a shaft. Discuss possible ways of fastening these two parts together in a production system. Disassembly is anticipated only when a major repair becomes necessary.

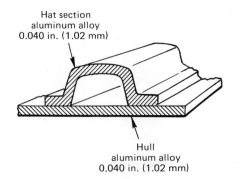

Hat section
aluminum alloy
0.040 in. (1.02 mm)

Hull
aluminum alloy
0.040 in. (1.02 mm)

FIGURE P1–3. *Hat sections are attached to the inside bottom of the boat hull for reinforcing.*

1–6. Shown in Fig. P1–6 is a corner joint made for adhesive assembly. Show, with the aid of sketches, how the strength of the joint, or a modified version of it, may be improved.

1–7. Shown in Fig. P1–7 is a cross-sectional view of two pieces of tubing to be joined by adhesives. Sketch three methods that would improve the joint for adhesive joining.

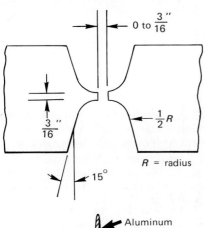

FIGURE P1–4. *A butt joint as prepared for welding in joining heavy plate.*

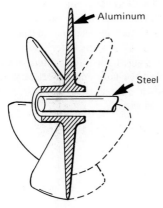

FIGURE P1–5. *Sectional view of a fan on a shaft.*

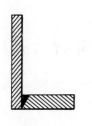

FIGURE P1–6. *A corner joint made by adhesive bonding.*

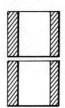

FIGURE P1–7. *Two pieces of tubing to be modified for joining with adhesives.*

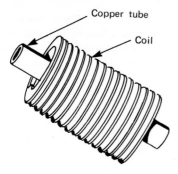

FIGURE P1–8. *An aluminum cooling fin is pressed on a copper tube and fastened in place.*

1–8. The aluminum cooling fin shown in Fig. P1–8 is pressed on the copper tube and fastened in place. The temperature range in use is from −30 °F (34 °C) to 220 °F (104 °C). What is the best method of joining the two parts?

1–9. Aluminized steel mufflers have been found to last 100 percent longer than the plain mild steel type. The ends must be made to join with exhaust tubing as shown in Fig. P1–9. Also, the seam must be securely fastened. What methods do you recommend for (a), the seam on top? And (b), the end?

1–10. Assume the lapped area of the joint shown in Fig. P1–10 is 2 in. (5.08 cm) long and 3 in. (7.62 cm) wide. It is to be used in an environment that often reaches 380 °F (193 °C). (a) Select an adhesive that will provide the highest shear strength at this temperature. (b) What are some advantages in making this joint with an adhesive rather than welding it? (c) How may the design be improved?

1–11. An electrical ground connection is shown in Fig. P1–11. It is made out of copper and the butt end must be fastened to the flat side of a 1/16 in. (15.87 mm) thick steel sheet on a production basis. Suggest two alternative methods to accomplish the fastening process and tell why each is chosen. NOTE: The terminal shown may be modified slightly.

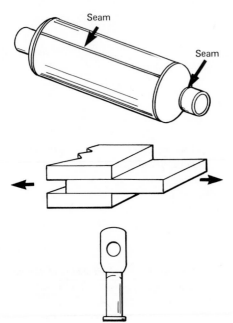

FIGURE P1–9. *An aluminized steel muffler.*

FIGURE P1–10. *A double lap joint subjected to shear stress.*

FIGURE P1–11. *A copper terminal that is to be connected at the butt end to the flat side of a 1/16 in. thick steel sheet.*

TABLE 1.A. *Metric Conversion Factors*

| Property | Unit | Symbol | Exact Conversion | | | Approximate Equivalency |
			From	To	Multiply by	
Length	metre	m	inch	mm	2.540×10	25 mm = 1 in
	centimetre	cm	inch	cm	2.540	300 mm = 1 ft
	millimetre	mm	foot	mm	3.048×10^{-4}	
Mass	kilogram	kg	ounce	g	2.835×10	2.8 g = 1 oz
	gram	g	pound	kg	4.536×10^{-1}	kg = 2.2 lbs (35 oz)
	tonne (megagram)	t	ton (2000 lb)	kg	9.072×10^{2}	1 t = 2200 lb
Density	kilogram per cu metre	kg/m³	pounds per cu ft	kg/m³	1.602×10	16 kg/M³ = 1 lb/ft³
Temperature	deg Celsius	°C	deg Fahrenheit	°C	(°F − 32) × 5/9	0 °C = 32 °F
						100 °C = 212 °F
Area	square metre	m²	sq inch	mm²	6.452×10^{2}	645 mm² = 1 in²
	square millimetre	mm²	sq ft	m²	9.290×10^{-2}	1 m² = 11 ft²
Volume	cubic metre	m³	cu in	mm³	1.639×10^{4}	16400 mm³ = 1 in³
	cubic centimetre	cm³	cu ft	m³	2.832×10^{-2}	1 m³ = 35 ft³
	cubic millimetre	mm³	cu yd	m³	7.645×10^{-1}	1 m³ = 1.3 yd³
Force	newton	N	ounce (Force)	N	$2,780 \times 10^{-1}$	1 N = 3.6 oz
	kilonewton	kN	pound (Force)	kN	4.448×10^{-3}	4.4 N = 1 lb
	meganewton	MN	Kip	MN	4.448	1 kN = 225 lb
Stress	megapascal	MPa	pound/in² (psi)	MPa	6.895×10^{-3}	1 MPa = 145 psi
	newtons/sq m	N/M²	Kip/in² (ksi)	MPa	6.895	7 MPa = 1 ksi
Torque	newton-metres	Nm	in-ounce	Nm	7.062×10^{3}	1 Nm = 140 in oz
			in pound	Nm	1.130×10^{-1}	1 Nm = 9 in lb
			ft pound	Nm	1.356	1 Nm = 75 ft lb
						1.4 Nm = 1 ft lb

BIBLIOGRAPHY

Books:

Adhesive Bonding Aluminum, Reynolds Metals Company, Richmond, Virginia.

BIKALES, N. M., *Adhesion and Bonding*, Wiley-Interscience, New York, N.Y., 1971.

Fastener Standards, 5th ed., Industrial Fasteners Institute, Cleveland, Ohio, 1970.

LAUGHNER, V. H., AND HAUGAN A. D., *Handbook of Fastening and Joining Metal Parts*, McGraw-Hill, New York, N.Y., 1956.

PARKER, R. S. R., *Adhesion and Adhesives*, Pergamon Press, New York, N.Y., 1966.

Report on Industrial Fastening and Assembly Conference, London Factory Publications, 1964.

SOLED, J., *Fasteners Handbook*, Reinhold Publishing Corp., New York, N.Y., 1957.

Periodicals:

Machine Design, *American Machinist Special Report No. 639*, 1971 Fastening and Joining Reference Issue, January 12, 1970.

SHARPE, L. H., "Assembling With Adhesives." *Machine Design*, **5**, August 1966.

SNOGREN, R. C., "Space Age Bonding Techniques." *Mechanical Engineering*, May 1970.

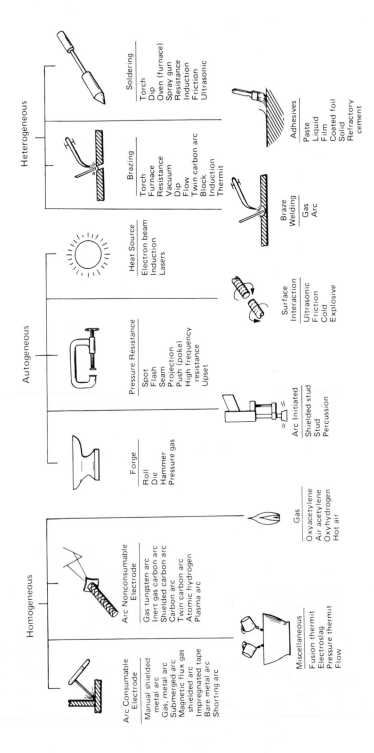

FIGURE 2-0. *Master chart of welding and other hot joining processes.*

2

The Physics of Energy Sources
for Materials Joining

INTRODUCTION

The art of joining materials has been in existence for thousands of years, yet the science remains in its infancy. For the general public the art of joining materials has often established an image that is less than glamorous. Most people still relate welding to a black helmet and showers of sparks. The advent of lunar travel and oceanographic exploration have changed this image (to some extent). The welding engineer must be knowledgeable in all branches of engineering. He must be able to apply engineering design and principles of thermodynamics, metallurgy, chemistry, stress analysis, heat transfer (Fig. 2–1), and many other specialities to determine the behavior of that relatively small mass of deposited metal involved in the welding operation.

The science of materials joining can best be understood if it is realized that, in theory, the mating surfaces of two pieces of metal, adequately cleaned and smoothed, will join together automatically when brought in intimate contact with each other. Metals are held together by internal bonds or attracting forces, which are the attraction of fractional positive ions for a negative electron gas. The ideal joint made between two pieces of metal would have this same attracting force across the boundary formed by the contacting surface. Unfortunately, the absence of advanced technology precludes this kind of union without the utilization of additional energies.

Environmental contaminants such as oxygen and nitrogen create several problems in joining materials. In addition to surface contamination, many technical

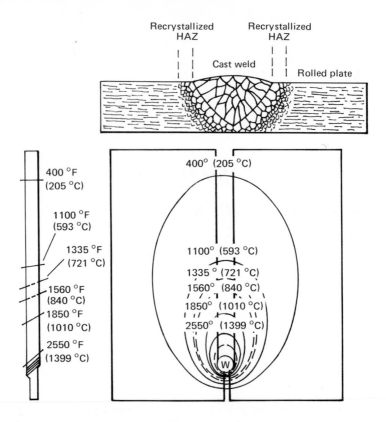

FIGURE 2–1. *The science of welding includes a knowledge of metallurgy and heat transfer. (Courtesy of American Welding Society.)*

problems remain in obtaining the required surface finish. The finest polishing and honing equipment available is not able to produce the surface finish required, which is within a few angstroms, for automatic atomic joining. When Surveyor IV landed on the moon, its predicted operational life expectancy was five years. However, within five days the unit was inoperative. The reason, as reported, was that friction between critical metal surfaces created a desirable degree of smoothness, but at the same time, the rarer atmosphere helped remove the oxide film. The natural partial vacuum prevented oxides from reforming and resulted in atomic joining of the mating parts.

To ensure successful bonding by welding, brazing, or soldering techniques, the following steps are necessary:

1. Produce an adequately smooth surface so that atomic bonding can take place.
2. Remove all possible contamination from the surfaces of the metal to be joined.
3. Prevent further contamination from taking place before, during, and immediately following the joining process.

An adequately smooth surface can be achieved through melting, by applying pressure, or by sandwiching a layer of alloying material between the surfaces to be joined. A metal surface can usually be cleaned by chemical, mechanical, or melting techniques. Perhaps the most difficult problem to overcome is contamination at the joint during the actual joining. This becomes a particularly difficult problem when the faying surfaces of the metals are at elevated temperatures. Almost without exception, every welding process must use some means to protect the metal's surface from the atmosphere during the time that joining takes place. The various techniques used will be discussed later in this text.

Classification of Energy Sources. A better understanding of the principles of materials joining can be gained by relating it to the basic type of energy used in the act of making the joint. Energy used for this purpose can be classified as follows:

1. Mechanical (kinetic).
2. Chemical (potential).
3. Electrical (potential).

Energy is defined broadly as the capacity to do work; it divides into two classes, *Kinetic Energy* (KE), the energy of motion; and *Potential Energy* (PE), the energy due to position or molecular arrangement. The energy sources used in material joining are of both types. Each energy source includes several modified joining techniques, which will be discussed in Chapters 3 and 4.

Most joining techniques have a very low efficiency, simply because the energy must be applied to the surface of the material and then be absorbed by thermal conduction. Because the thermal conduction of materials is relatively slow, a majority of the heat applied is lost to the atmosphere or is needlessly absorbed by the material mass. For example, the oxyacetylene torch is reported to be only 2% efficient. Because of heat losses, the weldments made by conventional processes have large heat affected zones and undesirable distortion problems. Heat affected zones (HAZ) are metal areas surrounding the weldment that have been heated to a temperature high enough to change the microstructure. The electron beam (EB) joining process is an exception in that due to the high current density it has a narrow HAZ. Thus, with excellent focusing systems, the energy penetrates only the metal being welded thereby eliminating the movement of heat by thermal conduction, Fig. 2–2.

Another medium for energy penetration, similar to that of the EB welder, is that of micro-ovens used for cooking. These systems are considered to be 95 % to 100 % efficient. The EB welder is capable of giving a 25 to 1 width-to-depth ratio compared to perhaps a 1 to 1 ratio or less with conventional welding processes, Fig. 2–3.

In welding, heat is conducted through gases, liquids, and solids. Heat conducted in gases and liquids is affected by the random motion of gas molecules, Fig. 2–4. A molecule that strikes a hot wall absorbs energy and rebounds traveling faster than when it arrived. Molecules share energy as a result of random collisions.

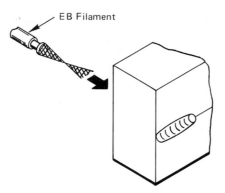

FIGURE 2–2. *The electrons transmitted from the filament of the EB welding gun completely penetrate the material being welded, giving a 95% to 100% efficiency of the transmitted energy.*

The heat diffuses by moving from the hot side to the cold side. The rate of heat transport depends chiefly on a molecule's mean free path between collisions. Thermal conduction in solids differs from that in gases because the atoms are tied to lattice positions. The neighboring atoms are bonded together and tend to vibrate in unison, Fig. 2–5.

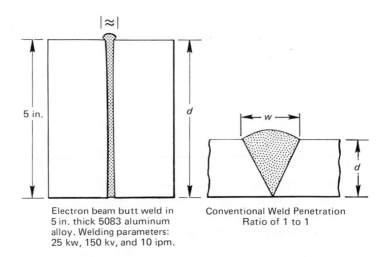

Electron beam butt weld in 5 in. thick 5083 aluminum alloy. Welding parameters: 25 kw, 150 kv, and 10 ipm.

Conventional Weld Penetration Ratio of 1 to 1

FIGURE 2–3. *The EB system is capable of giving a 25 to 1 penetration ratio as compared to a 1 to 1 ratio or less for the conventional welding processes. (Courtesy of American Welding Society.)*

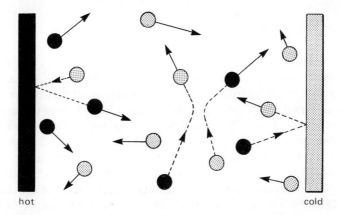

FIGURE 2-4. *Heat conduction in gas is affected by the random motion of gas molecules. A molecule that strikes the hot wall of a container (left) absorbs energy and leaves traveling faster than when it arrived. Within the container, molecules share energy as a result of random collisions. Thus, heat diffuses through the gas, carrying heat from the hot wall to the cold one. The rate of transport depends chiefly on a molecule's mean free path between collisions. (Courtesy of Scientific American.)*

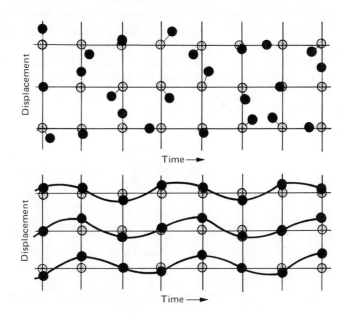

FIGURE 2-5. *Heat conduction in solids differs from that in a gas because the atoms are tied to lattice positions. In 1907 Albert Einstein devised a formula for heat conduction based on the assumption that atoms vibrate independently (top). In 1912 Peter J. W. Debye argued that neighboring atoms, being bonded together, tend to vibrate in unison (bottom). (Courtesy of Scientific American.)*

JOINING BY MECHANICAL ENERGY

The layer of oxides at the faying surfaces is ruptured by stretching and changing the shape of the material as it is joined using mechanical energy techniques. This principle is referred to as plastic deformation. Deformations can be accomplished through pressures from a punch and die, Fig. 2–6, blows from a hammer, friction, and explosive forces. The metals being joined are protected from further atmospheric contamination by the closely fitted surfaces of the material at the interface. When better protection is required, a flux can be used. Flux is a cleaning agent used to dissolve base metal oxides and to float out impurities. It also blankets the base metal from further atmospheric contamination. Another commonly used technique is to shield the surfaces being joined in an environment of inert gases or in a vacuum. An inert gas is one that does not combine easily with other elements. The most commonly used inert gases for metal joining are helium and argon.

The type of energy necessary for surface matching is provided through a variety of techniques, all of which use mechanical energy. The mechanical energy joining processes include forge, explosive, and friction welding, and ultrasonic and diffusion joining. The latter process is also referred to as solid state bonding because *diffusion* plays an essential role in both the practical and theoretical aspects of it. Eutectic bonding and solid state bonding are other terms for diffusion joining. Diffusion joining is dependent on time, pressure, and temperature. The metals to be joined are placed in contact with each other while an external pressure is applied. The unit is then heated to a predetermined temperature and held under pressure at this temperature for a specified length of time. The added energy (heat) increases the activity of the atoms at the faying edges of the metal. Eventually the atoms move across the joint interface, giving a solid state type bond. Each of these processes is discussed in detail in Chapter 3.

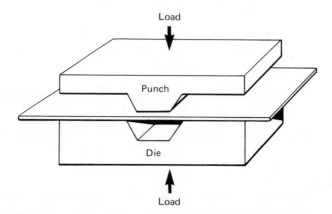

FIGURE 2-6. *Mechanical energy is used to stretch the material, changing its shape, the action of which ruptures the layer of oxides at the faying surface achieving atomic joining.*

Two theories explain the movement of atoms in metals; namely, the interstitial mechanism and the vacancy mechanism, Fig. 2–7. According to the interstitial mechanism, the atoms move from one location to another along the interstices of the crystal structure. The vacancy mechanism involves the matrix or substitution principle, meaning that a vacant location must exist before an atom can move or be diffused. Because there are fewer vacant locations than there are interstices, the vacancy mechanism is the slower bonding principle of the two.

Each of the mechanical joining processes has some unique principle for removing its surface contamination and achieving a metal to metal contact, the requisites for atomic bonding across the surface interface. The closeness of the metals being joined prevents further atmospheric contamination at the joint.

JOINING BY CHEMICAL ENERGY

The chemical energy group of joining processes is dependent upon heat generated by reactions of one material upon another, or upon itself. This heat is used to melt a filler metal, a flux, or the surfaces of the base metal to be joined. It is the melting action that destroys the undesirable surface films and provides for the surface to surface matching needed for atomic joining. The three processes generally classified as belonging to the chemical energy group are: combustible gases, atomic hydrogen, and thermit. Typical chemical reactions for each of these processes are:

1. Combustible Gases Energy (neutral flame)—inner cone reaction.

$$C_2H_2(\text{acetylene}) + O_2(\text{oxygen}) \rightarrow 2\,CO(\text{carbon monoxide}) + H_2(\text{hydrogen})$$

And for the envelope the reaction is:

$$CO + O = CO_2 \quad \text{and} \quad H_2 + O = H_2O, \qquad \text{(Fig. 2–8a)}$$

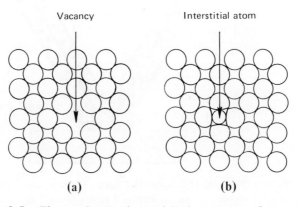

FIGURE 2–7. *The two theories that explain the movement of atoms in metals are the vacancy mechanism and the interstitial mechanism: (a) a vacancy type of defect, and (b) an interstitial type of defect.*

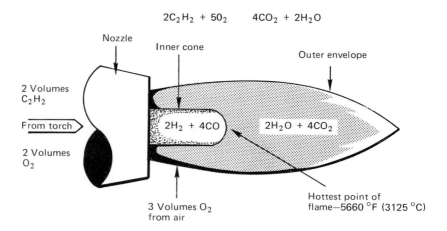

FIGURE 2–8. (a) *The chemical reaction within the oxyacetylene flame. When adequate oxygen is supplied, the two cones shrink into one and the two reactions collectively become* $2C_2H_2 + 5O_2 \rightarrow 4CO_2 + 2H_2O$ *or what is known as a neutral flame.*

2. Atomic Hydrogen Energy.

$$H_2 \rightarrow 2\,H \rightarrow H_2, \text{ (Fig. 2–8b)}$$

$$H_2(\text{hydrogen atoms}) \rightarrow 2\,H(\text{hydrogen atoms dissassociated})$$

$$\rightarrow H_2(\text{hydrogen atoms reassociated})$$

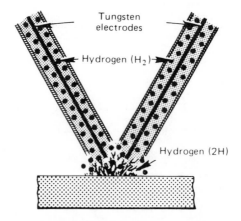

FIGURE 2–8. (b) *Hydrogen* (H_2) *passes through tubes around each of two tungsten electrodes. As the hydrogen passes through the arc between the two electrodes, the hydrogen atoms dissassociate* (2H), *absorbing energy. As the hydrogen comes in contact with the base metal, the hydrogen atoms reassociate* (H_2), *giving up the energy that is used to melt the metal being joined.*

3. Thermit Energy.

$$3\,Fe_3O_4(\text{iron oxide}) + 8\,Al(\text{aluminium}) \rightarrow 9\,Fe(\text{iron})$$
$$+ 4\,Al_2O_3(\text{aluminium oxide})$$

JOINING BY ELECTRICAL ENERGY

The electrical energy group is by far the largest and most popular source of energy used for joining metals. The joining processes included within this group are: gas resistance, liquid resistance, solid resistance, and a special group. Each of these processes, except for the last, depends upon some form of resistance to the flow of an electric current for generating heat, and each has its own unique way of protecting the materials being joined from atmospheric contamination.

Gas Resistance Theory. The gas resistance theory includes all processes whereby heat is generated from the resistance of a column of ionized gas or plasma to the passage of electrons from the cathode to the anode, Fig. 2-9. The processes included within this classification are: gas metal arc (GMA), gas tungsten arc (GTA), shielded metal arc (SMA), submerged arc, and electro-gas.

A typical welding power source (see Fig. 2–10b) set at 300 amperes dc and 25 volts will theoretically produce 7500 watts, or the equivalent of 1800 cal/sec (7.15 Btu/sec). From 20 to 85 % of the arc heat produced is utilized in the melting of the metal to be joined. The remainder is lost through conduction, radiation, spatter, and convection. Heat input from the arc is measured by joules and is expressed as:

$$H = \frac{E \times I \times 60}{S}$$

where H = joules (watt–sec) per linear inch.
E = arc voltage (V).
I = arc current (A).
S = arc travel, inches per minute.

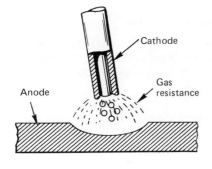

FIGURE 2-9. *The gas resistance theory includes all the processes whereby the heat is generated through the resistance of electrons in a column of ionized gas or plasma passing from the cathode to the anode.*

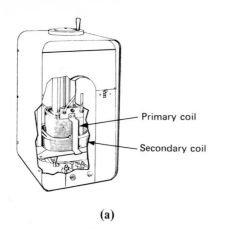

Primary coil

Secondary coil

(a)

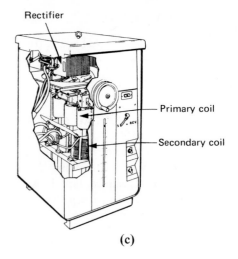

Rectifier

Primary coil

Secondary coil

(c)

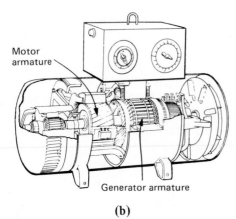

Motor
armature

Generator armature

(b)

FIGURE 2–10. *Typical constant current welding power sources. (a) The ac transformer used for arc welding. (b) A dual control, motor generated type welding machine. (c) Direct current can be supplied for welding through a rectifier.*

Example. A weld made at 250 A, 32 V, and at 10 in./min, the energy input (H) in joules per linear inch would be:

$$H = \frac{32 \times 250 \times 60}{10}$$

= 48,000 joules per linear inch

The 60 is used to convert time from seconds to minutes, as most torch speeds are measured in inches per minute. Welding arc temperatures vary from 5,000 °F (2777 °C) for a transferred arc (Fig. 2–9), to an estimated 50,000 °F (27770 °C) for the nontransferred arc plasma jet, Fig. 2–11.

Under certain conditions the welding arc becomes unmanageable. It has a tendency to lose shape and refuses to concentrate at the point of joining. This phenomenon is known as *arc blow*. Arc blow is the result of magnetic disturbances

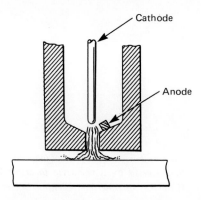

FIGURE 2–11. *A non-transferred arc means that the cathode and anode are both contained within the torch.*

that unbalance the symmetry of the self-induced field surrounding the welding arc. Arc blow results from one or both of two basic principles:

1. A direction change occurs in the flow of the current from the time it enters the work through the electrode and is conducted away to the ground clamp, Fig. 2–12. As noted, the lines of force are closer together on the inside bend, therefore giving unbalanced pulls.
2. The usual position of the electrode to the base material when welding fillets, or in the vee for thicker materials, tends to set up an ideal situation for arc blow. It is a well known fact that magnetic lines of force pass more freely through magnetic materials than through the air; therefore, it is only natural that most of the force lines follow the steel and create this arc disturbance, Fig. 2–13.

One sure way to reduce arc blow is to use ac current for welding. With ac current, eddy currents are induced into the work and thereby tend to neutralize the magnetic field, Fig. 2–14.

Liquid Resistance Theory. The liquid resistance theory applies to those processes whereby heat is generated when electrons pass from the cathode to the anode through a liquid. An example of this principle is the electroslag process. An arc is initially

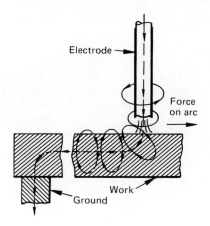

FIGURE 2–12. *Distortion of induced magnetic field caused by location of ground. (Courtesy of American Welding Society.)*

35

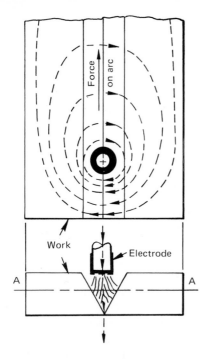

FIGURE 2-13. *Distortion of induced magnetic field caused by asymmetric location of iron. (Courtesy of American Welding Society.)*

used to fuse the granular flux into molten slag. The molten slag becomes a highly ionized mass, offering a path of relatively low resistance to the current flowing from the wire to the weld pool and adjacent plate edges, Fig. 2-15. Only the molten slag in the immediate region of the electrode tip is useful in generating the heat necessary for welding. The quantity of heat generated in the slag pool can be determined by the formula:

$$H = REI$$

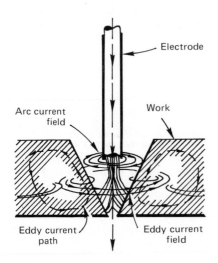

FIGURE 2-14. *Effect of eddy currents in neutralizing field induced by alternating arc current. (Courtesy of American Welding Society.)*

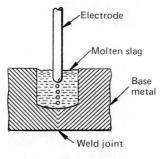

- Electrode
- Molten slag
- Base metal
- Weld joint

FIGURE 2–15. *The electro-slag welding process develops a molten slag into a highly ionized mass that has a path of relatively low resistance to the flow of electric current from the electrode wire through the weld pool and to the adjacent plate edge.*

where H = quantity of heat generated (cal/sec).
R = flux resistance in ohms.

Example. In electroslag welding where the flux being used has a resistance value of 0.24 ohms (Ω) and the power source is producing 40 volts and 950 amps, the energy input would be 9120 cal/sec.

$$\therefore H = 0.24\,\Omega \times 40\,\text{V} \times 950\,\text{A}$$

$$= 9120\,\text{cal/sec}$$

Solid Resistance Theory. The solid resistance theory implies that the energy necessary for the joining of metals is generated by the resistance to the flow of electrons through the base metal. Processes utilizing this principle are spot welding, seam welding, projection welding, percussion welding, upset welding, and high frequency resistance welding. The principle for each of these processes is closely related and can best be explained using the principle of a single spot weld, Fig. 2–16. The pieces being joined are in the secondary circuit of a transformer that converts a high voltage and low amperage ac power supply to low voltage and high amperage. The heat generated by current flow is expressed as:

$$H = I^2 R T$$

where H = heat generated in joules (watt-secs).
T = time of current flow, in secs.

Current and time are easily ascertained, but measurement of the various resistances is more difficult. In spot welding there are seven areas where the amount of resistance affects the quality of the weld; included are the areas between the electrode and the work, between the workpieces at the point of joining, and within the bodies of the workpieces and the electrodes, Fig. 2–16.

The principles of high frequency resistance welding differ from the other processes within its group in that ac currents with frequencies up to 500,000 Hz are employed as the heating source.

Special Group. The principles for generating energy in induction heating or electron beams and laser welding are different for each process and rely only indirectly on resistance to create heat. Therefore, each will be discussed independently.

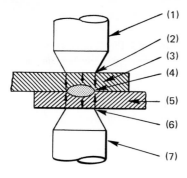

(1)
(2)
(3)
(4)
(5)
(6)
(7)

FIGURE 2–16. *The resistance spot weld has seven areas where the degree of resistance affects the quality of the weld. Areas 1 and 7 are the properties of the electrode material; 2 and 6 are the surface contamination and the area of the face for each electrode and the surface condition of the base metal at the point of electrode contact; 3 and 5 are the thermo properties and thickness of the metals being joined; and 4 is the surface condition of the base metal at the place of the nugget.*

Induction Heating. Induction heating is based on the theory that an electrical current passing through a wire sets up a magnetic field around the wire, Fig. 2–17. The intensity of the magnetic field is governed by the amount of current passing through the wire, the number of turns in the coil, and the medium surrounding the coil. As ac current flows through a wire, it causes the lines of magnetic force to expand and contract. The expanding and contracting magnetic field will penetrate any electrical conductor. When the frequency of the ac current exceeds 50,000 Hz, any metal object in the magnetic field will heat up rapidly. If the metal being heated is a magnetic material, the magnetized then demagnetized cycle produced by the ac current causes *molecular friction* to take place. Molecular friction is the continuous rearranging of the molecules in the magnetized material by the ac current used to generate heat within the object. Molecular friction can be defined further as electron

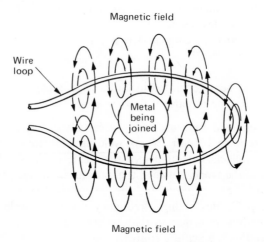

FIGURE 2–17. *As alternating current flows through a wire or bar, it causes the lines of magnetic force to expand and contract. The expanding and contracting magnetic field will penetrate any electrical conductor, which in turn heats up the metal.*

Vacuum Electron Beam Welding Machine

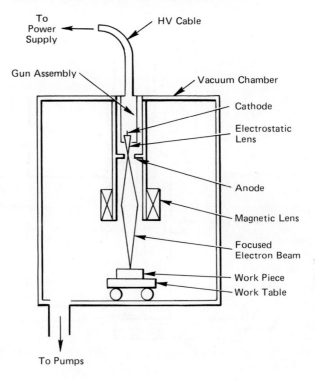

FIGURE 2–18.   *The cathode (filament) is heated by resistance to the electric current and emits electrons to the workpiece being welded. (Courtesy of American Welding Society.)*

collisions and hysteresis. This type of heat is the source of energy for welding or brazing. There is no arc or flame and the hardware of the power supply never comes in contact with the material being joined.

Electron Beam Welding.   The electron beam (EB) differs from other types of energy in this group in that high voltage is used to bombard electrons from a welding gun into the metal being joined, Fig. 2–18. The welding gun consists of a tantalum or tungsten filament heated by the resistance of an electric current to about 3600 °F (2000 °C) in a hard vacuum (10^{-4} or 10^{-5} mm Hg, or about one ten-millionth of atmospheric pressure). At this temperature, the filament emits electrons. The filament is positioned as the cathode in a high voltage field (500 to 150,000 volts), which accelerates the electrons to 30,000–120,000 miles (48,000 Km–192,000 Km) per second, creating tremendous kinetic energy. The electrons pass through the annular anode and are then focused by electrostatic or electromagnetic "lenses" into a fine, dense beam.

39

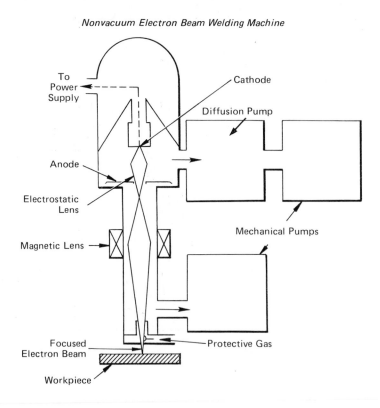

Nonvacuum Electron Beam Welding Machine

FIGURE 2–19. *Out-of-vacuum EB welding speeds up production but reduces the weld quality. (Courtesy of American Welding Society.)*

As the electron beam strikes an object that is opaque to electrons (such as metal), the electrons' kinetic energy is converted into heat, melting and vaporizing the metal to make a hole through the object. When the beam is advanced along the joint, molten metal flows around behind the beam and resolidifies, producing the weld (Fig. 2–2).

When the welding is done in the same vacuum that the beam is produced, extremely large depth-to-width penetration ratios (as high as 25 to 1) are possible in a single pass (Fig. 2–3). The high vacuum also provides for environmental control, and extremely clean welds can be made in materials that are very sensitive to atmospheric contamination. The depth-to-width ratio of the penetration is due to high power densities. Power concentrations of 10^6 to 10^7 watts/cm^2 are common for the electron beam process.

A modification of the equipment makes it possible for electron beam welding to be done in lower vacuums. Also gaining in popularity is the use of "out-of-vacuum EB welding", Fig. 2–19.

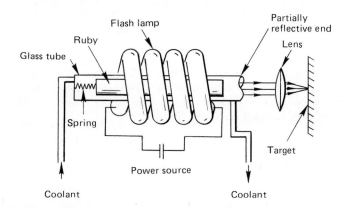

Figure 2–20. *The xenon lamp excites the chromium atoms until a self-supporting chain reaction occurs, triggering almost all of the chromium atoms at one time. Then a pulse of monochromatic unidirectional light is released from the partially reflective end of the crystal.* (*Courtesy of American Welding Society.*)

Laser Welding. Lasers (light amplification by stimulated emission of radiation) can be gas, liquid, or solid state. The most common types associated with materials joining are the solid state (ruby) or gas (carbon dioxide—CO_2) processes. For the purpose of this discussion, the physics of laser welding will be confined to the pulsed (intermittent firing) ruby, Fig. 2–20. It consists of a cylindrical single crystal of fused Al_2O_3 with 0.05 % Cr added as an intentional impurity. The ends are optically flat and parallel. One end is highly reflective while the other end is partially reflective. A xenon flash tube "pumps" green light into the ruby, exciting the Cr atoms to a higher energy level until:

1. They decay back to the dead state according to a statistical time distribution, emitting red light or . . .
2. They are struck by the characteristic red light.

The xenon lamp excites more and more Cr atoms until a self-supporting chain reaction occurs, triggering almost all of the Cr atoms at once. Then a pulse of monochromatic, unidirectional light is released from the partially reflective end of the crystal.

The time duration of the pulse is very short (a few milliseconds), but it may contain as much as 25,000 watts (comparable to EB power) and can be focused by optical lenses into a small pinpoint capable of melting and partially vaporizing its target. The duration of the pulse is, in fact, so short compared to the off-time that the molten material resolidifies before the next pulse. Since solidification is so rapid, exclusion of the atmosphere is often not needed, although inert gas may be used.

EXERCISES

Problems

2–1. Using the submerged arc process with 40 arc volts, 600 amperes, and a torch travel speed of 21 in/min (53.34 cm/M), how much heat in joules is being put into the base metal per inch? (Show your work.)

2–2. Why is it not considered feasible to weld aluminum to carbon steels with the conventional welding processes?

2–3. A typical stick electrode arc reportedly produces a temperature of 6300 °F. With the U.S. rapidly changing to the metric system, calculate what this temperature would be in degrees centigrade? Use the equation $C = \frac{5}{9}(°F - 32)$.

2–4. As a welding engineer, you are consulting with a German industry. They report a certain temperature to be 889 °C. What temperature would this be in Fahrenheit? $F = \frac{9}{5}(°C) + 32$.

2–5. A typical problem in welding is the prevention or reduction of distortion. The kind of process used readily influences the amount of distortion. Can you explain why this is true?

Questions

2–1. Theoretically, two like pieces of metals, if clean enough and smooth enough, will automatically join when brought together. If this is so, then why all the problems when welding or brazing?

2–2. To achieve the best possible joint when welding, brazing, or soldering, what rules should one observe?

2–3. Three different groups of energy sources were discussed in this chapter. Name the groups and the joining processes for each.

2–4. The amount of heat used for the joining of materials quite often affects the properties of the materials. How is joules heat input calculated? Design and solve a simple problem.

2–5. What equation is used to determine the amount of heat generated in solid resistance welding? Design and solve a simple problem.

2–6. Which of the many sources of energy discussed in this chapter performs its function without coming into direct contact with the pieces being joined?

2–7. Explain the principle of EB welding that makes possible nearly 100 % power efficiency.

2–8. Why is the 0.05% Cr added to the Al_2O_3 single crystal for a pulsed type laser?

2–9. What events in history did much to change welding from an "art" to one of "science?"

2–10. What are the qualifications for a "welding engineer?"

2–11. What would you consider to be the greatest enemy to successful welding?

2–12. Name and discuss the three kinds of energy used in welding.

BIBLIOGRAPHY

AMERICAN WELDING SOCIETY, "Resistance Welding", Miami, Florida, 1962.

AMERICAN WELDING SOCIETY, "Current Welding Processes", Miami Florida, 1965.

AMERICAN WELDING SOCIETY, "Modern Joining Processes", Miami, Florida, 1966.

AMERICAN WELDING SOCIETY, "Fundamentals of Welding", *Welding Handbook*, 6th ed., sec. 1, Miami, Florida, 1968.

AMERICAN WELDING SOCIETY, "Welding, Cutting and Related Processes", *Welding Handbook*, 6th ed., sec. 3A, Miami, Florida, 1970.

ARCOS CORPORATION, "Arcos Vertomatic Welding Machines." Murray Hills, N.J.

BRATON, N. R., DUCHON, G. A., AND LOPER, C. R., "Manufacturing Processes", University of Wisconsin, Madison, Wisc., 1966.

GURTNER, FRANCES B., "What is Welding?",

BIBLIOGRAPHY

Edgewood Arsenal Special Publication EASP-400-20, Department of the Army, August, 1969.

HOULDCROFT, P. T., "Welding Processes", *British Welding Research Association*, Cambridge University Press, London, England, 1967.

JONES, S. B., AND MEYERS, F. R., "Ultrasonic Welding of Structural Aluminum Alloys", *Welding Research Supplement*, Miami, Florida, 1958.

JUSTICE, JAMES F., "Design Control . . . Keys to Quality in Friction (Inertia) Welding", *Metal Progress*, Metals Park, Ohio, July, 1968.

KOZIARSKI, J., "Ultrasonic Welding", *Welding Journal*, Miami, Florida, April, 1961.

LEPEL HIGH FREQUENCY LABORATORIES, INC., "Review", *High Frequency Heating*.

LINDE DIVISION–UNION CARBIDE CORPORATION, "Electric Welding Progress", vol. 18, no. 2, New York, N.Y.

OBERLE, T. L., LOYD, C. D., AND CALTON, M. R., "Inertia Welding Dissimilar Metals", *Welding Journal*, June, 1967.

PATON, B. E., "Electroslag Welding", *Mashgiz*, Government Publishers of Scientific and Technical Literature for Equipment Manufacturing, Moscow-Leningrad, USSR, 1959.

SONOBOND CORPORATION, "Ultrasonic Welding Without Fusion", *Bulletin SW-87*, 1957.

STANFORD RESEARCH INSTITUTE, "Explosive Welding", *Research for Industry*, vol. 14, no. 1, January-February, 1962.

THOMAS, R. DAVID, "Electroslag Welding", *Welding Journal*, Miami, Florida, February, 1960.

TYLECOTE, R. F., "The Solid Phase Welding of Metals", St. Martins Press, New York, 1968.

VILL, V. I., "Friction Welding of Metals", *Mashgiz*, Government Publishers of Scientific and Technical Literature for Equipment Manufacturing, Moscow-Leningrad, USSR, 1959.

ZIMAN, JOHN, "The Thermal Properties of Materials", *Scientific American*, New York, N.Y., September, 1967.

3

Mechanical and Chemical
Energy Joining Systems

INTRODUCTION

The group of *mechanical energy joining systems* includes forge welding, cold welding, explosive welding, friction welding, ultrasonic welding, and diffusion bonding. Each process incorporates a unique principle for breaking up and dispersing the oxide layer so that a metal-to-metal contact can be achieved at the joint. The smoothness necessary for atom-to-atom joining at the interface is achieved either by pressure, heat, or a combination of the two, or, as in ultrasonic welding, an excitation of the atoms. Atmospheric contamination is reduced by intimately fitted joining surfaces.

FORGE WELDING

Forge welding is not only the first known welding process, dating back to 5,000 B.C., but it also is distinguished for joining metals for a longer period than any other welding process. Basically, forge welding involves heating the materials to be joined to a temperature approaching the onset of melting, but not reaching it, and then applying pressure, Fig. 3-1. The function of heating is only to make plastic deformation easier, as the yield strength of most materials decreases greatly as the temperature is raised.

The most commonly used joint for forge welding is the scarf. Other types of joints can also be forged welded, some utilizing pressures up to 150 tons/sq. in. (2068 MPa). The forging temperature for steel is usually in the range of 2100 °F

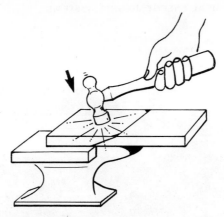

FIGURE 3-1. *Forge welding involves heating the materials to be joined to a temperature, which approaches the onset of melting, but does not reach it, and then applying pressure. A scarf joint is used.*

(1149 °C) to 2350 °F (1288 °C), whereas approximately 2700 °F (1482 °C) is necessary for melting. At these temperatures the surfaces oxidize rapidly; therefore, a flux of borax or silica sand is often used to aid in dissolving the oxides. A combination of pressure, flux, heat, and a clean surface provides for intimate metal-to-metal contact to permit atomic bonding. This process is generally limited to wrought iron and low carbon steels. Other metals that have been successfully forge welded are aluminum, magnesium, and copper.

The forge welding process produces good stress characteristics at the joint. This is because joining takes place at elevated temperatures, and working of the metal reduces the residual stresses that are generally prevalent in other kinds of weldments. Complete removal of oxides and other surface contamination at the faying edges is difficult to achieve. Also, this method is more time consuming than most welding processes.

COLD WELDING

Cold welding is the logical extension of forge welding at room temperature. Because the yield strength of a given material is higher at room temperature than at elevated temperatures, higher stresses are needed to obtain the required plastic deformation. Thus, the process is designed for more localized plastic deformation by means of punches, dies, or grooved rollers. At room temperature, the oxide film does not form as rapidly as at elevated temperatures; therefore, the two lapping surfaces provide for the required environmental control. However, complete degreasing and scrubbing for the removal of oxides and other surface contamination is necessary. The geometry of the matching punch and die sets provide for adequate deformation. As the metal deforms, surface oxides break down and permit metal-to-metal contact, which is so necessary for atomic joining. A cold weld can take place only if the mating surfaces are completely free of oxides and other surface contamination.

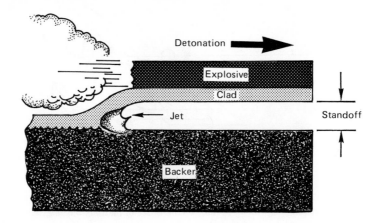

FIGURE 3-2. *In explosive welding, directional detonation sets off a wave train along the metal surfaces and plastically deforms them, stretching and rupturing the surface film to permit bonding.*

This process is often used in making electronic and refrigeration components. The ends of copper tubes on compressors are cold welded as they are pinched off for sealing. Lightweight metal containers used in the food processing industry are manufactured by intermittent cold welds for rigidity, followed by soldering for sealing purposes. The difficulty in removing all surface contamination and the need for highly localized pressures limit the use of this process. It is used on mass production lines.

EXPLOSIVE WELDING

Explosive welding, as the name implies, is the use of an explosive material as the source of energy for joining two or more pieces of material. Directional detonation (starting at one edge and proceeding in the opposite direction) causes the pieces of materials to collide, expelling most of the air between them, Fig. 3-2. The collision also sets off a wave train along the metal surfaces that plastically deforms them, stretching and rupturing the surface films to permit bonding.

While the collisions also generate some surface heating, melting seldom, if ever, occurs. The bond from explosive welding has good properties, but is different from other types of bonds in that it has an interlocking wavy pattern, Fig. 3-3. The bond is a combination of atomic and mechanical joining.

The explosive process has many applications. Its potential has not yet been fully explored. In excess of fifty different metal combinations have been welded by this technique. Thirty-five foot tantalum liners have been welded into stainless steel tubes. Slabs of copper and cupro-nickel 5 in. (12.7 cm) by 3 ft (91.4 cm) by 4 ft (121.92 cm) are welded by this process as one step in producing laminated silver and copper coins for the U.S. Government.

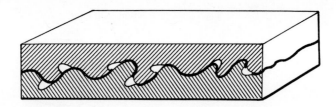

FIGURE 3–3. *The bond from explosive welding is a combination of atomic and mechanical joining.*

Explosive welding is an excellent process for joining dissimilar metals, especially over a comparatively large area. Since the process is extremely fast, melting of the metal is avoided. Explosive detonations make the process very noisy. Joints are limited to the lap type.

FRICTION WELDING

Friction welds are achieved through the principle of rubbing two metals together until the surfaces to be joined become smooth and clean enough for atomic joining. One of the two pieces about to be welded is rigidly held in a static position while the mating part is spun against it; friction at the mating surfaces generates the heat, Fig. 3–4. The elevated temperature of the two parts being joined permits

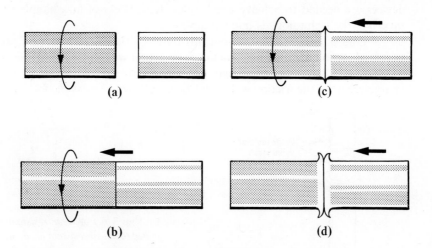

FIGURE 3–4. *Basic processes of friction welding: (a) Rotating member is brought up to desired speed. (b) Nonrotating member is advanced to meet the rotating member and pressure is applied (c). Heating phase—pressure and rotation are maintained for a specified period of time (d). Forging phase— rotation is either maintained or increased for a specified period of time. Total welding time—2 to 30 seconds. (Courtesy of American Welding Society.)*

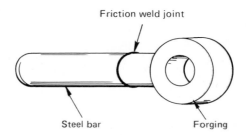

FIGURE 3–5. *Costly forgings can be welded to low carbon steel bars to produce hydraulic pistons with eyes.*

ease of deformation, which breaks up and expels the oxides and other surface contaminates requiring removal before atomic bonding can be accomplished. At the exact moment when the interface temperature of the metal being joined is just right, one of two actions must be taken; either the spinning part must be stopped or the static part permitted to spin. Both techniques are used successfully.

The Caterpillar Tractor Company has perfected a unique principle for friction welding. It is patented as the "Inertia Welding Process." For the braking system, Caterpillar uses a flywheel of the proper mass necessary to spin the revolving part just long enough to generate the exact amount of heat. The kinetic energy in the flywheel converts to heat energy when the rotating workpiece is brought in contact with the stationary piece under a computed thrust force. Various materials require different speeds, loads, and energy. Pressure, timing, and adequate braking devices are needed to achieve good welds. Regardless of how simple the principle involved, the technology is quite complex.

One of the most important attributes of the friction joining process is its potential for joining dissimilar metals. The actual weld time is often less than three seconds, and the short time span contributes to the narrow heat affected zone characteristic of the friction weld. This process makes it possible for a manufacturer

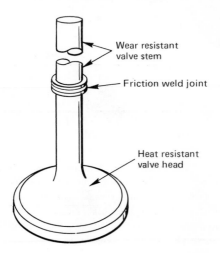

FIGURE 3–6. *A wear resistant stem can be joined to a heat resistant head, the combination gives an improved engine valve.*

to selectively place materials with desirable properties at critical points within a unit. For example, an expensive forging can be joined to a steel bar giving a hydraulic piston with eye; Fig. 3–5, or a wear resistant stem can be joined to a heat resistant head, the combination giving an improved engine valve, Fig. 3–6. Materials other than metals, such as plastics, can also be joined by friction welding.

The advantage of friction welding lies in its ability to join massive sections of dissimiliar metals in short periods of time. The metallurgical properties at the joints are similar to those of the base metal. The process is readily adaptable for automation, but due to the initial cost of friction welding equipment, it is not widely used in industry as yet. Friction welders sell for more than a half-million dollars.

ULTRASONIC WELDING

Ultrasonic welding is closely related to friction welding in principle. The pieces to be joined are clamped lightly together between two electrodes. Vibrational energy is fed into one of the electrodes. This electrode is permitted to oscillate in a plane parallel to the weld surface, Fig. 3–7. The motion develops elastic hysteresis

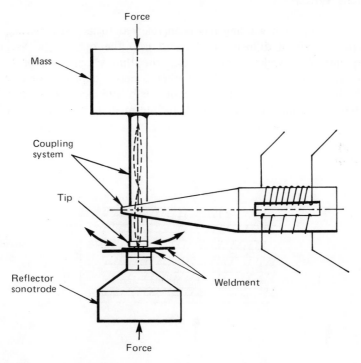

FIGURE 3–7. *Ultrasonic joining: vibration energy is induced into the upper electrode. The motion develops elastic hysteresis and plastic deformation takes place at the faying surfaces; the two pieces are thereby joined. (Courtesy of American Welding Society.)*

(internal friction) and plastic deformation at the faying surfaces of the materials to be joined. The ultrasonically produced activity generates interface temperatures 35 % to 50 % of the actual melting temperature of the base metals. Therefore, there is no melting or cast nugget, and little or no recrystallization, grain growth, or formation of brittle intermetallic compounds, each of which affects the properties of a weldment.

Ultrasonic welding is similar in nature to both friction and explosive welding in its ability to join dissimilar metals. The major difference is that friction and explosive welding are generally used for joining heavier materials, whereas ultrasonic is limited to foil thicknesses.

Ultrasonic joining techniques are not limited to welding. The application of ultrasonic principles to soldering and brazing is common. Many different metals or combinations of metals can be ultrasonically welded. Foils as thin as 0.00017 in. (.0043 mm) and as thick as 0.100 in. (2.5 mm) have been welded. Surface cleanliness is not critical, and fluxes are not required for ultrasonic application. Also, selective plastic materials are readily weldable with this process.

DIFFUSION WELDING

Diffusion welding has been referred to as "self-welding." The principle is simply that of diffusing one metal into the other. The variables affecting the quality of the weld are cleanliness, smoothness, time, and pressure. The coefficient of diffusion of one metal for another is better than the coefficient of self-diffusion. Also, the process is more applicable for brittle metals than ductile metals. When a 100 % metallic bond is required, an intermediate or sandwiched layer of metal that

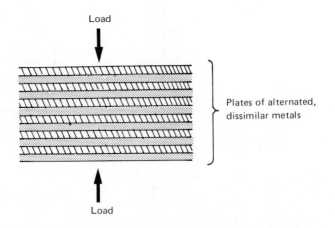

Load

Load

Plates of alternated, dissimilar metals

FIGURE 3–8. *A typical application of diffusion bonding is for joining large numbers of flat metal components into stacks. Time and pressure are necessary for diffusion to take place.*

goes through the molten phase becomes necessary. Without the intermediate layer, high pressures would be required to give a good area contact. The process is slow, but the rate of diffusion can be speeded by adding heat.

A typical application of diffusion bonding is for joining large numbers of flat components and for joining dissimilar metals, Fig. 3–8.

CHEMICAL ENERGY JOINING PROCESSES

The chemical energy group of processes include combustible gases (flame type), atomic hydrogen, and thermit. Energy produced by chemical reactions between materials has been used for joining and cutting metals for years. However, as new processes are developed, the uses for these processes become more specialized. For example, the joining of carbon steels by the gas torch was a common practice in the twenties and thirties. Today the same material is joined better and faster with some form of electric welding.

COMBUSTIBLE GASES

The combustible gas group of energy sources includes any fuel gas which, when supported by oxygen, creates a desirable flame, thus providing the heat necessary for materials joining, cutting, and surfacing. The fuel gases most often used are acetylene, hydrogen, natural gas, propane, butane, Mapp,* and Flamex.** The properties and characteristics for each of these gases are discussed in detail in Chapter 8. Oxygen is not a fuel gas; it only supports the combustion of fuel gases. From this list of fuel gases, acetylene has been used the longest time. Although still popular in the materials joining group, acetylene is rapidly being replaced by other gases in metal cutting. The decision of industry in choosing a particular gas is usually based on its properties, Table 3.1, availability, and cost per unit. For joining purposes the flame most often used is oxyacetylene, primarily because it gives the hottest combustion and the cleanest burning flame of all gas mixtures.

The ratio of oxygen to acetylene mixture determines the characteristics of the resulting flame, Table 3.2. The combustion of oxygen and acetylene occurs in three stages. The first reaction, C_2H_2 (acetylene) $\rightarrow$ 2C (carbon) + H_2 (hydrogen), occurs without the presence of oxygen, which makes acetylene a dangerous gas to handle—especially under pressure. The second reaction is $2C + O_2$ (oxygen) $\rightarrow$ 2CO (carbon monoxide). When adequate oxygen is supplied, the two cones shrink into one; and the two reactions collectively become $C_2H_2 + O_2 \rightarrow 2CO + H_2$, or what is known as a neutral flame, Fig. 3–9. This flame is used whenever melting of the base metal

---

* A product of Dow Chemical Corp., Division of Airco Co., Inc.
** A product of Flamex Corp.

TABLE 3.1. *Comparison of Properties for U.S. Industrial Fuel Gases (Courtesy of Dow Chemical, Division of Airco Co., Inc.*

	MAPP Gas	Acetylene	Natural Gas	Propane	Flamex
Shock sensitivity	Stable	Unstable	Stable	Stable	Stable
Explosive limits in oxygen, %	2.5–60	3.0–93	5.0–59	2.4–57	*––
Explosive limits in air, %	3.4–10.8	2.5–80	5.3–14	2.3–9.5	–*–
Maximum allowable regulator pressure, psi	Cylinder [225 psig (1.653 MPa) at 130°F (54°C)]	15	Line	Cylinder	Cylinder
Burning velocity in oxygen, ft/sec	7.9 (2.4 m/sec)	17.7 (5.4 m/sec)	8.2 (2.5 m/sec)	5.9 (1.8 m/sec)	14.5 (4.4 m/sec)
Tendency to backfire	Slight	Considerable	Slight	Slight	Slight
Toxicity	Low	Low	Low	Low	Low
Reactions with common materials	Avoid alloys with more than 67% copper	Avoid alloys with more than 67% copper	Few restrictions	Few restrictions	*––

* Information Not Available.

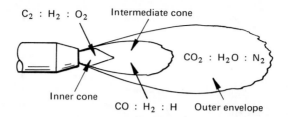

$C_2 : H_2 : O_2$ Intermediate cone

$CO_2 : H_2O : N_2$

Inner cone

$CO : H_2 : H$ Outer envelope

FIGURE 3-9. *The chemical reaction within the oxyacetylene flame. When adequate oxygen is supplied, the two cones shrink into one and the two reactions collectively become* $C_2H_2 + O_2 \rightarrow 2CO + H_2$ *or what is known as a neutral flame.*

is involved. The outer flame (envelope) is where the two volumes of carbon monoxide and one volume of hydrogen combine with oxygen from the air and burn to form two volumes of carbon dioxide and one volume of water vapor. The carburizing flame, second reaction, is used for soldering, brazing, and hard surfacing applications. The carbonizing flame, if used with carbon steels, adds carbon to the base metal, thereby changing the hardening properties. It is likewise possible to produce an oxidizing flame that has an excess of oxygen. This flame is seldom used except for removing paint and grease from the surface of metals.

The equipment for all flame energy systems is basically the same. It consists of a fuel gas cylinder or generator, oxygen cylinder, regulators, hoses, and torch or gun, Fig. 3–10. Fuel gases, except acetylene, can be liquidified. Gas in liquid form is used only for storage and shipping purposes. All liquid fuel gases, for combustion purposes, must be vaporized prior to mixing with oxygen. Gases that vaporize easily at room temperature are stored in cylinders (850 psi (5.962 MPa) and lower). Gases such as oxygen and hydrogen are stored in cylinders at high pressure (2,000 psig (13.89 MPa) and higher). Acetylene gas can be used directly from the generator with no storage necessary; however, it is often stored in a specially constructed cylinder.

Regulators are used to reduce gas pressures and to provide control over gas flows. The two gauges on gas systems indicate the tank pressures in psi (N/m^2 or MPa) the cubic foot (cubic meter) content, and the line pressures in psi. Regulators

TABLE 3.2. *Oxygen–Acetylene Flame Characteristics*

Ratio of (cylinder) Oxygen/Acetylene	Flame Temperature °F (°C)	Flame Characteristics
0.8 to 1.0	5,550 (3066)	Carbonizing
0.9 to 1.0	5,700 (3149)	Carbonizing
1.0 to 1.0	5,850 (3232)	Neutral
1.5 to 1.0	6,200 (3427)	Oxidizing
2.0 to 1.0	6,100 (3371)	Oxidizing
2.5 to 1.0	6,000 (3315)	Oxidizing

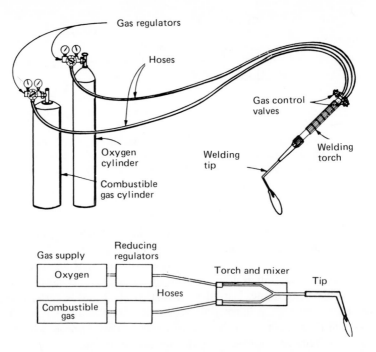

FIGURE 3-10. *The principle for the flame equipment used in metals joining techniques. (Courtesy of American Welding Society.)*

should be used only with the gas for which they were designed. Hoses should be lightweight, flexible, nonporous and constructed to withstand maximum pressures. Recommended hoses are those where the one carrying fuel gas and the other carrying the oxygen are manufactured as one, Fig. 3-11. This design is more convenient and safer to use than the traditional two hose hook-up. For ease of identification and safety, the color red is usually associated with acetylene and green with oxygen. The International Institute of Welding (IIW) specifies blue as the international color for oxygen hoses. To prevent accidents, all acetylene hose fittings are brass with left-hand thread; and for ease of identification, the acetylene fittings are circumferentially

FIGURE 3-11. *A cross-sectional view of molded hoses used for safety in gas processes.*

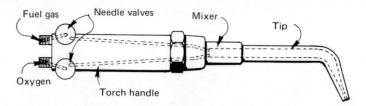

FIGURE 3–12. *The basic elements of a welding torch.* (*Courtesy of American Welding Society.*)

grooved in the center. Oxygen hose fittings are brass and have right-hand threads. Use of non-approved fittings or hoses should not be permitted.

The torch consists of two valves, a torch handle, a mixing chamber, and a tip or head, Fig. 3–12. The purpose of the torch is to mix gases, control volume, and direct the flame. Each torch is equipped for interchangeable heads and may use as many as fifteen different sizes.

The oxy-fuel gas flame is used for soldering, brazing, braze welding, fusion welding, metal spraying, hardsurfacing, cambering, heat treatment, and ferrous metal cutting. Of this group, fusion welding is the least used in industry. The reason is the low efficiency of the flame compared to that of other fusion welding processes. Due to the large number of processes using oxy-fuel as their source of energy, most metals can be formed and joined by the application of the flame.

The soft oxyacetylene flame offers excellent control for joining low melting point metal. The neutral flame provides adequate shielding for welding steel, cast iron, and some nonferrous metals. The oxy-hydrogen flame is useful for underwater cutting and joining. The other fuel gases are used in all applications except fusion welding. The required equipment for combustible gas processes can be made into portable units. Several disadvantages of the process are: low efficiency of energy transfer, hazardous conditions created by gas and its mixtures, and the demurrage charges which accumulate for cylinders kept beyond the specified rental time.

ATOMIC HYDROGEN ENERGY

Atomic hydrogen energy is developed within an electric arc; therefore, it would appear that this process should be classified with the electric group. However, because of the nature in which the energy is created, it seems more appropriate that it be referred to as a chemical source of energy. Molecular hydrogen (H) is passed through an arc that is maintained between two tungsten electrodes. At this time the molecules of hydrogen disassociate into atomic hydrogen. At the instant of disassociation, a great amount of heat from the arc is absorbed by the hydrogen atoms. As the atoms leave the arc stream and come in contact with the metal being welded, they

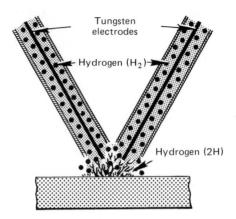

FIGURE 3–13. *In the atomic hydrogen welding process, molecular hydrogen is passed through an arc. Disassociation at the arc and re-association of the hydrogen atoms as the metal liberates sufficient heat to melt the base metal.*

recombine to form stable diatomic gas molecules (2H). This results in the liberation of the absorbed heat, which raises the temperature of the base metal to its melting point. Throughout the process, additional heat is generated by the burning of molecular hydrogen in the air to form water vapor. Water vapor effectively protects the molten weld metal from contamination. Therefore, the surfaces of the metals to be joined are cleaned and atomically joined through melting. At elevated temperatures, the metals are adequately protected from the atmosphere by the vapor formed during combustion. The temperature of the atomic hydrogen arc is approximately 10,000 °F (5540 °C).

The atomic hydrogen process can be used for both manual and automatic welding. The flame is wide, soft, and high in energy. Originally the atomic hydrogen process was used for the joining of thin gauge material. More recently developed processes are rapidly replacing atomic hydrogen in this area. However, it is often used for resurfacing the worn faces of dies and punches. The high flame temperature is ideal for depositing a thin layer of high melting material on heavy bases.

The atomic hydrogen process includes an ac constant-potential power source, a cylinder of hydrogen or an ammonia dissociator, a regulator, a connecting hose, an electrical control unit, and the torch. The ac power source is a high voltage machine, up to 300 open circuit volts (OCV). The high voltage is required to ignite and maintain an arc in the hydrogen atmosphere. The hydrogen is compressed into a cylinder at 2,000 psig (13.89 MPa). An ammonia dissociator can be used in lieu of the compressed hydrogen. The dissociator is a compressor mechanism capable of condensing ammonia gas (NH_3) at −28 °F (−33.3 °C), thus separating the hydrogen from the nitrogen. The regulator is used to reduce the cylinder pressure from 2000 psig (13.89 MPa) to a 3 psig (0.122 MPa) working pressure. The electric control unit, which is usually foot operated, gives the operator starting and stopping control over the flow of the electric current between the power source and welding gun. The torch is equipped with two tungsten electrodes, Fig. 3–13. The electrode ends form a gap just beyond the hydrogen orifice. The arc is ignited by bringing the ends of the two electrodes in contact with each other and then immediately separating them to a specified distance. Hydrogen is passed through two nozzles, one encasing each of the electrodes. As the hydrogen enters the arc stream it becomes a fan-shaped ionized flame extending between the tips of the two electrodes. The work piece is not a part of the electrical circuit.

THERMIT WELDING

Thermit welding is closely related to the casting techniques used in a foundry. To prepare the metal pieces to be welded, the surfaces to be joined must be thoroughly cleaned. The pieces are firmly and accurately jigged into position. A wax is used to fill in the joint and form the geometrics of what the completed weld is to look like. A sand mold is then built around the unit, Fig. 3–14. The features of traditional sand mold construction are then built into the mold; included are: the sprue, riser, air vents, gating, and a preheating gate. A gas flame is generally used to preheat the casting prior to the welding operation. The pieces to be joined are often preheated to a temperature of "red heat." The wax melts and flows away from the mold. The remaining cavities are eventually filled with the super-heated metal from the crucible. The preheating also dries out the mold and the metal to be joined. The high preheat temperature is also advantageous metallurgically; it aids in better

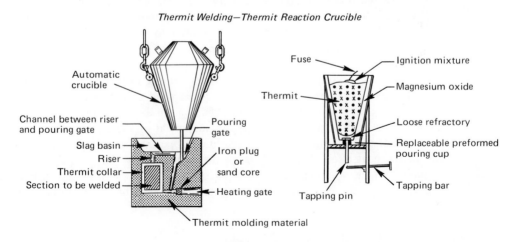

Thermit Welding—Thermit Reaction Crucible

FIGURE 3–14. *The mechanics of thermit welding. The energy of thermit welding is obtained from a powdered metal and a powdered metallic oxide, which produces highly superheated liquid metal. (Courtesy of American Welding Society.)*

TABLE 3.3. *Commonly Used Thermit Mixtures and the Temperatures Produced Through the Chemical Reaction*

$3\,Fe_2O_4 + 8\,Al \rightarrow 9\,Fe + 4\,Al_2O_3$	(5590 °F, 3066 °C)
$3\,FeO + 2\,Al \rightarrow 3\,Fe + Al_2O_3$	(4532 °F, 2500 °C)
$Fe_2O_3 + 2\,Al \rightarrow 2\,Fe + Al_2O_3$	(5360 °F, 2960 °C)
$3\,CuO + 2\,Al \rightarrow 3\,Cu + Al_2O_3$	(8790 °F, 4866 °C)
$3\,Cu_2O + 2\,Al \rightarrow 6\,Cu + Al_2O_3$	(5680 °F, 3138 °C)
$3\,NiO + 2\,Al \rightarrow 3\,Ni + Al_2O_3$	(5740 °F, 3171 °C)
$Cr_2O_3 + 2\,Al \rightarrow 2\,Cr + Al_2O_3$	(5390 °F, 2977 °C)
$3\,MnO + 2\,Al \rightarrow 3\,Mn + Al_2O_3$	(4400 °F, 2427 °C)
$3\,MnO_2 + 4\,Al \rightarrow 3\,Mn + 2\,Al_2O_3$	(9020 °F, 4993 °C)

fused joints and permits a longer cooling period for the separation of impurities from the molten metal by flotation. All these conditions provide for an improved joint weld. When many small units are to be joined, such as reinforcing bars, permanent molds made of steel or carbon are recommended. A number of metal oxides can be used to produce extra energy for chemically combining with aluminum (Table 3.3). A ferric oxide produces the highest temperature for joining steels, for example:

$$3\,Fe_3O_4 + 8\,Al \rightarrow 9\,Fe + 4\,Al_2O_3 \rightarrow 5590\,°F\ (3066\,°C)$$

A charge of 1000 grams (g) of thermit produces 476 g of slag, 524 g of iron, and 181,500 cal. The ferric oxide is obtained from mill scale.

A thermit mixture ignites at temperatures above 2370 °F (1300 °C); therefore, a magnesium strip is usually used as the ignitor. The length of time per weld is short, usually less than 30 seconds. Small pieces of scrap metal can be added to improve the quality of the weld metal. The aluminum in the thermit mixture is also used as a deoxidizer, which gives an improved weldment. The hardware used with the thermit joining process is limited to cupola and mold. The mold may be a once used sand mold or a semipermanent mold of metal or carbon used repeatedly.

Thermit welding is commonly used for joining rails for railroad tracks and for reinforcing bars to be used in building and road construction. It is also used for joining cables and bars of both ferrous and nonferrous materials. The thermit welding process is being replaced somewhat by the electroslag process.

Good metallurgical and engineering properties are obtained at the joint and in the heat affected zone. A large investment for welding equipment is not required and the welding time is short. However, the cooling period is long, and building the sand mold is time consuming. Moisture in the mold can cause violent explosions.

CHEMICAL ENERGY CUTTING PROCESSES

Flame Cutting. Flame cutting is a term used to describe the chemical reaction of an oxy-fuel flame with metal. Although not a joining process, rather, a separation process, it is included here because it is used extensively in welding fabrication.

Flame cutting refers almost exclusively to ferrous metals. In fact, plain carbon steels are the only metals that oxidize before they melt. When steel is brought to a temperature of between 1400 °F and 1600 °F (760–871 °C), it reacts rapidly with oxygen to form iron oxide, (Fe$_3$O$_4$). The chemical reaction may be shown as $3\,Fe + 2\,O_2 = Fe_3O_4$. The melting point of the oxide is somewhat lower than that of the steel. The heat generated by the burning iron is sufficient to melt the iron oxide and some free iron, which runs off as molten slag, exposing more iron to the oxygen jet.

Theoretically, 1 cu ft of oxygen is required to oxidize about 3/4 cu in. of iron. Actually, the *kerf*, or cut, is not entirely oxidized, but 30 to 40 percent of the metal is washed out as metallic iron.

Once the cut is started, enough heat should be present to continue cutting without the need for preheating flames, using only the oxygen. But, in practice, this does not work out. Excessive radiation at the surface, small pieces of dirt, and paint or scale make it necessary that the oxygen be surrounded with preheating flames throughout the cutting operation.

Although acetylene is used widely as the fuel gas, other gases that can be used include hydrogen, natural gas, propane, propylene, MAPP gas (methylacetylene plus propadiene), and Flamex.

The cutting processes can be broadly classified as either manual or by machine.

Manual Cutting. The cutting torch somewhat resembles the welding torch, but it has provisions for several neutral flames and a jet of high pressure oxygen in the center, Fig. 3–15. The neutral preheating flames are adjusted and controlled as described previously in gas welding. The oxygen needed for cutting is controlled by a quick acting trigger or lever-type valve.

The thickness and type of material to be cut will determine the tip size. Best results are obtained when the cutting oxygen pressure, cutting speed, tip size, and preheating flames are controlled to give a narrow, clean cut. Improperly made

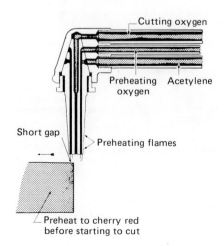

FIGURE 3–15. *The cutting torch has an oxygen jet surrounded by preheating flames.*

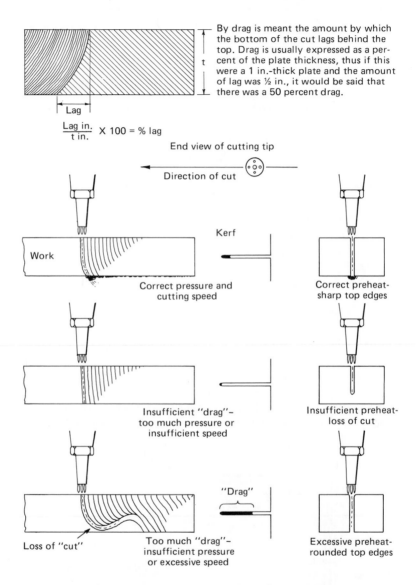

By drag is meant the amount by which the bottom of the cut lags behind the top. Drag is usually expressed as a percent of the plate thickness, thus if this were a 1 in.-thick plate and the amount of lag was ½ in., it would be said that there was a 50 percent drag.

$$\frac{\text{Lag in.}}{t \text{ in.}} \times 100 = \% \text{ lag}$$

End view of cutting tip

Direction of cut

Kerf

Work

Correct pressure and cutting speed

Correct preheat-sharp top edges

Insufficient "drag"—too much pressure or insufficient speed

Insufficient preheat-loss of cut

"Drag"

Loss of "cut"

Too much "drag"—insufficient pressure or excessive speed

Excessive preheat-rounded top edges

Sight method for determining correct cutting procedure.

FIGURE 3–16. *Correct cutting conditions are indicated by the correct amount of drag.*

FIGURE 3–17. *Faults to avoid in torch cutting. (Courtesy of Linde Co., Division of Union Carbide Corp.)*

cuts will produce ragged and irregular edges, with slag adhering at the bottom of the plates. One indication of proper speed is the drag lines. The drag of a cut is the distance between the point where the oxygen stream enters the top of the metal and the point where the slag emerges at the bottom of the cut, Fig. 3–16. The entrance and exit points of the cutting oxygen are expressed as a drag ratio or percentage.

$$Drag = \frac{d}{t} \qquad Drag \% = \frac{d}{t}(100)$$

where d = amount of horizontal lag in in.
t = thickness of the metal being cut in in.

Most shops that do considerable cutting attempt to achieve an economical drag ratio; this is the longest drag that will produce the accuracy required with a minimum of uncut corners. Generally, a drag ratio of 10 % is the uppermost limit, but some straight cutting may go as high as 20 %. Defects that arise in the flame cutting of steel are shown and explained in Fig. 3–17.

Machine Cutting. Although manual cutting is used extensively for salvage and repair work, it lacks the speed, accuracy and economy of machine cutting. With a cutting torch mounted on a *radiograph* (a variable speed electric motor drive),

FIGURE 3–18.   *A photoelectric tracer used to guide a multiple torch setup for shape cutting. (Photo courtesy of Airco Welding Products, Div. of Airco, Inc.)*

which in turn may be mounted on a track, very accurate cuts can be made. A compass attachment is used for circular cuts.

Other methods used in guiding the cutting torch are templates, photoelectric tracers, and numerical controls. Shown in Fig. 3–18 is a photoelectric tracer unit being used to guide several cutting torches at once.

Stack Cutting.   To increase production, sheets may be stacked and cut simultaneously. However, the sheets must be clamped tightly so that no air passes between them, this would cause the cut to be lost. Clamping also keeps thin parts from warping during the cut.

Cutting Tolerances.   Shown in Table 3.4 is a list of ASM tolerances that can be expected on flame cut parts. Close cutting tolerances start at 0.010 in. (0.25 mm) for light materials and increase to 0.030 in. (0.75 mm) for $\frac{1}{4}$ to 2 in.

TABLE 3.4.   *ASM Cutting Tolerances*

Plate Thickness		Tolerance	
(in.)	(mm)	(in.)	(mm)
to $\frac{1}{2}$	12.7	1/16	1.58
$\frac{1}{2}$ to $1\frac{1}{3}$	12.7 to 33.78	3/32	2.38
$1\frac{1}{3}$ to 3	33.78 to 76.20	1/8	3.175
3 to 5	76.20 to 127.00	5/32	3.96
5 to 6	127.00 to 152.40	3/16	4.76
6 to 8	152.40 to 203.20	7/32	5.55
8 to 10	203.20 to 250.40	1/4	6.35

(6.35–50.80 mm). Close cutting costs more than cutting to ASM tolerances, and it requires excellent equipment, more planning, supervision, and preparation.

Physical and Metallurgical Effects of Torch Cutting. Flame cutting of mild steel has very little effect on the metal adjacent to the cut. However, as the carbon or alloys content is increased, some hardening of the edge will occur due to the quenching action of the adjacent metal and the atmosphere. The hard edges may be difficult to machine, and the lowered ductility of this hard layer may cause cracking under load. The best method of avoiding this condition is to preheat the metal. Medium carbon steels should be heated from 350 °F to 700 °F (177–371 °C); low alloy, high tensile strength steels require from 600 °F to 900 °F (316–482 °C).

Heavy plates do not warp when flame-cut, but plates $\frac{1}{2}$ in. or less in thickness may have to be clamped or the amount of cutting done at one time may have to be restricted. Judicious postheating with a torch can be effective in reducing or eliminating warpage.

CUTTING ALLOY STEELS, CONCRETE, AND CAST IRON

Alloy Steels. As stated previously, steel is readily cut by the oxyfuel flame because the oxide film forming over the molten metal is of a lower melting point than the base metal. This is not true in cutting some alloy steels. For example, stainless steel, aluminum, bronzes and high nickel alloys form oxides that have a higher melting point than the base metal. To cut these metals, an iron powder is fed into the oxygen stream. This method is termed *powder cutting.*

Powder Cutting. The powder cutting method, first introduced in 1943, employs a cutting torch with a tube feed unit. The tube contains iron rich powder that is fed to the torch tip under the pressure of either compressed air or nitrogen. When the powder reaches the oxygen cutting stream, it burns with considerable heat. With this increased heat, the oxygen flame cutting action is quickened, and the number of materials to which it can be applied is substantially increased.

The powder cutting process was developed originally to cut stainless steel, but now is used successfully on other alloy steels, cast iron, bronze, nickel, aluminum, certain refractories, and concrete. Nonferrous materials require that a mixture of aluminum and iron powder be used instead of the straight iron powder. Powder cutting is still being used, but is being replaced by plasma arc cutting, a method discussed in Chapter 4.

Concrete Cutting. Concrete and reinforced concrete can be efficiently cut with a powder cutting torch for thicknesses up to 18 in. This is a mechanized operation and average cutting speeds range from 1 to $2\frac{1}{2}$ ipm. For concrete thicker than 18 in., the powder lance is used.

Cast Iron Cutting. Cast iron also has a higher melting point oxide than the base metal. When the cast iron melts, the oxides and impurities mix together, making it very difficult to cut. It requires a much higher preheat temperature and an oxygen pressure from 25 to 100 percent greater than for steel. The preheating flames should be adjusted for an excess of acetylene. The motion used in cutting is to swing the torch from side to side in a small arc. The diameter of the arc is equal to the width of the kerf, which is determined by the thickness and quality of the metal. Thick or poor quality metals require some working of the molten metal with a steel rod.

Lance Cutting. An oxygen lance is a simple device consisting essentially of a length of steel pipe [usually 1/8 in. or 1/4 in. size (3.175 or 6.35 mm)], a length of hose, some couplings, a control valve, and an oxygen tank complete with regulator. With this equipment no preheating flame is provided, and an auxiliary torch is needed. After heating the metal to the kindling temperature, the lance is brought over to start the cut. Other methods also are used to obtain the heat necessary to start the cut, such as placing a red-hot piece of steel on the starting point or heating the end of the lance until it is red hot. When the lance is brought in contact with metal and the oxygen is turned on, the end of the pipe will burn brilliantly, furnishing enough heat to start the cut.

The oxygen lance is an excellent tool for piercing holes in steel. For example, a hole $2\frac{1}{2}$ in. in diameter (63.50 mm) can be cut in one foot thick (30 cm) steel in a matter of two minutes. It is routinely used in tapping blast and open-hearth furnaces.

For continuous cuts, the lance and cutting torch are frequently used together. The torch furnishes the heat needed to keep the cut going along the top surface, and the lance carries the cut through the piece. When the lance reaches the bottom of the metal, it is brought back to the top and again lowered, each time making an advance. Masses of steel eight feet thick have been cut by this method.

Powder cutting is also done with lance equipment. This consists of a lance handle attached to one or more lengths of black iron pipe. Iron and aluminum powders are mixed with oxygen in the lance handle, and they burn at the end of the pipe.

The powder-cutting lance has proved successful in cutting fire-brick, aluminum billets, bronze, both steel and cast iron containing inclusions, and concrete. The process is particularly good for cutting thick concrete. Examples of this work are cited to give an understanding of the method's usefulness: The powder lance was called on to cut a hole for a pipeline in a concrete wall 12 ft thick (360 cm). A clean eight inch (20.32 cm) diameter hole was pierced in a matter of $1\frac{1}{4}$ h. Powder cutting of concrete walls 4 ft (120 cm) thick have been made at the rate of $1\frac{1}{2}$ ft per h.

EXERCISES

Problems

3-1. You are the welding engineer for the "X" company. A customer comes to you with a joining problem for 500,000 cast steel halves of track idlers. Each half weighs 15 lb (6.8 Kg) and is 8 in. (20 cm) in diameter and 6 in. (15 cm) across the face. The customer specifies a narrow heat affected zone. The halves are shown below.

What welding process would you propose for this job? Why?

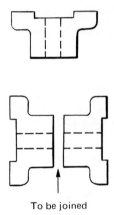

To be joined

FIGURE P3–1.

3-2. Your plant is expected to provide a lamination by welding a number of plates (2′ × 6′ × 2″) (0.6 m × 1.8 m × 5 cm) of dissimilar metals (copper to steel). The end product will consist of one steel plate sandwiched between two layers of copper. Which process would you choose? Why?

3-3. Assume you have an assignment similar to that in problem 3–2, except many strips of thin, hard, and brittle metals are to be laminated. The property requirements for the end product eliminate the use of arcs, flames, or any movements of the pieces during the joining process. Which process would you suggest? Discuss any variations of the technique.

3-4. With the information provided in Table 3.1, name and discuss the ideal fuel gas.

3-5. After you have studied the sketches for thermit welding in Fig. 3–14, discuss the following statement: "The thermit welding process provides excellent metallurgical properties."

3-6. After studying Fig. 3–9, discuss the following statement. To achieve complete combustion (neutral flame) with the oxyacetylene torch, you must have a ratio of 2-1/2 parts of oxygen to 1 part of acetylene. Figure 3–10 shows only one part of each entering the torch. Explain.

3-7. What is the maximum and minimum size that could be expected in cutting a 10 in. diam circle from 1 in. thick steel plate?

3-8. A 2 in. steel plate is flame cut. In examining the edge after the cut, the drag is determined to be 1/8 in. Is this excessive?

Questions

3-1. Name the six principles of mechanical energy joining systems.

3-2. Explain exactly how the faying surfaces of the metal being joined are cleaned free of oxides and other contaminates with the mechanical joining process.

3-3. Explain exactly how the faying surfaces of the metal being joined are adequately smoothed to achieve atom-to-atom joining.

3-4. What is the oldest known welding process, and with what kind of metal was it generally used?

3–5. It is well known that as the temperature of steel rises the metal becomes more plastic and works more easily. What is the greatest disadvantage of heating metals for welding?

3–6. With several welding procedures a flux is used. Explain the purpose of a flux when welding.

3–7. Forge welding produces welds with better stress distribution than most other welding processes. Explain why this happens.

3–8. At the first opportunity, take a close look at the ends of the copper tubes near the compressor of your refrigerator and determine what kind of welding process was used to seal them.

3–9. Explain the principle for joining material by explosive welding.

3–10. Control of what three variables determines the difference between a good or poor friction weld?

3–11. What is the difference between a conventional friction welding process and Caterpillar's patented "Inertia Welding Process?"

3–12. What is meant by "elastic hysteresis" as applied to ultrasonic welding?

3–13. State the four variables necessary for quality joining by the diffusion process.

3–14 Name the nine most common applications for the oxy-fuel flame.

3–15. For what application would the oxy-hydrogen flame be used?

3–16. Name the seven most commonly used fuel gases for the oxy-fuel flame.

3–17. Why is oxygen not considered a fuel gas?

3–18. Of the several fuel gases, only one is capable of giving a good fusion weld with carbon steels. Name the gas and the reason for this being true.

3–19. Explain an oxidizing flame and its applications.

3–20. Explain a reducing (carburizing) flame and its applications.

3–21. What is the function of a gas regulator?

3–22. What is the most common application for the atomic hydrogen process?

3–23. The thermit welding process offers a number of desirable features. Please discuss.

3–24. What is meant by drag in the flame cutting of metals?

3–25. Why would it be advisable to preheat medium carbon or high tensile strength steels before cutting them?

3–26. Why is it harder to flame cut cast iron than steel?

3–27. How can nonferrous metals and stainless steel be successfully flame cut?

3–28. (a) When is it particularly advantageous to use the oxygen lance? (b) When is the powder lance used to advantage?

BIBLIOGRAPHY

ALTHOUSE, A. D., TURNQUIST, C. H., AND BOWDITCH, W. A., *Modern Welding*, Goodheart-Wilcox, Homewood, Ill., 1965.

PHILLIPS, A. L., *Current Welding Processes*, American Welding Society, Miami, Florida, 1964.

GIACHINO, J. W., WEEKS, W., JOHNSON, G. S., *Welding Technology*, The American Technical Society, Chicago, Ill., 1968.

KISIELEWSKI, ROBERT, *Statistical Analysis of Fuel Gases for Flame Cutting*, MSc Thesis, University of Wisconsin, Madison, Wi., 1972.

HOULDCROFT, P. T., "Welding Processes," British

BIBLIOGRAPHY

Welding Research Association, Cambridge University Press, London, Eng., 1967.

LINDBERG, R. A., *Processes and Materials of Manufacture*, Allyn and Bacon, Inc., Boston, Mass., 1964.

PHILLIPS, A. L., *Modern Joining Processes*, American Welding Society, Miami, Fla., 1966.

VILL, V. I., *Friction Welding of Metals*, American Welding Society, Miami, Fla., 1962.

"Welding Handbook," 6th ed., sect. 1, American Welding Society, Miami, Fla., 1968.

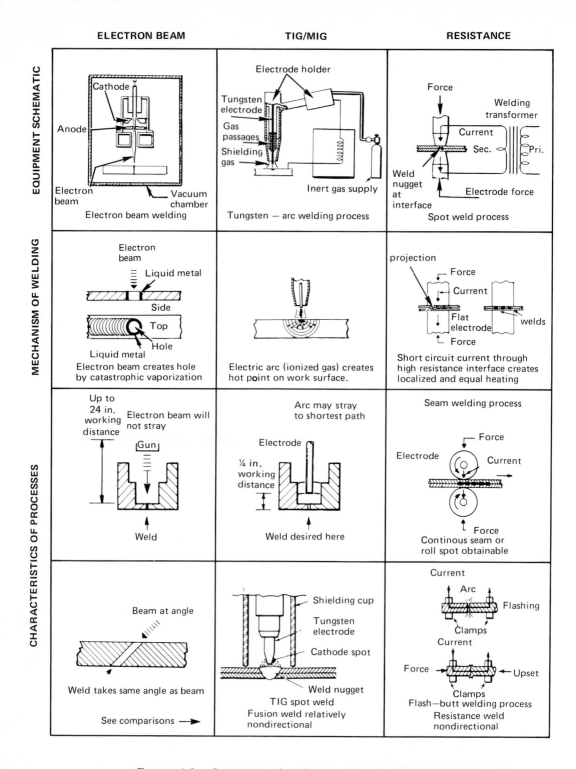

FIGURE 4-0. *Comparison of similar operations for EB, GTA, GMA, and resistance welding. Comparison of equipment, mechanisms and characteristics. (Courtesy of Sciaky Brothers, Inc.)*

4

Electrical Energy
Joining Processes

INTRODUCTION

Electrical energy is readily available, easy to control, and adaptable as the source of heat in many welding processes. These characteristics make electricity the most frequently used source of energy in the field of material joining. Because resistance to the flow of electrons in one form or another generates heat, this section will discuss resistance heating under three natural divisions. They are *gas resistance*, *solid resistance*, and *liquid resistance*. Two exceptions apply to the above classifications, whereby electricity is the source of energy but the heat generated is not due to resistance. The exceptions are electron beam welding and laser welding. However, for discussion purposes these two processes are included with the solid resistance group.

A clear understanding of the terminology is necessary for good learning. The electrical energy group has a number of terms common to several of the processes. To assure continuity and clarity, these terms are discussed in the next few pages of the chapter.

Constant Potential Power Source (CP). The CP power source, sometimes referred to as constant voltage (CV), has a flat curve (theoretically), Fig. 4–1. A control sets the open circuit voltage (OCV), and the amperage output is dependent on the wirefeed of the wiredrive unit. The CP power source can be compared to the ordinary wall receptacle found in the home; the voltage is constant, and the amperage is provided to match the load requirements, which in this case is the

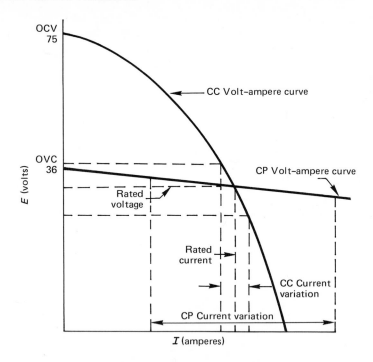

FIGURE 4–1. *Volt-ampere curve for the CC power source superimposed on the volt-ampere curve for the CP power source. (Courtesy of Charles R. Stevens, Senior Manufacturing Engineer, Hyster Co.)*

wire melt-off rate (MOR). The CP power source is never used with stick electrodes. The National Electrical Manufacturers Association (NEMA) long ago developed standards for electric welding machines. These standards cover established practice in ratings, basis of ratings, methods of tests, and broad overall principles of construction contributing to safety.

Constant Current Power Source (CC). The CC power source is often referred to as the "drooper", Fig. 4–1. The only control the operator has over the voltage on a CC power source is through the length of the arc. As the arc becomes longer, the voltage rises and the amperage drops slightly. Therefore, the operator can knowingly or unknowingly control the voltage and, to a lesser degree, the amperage. A change in voltage will change the geometry of weld penetration. A change in amperage affects the MOR of the electrode.

Direct Current (dc). Direct current is electrical current that flows in only one direction and exhibits both a positive and a negative polarity. Thus, the welding engineer has a choice of polarity, straight (DCSP) or reverse (DCRP). The correct polarity is determined by the process, type of metal, or joint geometry. With DCSP, the electrode is negative and the ground positive. The electrons flow from the electrode (cathode " − ") to the work piece (anode " + "), and the ions

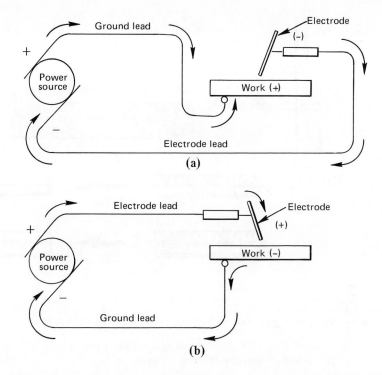

FIGURE 4–2. (a) *DCSP for a welding power source, electrode is negative and the work piece is positive,* (b) *DCRP for a welding power source, electrode is positive and the work piece is negative.*

flow from the work piece to the electrode, Figs. 4–2a and 4–2b. Two-thirds of the heat from an arc is estimated to be on the positive side. Therefore, a rule of thumb is that straight polarity will produce less heat at the electrode and should be used for easy-to-weld materials in easy-to-weld positions. Conversely, reverse polarity concentrates approximately two-thirds of the total heat into the small cross-sectional area of the electrode and produces a high current density. This is helpful for the more difficult-to-weld metals and for *out-of-position welding*. Out-of-position welding refers to any position where the piece being welded is not parallel to the surface of the earth.

Alternating Current (ac). Alternating current is a compromise from dc in that 50% of the time it is straight polarity, and 50%, reverse polarity. Generally, this is ordinary 60 Hz current, or 120 reversals of direction per second.

Open Circuit Voltage (OCV). Open circuit voltage is the potential voltage of a power source at no load ($I = 0$). It occurs when the machine is idling and is usually in the range of 65–100 volts. The OCV for a transformer type power source is generally 80 volts and near 100 volts for a motor generator type power source.

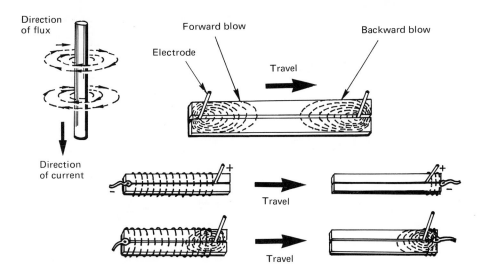

Backward blow: When welding towards ground connection end of a joint, or in a joint, or in a cover.

Forward blow: When welding away from the ground at the start of a joint.

FIGURE 4–3. *The phenomenon of arc blow when welding with dc current. (Courtesy of American Welding Society.)*

Short Circuit Voltage (SCV). Short circuit voltage represents a condition of maximum current draw with the voltage approaching zero. When welding with a constant current power source, the load voltage (LV) or arc voltage is the total voltage between the electrode holder and the base metal immediately adjacent to the arc terminals.

Arc-Blow. The phenomenon of arc blow is observed almost exclusively with dc welding. The dc welding current flowing through the electrode and plate sets up magnetic fields around the electrode; the fields tend to deflect the arc either to the sides or more often backwards and forwards from its intended path, Fig. 4–3.

Contact Tube Height

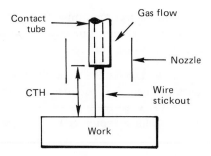

FIGURE 4–4. *The contact tube is the only connection between the power source and the wire being fed into the arc. The average contact tube height (CTH) is 3/4 inch (1.905 cm). (Courtesy of Charles R. Stevens, Senior Manufacturing Engineer, Hyster Co.)*

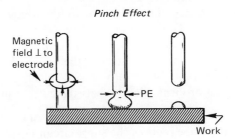

Pinch Effect

Magnetic field ⊥ to electrode

PE

Work

FIGURE 4–5. *A magnetic field forms perpendicular to the wire and acts to overcome the surface tension of the molten metal, pinching it off. (Courtesy of Charles R. Stevens, Senior Manufacturing Engineer, Hyster Co.)*

Burn-Back. Burn-back results when the wire (consumable electrode) melts off faster than it is being fed from the gun. The end of the filler wire fuses with the copper contact tube.

Stubbing. Stubbing is just the opposite of burn-back, in that the wire is being fed out at a faster rate than it is being melted off. This often causes the electrode to freeze in the base metal when the flow of current is shut off.

Contact Tube. The contact tube makes the electrical connection between the power source and the wire being fed into the arc, Fig. 4–4.

Pinch Effect. Pinch effect is the result of a temporary short circuit between the wire electrode and the work. (See Fig. 4–11.) During this brief period, the current reaches a high value and the wire begins to melt. A magnetic field forms perpendicular to the wire and acts to overcome the surface tension of the molten metal, pinching it off, Fig. 4–5. The rate at which the pinch effect is applied to the current carrying conductor will control the amount of spatter, Fig. 4–6.

Spatter. Spatter is small, hot, meteorite-like material that has wandered or been forced outside of the shielding cone of an arc stream. Once outside the shielding median, the material combines with the atmosphere and may burn up or deposit itself on any surface that happens to be in its path.

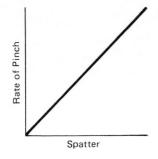

FIGURE 4–6. *The rate at which the pinch effect is applied to the current carrying conductor will control the amount of spatter. (Courtesy of Charles R. Stevens, Senior Manufacturing Engineer, Hyster Co.)*

Slope Control. Slope makes reference to the shape of the static volts (E)-to-amperes (I) curve of the CP power source. The curve slope is derived from an impedance (magnetic resistance) produced by an inductor on the ac side (secondary coil) of the power source. The steeper the curve, the greater the slope. Indeed, so much slope can be built into a CP machine that it will take on the characteristics of a drooper. Slope affects the current in the following ways:

1. Slope controls the current surge at the time of electrode short circuiting.
2. Increasing the slope reduces the short circuit current, thus limiting pinch force and reducing spatter.
3. Increasing the slope reduces the output potential of the power source. However, there must be adequate pinch force so that the arc is clearly reignited after each short circuit.
4. To maintain the same welding conditions, as the slope is increased the OCV must also be increased a proportional amount, Fig. 4–7.

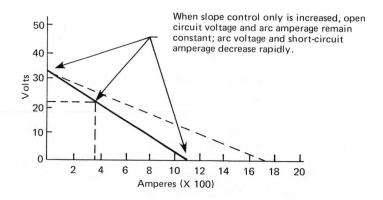

When slope control only is increased, open circuit voltage and arc amperage remain constant; arc voltage and short-circuit amperage decrease rapidly.

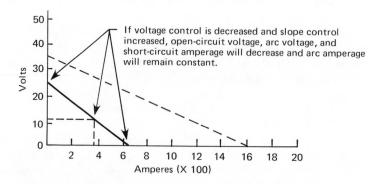

If voltage control is decreased and slope control increased, open-circuit voltage, arc voltage, and short-circuit amperage will decrease and arc amperage will remain constant.

FIGURE 4–7. *A change in slope can effect OCV, arc amperage, arc voltage, and short circuit amperage. (Courtesy of Charles R. Stevens, Senior Manufacturing Engineer, Hyster Co.)*

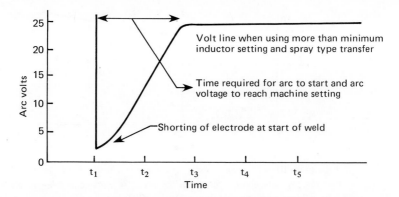

FIGURE 4–8. *Inductance is most useful for the starting of the arc as it may be used to prevent a blasting of the electrode when the current surges. (Courtesy of Charles R. Stevens, Senior Manufacturing Engineer, Hyster Co.)*

Inductance. Inductance is the electrical influence exerted by current flow in a magnetic field. The inductor is connected to the dc side of the rectifier and provides additional refinement to the slope control. The same effect can be produced simply by wrapping the welding cable around a box of coated electrodes or bars of steel. Inductance affects the current in the following ways, Fig. 4–8:

1. Inductance controls the rate of current rise without affecting the final amount of current used. Therefore, it sets the frequency of pinch force.
2. Increasing inductance decreases the number of short circuit metal transfers and increases the arc-on time per second, Fig. 4–9. This creates a more fluid and wetter puddle. A decrease in inductance works in the opposite manner, creating a smaller pool.

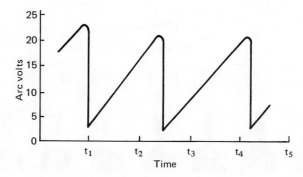

FIGURE 4–9. *Excessive reactance control increases reactance time until full arc voltages cannot be reached between shorts of the electrode to base metal. (Courtesy of Charles R. Stevens, Senior Manufacturing Engineer, Hyster Co.)*

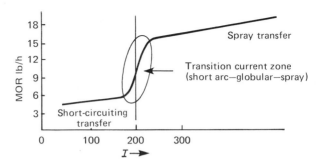

FIGURE 4–10. *The relationship of the current to the modes of metal transfer and the deposition rates for the filler metal. (Courtesy of Charles R. Stevens, Senior Manufacturing Engineer, Hyster Co.)*

Modes of Metal Transfer. Three different principles govern the transfer of metal from the electrode to the weld joint. These are short-circuiting, gobular, and spray, Fig. 4–10.

Short-Circuiting Transfer. Short-circuiting transfer is used with low voltage, low wire feeds (amperes), and small diameter filler wires. Note that wire feeds and amperage become synonymous in the gas metal arc (GMA) process. The short-circuiting transfer process never transfers the metal across an arc but rather through resistance as the wire is being fed against the base metal, Fig. 4–11. The pinch rate of the filler metal is from 20 to 200 times per second and is controllable by the slope adjustment found on most CP power sources.

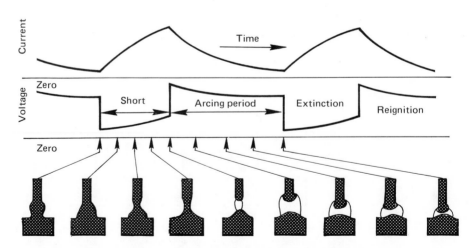

FIGURE 4–11. *Voltage and amperage waveforms associated with a short-circuiting electrode. (Courtesy of Charles R. Stevens, Senior Manufacturing Engineer, Hyster Co.)*

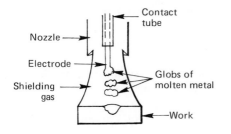

FIGURE 4–12. *Globular metal transfer.* (*Courtesy of Charles R. Stevens, Senior Manufacturing Engineer, Hyster Co.*)

Globular Transfer. Globular type metal transfer occurs across the arc in large, irregularly shaped drops or globs. This type of metal transfer is indicative of CO_2 shielded welding processes. Globular transfer can also exist because of low arc voltage, or low amperage when using other shielding gases, Fig. 4–12. A great amount of spatter is associated with this process.

Spray Transfer. Spray transfer is accomplished by the movement of a stream of tiny droplets of molten metal across the arc from the electrode to the weldment, Fig. 4–13. This process uses large diameter electrode wires, .045 (1.14 mm) to .062 in. (1.57 mm) and employs relatively high arc voltages and currents (24–40 arc volts, 250–500 amps). The current densities are much greater and result in high deposition rates and deep penetration. Spray transfer is most suitable for making heavy welds on thick material in the flat or horizontal position. The shielding gases used are generally inert gases, and frequently a mixture of inert gas and O or CO_2.

High Frequency Stabilization (HF). High frequency ac is used to help start and to stabilize the arc. A spark gap oscillator is used to produce the harmless high frequency voltage at radio frequencies. This voltage sometimes approaches 3000 V or more. Tube type oscillators may be substituted as they provide more definite control over the HF. High frequency oscillators may also be used in conjunction with dc welding currents but generally only for the initial starting of the arc.

Duty Cycle. Since welding loads are not as demanding in length of operation time as other loads, a power source that supplies a given current (I) for a short time need not be as large or rugged as one required to supply the same current

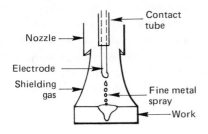

FIGURE 4–13. *Spray metal transfer.* (*Courtesy of Charles R. Stevens, Senior Manufacturing Engineer, Hyster Co.*)

continuously. Thus, a welding power supply is allowed to idle during part of its operating time. This is especially true for manual or semiautomatic welding. Therefore, a useful rating is the duty cycle of the welding machine, expressed as:

"A percentage of the portion of time that the power supply must deliver its rated output in each of a number of successive ten-minute intervals."

The duty cycle is based on the output current and not on the KVA or KW rating of the power supply.

Thus, 60% duty cycle (a standard industrial rating) means that the power supply can deliver its rated output for six minutes out of every ten minutes without harming the machine. Limited service and limited input type machines are rated at 20% duty cycle. Other industrial type machines are rated at 100% duty cycle.

$$\therefore \quad \% \text{ duty cycle} = \frac{(\text{I rated})^2}{(\text{I load})^2} \times (\text{rated duty cycle})$$

Example. Should one wish to use an ac power source rated at 20% duty cycle and 180 ampere output at 100% duty cycle, the maximum current output under the new duty cycle would be calculated in the following manner:

$$(\text{I load})^2 = \frac{(180)^2}{100} \times 20$$

$$\frac{32400 \times 20}{100} = \frac{648000}{100} \text{ or } 6480$$

$$(\text{I load})^2 = \sqrt{6480} \text{ or } 80 \text{ amperes at } 100\% \text{ duty cycle}$$

GAS RESISTANCE PROCESSES

As stated previously, the resistance to the flow of electrons across the arc generates the heat necessary to melt the metal for welding. Therefore, all processes producing an arc will be classified as being gas resistant. The list includes shielded metal arc (stick electrode), gas metal arc (continuous electrode), gas tungsten arc (nonconsumable electrode), submerged arc, electro-gas, stud welding, arc spot, carbon arc, and plasma arc.

Shielded Metal Arc. The shielded metal arc process is most often thought of when the term arc welding is used. It is characterized by a manually manipulated stick (coated wire) electrode. Both the rod and the base metal melt due to the intense heat of the arc. The gases formed by the burn and vaporization of the electrode coating act to form a shield from the atmosphere for both the arc and the molten metal of the weld, Fig. 4–14. The energy to maintain the arc is obtained from one of several power sources; e.g., a generator, a transformer, a transformer plus rectifier, a bank of capacitors, or storage batteries. Arc welding (dc) was first used in 1890,

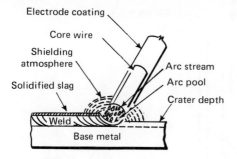

Electrode coating
Core wire
Shielding atmosphere
Solidified slag
Arc stream
Arc pool
Crater depth
Weld
Base metal

FIGURE 4-14. *Schematic representation of the shielded metal arc.* (*Courtesy of American Welding Society.*)

when the source of power was a bank of storage batteries. Batteries are no longer used for welding purposes, except in research where close control over voltage/amperage is important.

Generator. The generator can be powered by an engine or electric motor. If an engine is used as the source of power, it would be equipped with an automatic throttle and governor. The advantage of the engine driven generator is that it can be used in remote areas independent of electric power lines. The disadvantage of this kind of power source is its many moving parts, which increase the need for maintenance and result in greater wear of the parts. Moving parts also create more noise. The opening and closing of the throttle each time an arc is struck is particularly noisy.

Transformer. The basic transformer type power source delivers only ac current, Fig. 4-15. This machine has no moving parts except a cooling fan motor; therefore, initial investment and maintenance costs are generally low. The amperage may be controlled by tapping the secondary coil, using a shunt, or by means of a movable coil to cut the magnetic lines of force, Fig. 4-16. Other means to adjust the amperage are shown in Figs. 4-16a and 4-16b.

Transformer-Rectifier Power Source. The principle for this kind of dc current came after the development of ac current for welding. The function of a

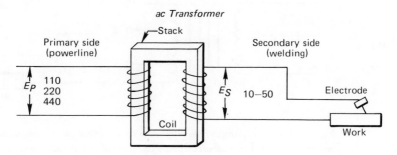

ac Transformer
Stack
Primary side (powerline)
Secondary side (welding)
E_P 110 220 440
E_S 10–50
Electrode
Coil
Work

FIGURE 4-15. *The transformer changes the values of the volts and amperes.* (*Courtesy of American Welding Society.*)

79

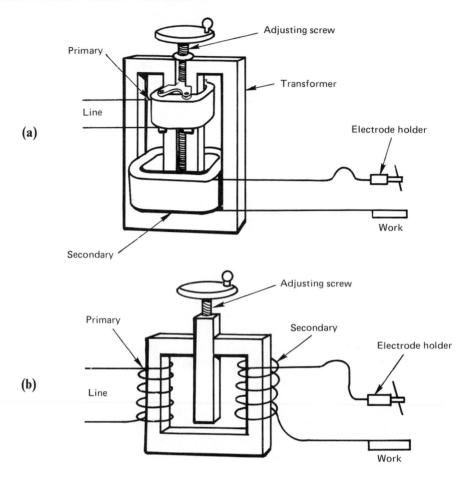

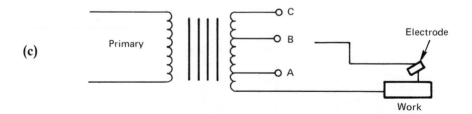

FIGURE 4-16. *Techniques used for amperage control on ac transformer type power sources:* (a) *Moveable coil amperage control.* (b) *Movable shunt amperage control.* (c) *Tapped secondary amperage control.* (*Courtesy of American Welding Society.*)

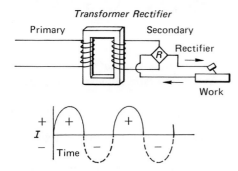

FIGURE 4-17. *The rectification of ac current to dc current. (Courtesy of American Welding Society.)*

rectifier is to change alternating current to direct current, Fig. 4–17. To reverse the flow of current, or to change the polarity, the second half of the rectifier is used. There are two types of rectifiers in use today, vapor and metal.

The vapor-arc principle, better known as ignitron and thyrotron tubes, is the oldest of the systems and is still in use in many of the older resistance welding power sources. The metallic rectifiers are of two types, namely, the stack and the diode. The stack rectifier, Fig. 4–18, is made from many assembled aluminum base plates, referred to as back electrodes, surfaced only on one side with a 0.004 in. (0.102 mm) thick selenium deposit. To complete the cell, a part of the selenium surface is sprayed with a coat of low melting point alloying material. The sprayed material is called the front electrode. Each selenium cell forms a half-wave rectifying element. The individual cells must be assembled in proper orientation so that the current will flow freely in one direction and be blocked from flowing in the opposite direction.

The diode rectifier, Fig. 4–19, uses a wafer from a silicon crystal as the half-wave rectifying element. The wafer is sliced from a pole oriented, single

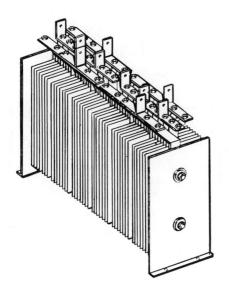

FIGURE 4-18. *A complete selenium main power rectifier. (Courtesy of Power Publications Co., Appleton, Wis.)*

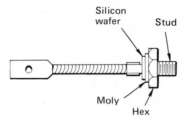

Silicon
wafer

Stud

Moly

Hex

FIGURE 4-19. *A silicon diode rectifier used to convert ac to dc current. (Courtesy of Power Publications Co., Appleton, Wis.)*

crystal ingot of silicon. At the time of assembly, it is important that the wafers of the diode be correctly oriented as to the direction of current flow. The crystal, measuring approximately 12 in. long and 1-1/2 in. (3.81 cm) in diameter, is grown in laboratories from a small, super-pure silicon seed and becomes a two pound crystal in about four hours. The pole orientation for each crystal is determined by the kind of impurity used to initiate the growth.

Gas Metal Arc (GMA). The GMA process dates from approximately 1946. This process consists of a continuous hot wire electrode fed into the arc, usually shielded by some type of gas and possibly by a flux from within the wire, Fig. 4-20. It presents a major challenge to the shielded metal arc process for mild steel and stainless steels, and has been used extensively for large weldments in aluminum and nickel alloys.

When this process was first introduced, it was referred to as the metal inert gas process or MIG, primarily because the process was limited to welding nonferrous metals where inert shielding gases were required. The GMA process uses DCRP, and where the equipment is available, CP power sources are recommended. A CC

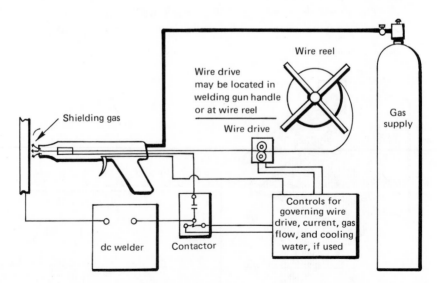

FIGURE 4-20. *A schematic diagram of the gas-metal-arc welding process. (Courtesy of American Welding Society.)*

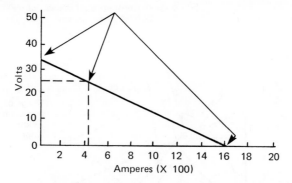

FIGURE 4–21. *A normal machine and wire feed setting for welding with the constant voltage power source. (Courtesy of Charles R. Stevens, Senior Manufacturing Engineer, Hyster Co.)*

power source can be used; however, most wire drive units have a constant feed wire unit, and if the amperage is set too high for the wire melt-off rate, burnback would occur. When this happens, all welding is stopped until the contact tube is replaced or redrilled. The contact tube height (CTH), Fig. 4–4, affects wire alignment, contact tube heating, resistive heating of the filler wire, and gas shielding. The CP power source balances the MOR of the wire with the wire feed rate (amperes).

A study of the normal E/I curve for GMA welding, Fig. 4–21, will provide a brief review of the two most common variables, E and I, as they affect welding characteristics of the CP power source. Fig. 4–22 shows what effect a change in wire feed speed (wfs) has on the arc amperage. It was pointed out earlier in this chapter that the amperage is controlled by the wfs when using a CP power source and a wire drive unit.

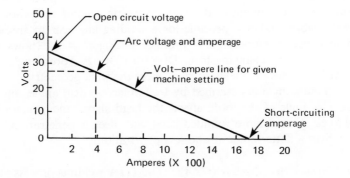

FIGURE 4–22. *When the wire feed speed is changed, the open circuit voltage and short-circuiting amperage remain constant, but the arc amperage and arc voltage change inversely. (Courtesy of Charles R. Stevens, Senior Manufacturing Engineer, Hyster Co.)*

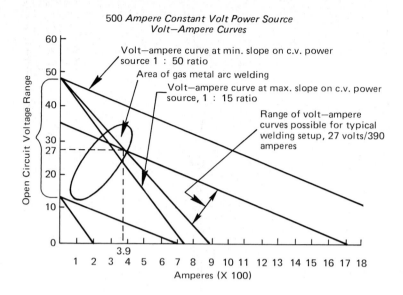

FIGURE 4-23. *The CP volt–ampere curve with varying amounts of slope and the volt–ampere ratios. The oval shows the proper range settings for good GMA welding. (Courtesy of Charles R. Stevens, Senior Manufacturing Engineer, Hyster Co.)*

In summary, for successful GMA welding it is desirable to coordinate the E/I settings on the power source and wire drive unit to fit within a range, as shown in the oval circle in Fig. 4–23. Voltage, amperage, wire size, and type of gas shielding all have a part in determining the kind of metal transfer an operator gets.

The above discussion points out that the GMA process can become rather involved. It also shows that this process has removed many of the uncontrolled variables from welding with the manual stick electrode process. The GMA process adapts easily to semiautomatic and automatic welding operations. The control of these variables makes it possible for a welding engineer to specify *weld geometry* and to expect the operator to produce it. Figure 4–24 shows two extremes of weld geometry by changing from DCRP to DCSP. Torch angles, voltage settings, and various types of shielding gases affect the weld geometry.

Torch angle is described by lead angle, vertical angle, and trail angle, Fig. 4–25. The angle of the torch affects the bead shape and penetration pattern. The trail angle gives a more rounded bead with deeper penetration, whereas the lead angle gives a flatter contour and less penetration.

Gas Tungsten Arc Process (GTA). The GTA welding process was developed in 1942 for welding magnesium alloys for aircraft parts. The arc is maintained between a nonconsumable tungsten electrode and the base metal. A nonconsumable electrode is one that does not become part of the weld metal. The environment is controlled by an atmosphere of inert gas, usually argon, helium, or a mixture of the

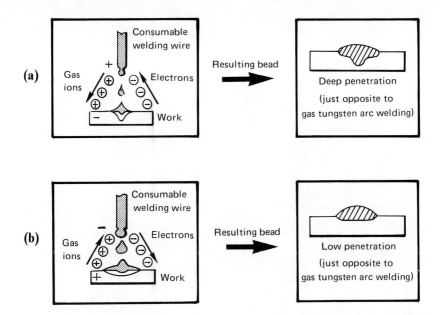

FIGURE 4-24. *The weld geometry can be varied by changing the polarity. (a) DCRP gives deep penetration—GMA process. (b) DCSP gives a wider and flatter bead with less penetration—GMA process. (Courtesy of American Welding Society.)*

two. The process is used extensively for the joining of stainless steel, aluminum, magnesium, copper, nickel, alloy steels, and titanium. Prior to its development, nonferrous metals were welded with the oxyacetylene flame or the stick electrode, neither giving the quality of weld achieved with the GTA process. This process uses DCSP or ac high frequency (ACHF) power sources. The torch may be air or water cooled, as shown in Fig. 4–26. The tungsten electrode is nonconsumable and is used only with DCSP or AC current to prevent it from melting. The melting temperature for tungsten is approximately 6300 °F (3482 °C); therefore, when using DCSP a higher amperage can be used without melting the electrode, thereby obtaining deep penetration, Fig. 4–27c. A combination ac/dc–CC power source is generally used with GTA welding.

Welding by ACHF is particularly adaptable for aluminum. The reverse cycle of the ac syne wave has the phenomenal characteristic of removing the high temperature melting layer of aluminum oxide (ALO_3) found on the surface of all aluminum. The slow moving but hard hitting gas ions bombard this layer of ALO_3 and break it up. The fast moving electrons then lift these ALO_3 scales leaving a clean surface conducive to a good weld. The DCSP side of the cycle provides excellent penetration and cooling for the electrode. The inert shielding gas prevents further oxides from reforming.

DCSP current is more adaptable for the welding of stainless steels than ACHF because the cleaning action is not necessary for stainless steel. Filler metal

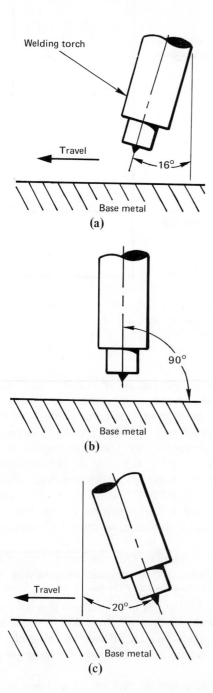

FIGURE 4-25. *The torch angle effects the penetration geometry of welds. (a) Relationship of welding torch with 16 degrees leading angle to the base metal. (b) Relationship of vertical welding torch to the base metal. (c) Relationship of welding torch with 20 degree trailing angle to the base metal.*

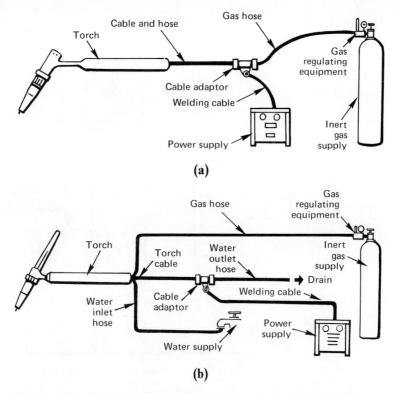

FIGURE 4–26. *Schematic diagram for gas tungsten arc, air-cooled and water-cooled guns. (a) The air-cooled GTA welding torch. (b) The water-cooled GTA welding torch. (Courtesy of Linde Division, Union Carbide Corp.)*

can be added either by hand feeding or as a cold wire with a wire drive unit. For very high deposition rates and welding speeds, hot wires can be fed into the GTA arc, Fig. 4–28. This requires a second power source and a wire drive unit. Actually, this technique combines GTA and GMA processes in one operation.

Tungsten Arc Spot Welding. To arc-spot thin gauge metals, the common practice is to use a tungsten torch with no filler metal. The principle is simply to fuse a spot of the two metals together. This technique works well on both ferrous and nonferrous metals.

Submerged Arc Welding. Submerged arc welding derives its name from submerging the entire welding action beneath a mineral base material known as flux, Fig. 4–29. The arc is ignited either by bringing the continuous type electrode in contact with the base metal, by using a steel wool fuse ball, or by using a high frequency spark. Once the current begins to flow, the arc becomes submerged in a sea of molten flux. Both the molten and granular fluxes aid in shielding the arc and molten metal from the atmosphere. The weld bead and heat affected zone (HAZ)

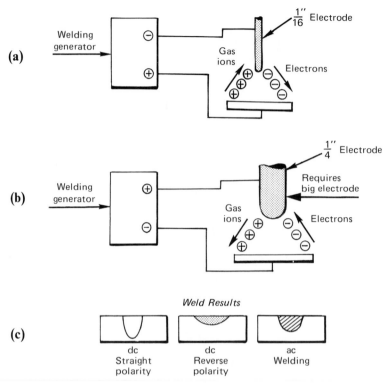

FIGURE 4–27. *The effects of polarity on weld penetration geometry when using the GTA process. (a) DC straight polarity machine connection. (b) DC reverse polarity machine connection. (c) Weld result. (Courtesy of American Welding Society.)*

remain covered with a layer of granular flux that retards the cooling rate and improves the weld properties. The flux is later removed either by hand or by a vacuum device that can be attached to the torch carriage. The granular flux is recycled and used over again, while the fused, glasslike substance resulting from the liquid flux is generally discarded. The submerged arc process is primarily suited for welding in the flat or nearly flat position since the flux would otherwise fall off, Fig. 4–30. The process produces welds of excellent mechanical properties in mild steel and stainless steel and is also used extensively for overlay operations.

The electrode is bare wire, and for a CC power source, the speed of the motor feeding the electrode is governed by the arc voltage. As the arc lengthens, the voltage rises and the motor driving the feed rolls speeds up; as the arc becomes shorter, the voltage drops and the driving motor slows down. This principle gives excellent arc voltage control and results in repetitive high quality welds. The CC power source is best suited for the welding of plate. More than one electrode can be used to weld heavy plate at high speeds, Fig. 4–31. When welding thin gauge material with the submerged arc process, a CP power source is recommended.

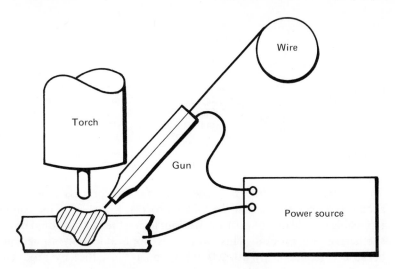

FIGURE 4–28. *Filler wire can be supplied to a GTA puddle by either a cold or hot wire. The figure shows the use of a hot wire.*

The submerged arc process uses extremely high current densities that result in deep penetration and high melt off rates. To find the current density for a given electrode, simply divide the cross-sectional area of the electrode into the amperage being used. The resulting figure is the amperage per square inch of welding wire.

Example. A 3/32 in. (2.38 mm) diameter welding wire has a cross-sectional area of 0.0069 in^2. (0.0445 cm^2). Assume that 600 A is average for this size wire, therefore:

$$D_c = 0.0069 \sqrt[\displaystyle 86985.0]{600.0000}$$

or a current density of 86,985 amperes per square inch of weld wire. The current density for the same size electrode with the shielded metal arc process and 75 A would be 10,869 A/in^2.

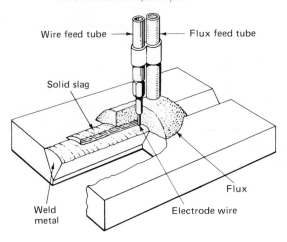

Wire feed tube —

— Flux feed tube

Solid slag

Weld metal

Electrode wire

Flux

FIGURE 4–29. *Submerged arc welding is used with either alternating or direct current for heavy welding and high deposition rates. Either CC or CP power sources can be used. (Courtesy of Linde Division, Union Carbide Corp.)*

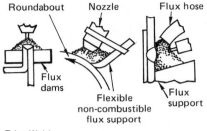

Roundabout   Nozzle   Flux hose

Flux dams

Flexible non-combustible flux support

Flux support

Edge Weld

FIGURE 4–30. *Suggested techniques for welding positions with submerged arc welding. (Courtesy of Linde Division, Union Carbide Corp.)*

The Magnetic Flux Process.   The magnetic flux process is a combination of the submerged arc process and the GMA process, Fig. 4–32. This combination makes it possible to weld out-of-position without losing the flux, as would be the case with the submerged arc process. This process is applicable only to ferrous filler metals because it magnetically attracts the granular flux, thereby producing the desired electrode coating an inch or two before the flux is fed into the arc. The magnetic flux process has an exposed arc rather than the hidden arc of the submerged process and, as a consequence, does not provide adequate protection from the atmosphere for quality welding. Therefore, a shielding gas is used along with the magnetic flux. The

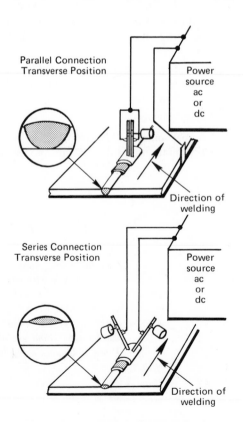

Parallel Connection
Transverse Position

Power source ac or dc

Direction of welding

Series Connection
Transverse Position

Power source ac or dc

Direction of welding

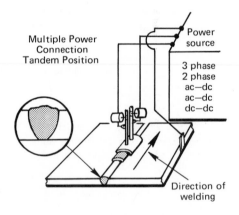

Multiple Power Connection
Tandem Position

Power source

3 phase
2 phase
ac–dc
ac–dc
dc–dc

Direction of welding

FIGURE 4–31. *Recommended electrode arrangements for the welding of heavy plate at high speeds when using the submerged arc process with multiple electrodes. (Courtesy of Linde Division, Union Carbide Corp.)*

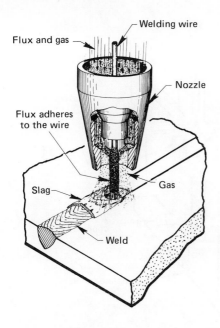

Flux and gas

Welding wire

Nozzle

Flux adheres to the wire

Slag

Gas

Weld

FIGURE 4–32. *The magnetic flux process has an automatically fed consumable electrode shielded by carbon dioxide and a magnetic flux. (Courtesy of Linde Division, Union Carbide Corp.)*

development of the "flux-cored" wire has decreased the attractiveness of the magnetic flux process. Flux-cored wires are discussed in detail in Chapter 8 on filler wires.

Electro-Gas Process. The electro-gas welding process is fully automatic and is readily used for vertical fusion welding of butt, corner, and tee joints. The process uses either solid or cored wire fed into an open arc shielded by a gas, Fig. 4–33. The electro-gas process is desirable for the welding of plate from $\frac{1}{2}$ in. (1.27 cm) to 3 in. (7.62 cm) thick in one pass. This is similar in principle to the electro-slag process, which will be discussed under the heading Liquid Resistance. The electro-gas process uses a CP power source, a wire drive unit, and a pair of water-cooled copper shoes. The control system automatically directs the lifting of the tubes guiding the filler wire and the copper shoes that are used for damming the molten metal between the two pieces to be joined.

This process is used mostly for field welding of fuel storage tanks, water tanks, blast furnace shells, and ship bulkheads. The initial investment is costly; however, the high electrode deposition rate of 35-to-45 lb of filler metal per hour per electrode for the appropriate applications makes the cost quite appealing to the manufacturer.

Stud Welding. Stud welding consists of placing a stud in a spring loaded collet of a welding gun. At the end of the stud is either a drilled cavity filled with a flux and steel cap, a pressed aluminum disc, or a machined projection Fig. 4–34. The flux or aluminum disc is vaporized as the arc is initiated. The vaporization provides a desirable gas shield, acts as deoxidizer, and aids in cleaning the metal surface. A ceramic ferrule placed over the end of the stud confines the arc to a limited

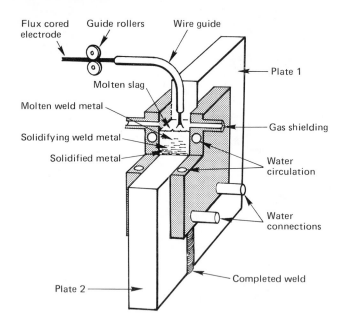

FIGURE 4-33. *A schematic of the electrogas welding process.* (*Courtesy of American Welding Society.*)

area, molds a neat fillet, and provides some shielding of the molten metal from the atmosphere, Fig. 4-35a. This process produces excellent penetration and complete fusion between the parts being joined, Fig. 4-35b. Because automation eliminates most of the human error, it is an extremely dependable technique.

Studs of any design and many different sizes can be welded by this process. Time and material can be saved by stud welding rather than drilling holes in steel and using bolts. The drilling of a hole weakens the structure; therefore, heavier sections should be specified. The average time required for welding is approximately 20 cycles or one-third second per stud. However, loading the gun is far more time consuming because this must be done singly and by hand with the arc drawn principle. Another stud welding principle uses stored energy to do the welding. Capacitors store energy up to 200 volts, Fig. 4-36. When the high voltage is released, current is forced through a small diameter projection that has been machined into each stud, and the projection melts. An arc is created and fusion results for welding. No ferrule or flux is used with the capacitor discharge stud welder. When welding nonferrous studs, an inert gas shield is sometimes used. In some automobile industries, air pressure is used to load the stud into the gun automatically.

Arc Spot Welding. Arc spot welding is similar in appearance to a rivet, but its properties are like those of a resistance spot weld, Fig. 4-37. This process has replaced the resistance spot weld in many shops. The equipment is the same as that used for GMA welding, with the addition of an automatic control system.

Electric Arc   Capacitor Discharge   Drawn Arc
                                     Capacitor Discharge

FIGURE 4–34. *Three common methods used for stud welding.* (*Courtesy of TRW Nelson Division.*)

The control system regulates the "arc on" time and the "arc off" time in cycles. It also controls the *wire stick out* or *burn-back* of the filler wire as the arc is extinguished. The control unit removes most of the human error from the operation; therefore, the arc spot process can be relied upon to duplicate weld characteristics consistently. Arc spot welding is used to join sheet material or plate up to $\frac{1}{2}$ in. (1.27 cm) thick without drilling or other costly preparations. Thin materials can be welded to heavy plates, or heavy plates to thin materials. Dissimilar metals can be welded to each other when the filler material is

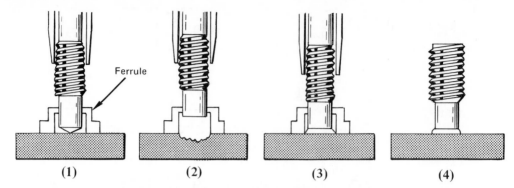

FIGURE 4-35. (a) *The four steps for welding when using the arc-drawn system.* (1) *Fluxed end of stud is placed in contact with work.* (2) *Stud is automatically retracted to produce an arc.* (3) *Stud is plunged into pool of molten metal.* (4) *Weld is complete.*

compatible with the lower piece. Once the parameters have been determined, the operator has only to place the gun where the weld is required and trigger the gun. Most welds can be completed in one to five seconds.

Carbon Arc Process. Carbon arc welding is similar to the shielded metal arc process except that the electrodes are nonconsumable, they do not become a part of the weld. When two carbon electrodes are used, one acting as the cathode and the other as the anode, it is a *carbon arc torch.* The equipment for a single carbon electrode is identical to that used for the stick electrode. However, when two carbon electrodes are used, a special holder is required. The carbon arc is best used for cutting applications, whereas the carbon arc torch is used for soldering, brazing, and braze welding.

The arc of a carbon arc torch is similar to certain kinds of gas flame and is often used in lieu of the gas flame, e.g., where flame requirements are small and cylinder demurrage is costly. With the carbon arc torch, the initial investment is low and demurrage charges are nonexistent.

FIGURE 4-35. (b) *The schematic shows the fillet and the complete penetration of the weld stud. (Courtesy of TRW Nelson Division.)*

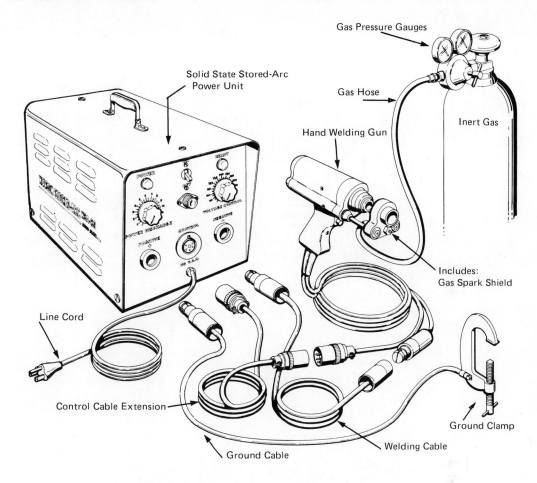

FIGURE 4–36. *The capacitor type stud welder complete with inert shielding gas. (Courtesy of TRW Nelson Division.)*

Plasma Arc Welding. The plasma arc is used for cutting ferrous and nonferrous metals, fusion plating, and welding. The plasma arc welding torch uses two separate gas flows. One of the gases surrounds the electrode and becomes ionized as it passes through the arc. The flow of the ionized gas is constricted by a small orifice directly below the point of the electrode. Due to the constriction and the sonic speed of the gas as it passes through the torch orifice, temperatures of 30,000 °F (16648 °C) are generated. A column of slower moving, less constricted cool gas shields the arc and molten metal from the atmosphere, Fig. 4–38. Argon is generally used for the plasma and helium for shielding; however, argon has been used for both purposes.

The narrow column of ionized gas cuts a "keyhole" shaped hole through

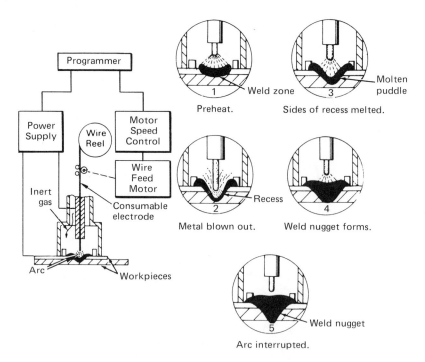

FIGURE 4–37. *The arc spot weld resembles a rivet, however, it gives a rigid joint.*

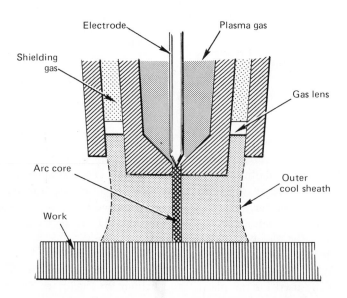

FIGURE 4–38. *Plasma welding uses a central core of extreme temperature surrounded by a sheath of cool gas. (Courtesy of Amerian Technical Society.)*

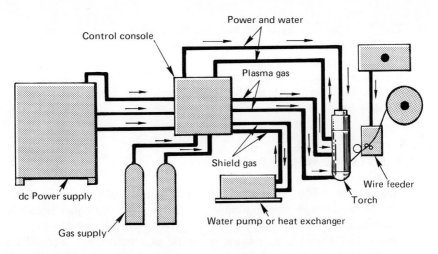

FIGURE 4–39. *Schematic of a plasma welding installation.* (*Courtesy of American Technical Society.*)

the metal. As the plasma torch moves over the abutting edges of the metal being joined, the molten metal in front of the arc flows around the arc column to the rear, where fusion and solidification take place. The power supply for plasma welding is a dc rectifier. A heat exchanger cools the water, which in turn cools the torch and electrode. The control unit regulates the flow of current, water, and gases. When a filler metal is needed, a cold wire is fed into the arc by a conventional wire drive unit, Fig. 4–39.

Schematics of the two basic designs for plasma torches are shown in Fig. 4–40.

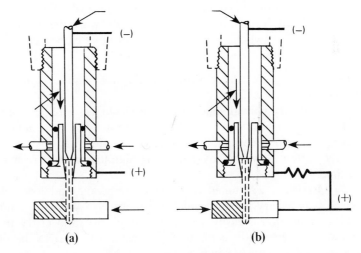

FIGURE 4–40. *Two kinds of plasma torches are: nontransferred and transferred. (a) Nontransferred torch has an internal anode. (b) The transferred torch has the work piece as its anode. (Courtesy of American Welding Society.)*

The "nontransfer" type has the anode as an integral part of the torch and is used for plating purposes. The "transferred" type torch has the anode as the work piece. This type is used for metal cutting and joining. The plasma arc welding process is applicable to most metals and their alloys. The initial cost of the equipment is high. However, the high arc temperatures permit fast travel speeds that reduce the cost per weld when properly programmed.

SOLID RESISTANCE WELDING

Introduction. The initial development of solid resistance welding was rather slow due to the limited power supplies available. Not until just before WWII did it become a commercial proposition. Its development is parallel to that of the press tool industry, and it is now indispensible as a convenient method for fabricating a host of items. Most of the solid resistance welded items are manufactured from metal stampings and include automobiles, aircraft, household appliances, and products of allied industries.

All welding processes that generate heat by the resistance of a solid to a current flowing through it are classified as *solid resistance processes*. The processes include spot welding, seam welding, projection welding, flash butt welding, percussion, upset welding, high frequency electrical resistance welding, electron beam, and laser beam welding.

The basic equipment required for all solid resistance welds consists of three major units:

1. A transformer.
2. A mechanism (mechanical, pneumatic, or hydraulic) to provide a means to exert force through the electrode onto the work piece, for quality welds.
3. A precision switching system (mechanical or electrical) to control the flow of electric current to the work piece.

Electron beam and laser processes do not require a mechanism to control forces on the work piece.

The theory behind solid resistance welding is forcing electric current to flow with minimum heat loss. Therefore, the heat must be confined to the exact area where the weld is desired. This is accomplished by providing a concentrated area for the current flow, thereby generating high temperature over a very short period of time.

Earlier it was noted that a high amperage and low voltage combination, which is necessary for most solid resistance type welding, requires a transformer. This combination is easily achieved through the ratio of the turns of the primary (input side wire windings about the transformer core) to the number of turns of the secondary (output) winding. Thus:

$$\frac{\text{primary turns}}{\text{secondary turns}} = \text{turns ratio}$$

For a turns ratio of 40 : 1, it is feasible to convert 45 amperes at 440 volts on the primary side to 1800 amperes at 11 volts on the secondary side. To calculate the amount of heat generated in an electrical circuit, Ohm's law is used, for which the equation is:

$$I = \frac{E}{R} \qquad \text{or} \qquad R = \frac{E}{I}$$

where I is the current in amperes, E is the emf (electromotive force) or voltage, and R is the resistance through the material in ohms. The total energy can be expressed in joules with the equation:

$$\text{joules} = IET$$

where T is the time in seconds that the current flows.

The amount of heat generated is related to the magnitude of the electric current, the resistance of the current conducting path, and the time the current is allowed to flow. Of course, the metal makes a much better path for the current than the arc in arc welding; therefore, the current must be much higher. Since the area of greatest resistance should be at the interface of the two joining surfaces, it is also the area of greatest heat. The heat generated is directly proportional to the square of the current times the resistance and is expressed by the formula:

$$Q = I^2RT$$

Where Q equals heat generated in watt hours, T equals time in hours, I equals current in amperes, and R equals resistance in ohms. The heat may be changed from watt hours to BTU by multiplying the current by 3.412, since there are 3412 BTU in one killowatt hour.

There are actually seven critical areas where the resistance factors can easily affect the weld quality, Fig. 4–41. Areas 1 and 7 are affected by the electrical resistance properties of the electrode material. The areas of contact resistance between the electrode and base metal (2 and 6) can be affected by several variables. The magnitude of resistance depends upon the surface condition of the base metals and that of the electrode face; the area of electrode contact and the electrode force is similarly influential (resistance is roughly inversely proportional to the contacting force). Even though this area is a point of high heat generation, the fact that the electrode material exhibits high thermal conductivity and is usually water cooled keeps the surface temperature of the base metal below its melting temperature.

Solid resistance welding applications can be divided into two main groups: lap welds—including spot, seam, and projection, and butt welds—including upset butt and flash butt, Fig. 4–42.

Spot Welding. Spot welding is a basic type of solid resistance welding. It will be discussed in greater detail since many of the points covered are common to the other types, Fig. 4–43. Squeeze time, weld time, hold time, and off time are the fundamental variables of spot welding. For welding most metals, but

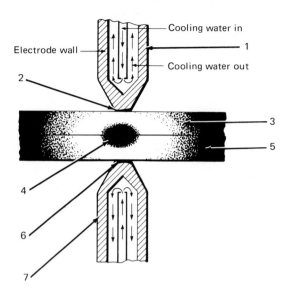

FIGURE 4–41. *Critical points of resistance for solid resistance welding. (1) Upper electrode. (2) Point of contact between upper electrode and base metal. (3) Base metal–thickness, density, and chemistry. (4) Point of contact of base metals in area of nugget. (5) Base metal–thickness, density, and chemistry. (6) Point of contact between lower electrode and base metal. (7) Lower electrode.*

especially the nonferrous types, these variables must be controlled within very close limits.

Squeeze Time. Spot welds are made by first cleaning the two pieces of metal to be lapped, then placing them between the copper electrodes of the spot welding machine. The squeeze time brings the two workpieces together in intimate contact just prior to the current flow. The pressure also localizes the current flow.

Weld Time. The weld time is the time when the current flows causing a nugget to form at the interface of the two pieces of metal.

Hold Time. Hold time is basically a cooling period. It is the interval from the end of the current flow until the electrodes part. The water-cooled electrodes transfer the heat away from the weld rapidly.

Off Time. The off time is the interval that the electrodes are apart before the cycle automatically repeats for the next weld. If this portion of the control is switched out, the machine will stop after each weld. The sequence just described is shown in Fig. 4-44. Also shown is a *pulsation-welding* cycle. The main difference between this and the conventional cycle is that, instead of the weld time being one

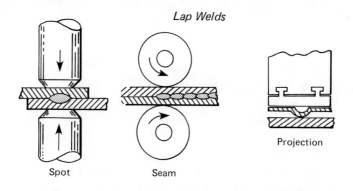

Lap Welds

Spot   Seam   Projection

Butt Welds

Flash   Upset

FIGURE 4–42.   *Resistance welds may be classified as lap and butt welds.*

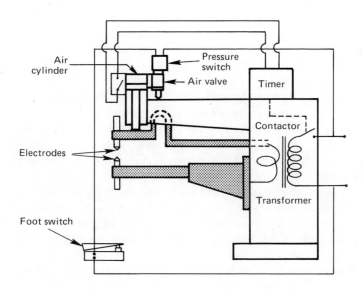

FIGURE 4–43.   *Press type spot welder and simplified electrical circuit.*

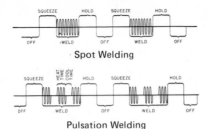

Spot Welding

Pulsation Welding

FIGURE 4-44. *The normal timing sequence for spot welding compared with pulsation welding.*

period of current flow, there is an intermittent flow with cooling no-current intervals between the current flow periods. This method is frequently used on multiple layer welds, projection welds, and welds on two pieces of steel thicker than $\frac{1}{8}$ in. It also provides increased electrode life.

Other variables that may be sequenced along with the four control steps already mentioned are: low current for preheating, either high or low current as required for heat treatment, and a forging action to refine the grain. Grain refinement is accomplished by increasing the electrode pressure during the cool time. Some of these added functions are particularly useful in joining hardenable alloy steels, which may crack if welded with one surge of current and then allowed to cool.

The squeeze, hold, and off portions of the sequence control are relatively simple timing devices that need not be precise. Usually plus or minus two cycles ($\frac{1}{30}$ sec) is sufficient. However, the *weld* portion of the sequence is the most critical and also the most complicated. The primary welding current must switch on and off in such a manner as to produce the desired waveshape in the secondary current. This requires the switching of currents on the magnitude of thousands of amperes, and all the related problems.

Spot Welding Power Supplies. There are two basic types of resistance power supplies, *stored-energy* and *direct-energy*. The stored-energy machine draws current at a relatively low rate and stores it for instantaneous discharge when needed to make a weld. The low current input rate has the advantage of a more nearly balanced load on the three phase electrical system of most shops. Also, variations in line voltage have relatively little effect on the efficiency of the machine.

The main disadvantage of the stored-energy machine is the physical size and the economic limitations of the storage system. Batteries have proven to be the best power supply but still are not commercially successful except for welding light gauge metals, where they render a reasonable service life.

The direct-energy machine is a transmission agent, discharging the power to make the weld as soon as it is received. The two main advantages are the lower initial cost and the lower maintenance cost. The waveshape of the current is also more controllable, making it quite versatile for a greater variety of metals. The main disadvantage of the direct energy machine is the need for more complicated controls. This is especially true for aluminum and alloy steel welding. Direct-energy machines may be either single phase or three phase.

The single phase direct-energy machine is quite satisfactory for welding

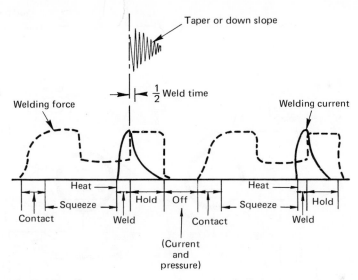

FIGURE 4-45. *Representative waveform of a single-phase ac machine using slope control.*

steel and may be used on aluminum when the waveform is properly modulated and the voltage conditions are stable. It must be remembered, however, that the current draw for aluminum is very high compared to that for steel. A heavy draw on one phase of a three phase line can cause an excessive voltage drop and affect the performance of other equipment in the shop.

The modulated waveform mentioned previously is a development that permits variation in the rate of rise and fall of welding current and is called *slope control*. Slope control consists of having the current build up gradually as in *up-slope* or decrease gradually as in *down-slope, current decay*, or *taper*, as shown in Fig. 4-45. *Up-slope* uses a few pulses of 60 Hz current to build up. This allows the electrodes to sink into the softening aluminum and seat themselves before they have to carry the full welding current. If this were not done, overheating would occur at the electrode tip-to-aluminum contact surface and result in an unwanted welding of aluminum and copper.

In order to keep the current flowing, using both polarities of the ac current, two ignitrons are used in a "back-to-back" arrangement. This is, the electrons flow from cathode to ignitor in one ignitron, and in the reverse direction in the other ignitron. Since this requires the current to flow in the wrong direction (towards the cathode) in the second half of the cycle, solid state rectifiers are placed in the circuit. The current can flow only in the direction indicated by the arrow on the symbol, Fig. 4-46, or only from the cathode to the ignitor. A water switch is included in the circuit to stop operation if the cooling water fails to flow to the ignitrons.

The thyratron also prevents reverse current flow. It is a three electrode tube consisting of anode, grid, and cathode. These electrodes are enclosed in an evacuated glass envelope containing a few drops of mercury. The mercury provides a vapor

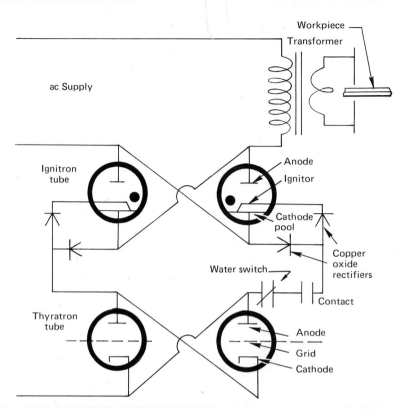

FIGURE 4-46. *The ignitron–thyratron switching circuit with rectified protection.*

that sustains the arc of current flowing through the tube. The electron flow is from cathode to anode, the electrons being attracted by the potential of the anode, which is positive in respect to the cathode. A negative voltage on the grid will prevent current flow in the anode–cathode circuit. Assuming the anode to be positive, for a given potential there exists a "critical" grid voltage. That is, if the grid voltage drops below a certain level, no current will flow.

If the anode voltage is reduced to zero or made negative, the current flow stops and the grid regains control. Thus with ac current fed to the thyratron, the arc is extinguished when the current wave goes through zero, or once each cycle. The grid then takes control and prevents restarting another current flow when the anode becomes positive again on the next cycle. Two thyratrons then, connected back to back, i.e., the anode of one connected to the cathode of the other, can control an ac load.

Thyristors and silicon controlled rectifiers (SCR's) are now replacing thyratron tubes. They are semiconductive solid state devices capable of switching current from a few milliamps to several hundred amperes. The advantages of the thyristor over the thyratron tube are: space savings, long life, elimination of water cooling (on small devices), and greater efficiency. Thyristors, however, may be easily destroyed if subjected to voltages in excess of their ratings.

Many three phase welders use combinations of ignitron and thyratron tubes. This system can be made *synchronous* (eliminating electrical transients) to provide a basic method of control for timing the various sequences in resistance welding.

Spot Weld Design Considerations. There is a direct relationship between the weld strength of a given joint and its design. The factors that must be taken into consideration for a given material are the amount of overlap, the spot spacing, the spot strength, and the accessibility. The amount of overlap required for a good spot weld in lap joints is determined by the weld size which, in turn, is determined by the metal thickness. An acceptable weld size for a thickness range of 0.032 to 0.188 in. (0.08128 cm to 0.47752 cm) is roughly estimated at 0.10 in. (0.254 cm) plus two times the thickness of the thinnest member. The overlap should be equal to two times the weld size plus $\frac{1}{8}$ in. (0.3175 cm) ($\frac{1}{8}$ in. is the tolerance for positioning the weld). If fixturing is used to center the spot in the overlap, the $\frac{1}{8}$ in. (0.3175 cm) tolerance can be disregarded.

Welds placed too close to the edge will often squirt to the previous weld, and thus reduce the size of the weld being made. A general rule is to allow $16t$ between welds, t = the thickness of the material. However, if eliminating distortion is more important than strength, the figure should be increased to $48t$. When it is necessary to place spot welds where there is apt to be current shunting, the current must be increased to compensate.

Spot weld strength varies directly with area. It is generally safe to assume that the strength will be equal to the weld area multiplied by the tensile strength of the metal in the annealed state. Offset electrodes can be used, Fig. 4–47, to make spot welds in places that are inaccessible to the conventional electrode. If the offset is great, excessive tip deflection, skidding, and surface deformation are likely. When the size of the electrode must be restricted to accommodate the joint, it will tend to overheat.

Tables 4.1, 4.2, and 4.3 contain spot welding data for mild steel, stainless steel, and aluminum. Current settings are not intended to be exact, but rather to serve as a starting point. All dimensions are given in inches. The tip diameter refers to the outside diameter that is dressed down by a 20 degree bevel. A general

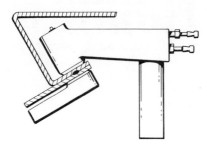

FIGURE 4–47. *An offset electrode is needed for the less accessible areas.*

TABLE 4.1. *Spot Welding Data* for Mild Steel*

Sheet Thickness (cm)	Tip Diameter (cm)	Electrode Force (lb) (Newton)	Weld Cycles	Amperes	Diam of Fused Zone (cm)	Minimum Spot Weld Spacing (cm)
0.020 (0.0508)	3/8 (0.9525 cm)	300 (1334)	6	6,500	0.13 (0.3302)	3/8 (0.9525)
0.035 (0.0889)	3/8	500 (2224)	8	9,500	0.17 (0.4318)	1/2 (1.27)
0.047 (0.11938)	1/2 (1.27 cm)	650 (2891)	10	10,500	0.19 (0.4826)	3/4 (1.905)
0.059 (0.14986)	1/2	800 (3559)	14	12,000	0.25 (0.635)	1 (2.54)
0.074 (0.18796)	5/8	1,100 (4893)	17	14,000	0.28 (0.7112)	1-1/4 (3.175)
0.089 (0.22606)	5/8 (1.5875 cm)	1,300 (5783)	20	15,000	0.30 (0.762)	1-1/2 (3.81)
0.104 (0.26416)	5/8	1,600 (7117)	23	17,500	0.31 (0.7874)	1-5/8 (4.1275)
0.119 (0.30226)	7/8 (2.2225 cm)	1,800 (8007)	26	19,000	0.32 (0.8128)	1-3/4 (4.445)

* Dimensions are in inches unless otherwise indicated.

TABLE 4.2. *Spot Welding Data* for Stainless Steel*

Sheet Thickness (cm)	Tip Diameter (cm)	Electrode Force (lb) (Newton)	Weld Cycles	Amperes	Diam of Fused Zone (cm)	Minimum Spot Weld Spacing (cm)
0.013 (0.03302)	1/4 (0.635)	300 (1334)	3	3,200	0.10 (0.254)	1/4 (0.635)
0.023 (0.05842)	1/4	520 (2313)	4	4,100	0.12 (0.3048)	3/8 (0.9525)
0.032 (0.08128)	3/8	750 (3336)	5	5,500	0.14 (0.3556)	1/2 (1.27)
0.035 (0.0889)	3/8 (0.9525)	900 (4003)	6	6,300	0.16 (0.4064)	5/8 (1.5875)
0.047 (0.11938)	3/8	1,200 (5338)	8	7,500	0.18 (0.4572)	3/4 (1.905)
0.059 (0.14986)	1/2 (1.27)	1,500 (6672)	10	9,000	0.20 (0.508)	1 (2.54)
0.074 (0.18796)	5/8 (1.5875)	1,900 (8452)	14	11,000	0.24 (0.6096)	1-1/4 (3.81)
0.104 (0.26416)	3/4 (1.905)	2,800 (12455)	18	14,000	0.28 (0.7112)	1-1/2 (4.1275)
0.119 (0.30226)	3/4	3,300 (14679)	20	15,500	0.30 (0.762)	2 (4.445)

* Dimensions are in inches unless otherwise indicated.

TABLE 4-3. *Spot Welding Data* For Aluminum Alloys*

Sheet Thickness (cm)	Diam (cm)	Tip Radius (cm)	Electrode Force (lb) (Newtons)	Up-Slope Preheat Current (A)	Weld Heat Cycles	Amperes (A)	Cycles Up-Slope Weld Time	Cycles Down-Slope Post Heat	Final Down-Slope To 60% Weld Current (A)	Spot Spacing (cm)
(0.08128) 0.032	5/8	3 (7.62)	500 (2224)	7,500	8	25,000	12	10	15,000	3/4 (1.905)
(0.08128) (0.16002) .032 to .063	5/8 (1.5875)	3	600 (2669)	8,500	10	28,000	14	12	17,000	3/4
(0.08128) (0.23114) .032 to .091	5/8 or 7/8	Flat or 3 to 8 (7.62 to 20.32)	750 (3336)	10,000	12	33,000	16	16	20,000	1 (2.54)
(0.08128) (0.3175) .032 to .125	7/8 (2.2225)	Flat or 3	1,000 (4448)	11,500	12	38,000	16	16	23,000	1
(0.16256) (0.08128) .064 to .032	5/8 (1.588)	3	600 (2669)	8,500	10	28,000	14	12	17,000	3/4 —(1.905)
(0.16756) (0.16256) .064 to .064	5/8 or	3 (7.62)	750 (3336)	10,500	12	33,000	16	16	20,000	1
(0.16256) (0.23114) .064 to .091	7/8	8(—20.32) Flat or	1,000 (4448)	11,500	12	38,000	16	16	23,000	1 (2.54)
(0.16256) (0.3175) .064 to .125	7/8 (2.2225)	Flat or 8 (20.32)	1,200 (5338)	12,000.	12	40,000	16	18	24,000	1

* Dimensions are in inches unless otherwise indicated.

formula for the resulting small diameter is: $d = 2t + 0.1$ in. (0.254 cm) or $d = \sqrt{t}$. The latter formula has been found more applicable to thicker materials. If a dome-shaped electrode is used, the spherical radius used for mild steel and stainless steel is 3 in. (7.62 cm). The radius recommended for welding aluminum varies as shown in Table 4.3.

Pre-weld surface preparation is necessary to remove two basic types of contaminants: foreign matter such as grease, oil, dust, dirt, paint, films, etc., and oxide films. A uniform oxide film on steel is not detrimental, but rusty metal should not be spot welded without cleaning. Oil will not affect weld quality as such, but particles of dirt in the oil may cause erratic welding. Oil and oxides in contact with the tips causes carbonization, decreasing tip life.

The oxide film on aluminum is more critical than on steel. In order to produce consistently high quality welds and maintain long electrode life, the oxide should be reduced to a minimum either by mechanical or chemical methods. Mechanical oxide removal is done by means of abrasive powders or cloths. The action should be carefully regulated as it must be severe enough to cut through the oxide layer yet not make harmful scratches on the metal. Brush bristles should be less than 0.004 in. (0.01016 cm) in diameter and abrasive cloths should not be coarser than no. 240. Abrasive residue must be removed before welding or inclusions and electrode contamination will result.

Chemical oxide removal is the approved method used by aluminum fabricators for consistently high quality seam and spot welds. Although each alloy forms its

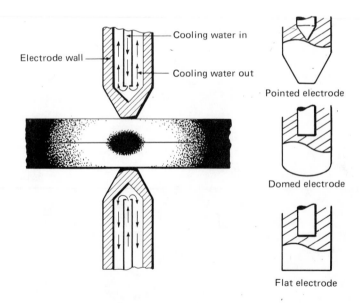

Electrode wall

Cooling water in

Cooling water out

Pointed electrode

Domed electrode

Flat electrode

FIGURE 4-48.   *Electrode tips must provide passageway for cooling water when high currents or high duty cycles are required.*

oxides at a different rate and reacts differently to various deoxidizing treatments, group treatments are satisfactory. Various commercial preparations are available for this purpose. Sandblasting should not be used as a method of cleaning the metal in preparation for spot or seam welding since the silica that becomes embedded on the surface causes erratic results.

Several types of spot welders are available, ranging from small manually controlled machines up through large three-phase, direct-energy machines with solid state controls. Two distinct methods are used to bring pressure on the workpiece, namely, *rocker-arm* and *press method*. The rocker-arm welder has a pivoting arm that moves the electrodes to the workpiece. The operation can be done either manually through mechanical linkage or by a pneumatic cylinder. The press type spot welder exerts pressure on the electrode with an air operated cylinder and ram. Hydraulically operated cylinders are used for welding heavy gauge metals.

It is not always convenient to move units into many different positions when it is being fabricated by spot welding. Often the surfaces are not accessible with standard equipment; thus, portable guns are now extensively used, as for example, on truck and auto-body construction. The guns are mechanically, pneumatically, or hydraulically operated. Portable welding equipment requires the same cooling and controls used on stationary machines for equivalent work. Air operated guns are usually equipped with remote solenoid valve controls.

The welding electrodes are made of copper alloys, and, when the duty cycle is high, they are constructed to provide for water cooling, Fig. 4–48. This prevents spreading or mushrooming of the tips. As a general rule the harder copper alloy will have a longer life but will be less conductive. Consequently, hard alloys are restricted to uses where they can be adequately cooled.

Tapered tips are used, since they mushroom uniformly with continued wear. Domed tips, used for welding many non-ferrous materials, are characterized by their ability to withstand heavy pressure and severe heating without mushrooming. The radius of the dome varies, but a 2 in.-to-4 in. radius is most common. Filing electrodes while on the machine should not be permitted because it is too difficult to restore the original contour. When invisible or inconspicuous welds or a minimum weld indentation are desired, a flat tip is used. A domed tip is used for the mating electrode.

Offset electrodes are made for corner welds, for parts with overhanging flanges, and for the less accessible areas. Their limitations are mentioned previously. Normally, the opposing electrodes will be of similar size and shape. In certain cases, however, it is necessary to have unequally sized electrodes. This is true in welding dissimilar metals or in welding similar materials of unequal thickness. A larger electrode will be needed on the side of the thicker material, or the material with the least resistivity, in order to center the spot between the two materials.

Seam Welding. Seam welding is a continuous lap type of spot welding. Instead of using pointed electrodes that make one weld at a time, the work is passed between copper wheels or rollers that act as electrodes. Thyratron and ignitron tubes are used to "make and break" the circuit. The completed weld is a series

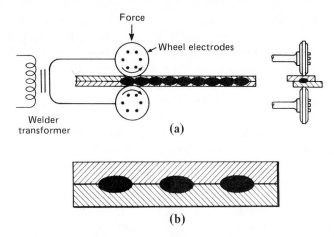

FIGURE 4-49. *Seam and roll spot welding. Seam welding can be done as shown, or one of the wheels can be replaced with a flat backing electrode that supports the work for the entire length of the seam. (a) Seam welding. (b) Roll spot.*

of overlapping spot welds that resemble stitches, hence the vernacular name, *stitch welding*, Fig. 4-49a.

Seam welding can be used to produce highly efficient water and air tight joints. A variation of seam welding, called *roll welding*, is used to produce a series of intermittent spots, Fig. 4-49b. Seam welds tend to cause warping due to the high heat concentration. There are two standard methods of minimizing this effect. Intermittent spot welds can be made as in roll welding. A second pass on the same seam will fill in the alternate welds to make a solid seam. Another method of keeping the heat to a minimum is to use a flow of water over the electrode. Refrigerants inside the electrode are used by some manufacturers to eliminate the inconvenience of external water. Cooling not only reduces distortion of the work, but lowers the maintenance costs of the wheel electrodes. The speed at which various thicknesses of mild steel can be seam welded are given in Table 4.4. The electrode width refers to the width of the wheel. The edge of the wheel is beveled on each side at 20 degrees. The contact area is similar to that of the spot weld explained previously. The net electrode force refers to the exact force between the contact surfaces.

Projection Welding. Projection welding is another variation of spot welding. Small projections are raised on one side of the sheet or plate with a punch and die, Fig. 4-50. The projections localize the heat of the welding circuit. During the welding process, the projections collapse due to heat and pressure, and the parts to be joined are brought in close contact, Fig. 4-51.

Projection welds can be made on a spot welding machine that is equipped with the proper electrodes. Electrode life is long, since there is no direct contact

TABLE 4.4   *Seam Welding Data* for Mild Steel*

Thickness of Thinnest Outside Piece (cm)	Electrode Width (cm)	Net Electrode Force (Newtons)	On Cycles	Off Cycles	Weld Speed (ipm) (cm/min)	Welds Per Inch (per cm)	Amperes	Minimum Overlap (cm)
0.021 (0.05334)	3/8 (0.9525)	550 (2447)	2	2	75 (190.5)	12 (4.7)	11,000	7/16 (1.111)
0.031 (0.07874)	1/2	700 (3114)	3	2	72 (182.9)	10 (3.9)	13,000	1/2 (1.270)
0.040 (0.1016)	1/2 (1.27)	900 (4003)	3	3	67 (170.2)	9 (3.5)	15,000	1/2 (1.270)
0.050 (0.127)	1/2	1,050 (4670)	4	3	65 (165.1)	8 (3.1)	16,500	9/16 (1.429)
0.062 (0.15748)	1/2	1,200 (5338)	4	4	63 (160.0)	7 (2.8)	17,500	5/8 (1.588)
0.078 (0.19812)	5/8 (1.5875)	1,500 (6672)	6	5	55 (139.7)	6 (2.4)	19,000	11/16 (1.746)
0.094 (0.23876)	5/8	1,700 (7562)	7	6	50 (127.0)	5.5 (2.2)	20,000	3/4 (1.905)
0.125 (0.3715)	3/4 (1.905)	2,200 (9786)	11	7	45 (114.3)	4.5 (1.8)	22,000	7/8 (2.223)

* Dimensions are in inches unless otherwise indicated.

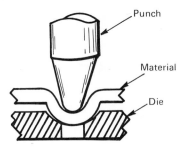

FIGURE 4–50. *A punch and die used in forming projections.*

with the hot metal. Also the electrodes can be made from harder, more wear resistant material. The prelocated spots permit welds that are impractical with other resistance methods. Outer or top surfaces can be produced without electrode marks, making it possible to paint or plate without grinding or polishing. The process is fast since a number of welds can be made at the same time. Some projection welding design data for mild and stainless steel is given in Table 4.5.

The main limitation of projection welding is its practical use for a comparatively small group of metals and alloys. These are low carbon steels, high carbon and low alloy steels, stainless and high-alloy steels, zinc die castings, terneplate, and some dissimilar and refractory metals. With brasses and coppers the method has not been very satisfactory. Aluminum applications are rare as the metal does not have sufficient hot strength. Free cutting steels, high in sulphur and phosphorus, should not be used because the welds are porous and brittle.

One of the most common applications for projection welding is the attachment of small fasteners, nuts, special bolts, studs, and similar parts to larger components. A wide variety of these parts are available with preformed projections. Wire grids and fencing materials form natural projections for welding. Projection welded joints are not generally water or air tight, but can be made so by sweating solder into the seam. This is usually satisfactory unless the parts are exposed to substances

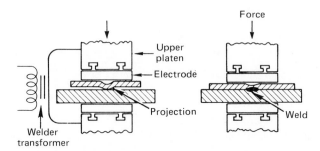

FIGURE 4–51. *In projection welding the current is concentrated in the areas of the raised projections. During the welding process the projections are flattened by heat and pressure. Several projections can be welded at one time.*

TABLE 4.5. *Projection Weld Design* Recommendations*

Outside Piece (cm)	Projection Diam at Base (cm)	Projection Height (cm)	Min Diam of Fused Zone (cm)	Min Shear Strength/Spot Tensile Strength psi	
				Below 70,000 (482.6 MPa) (MPa)	70,000 to 150,000 (482.6 to 1034 MPa) (MPa)
0.010 (0.0254)	0.055 (0.1397)	0.015 (0.0381)	0.112 (0.2845)	130 (0.8274)	180 (1.241)
0.021 (0.0533)	0.067 (0.1702)	0.017 (0.0432)	0.112 (0.2845)	320 (2.206)	440 (3.034)
0.031 (0.0787)	0.094 (0.2388)	0.022 (0.0559)	0.169 (0.4293)	635 (4.379)	850 (5.861)
0.062 (0.1575)	0.156 (0.3962)	0.035 (0.0889)	0.225 (0.5715)	1,950 (13.44)	2,250 (15.51)
0.125 (0.3175)	0.281 (0.7137)	0.060 (0.1524)	0.338 (0.8585)	4,800 (33.10)	5,700 (29.30)

* Dimensions are in inches unless otherwise indicated.

that will attack the solder. A tension-shear test of production run samples is the accepted method of checking projection welds.

Upset Butt Welding. The material to be welded is clamped in suitable electrode clamps. The ends to be welded are brought in contact before the current is turned on. Current densities ranging from 2000-to-5000 A/in.^{2} (310 to 775 A/cm^2) are used. The high resistance of the joint causes fusion at the interface. Just enough pressure is applied to keep the joint from arcing. As the metal becomes plastic, the force is enough to make a large, symmetrical upset that expels oxidized metal from the joint area. The current may be interrupted several times for large areas. Final pressures are applied after heating is completed and range from 2500 to 8000 psi (17.24 to 55.16 MPa) depending upon the material. The metal is not melted, and no spatter results. The smooth symmetrical upset must, however, be machined before use for most applications. Common uses are joining bars and wire.

Flash Butt Welding. In flash butt welding, the ends of the stock are clamped with a slight separation. As the current is turned on, it causes flashing and great heat. The metal burns away, and the pieces move together in an accelerated motion, maintaining uniform flashing action. When the inner faces reach the proper temperature, they are forced together under high pressure, and the current is cut off, Fig. 4–52.

Flash butt welding offers strength factors up to 100 percent of the base metal. No extra material, such as welding rod or flux, is required. Generally, no special preparation of the weld surface is required. Dissimilar metals with varying melting temperatures can be flash butt welded. The size and shape of the parts should be similar, but a 15 percent variation in end dimensions is permissible for commercial use. The process is regularly used for end joining of rods, tubes, bars, forgings, fittings, and so forth. Heavy forgings can sometimes be eliminated by welding small forgings to bar stock.

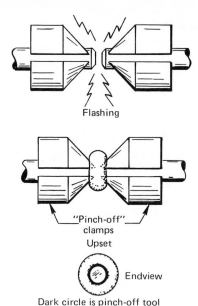

Flashing

"Pinch-off"
clamps
Upset

Endview

Dark circle is pinch-off tool

FIGURE 4–52. *Flash butt weld with "pinch-off" clamping.*

Flash and upset butt welding are purely high production methods. The preparation of work holding equipment and the set-up of the welding machine make the process too costly unless there are a large number of similar welds to be made. Production rates can be quite high since the weld itself takes only 2 to 3 seconds to make.

Percussion Welding.   Percussion welding is a variation of flash welding in that two workpieces are brought together at a rapid rate. Just before they meet, a flash or arc melts both of the colliding surfaces. The molten surfaces are then squeezed together by percussive hammer blows, and some of the metal is forced out to the sides of the joint.

Percussion welding is particularly good for joining small diameter wires, for example, welding 0.002 (0.00508 cm) and 0.015 in. (0.0381 cm) diameter wires in electronic applications, and for materials of widely differing properties. Wires can be joined by soldering or other processes, but the advantage of percussion welding is that it is almost instantaneous.

The welds are produced either by stored-energy type machines or by rapid dissipation of current from a standard 60 Hz ac source. Parts other than fine wires, if they have pinpoint type projections formed or coined into the part, can be percussion welded. These points localize the heat so that the small projection is instantaneously vaporized upon contact. Some metal combinations that have been welded by this process, with excellent results, are copper to Nichrome, copper wire to type 304 stainless plate, thorium to thorium, and thorium to Zircaloy.

Stud welding, discussed earlier in this chapter, may also be classified as

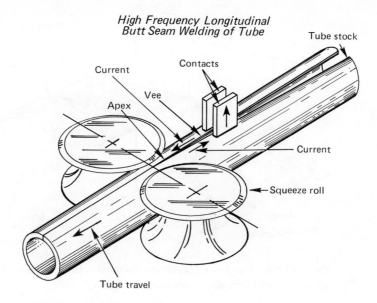

*High Frequency Longitudinal
Butt Seam Welding of Tube*

FIGURE 4–53. *High frequency resistance welding of tubing. Electrical energy entering the tube via the contacts travels to the apex of the vee by the skin and proximity effects. Since the apex is at the weld point, and since the work is continuously advancing through the squeeze rolls, a continuous weld seam is made. (Courtesy of Thermatool Corp., An Inductotherm Company.)*

percussion welding. After the arc melts a superficial layer of metal on the end of the stud, it is propelled to the work by springs that give it a percussive action.

High Frequency Electrical Resistance Welding. In high frequency electrical resistance welding, the energy necessary for raising the temperature of the surfaces about to be joined is provided by the high frequency of current reversals per second. To weld successfully, a minimum of 3000 Hz is necessary, and up to 500 kHz is not uncommon. High frequency ac current does not penetrate deeply into the body of metals. Therefore, only a few thousandths of an inch of surface metal becomes plastic for the actual welding joint. The heat affected zone is usually less than 1/8 inch (0.3175 cm) deep. This process is shown in Fig. 4–53.

Flat thin gauge metal is formed into pipe by a series of forming rolls. The current necessary to heat the edges to a plastic temperature travels from one copper electrode (contact) through the tube apex to the second electrode, thereby completing the electrical circuit. The highest energy concentration is at the notch of the apex and is therefore the most effective for the joining process. Forging pressure is always necessary to complete the weld. Dissimilar metals, such as stainless steel and low carbon steels, have been joined successfully by this process. In addition to joining thin wall tubing, Fig. 4–54, this process is also used for making "T" beams, Fig. 4–55, and for welding spiral fin-to-tube, Fig. 4–56. Speeds above

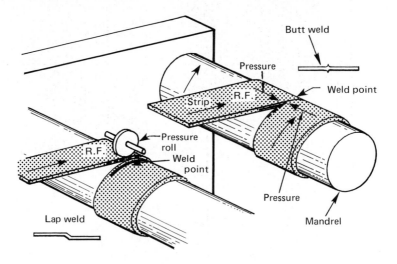

FIGURE 4–54. *Two forms of high frequency spiral welding are shown. (Courtesy of Thermatool Corp., An Inductotherm Company.)*

High Frequency Current Flow in Typical "T" Weld

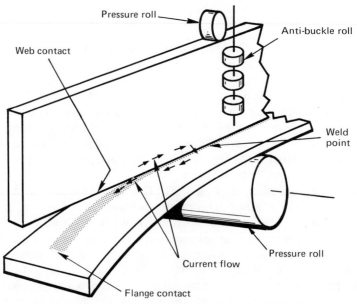

FIGURE 4–55. *T-welds can also be made using high frequency resistance welding. (Courtesy of Thermatool Corp., An Inductotherm Company.)*

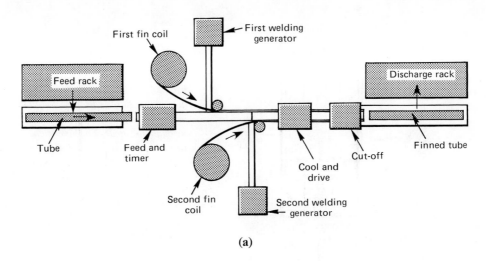

(a)

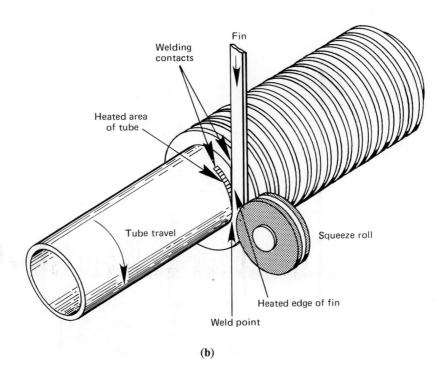

(b)

FIGURE 4–56. *The continuous welding of fins to tubing. (a) Longitudinal fin mill arrangement. (b) Fins being welded to tube. (Tubing up to 96 inches in diameter and fin thicknesses up to 1/8 inch have been successfully welded on a production basis on other units.) (Courtesy of Thermatool Corp., An Inductotherm Company.)*

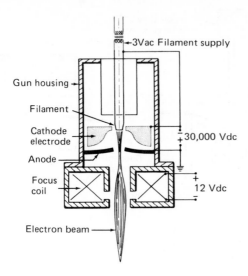

Gun housing

Filament

Cathode
electrode

Anode

Focus
coil

Electron beam

←3Vac Filament supply

30,000 Vdc

12 Vdc

FIGURE 4–57. *The electron beam gun operates in a hard vacuum. The hot filament generates the electrons that are directed through the anode. The kinetic energy converts to thermal energy as the electrons strike the workpiece, melting and/or fusing any known metal. (Courtesy of American Technical Society.)*

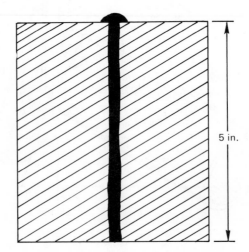

5 in.

FIGURE 4–58. *The electron beam welder is capable of giving a 25 to 1 width to depth ratio for the heat affected zone when welding in a hard vacuum. Shown is an electron beam weld in 5 in. thick 5083 aluminum alloy, with welding parameters of 25 kW, 150 kW, and 10 cpm. (Courtesy of American Welding Society.)*

200 ft (60 meters) per minute are often achieved. The greatest advantages of the high frequency electrical resistance welding process are its speed and its joint quality. The biggest disadvantage of the process is the installation cost of the basic equipment. The process is capable of welding butt joints on pipes with wall thicknesses ranging from .025 to 1.0 in., with diameters up to 96 in. o.d.

Electron Beam Welding. Electron beam (EB) welding is not often thought of as being a resistance process; however, when the principles for emission and penetration are considered, it fits naturally into the resistance group. Millions of electrons are transmitted from a filament, Fig. 4-57, at speeds up to hundreds of miles per second. The electron beam can be focused to a spot less than 0.010 in. (0.0254 cm) in diam. Heat concentrations five hundred times as intense as those available with conventional arc welding are possible. Such high heat concentration makes for relatively narrow welds; consequently, the heat affected area and corresponding stress is relatively low. The penetration, width to depth ratio, Fig. 4-58, is of course dependent on the voltage and power density capabilities of the machine.

Studies made of the electron beam process reveal that a multiple scattering occurs as the electrons strike the surface. They do not lose their energy immediately but experience multiple collisions with atomic nuclear and lattice electrons. When the power reaches a certain threshold, the metal vaporizes. This vaporized metal has no trouble escaping in a vacuum and in so doing pushes the molten metal in the pool aside, exposing the solid metal underneath. As this action is repeated, the beam bores a hole through the metal, as shown in Fig. 4-59.

The beam may also be used to clean the metal prior to welding. This is done by using a broad, out-of-focus beam that sweeps the surface. This vaporizes the oxides and rids the material of impurities and gaseous products. Precise heat control of the beam can be obtained by inducing a pulsed current into the control grid. A wide range of frequencies can be switched by a timing circuit similar to that used on the resistance welding machine to give short and consistent weld times.

The electron beam welder is comprised of three main components: the electron gun, a vacuum chamber, and equipment for positioning and feeding. The electron gun (as shown in Fig. 4-57) consists of a filament, a control electrode, and an anode. A current of sufficient capacity is passed through the filament to heat it to about 4000 °F (2204 °C). After heating, the filament releases a cloud of electrons. A high potential, up to 150,000 V, is established between the filament and the anode, which accelerates the electrons toward the anode at high velocities. The control electrode, a stainless steel cap placed around the filament and insulated from it, is maintained at a negative voltage between 0 and 1000 V. The function of the control electrode is to repel and direct the flow of electrons to pass through the orifice in the anode. The voltage on the control electrode can be pulsed to provide additional control of the beam. Varying the current in the focusing coil allows the beam to become concentrated on a very small spot on the workpiece.

The heat generated by the EB power source is equal to the beam voltage times the beam current times the welding time. To illustrate, a 60,000 V, 500 milli-ampere beam supplies 30,000 watt seconds, which is equal to 28.5 BTU. (It takes

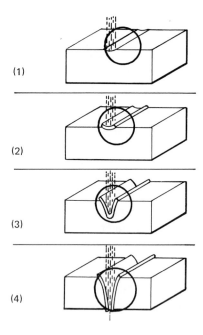

(1)

(2)

(3)

(4)

FIGURE 4–59.   *The successive stages of a deep penetration electron beam weld.*

600 BTU to melt one pound of steel.) The EB gun is by far the most efficient welding system known, giving the highest current density of any known welding system.

The vacuum chamber is built according to the size and specifications required for production. It is usually constructed of mild or stainless steel plate 1/4 in. (0.635 cm) thick or more. One or two sides have doors to give full access to the chamber. Leaded glass viewing windows are also provided. Thin wall chambers, 1/4 in. (0.635 cm) thick or less, require lead shielding to eliminate the danger of X-ray radiation. Work handling fixtures with lengthwise, crosswise, and rotational movements are available in various sizes. Drives may be manual but are usually powered by continuous variable speed motors.

Three words may be used to characterize the electron beam welding process— purity, concentration, and control. Making the welds in a vacuum results in high purity. This is particularly valuable in metals and alloys that are very susceptible to atmospheric contamination, such as zirconium, molybdenum, niobium, and titanium.

Applications.   The versatile electron beam welding process is used for a wide variety of applications ranging from joining tiny precise parts to high-speed seam welding of pipe. Its best applications are for difficult to weld materials and repair work. The electron beam process can be used to join dissimilar metals; however, as in all such joints, hard and brittle zones are often produced. It is possible to introduce

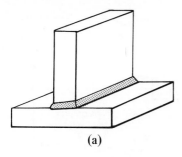

(a)

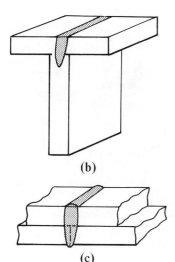

(b)

(c)

FIGURE 4–60. *Joints made by the electron beam process. The tee joint is shown properly made at (a) and with poor strength qualities at (b). A hidden joint can be welded as shown at (c).*

a third metal into the joint by placing a shim at the interface. Although this technique can be very beneficial, the possibilities for problems to arise do exist. Some typical joints are shown in Fig. 4–60. A tee-joint may be made as shown at (a) or (b). The joint at (b) will have complete penetration and be much stronger. Hidden joints may also be welded as shown at (c).

The electron beam process is especially well suited to repair work. Many parts relegated to the scrap bin can be salvaged. An unusual application of the process is to repair cracks. If the crack is small and tight, it can be welded with no additional filler metal. If additional filler metal is needed, it can be added by laying a strip of metal over the crack or by feeding in a cold wire. Salvage and rework usually have the advantages of saving time and material and meeting production schedules. Electron beam welding has been effective in reducing the amount of expensive and difficult to machine materials used in fabricated products. The desired alloy material can be fabricated in small segments, fitted into the strategic location in the assembly, and welded.

Due to the small area of heat concentration, stresses are much lower and the total material undergoing phase changes is relatively small. Even though welds

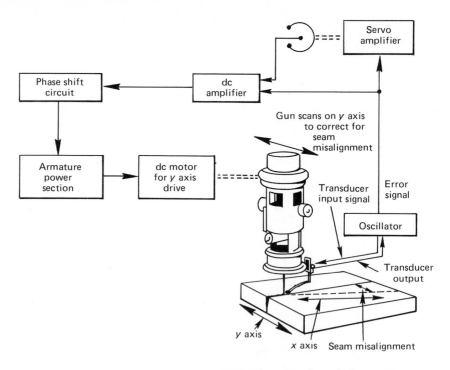

FIGURE 4-61. *A prober type seam tracking system for electron beam welding equipment. (Courtesy of Sciaky Brothers, Inc.)*

of this type are often made without cracks even in hardenable metals, it is best to perform a post weld heat treatment wherever possible to improve the fatigue properties. The EB welding process is readily automated. Automatic systems are employed for welding the collapsible automobile steering column at 1,200 units per hour and for joining strips of teeth to steel backing for band saw blades. Several different concepts for seam tracking adaptable to EB guns, as well as guns of other processes, have been developed, Fig. 4-61. Automatic control over the movement of the energy from its source to the work improves weld quality and speed.

Soft Vacuum Electron Beam Welders. Electron beam welders can be classified as "hard" and "soft" and "nonvacuum" machines. Most commercial machines are in the "hard" classification with the gun and work in a chamber at a vacuum of approximately 1×10^{-4} torr during welding. The soft vacuum welder consists of two chambers. The gun housing is made as small as possible and is evacuated by both a mechanical and a diffusion pump. The welding chamber is evacuated by a mechanical pump only. Connecting the two chambers is a passage that can be closed off by a special valve. With a passage size of 100 microns, the air molecules have a long free path, so that only a few of them randomly find their way into the gun chamber and are removed by the diffusion pump. After welding, the valve

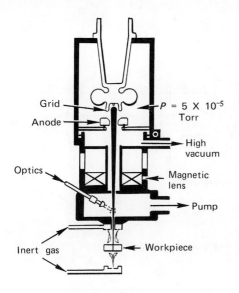

Grid

Anode

$P = 5 \times 10^{-5}$ Torr

High vacuum

Optics

Magnetic lens

Pump

Inert gas

Workpiece

FIGURE 4–62. *An out-of-vacuum electron beam welding system. (Courtesy of Manufacturing Engineering and Management.)*

between the two chambers is closed so that the gun chamber can remain at 0.1 micron while the welded part is removed and the next part loaded.

Shown in Fig. 4–62 is an out-of-vacuum system. That is, the welding is done without a vacuum chamber for the work. This system very much simplifies the design of the work handling equipment. There are limitations on the size of the part that can be welded. There is also a limitation on the working distance, which cannot be greater than one inch. Greater distances result in excessive scattering of the beam. An interesting application of the nonvacuum system is in seam welding high strength 3/4 in. (1.905 cm) thick longitudinal pipe welds in a single pass at speeds in excess of 100 in./min (254 cm/min). Another out-of-vacuum application is the welding of automobile frames by the A. O. Smith Co., of Milwaukee, Wisconsin. The main limitation of conventional electron beam welding is that it must be performed in a vacuum. This disadvantage, however, is being overcome at least in part by soft vacuum machines and guns that operate outside of the chamber in an area of protective gas. A second limitation is that the process is limited to joints that have good fit-up. This often requires careful machining and fitting before welding. Whenever EB welding is done out of vacuum, leaded shields are required to protect the operator and others workers from the hazardous X rays of the process. A schematic comparison of the principles and applications for the EB, GMA, GTA, and Solid Resistance Processes is given in the frontispiece to this chapter.

LASER WELDING

Lasers (**l**ight **a**mplification by **s**timulated **e**mission of **r**adiation) are classed into three categories, namely: solid state (ruby, neodymium, yttrium, aluminum, garnet), gas (helium-neon, carbon dioxide, argon), and injection (galium arsenide,

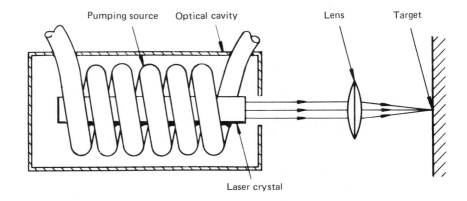

FIGURE 4–63. *A schematic for a pulsed ruby laser welder. (Courtesy of American Welding Society.)*

indium, antimonide). Until late 1971, the laser welding process was confined to the joining of materials of foil thickness. The early equipment was primarily the pulsed ruby type, Fig. 4–63. The more recent CO_2 lasers are capable of joining 3/8 inch (0.9525 cm) thick steel with a depth-to-width ratio of 10 : 1 and welding speeds of 50 ipm (127 cm/min). This breakthrough came with the idea of oscillating the base material under the CO_2 laser beam. The laser process is now comparable with the electron beam process for many applications. One company is experimenting in piping lasers throughout their plant with mirrors and tapping the laser energy at welding and cutting stations.

Energy transfer in the laser process permits it to be classified as a "solid resistance welding process." The resistance of the base metal to the thermal diffusivity determines the depth of penetration pattern. Thermal diffusivity is defined as the property of the material to accept and conduct thermal energy. For example, stainless steel and Rene 41 have low heat diffusivity and are difficult to weld with the laser process. On the other hand, aluminum, copper, and gold have a high thermal diffusivity and can be joined more easily by the laser process. For other comparisons, see Table 4.6. The laser and EB sources of energy provide the greatest power densities of all welding processes. The properties of intensity and spatial coherence from a laser give a source of radiant energy that can be concentrated to achieve these extremely high power densities. This energy is often sufficient to overcome the binding forces associated with the atomic and molecular structure of materials and melting or vaporization takes place.

The theory of laser welding is that the light strikes the metal's surface causing some vaporization. The vaporized metal prevents the remaining energy from being reflected by the metal; and if the metal being joined has the property of excellent thermal diffusivity, then greater penetration can be expected. The presence of the atmosphere does not necessarily change the quality of the weld. The properties

TABLE 4.6. *Thermal Diffusivity Properties of Metals and Their Alloys*

Material	Thermal Diffusivity Properties at 68 °F (20 °C) ft^2/h
Aluminum (pure)	3.665
Copper (pure)	4.353
Lead (pure)	0.924
Iron (pure)	0.785
Invar (36 % Ni)	0.108
Magnesium	3.762
Nickel (pure)	0.882
Silver (pure)	6.418
Bronze (75% Cu./25% Sn)	0.333

E. R. G. Eckert and R. M. Drake, *Heat and Mass Transfer*, 2nd ed. McGraw-Hill Book Company, New York, N.Y., 1959.

of metal that may affect the weld ability by the laser process are thermal, mechanical, surface condition, metallurgical and chemical. Highly reflective surface conditions are not generally considered condusive to good laser welding. This beam is capable of vaporizing or melting holes in opaque materials. Likewise, it passes freely through transparent materials without affecting them. This characteristic makes it possible to weld metals previously placed inside a vacuum tube. The feasibility of this application can be demonstrated: a pulsed ruby laser can break a blue balloon inside a clear outer balloon without harming the latter. The blue balloon bursts

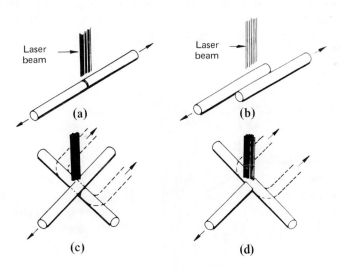

FIGURE 4–64. *Joint configuration for wire-to-wire welds made with a laser: (a) Butt. (b) Lap. (c) Cross. (d) Tee. (Courtesy of Linde Division, Union Carbide Corp.)*

because it absorbs the red light concentrated at one place. Since the outer balloon was clear, it could not absorb red light. A second illustration is laser beam erasing of typewritten characters by evaporating the ink without burning the white paper.

Because filler metal is not usually used when welding with the laser, good fit-up is needed between the pieces being joined. Joint configuration for joining wire to wire by the laser process is shown in Fig. 4–64. Laser welding has a low heat input into the metal being welded and is usually done in the atmosphere. The pulsed ruby is limited to foil thickness material and has a very low efficiency. Due to the opacity/high energy principles of laser welding, safety precautions are very important.

LIQUID RESISTANCE WELDING

Liquid resistance welding is known as the "electro-slag process," whereby the energy necessary to melt the base metal and filler metal is generated by the resistance of the molten metal to the flow of electric current. The electro-slag process is similar in nature but different in terms of physics to the electro-gas process. Both electro-gas and electro-slag processes are used mainly for one pass weldments on heavy plate positioned in a vertical plane. The electro-gas process is usually used with plate thickness of 1/2 to 3 inches (1.27 to 7.62 cm), while the electro-slag process is used on plate thickness from one inch to fifteen inches (2.54 to 38.1 cm) or thicker. The development of the electro-slag process is credited to personnel working at Paton Institute of Kiev, Russia, in 1953. The principle of the electro-slag process is unique in itself. The edges of the metal to be joined are non precision squared by any available process and placed in a vertical position gapped by one or two inches depending on the plate thickness. The thicker the

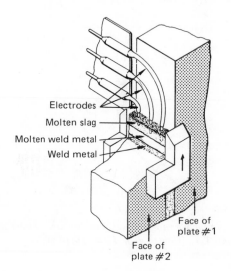

Electrodes
Molten slag
Molten weld metal
Weld metal

Face of plate #1

Face of plate #2

FIGURE 4–65. *Electro-slag welding is used for joining heavy plate in the vertical position.*

plate, the wider is the gap up to approximately two inches. The plate surfaces to be joined are positioned with a slightly wider opening at the top or finishing end of the weld compared to that dimension at the beginning of the weld. This difference compensates for a greater contraction of metals due to the expected increase of plate temperature as the weldment progresses, as well as the fact that heat normally rises and the metal nearing the end of a weldment will be hotter than when the weld was begun.

Figure 4–65 shows the principles for electro-slag welding. Once the pieces to be joined are properly positioned, copper shoes are used to close the gaps at each of the two sides. The shoes are often water cooled and are used to contain the large pool of molten metal characteristic of the process. Each copper shoe may be a single slab of copper, the length of the joint held in a stationary position, or a shorter piece of copper containing magnetic feet that automatically walk upward as sections of the weld are completed. A temperature sensing device signals the climbing mechanism that the weld metal has solidified and that the shoes should move upward along with the molten pool. With the base material properly fixtured and the shoes in place, the next steps are to place a back-up plate at the bottom of the gap and pour several inches of an approved granular flux into the opening. The remaining equipment consists of a conventional CP power source, a wire drive unit, and an electrical control panel. The filler metal wire is fed into the granular flux, which acts as an electrical conductor of high resistance. Turning on the current ignites an arc that, in turn, melts the flux. The molten slag is an electrical conductor that generates heat by resistance to the current as it passes from the electrode to the weld pool. The heat generated is sufficient to melt not only the wire but also the surfaces of the base metal being joined. As the electrode wire is melted away, the level of the weld pool rises and the water-cooled shoes creep upward at the appropriate rate. The liquid slag not only serves to provide heat by its resistance to the current, but it also forms a protective blanket over the molten metal. Additions of fluorides to the flux reduce the viscosity and improve the electrical conductivity. The depth of the slag averages from 1-1/2 to 2-1/2 inches (3.81 to 3.175 cm).

To provide equal penetration and consistent metallurgical properties throughout, the electrode is often oscillated when joining metals greater than two inches thick. For metals thicker than four inches, multi-electrode systems are used.

As mentioned earlier the electro-slag process is seen in several varieties, depending on the manufacturer. Electro-slag welding is used in constructing heavy walls such as those found in pressure vessels, press frames, and water turbines. It is also used for thinner wall materials in ships and storage tanks. In addition to welding, electro-slag techniques have been used for surfacing, ingot refining, and ingot hot topping. The electro-slag process is fast and reduces distortion to a minimum. Weld joints are completed in a single pass, thus reducing heat (energy) losses through interpass cooling. The deposit is cleaner and there are fewer weld defects than in other conventional welding processes. The process has been used successfully to weld mild steels, low alloy steels, high alloy steels, stainless steels, and titanium. Joining is limited to heavy plate material in the vertical position. Until

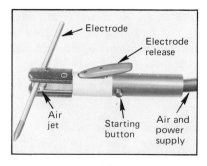

Electrode

Electrode release

Air jet

Starting button

Air and power supply

(a)

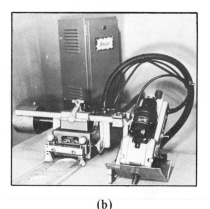

(b)

FIGURE 4-66.   (a) The arc-air torch is used in cutting, gouging, and beveling metals. (b) The metal is melted and simultaneously removed with compressed air. (Courtesy of Arcair Company.)

recently, both the weld metal and the heat affected zone of electro-slag welding were expected to have a low impact property. That property is caused by the coarse grain sizes which are inherent in this kind of joint. More recently, however, companies using the process report that the use of improved filler wires greatly improves the impact properties.

ELECTRIC ENERGY CUTTING PROCESSES

An ordinary mild steel electrode such as E-6010 or E-6011 can be used for cutting when a higher than usual current setting with straight polarity (electrode negative) is applied. This method is mostly for rough work such as cutting scrap, rivet cutting, and hole piercing. For most general purpose cutting, a 3/32 in. (2.38 mm) electrode is satisfactory for metal thicknesses up to 1/8 in. (3.75 mm). A 5/32 in. (3.96 mm) electrode is required for materials 1/4 in. (6.35 mm) thick or thicker.

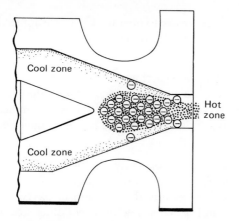

FIGURE 4-67. *The thermal pinch effect. Electrons move more freely in the hot zone, which is in the center of the plasma, than in the cool zone of boundary layer gases. Therefore, the majority of the electrons and the discharge current are concentrated in the hot zone.*

Arc cutting methods have been improved and now include arc-air, oxygen arc, and plasma arc.

Arc-Air Cutting. The arc-air method of cutting metals melts the metal with an electric arc and then removes the molten metal by means of a high velocity air jet. The equipment consists of a torch or electrode holder and a concentric cable carrying both compressed air and current. A lever at the bottom of the holder controls the flow of air that passes through a hole adjacent to the electrode, Fig. 4-66. The electrode usually is a combination of carbon and graphite, either plain or copper coated.

The arc-air process is not only used for cutting both ferrous and non-ferrous metals but also as an effective grooving tool. As an example, grooves 3/8 in. wide and 1/4 in. deep usually are cut at about 3 fpm (91.4 cm/min).

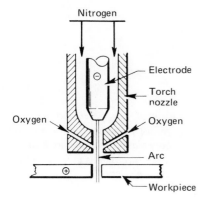

FIGURE 4-68. *The newer plasma cutting-torch design permits the addition of oxygen into the plasma below the electrode, which makes it last much longer.*

The arc-air torch is relatively inexpensive to operate. Costs are approximately $1.50 per hr, exclusive of labor, using 3/8 in. electrodes.

Oxygen Arc Cutting. Oxygen arc cutting is performed with a special electrode, a coated rod with a small diameter hole through the center, and a special electrode holder equipped with an oxygen connection. Cutting is accomplished by the heat of the arc and the oxygen that flows through the center of the electrode.

Plasma Arc Cutting. Plasma, the fourth state of matter, is formed by passing gas through a constricted arc or thermal pinch, as shown in Fig. 4–67. A schematic view of a newer type plasma torch that permits the addition of oxygen into the plasma below the electrode is shown in Fig. 4–68. The principal difference between the torch used in plasma cutting and plasma welding is the absence of a shielding gas. Open arc plasma temperatures are relatively low, about 10,000 °F (5465 °C), owing to the freedom with which the plasma can expand laterally in the open air; however, in the constricted portion of the arc they may be as high as 24,000 °F.

The plasma flow is initiated through a pilot arc, which is established by a high frequency generator. Current to maintain the arc is provided by a direct current power supply as shown schematically in Fig. 4–69.

The orifice gas, fed to the torch at the rate of 50 to 350 cfh (1.4–9.8 cu m/h), is determined by the particular application. Aluminum and magnesium are cut with nitrogen, nitrogen–hydrogen, or argon–hydrogen. Stainless steel thicker than 2 in. (5.08 cm) is best cut with argon and hydrogen.

When the torch is used for heavier cuts, cooling water is required to dissipate the 35,000 BTU per hour heat. The water pressure used is 80 psi (5.6 kg per sq m) with a flow rate of 3 gpm (11.4 liters per min).

Advantages and Limitations. The plasma arc torch has several advantages over the conventional oxyacetylene cutting torch:

1. It can be used easily on all types of conductive metal. (With the transferred type arc, the workpiece becomes the conductor).

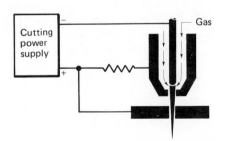

FIGURE 4–69. *Schematic view of the transferred arc plasma torch. The work is subjected to both plasma heat and arc heat. A thoriated tungsten electrode is used with direct current, straight polarity.*

2. The cut can be made faster and with a narrower kerf. Mild steel, 3/4 in. (19.05 mm) thick, can be cut at 70 ipm (177.8 cm/min) as compared to oxy-acetylene cutting rates of about 20 ipm (50.8 cm/min).

3. Operating costs for the plasma torch are about 60 percent that of operating an oxyacetylene torch of comparable size.

4. Cutting is quite automated. The power control unit performs all on–off and sequencing functions. The cooling water must be turned on or the waterflow interlock will block the starting circuit. Once the torch is located over the center of the cut, the starting button is pushed and current from the high frequency generator flows to establish the arc between the work and the nozzle. As soon as the cutting arc is established, the high frequency current shuts off and the carriage starts to move. After the cut is complete, the arc goes out since it has no ground, and the carriage is stopped. The main contactor opens and shuts off the gas flow.

Disadvantages. The main disadvantage of plasma-arc cutting is the high initial cost of the equipment.

EXERCISES

Problems

4–1. You live in Wisconsin, and you awake one January morning to find that the water pipes in your house are frozen. You have a 200 ampere welder with a 60 % duty cycle to thaw your pipes. You need 100 % duty cycle. What is the maximum amperage you can draw continuously from this machine without burning it out?

4–2. With the same machine as in problem 4–1, determine the duty cycle for the power source when drawing 400 A.

4–3. You are the welding engineer for the Better Fabricators Co. A customer brings you several zirconium parts that he wishes welded. When completed the joints are to undergo X ray and ultrasonic inspection. Which process would you recommend and why?

4–4. The current density at the tip of the electrode affects the depth of penetration when welding. What is the current density for a 1/8 in. (0.3175 cm)

electrode when using 125 amperes? [1/8 in. = 0.01227 area inches^2 (0.3175 cm = 0.07916 area cm^2)].

4–5. The heating principle for the EB welder is similar to that of the micro-wave oven used in the home. Explain the difference between this heating principle and the conventional method.

4–6. What would be the time difference for flame cutting a 5 ft. diam circle in $\frac{3}{4}$ in. thick steel plate and plasma arc cutting it?

Questions

4–1. Name the three natural divisions for the electrical energy joining processes.

4–2. Schematically show the volt–ampere curves for both the CP and CC power sources.

4–3. Schematically show the penetration patterns for each of the following: DCSP, AC, and DCRP when using a stick electrode.

131

4–4. You are a "rod burner," and your job is overhead welding with the stick electrode. Which current would you select?

4–5. How does the operator control the voltage when welding with a CC power source?

4–6. With arc welding, the operator uses one of two voltages. Name each and describe it.

4–7. Explain arc blow and when is it encountered.

4–8. Explain pinch effect in GMA welding.

4–9. Explain what is meant by steep slope in a CP power source.

4–10. What are the modes of metal transfer for GMA welding?

4–11. Discuss the duty cycle of a welding power source.

4–12. What is the purpose of a rectifier? Name two kinds of rectifiers used in welding power sources.

4–13. With the GMA welding process using the CP power source, how is the amperage controlled?

4–14. Why is ACHF current always used with the GTA process when welding aluminum?

4–15. How is the arc length controlled in the submerged arc process using a CC power source?

4–16. What are the unique differences of the electro-slag and the electro-gas processes over most other processes?

4–17. Name the seven critical points of resistance in the spot welding process.

4–18. Describe the two kinds of guns used in plasma welding.

4–19. You wish to convert high voltage/low amperage to low voltage/high amperage. What mechanism would you use? In what welding power source is this principle applied?

4–20. How does the thermal efficiency of the EB welder compare to that of the shielded metal-arc process?

4–21. What do the letters spelling "LASER" stand for?

4–22. Name the two kinds of welding lasers in use today.

4–23. What can be used to help automate the arc-air cutting and grooving process?

4–24. Why is plasma arc cutting limited to conductive materials?

BIBLIOGRAPHY

ALTHOUSE, A. D., TURNQUIST, C. H., AND BOWDITCH, W. A., *Modern Welding,* The Goodheart-Wilcox Co., Inc., Homewood, Ill., 1967.

AWS, *Resistance Welding, Theory, and Use,* Miami, Fla., 1956.

AWS Welding Handbook, sec. 1, 6th ed., Miami, Fla., 1968.

———, sect. 2, 6th ed., Miami, Fla., 1969.

———, sect. 3A, 6th ed., Miami, Fla., 1970.

DE GARMO, PAUL E., *Materials and Processes in Manufacturing,* The Macmillan Company, Toronto, Canada, 1969.

GALIANO, F. P., LIMBLEY, R. M., AND WATKINS, L. S., *Lasers in Industry,* the Proceedings Institute of Electronic and Electrical Engineering, vol. 57, no. 2, February 1969.

Gas Metal Arc Welding, Miller Electric Manufacturing Co.

GIACHINO, J. W., WEEKS, W., AND JOHNSON, G. S., *Welding Technology,* American Technical Society, Chicago, Ill., 1968.

HOULDCRAFT, P. T., *Welding Processes,* Cambridge University Press, London, England, 1967.

KING, K. D., *Solid State Controls for Resistance Welding,* British Welding Journal, August, 1967.

LINDBERG, R. A., *Processes and Materials of Manufacturing,* Allyn and Bacon, Inc., Boston, Mass., 1964.

MOTL, B. C., *Welding with the Carbon Dioxide Welding Processes,* A. O. Smith Corp., Milwaukee, Wis., 1962.

PHILLIPS, A. L., *Current Welding Processes,*

American Welding Society, Miami, Fla., 1964.

———, *Modern Joining Processes*, American Welding Society, Miami, Fla., 1966.

PIERRE, EDWARD R., *Welding Processes and Power Sources*, Pierre Publishing Company, Appleton, Wis., 1966.

Resistance Welding Aluminum, Reynolds Metals Co., 1963.

SAYER, L. N., AND BURNS, T. E., *Practical Aspects of Electron Beam Welding*, British Welding Journal, vol. 11, 1964, p. 163.

TYLECOTE, R. F., *The Solid Phase Welding of Metals*, St. Martin's Press, New York, N.Y., 1968.

Welding Encyclopedia, 13th ed., Jefferson Publication, Morton Grove, Ill., 1951.

Welding Processes, Hobart Welding School, Brochure EW–294, 1969.

WROS, E. M., AND BRATON, N. R., *The Weldability of Ductile Iron*, Mechanical Engineering Department, University of Wisconsin, 1968.

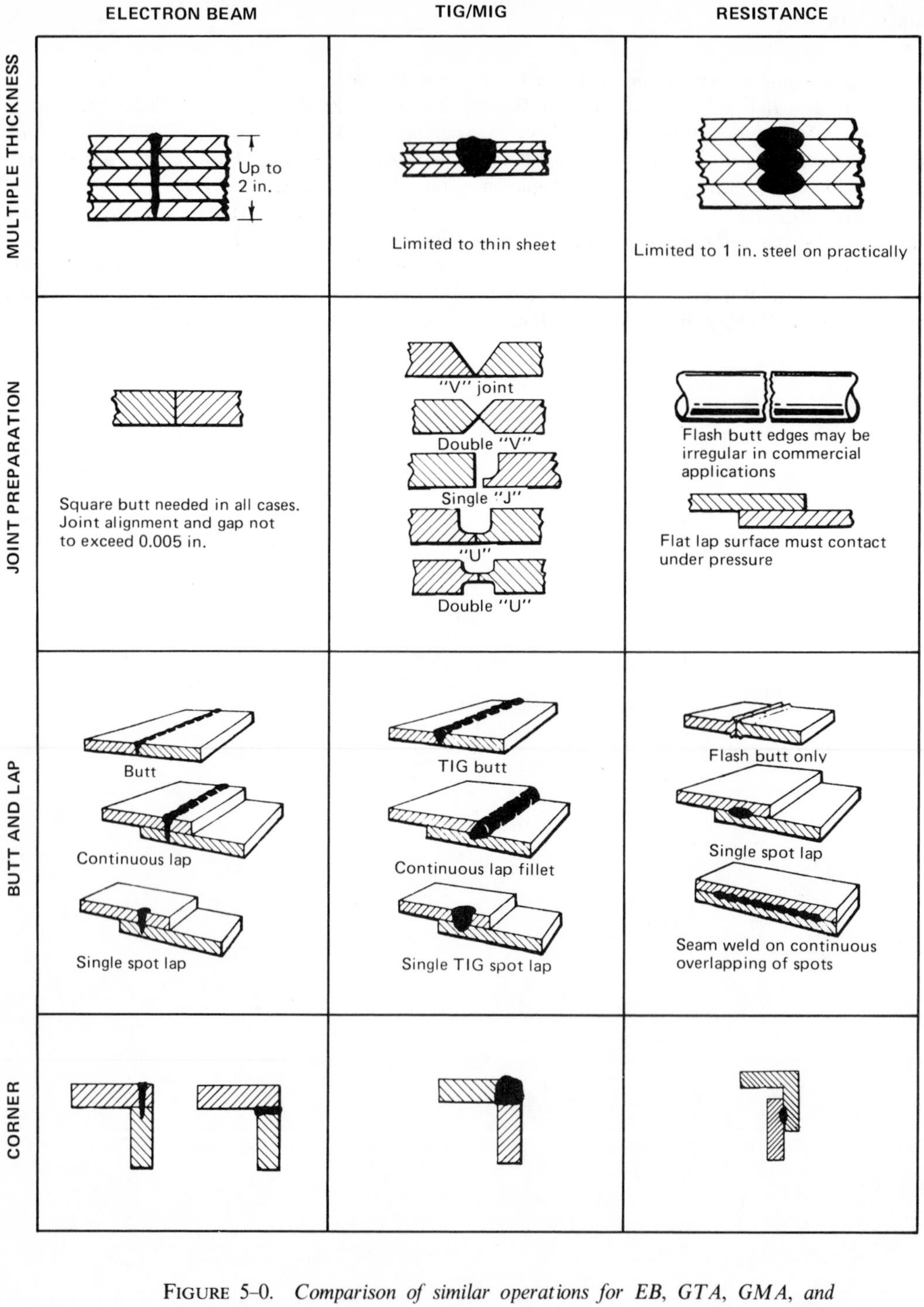

	ELECTRON BEAM	TIG/MIG	RESISTANCE
MULTIPLE THICKNESS	Up to 2 in.	Limited to thin sheet	Limited to 1 in. steel on practically
JOINT PREPARATION	Square butt needed in all cases. Joint alignment and gap not to exceed 0.005 in.	"V" joint / Double "V" / Single "J" / "U" / Double "U"	Flash butt edges may be irregular in commercial applications / Flat lap surface must contact under pressure
BUTT AND LAP	Butt / Continuous lap / Single spot lap	TIG butt / Continuous lap fillet / Single TIG spot lap	Flash butt only / Single spot lap / Seam weld on continuous overlapping of spots
CORNER			

FIGURE 5–0. *Comparison of similar operations for EB, GTA, GMA, and resistance welding. Comparison of applications. (Courtesy of Sciaky Brothers, Inc.) (Cont. from p. 68.)*

5

Heterogeneous Joining, Hard Surfacing, and Thermal Spraying

Heterogeneous joining processes as discussed in this chapter are processes in which metals are joined by a dissimilar lower melting point metal, as in brazing, braze welding, and soldering. Also discussed in this chapter are hard surfacing, the process of fusing a hard outer layer of metal over a softer one; and thermal spraying, the process of atomizing metal or ceramic materials to apply as a coating for build-up purposes or to change the characteristics of the base material.

BRAZING

Brazing is a process of joining metals with nonferrous filler metal that has a melting point below that of the metals being joined but by the American Welding Society's (AWS) definition above 800 °F (427 °C). The filler metal must *wet* the surfaces to be joined; that is, a molecular attraction must exist between the molten filler material and the materials being joined. The brazing alloy, when heated to the proper temperature, flows into small joint clearances by capillary attraction. A limited amount of alloying occurs between the filler and base metal at elevated temperatures, Fig. 5–1. As a result, the strength of a brazed joint, when properly made, may exceed that of the base material. The strength is attributed to three sources: atomic forces between the metals at the interface; alloying, which comes from diffusion of the metals at the interface; and intergranular penetration.

Although there are eight basic forms of brazing listed in the AWS master chart of welding processes, only the following four are of commercial importance today: torch, induction, furnace, and dip brazing.

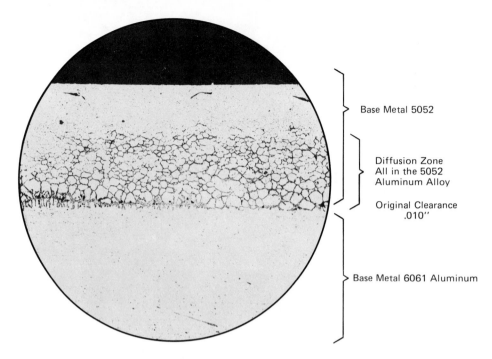

Base Metal 5052

Diffusion Zone
All in the 5052
Aluminum Alloy

Original Clearance
.010″

Base Metal 6061 Aluminum

FIGURE 5–1. *A brazed aluminum joint showing the diffusion zone between the two metals. The brazing filler metal was BAlSi-3 (10% Si, 4% Cu, Al.). Brazed in molten flux bath at 1100°F (593°C). (Courtesy of Wall Colmonoy.)*

Torch Brazing. Torch brazing is usually done with the regular oxyacetylene torch, however some modifications may be desirable. A tip with several orifices will permit a more general flame distribution. For larger work, bar type nozzles 6 or 8 in. (15.24–20.32 cm) long may be used. A variety of gases may be substituted, such as oxyhydrogen, oxypropane, and natural gas with compressed air. Regulator pressures are about the same as for gas welding, however, the flame must be soft to prevent blowing the filler metal. If an oxyacetylene flame is used, it should be made slightly reducing (excess acetylene) to help avoid any possibility of joint oxidation or overheating of the bare metal. Only the flame envelope is used for heating the joint.

Induction Brazing. The heat energy for induction brazing is furnished by an ac coil placed in close proximity to the joint. High frequency current is usually provided by a vacuum tube or solid state oscillator to produce frequencies of 200,000 to 5,000,000 Hz. This method has the advantage of providing (a) good heat distribution, (b) accurate heat control, (c) uniformity of results, and (d) speed. It is excellent for certain types of repetitive work that requires close control.

Furnace Brazing. Parts to be furnace brazed have the flux and brazing material preplaced at the joint. If the furnace has a protective atmosphere, the

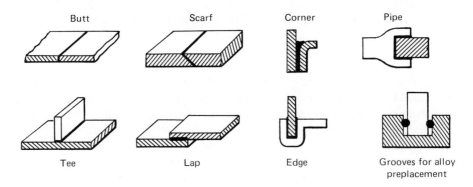

FIGURE 5–2. *Common types of joints used for brazing.* (*Courtesy of Welding Engineer.*)

flux may not be necessary. The furnace may be the batch or conveyor type. Conveyor furnaces have both heating and cooling chambers. A system of automatic controls regulates the time, temperature, and, where applicable, the atmosphere.

Dip Brazing. Dip brazing derives its name from the fact that the parts to be brazed are jigged and placed in a chemical or molten metal bath, which is maintained at the correct brazing temperature. The parts are first thoroughly cleaned, fluxed, and preheated, and the alloy preplaced. A flux is placed over the top of the molten mass of metal to prevent oxidation. Molten chemical (salt) baths are also used. The salts are generally of the cyanide or chloride type to provide a mild fluxing action. If heavier fluxing is desired, other salt mixtures may be used. After immersion is completed, cleaning operations are begun immediately to remove the flux.

Brazing Joint Design. Brazing alloys do not usually possess strength equal to that of the base metal. However, by proper joint design, the strength can be made equal to that of the base metal. Typical braze joints are shown in Fig. 5–2.

The butt joint presents a limited area to which brazing alloy can bond. So, unless product appearance or other considerations dictate its use, the butt joint should be bypassed in favor of the sturdier lap joint or scarf joint. The scarf joint (shown in Fig. 5–2) is designed to give a joint area more than three times that of the butt joint. Joints of this type are difficult to prepare and even more difficult to hold in alignment.

Lap joints should be used in preference to other types of joints where strength is the primary consideration. An overlap of three times the thinnest member will usually give maximum efficiency. Overlaps greater than this lead to poor brazing due to insufficient penetration, inclusions, and so forth. This joint is also recommended when leak tightness or good electrical conductivity are required.

Brazed Joint Clearance. The proper clearance for a brazed joint is dependent upon the following: type of base and filler metals, base metal thickness, mechanical strength desired, surface condition of parts, and the distance the alloy must flow. In general the clearance at room temperature is specified. It represents an optimum value considering the foregoing conditions. Generally, it can be said that the smallest possible thickness of brazing filler material metal in the joint after it has been completed gives the joint its best strength. Production brazing unfortunately cannot meet these theoretical requirements. When a mineral type flux is used, the joint clearance must allow it to flow and be pushed ahead of the filler metal as it is drawn into the joint by capillary action. In general, for flux brazing with silver alloy filler metals, the clearance should be between 0.002 and 0.005 in. (0.050–0.127 mm). Excessive clearance may result in the loss of capillary action and a weak joint. Insufficient clearance will not allow the flux or filler material to enter properly and thus cause voids and poor joint strength.

In the case of dissimilar metals, the coefficient of expansion must be considered. As an example, if an inner member has a greater rate of expansion than the outer member, the clearance will be lost and capillary action will be prevented. In this case, a clearance near the maximum should be used. The clearance should be carefully calculated in advance since either too much brazing material or too little will result in a weak joint.

Brazed Joint Strength. There is, unfortunately, no industry-wide standard test specimen for use with brazed joints. The published data available are applicable only to specific test specimens and are not comparable to other designs. Thus it is necessary to resort to empirical information in order to design assemblies using shear joints. The following simple formula is useful in determining the lap length (depth of shear) on either tubular or flat joints.

$$L_1 = \frac{S_t \, t F}{S_s}$$

where L_1 = lap length.
S_t = tensile strength of weakest member.
t = thickness of weakest member.
F = factor of safety.
S_s = shear strength of brazing alloy.

Example. A lap joint is made in copper tubing. A copper brazing alloy is used that has an S_s of 36,000 psi (248 MPa). The desired factor of safety is 3. The tubing is 16 gauge or 0.0625 in. (1.5875 mm) thick. How much overlap is required if the tensile strength of the copper tubing is 33,000 psi (227 MPa)?

Solution:

$$L_1 = \frac{33000 \times 0.0625 \times 3}{36000} = 0.171 \quad \text{or} \quad \frac{3}{16} \text{ in.}$$

$$= \left(\frac{227 \, \text{MPa} \times 1.5875 \, \text{mm} \times 3}{248 \, \text{MPa}} = 4.36 \, \text{mm} \right)$$

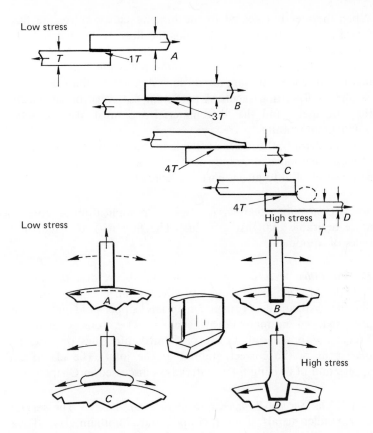

FIGURE 5–3. *Examples of how brazed joints can be designed, ranging from being able to accept only low stress loads due to concentration on the joint, to being able to accept high stress loads due to transference to the base metal.*

A rule of thumb is to use an overlap 3t for the shear type joints.

A good braze joint design will act to remove high stress concentrations from the edges of the joint and distribute them into the base metal. A good example of progressively transferring the stress to the base metal through design improvement is shown in Fig. 5–3.

In some joint designs, a relatively large volume of air is heated and may block the flow of flux and filler material. When possible, such cavities should be vented. As an example, a rod brazed into a dead end hole can be vented by making a small flat on the shaft equal in length to slightly more than the depth of the hole.

Brazing Fluxes. The primary purpose of brazing fluxes is to dissolve and absorb oxides that heating tends to form. Fluxes are not intended to act as primary cleaning agents. Therefore, the parts to be brazed must first be cleaned of all existing oxides. This includes oil, grease, and other foreign matter, by either degreasing, grinding, or pickling—or a combination of these cleaning processes.

When the metal is heated to the brazing temperature, the flux becomes a clear liquid that wets the surface and aids the flow of the filler metal. Several chemicals are used singly or in combination to achieve the desired properties. Some of the most common ingredients are sodium, potassium, and lithium. They are used in making up chemical compounds such as borates, fluorides, chlorides, boric acids, alkalies, and wetting agents. A particular flux must be chemically compatible with the base metal and the filler material and must also be active throughout the brazing temperature.

Generally, fluxes are available in three forms, powder, paste, or liquid. Of the three, paste is most commonly used although powdered flux is frequently mixed with water or alcohol to give a pastelike consistency.

Controlled Atmosphere. In certain applications, a controlled atmosphere is more desirable than using a flux. The flux may damage or contaminate some types of equipment.

Brazing Filler Metals. Brazing filler metals are divided into seven classifications. In order of popularity, these are: silver, copper, copper zinc, copper phosphorus, aluminum silicon, heat resisting materials, copper-gold and magnesium. These alloys are produced in many forms such as wire, rods, coated sheets, and powder. Frequently, brazing alloys are preformed into rings or special shapes to simplify placement of the correct amount at the joint. The classification, content, and application of brazing filler materials is discussed in Chapter 8.

The Service Temperature of Brazing Alloys. The service temperatures for brazing filler metals of silver, copper, and aluminum class alloys vary over a wide range, -100 to $600\,°F$ (-73 to $316\,°C$), but find most usage around room temperature.

A nickel class of brazing alloys has been introduced that can be used at temperatures from below $-300\,°F$ ($-184\,°C$) to above $2200\,°F$ ($1204\,°C$). The nickel-chromium-boron filler metal of this series is particularly good for stainless steel brazing. It diffuses and alloys with the base metal producing a joint with higher remelting temperatures, above $2500\,°F$ ($1371\,°C$), and a joint ductility that is better than the "as received" filler metal.

Advantages and Limitations of Brazing. Brazing is well suited to mass production techniques that join both ferrous and nonferrous metals. Some of the advantages of this joining method are:

1. Dissimilar metals can be easily joined.
2. Assemblies can be brazed in virtually a stress-free condition.
3. Complex assemblies can be brazed in several steps by using filler metals with progressively lower melting temperatures.
4. Materials of different thicknesses can be joined easily.
5. Brazed parts require little or no finishing other than flux removal.

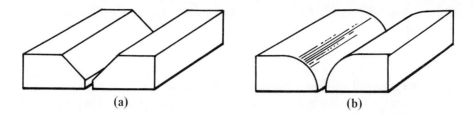

FIGURE 5-4. *A comparison of butt joints:* (a) *prepared for fusion welding and* (b) *braze welding.*

Some limitations are:

1. Joint design is somewhat limited if strength is a factor.
2. Joining is generally limited to sheet metal gauge material and relatively small assemblies.

BRAZE WELDING

Braze welding is similar to brazing in that the base metal is not melted but is joined to another piece of metal by an alloy of a lower melting point. The main difference is that in braze welding the alloy is not drawn into the joint by capillary action. A braze welded joint is prepared very much like one for welding; however, an effort should be made to avoid sharp corners that are easily overheated and may also be points of stress concentration, Fig. 5-4.

Braze welding is used extensively for repair work and some fabrication on such metals as cast iron, malleable iron, wrought iron, and steel. It is also used, but to a lesser extent, on copper, nickel, and high melting point brasses and bronzes. Some of the brasses and bronzes melt at a point so near that of the filler metal that fusion welding takes place.

Strength. The strength of a braze welded joint is dependent on the quality of the bond between the filler metal and the base metal. It is also dependent on the quality of the filler metal after it is deposited. Quality depends on freedom from blowholes, slag inclusions, and other physical defects. Some of the same forces at work in brazing, namely, interalloying and intergranular penetration, are at work here also. The average tensile strength of a braze weld joint on steel or cast iron is approximately 50 ksi (345 MPa).

SOLDERING

Soldering is a process of joining two or more pieces of metal by means of a fusible alloy or metal, called solder, which is applied in the molten state. Soldering can be distinguished from brazing in that it is done at a lower temperature, below 800 °F (427 °C), and there is less alloying of the filler metal with the base metal.

Solder Alloys. The four metals used as the principal alloying elements of all non-proprietary solder formulations are: tin, lead, cadmium, and zinc. These alloying elements fall into three thermal groups. Those that melt below 596 °F (312 °C) are called low melting-point solders. Those that melt between 596 and 700 °F (312–371 °C) are called intermediate solders, and those that melt between 700 and 800 °F (371–427 °C) are termed high-temperature solders. The strongest and most corrosive resistant are the high-temperature solders, and vice versa. The shear strength of low temperature solders is a bit in excess of 5000 psi (34.47 MPa).

The low melting point solders, 378–596 °F (190–312 °C), are the most widely used and are of the tin-lead or tin-lead-antimony types. The tin-lead alloys provide good strength at low temperatures and require a minimum of preparation. A combination of 63% lead and 37% tin produces the lowest melting point. In practice, this is referred to simply as a 60–40 solder. Being a eutectic alloy, it behaves like a pure metal with only one melting temperature, 378 °F (190 °C), at which it is completely liquid, rather than over a range of temperatures. Increasing the tin content produces better wetting and flow characteristics. Thus when less tin is used, greater care should be exercised in surface preparation of the metals to be joined.

The tin-lead-antimony solders are generally used for the same types of applications as the tin-lead alloys except they are not recommended for use on aluminum, zinc, or galvanized steel. The addition of antimony, up to a maximum of 6% of the tin content, increases the mechanical properties without seriously affecting the wetability or flow characteristics. The joint clearance for both the tin-lead and the tin-lead-antimony solders is 0.003 to 0.005 in. (0.076–0.127 mm).

The tin-zinc group of alloys is used primarily for soldering aluminum, particularly where a lower soldering temperature than that of a zinc-aluminum solder is required. Zinc aluminum solders are designed specifically for soldering aluminum. They provide high joint strength and good corrosion resistance. Indium solders, 50% indium and 50% tin, is particularly suitable for products subjected to cryogenic temperatures and for glass-to-metal bonds.

Design Considerations for Soldering. Soldered joints should not be placed under great stress. In sheet metal, the joint should take advantage of the mechanical properties of the base metal by using interlocking seams, edge reinforcement, or rivets. Butt joints should be avoided. Lap joints loaded in pure shear may be used if the proper clearance is allowed. Clearances of 0.003 to 0.010 in. (0.076–0.254 mm) are permissible, but best strength and ease of soldering are obtained with about 0.005 in. (0.127 mm) clearance. Care should be exercised to avoid placing the soldered joint where it will be subject to a peeling action, Fig. 5–5.

Soldering Methods. Many techniques are used to transmit the heat energy to the joint area. These include the well known soldering "irons" or "coppers" torches, resistance heating, hot plates, ovens, induction heating, dip soldering, and wave soldering.

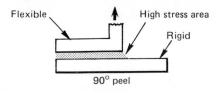

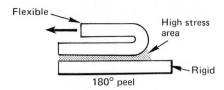

FIGURE 5–5. *A soldered joint should not be subjected to peel.*

Torches. In addition to the air-acetylene torch, the disposable tank propane torches are very popular, since they are easily portable and are a source of instantaneous heat. Various types of tips are used to control the flame from a "pencil point" for fine work to a flared tip for preheating. A newer type flameless soldering torch is shown in Fig. 5–6. It is capable of producing an air stream of up to 1100 °F (593 °C).

Resistance Soldering. The work is heated by its own resistance to applied electrical energy. A resistance unit can be made to work with approximately 6 volts, stepped down from that of regular plant voltage for reasons of safety and to prevent arcing.

Oven or Furnace Soldering. This is the most widely used method for high production soldering. Automobile radiator cores and other parts where a number of joints are desired at one time are assembled by this method. The parts are prepositioned and held in fixtures. Close tolerances are required as are the usual cleaning and fluxing. Clean warm water can be used to remove the flux residue after soldering.

Wave Soldering. Wave soldering is a technique that is particularly applicable to printed circuit boards. After all the components have been assembled on the board, the joining side is passed over a bath of molten solder as shown schematically in Fig. 5–7a. The molten solder is pumped under pressure and forced through a nozzle to form the wave, Fig. 5–7b. Of course, successful operation of what appears to be a very simple process requires well cleaned surfaces, proper selection and application of flux, and proper soldering temperature.

The flux is also applied in a wave form. The force of the wave and capillary action of the liquid flux combine to give good through-hole penetration on double-sided circuit boards. After fluxing, the circuit boards are preheated to

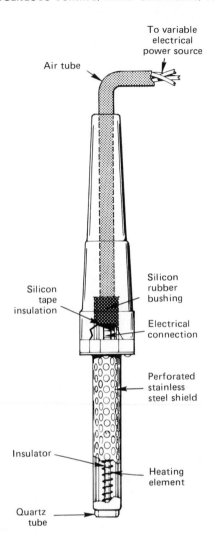

FIGURE 5-6. *The flameless soldering torch eliminates soldering iron scratches on thin film substrates. It provides an airstream of up to 1100°F (593°C) one-half inch (12.70 mm) beyond the tip. (Courtesy of The Western Electric Engineer.)*

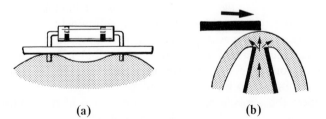

(a) (b)

FIGURE 5-7. (a) *Wave soldering a resistor in place.* (b) *The wave may be controlled by the pressure on the solder as it flows through the nozzle.*

bring them up to soldering temperature. The heat also serves to drive off the flux solvent or flux vehicle. The boards are then moved by a conveyor over the wave of solder. Since the height is critical, the conveyor mechanism is equipped with vernier height controls.

Aluminum Soldering. Aluminum is often thought of as being very difficult to solder. One reason for this is that an oxide film forms quickly over the surface after cleaning. Now, however, several methods have been developed that remove the oxide as the soldering is being done. These methods are flux soldering, friction soldering, and ultrasonic soldering, as shown in Fig. 5–8.

Of the low temperature solders used for aluminum, a tin-zinc eutectic, which melts at 400 °F (204 °C), has an advantage over brazing. It requires less heat, the result

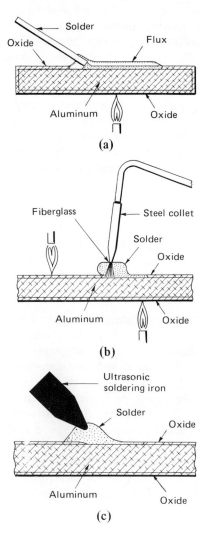

(a)

(b)

(c)

FIGURE 5–8. *Various methods of removing the oxide film from aluminum during the soldering operation. (a) Removing aluminum oxide by fluxing. Chemical action of the flux removes the oxide. (b) Abrasive removal of oxide. Brushing action abrades oxide from the aluminum surface, allowing solder to tin the surface. (c) Ultrasonic soldering. Ultrasonic waves cause cavitation, breaking up the oxide and floating it to the top of the solder puddle. A frequency of 20 kHz is often used to cause active cavitation. (Courtesy of Reynolds Metals Company.)*

of which is a minimum loss of strength and hardness and the least distortion. This is particularly important in complex structures. If higher strengths are important, the intermediate or high temperature solders may be used. The high temperature solders contain from 90 to 100% zinc and are the least expensive and strongest of all the aluminum solders.

Corrosion in aluminum soldered joints can be particularly troublesome. The corrosion may be either chemical or galvanic in nature. In any case moisture must be present. By coating the joint with a waterproof coating, the problem can be eliminated.

HARD SURFACING

One of the most important and yet frequently the least understood method of minimizing wear is by applying a wear resistant overlay. Wear resistant deposits are often associated with repair work, however, the process can prove to be very economical even in the original design and fabrication. Hard surfacing materials can be used where, otherwise, massive sections of special wear resistant alloys may be required. The applied hard surface materials resist wear, cavitation, corrosion, heat, or impact. Hard surfacing alloys can be applied by most of the welding processes already discussed, such as oxyacetylene and various forms of arc welding. Of these, the most common and most economical for small operations is the manual stick electrode method. For large and repetitive jobs that are of relatively simple geometry, the most economical in terms of per pound overlay is usually automatic arc welding. This may be either open arc or submerged arc.

Classification of Hard Surfacing Materials. The wide variety of hard surfacing materials available present a confusing array to choose from, particularly when they are best known by the manufacturers' trade names rather than by a standardized code system common to most metals. The classification presented here will be based on the properties produced by the alloying elements used and then the type of service that can be expected.

Alloying Elements. Nearly all available surfacing alloys have a base of iron, nickel and cobalt, or copper. Carbon is the most important auxiliary element since it alloys with chromium, molybdenum, tungsten, manganese, and silicon. The carbon joins with these elements to form hard and brittle compounds such as chromium carbides, molybdenum carbides, and so forth. These carbides in order of decreasing hardness are: tungsten, chromium, molybdenum, and iron. High percentages of tungsten or chromium with 2 to 4% carbon form distinctive carbide crystals that are harder than quartz. A fused mixture of tungsten carbide, WC or W_2C, is the most abrasive-resistant type constituent commonly used in hard surfacing. Chromium carbides (Cr_7C_3) found in high chromium alloys (20–30%) are softer than WC and less expensive. Iron carbides, Fe_3C or cementite, are used in many hard surfacing alloys and are usually modified by moderate percentages of Cr, Mo, or W as alloying elements.

Tungsten, molybdenum, vanadium, and chromium also contribute to high temperature strength in the 900 to 1200 °F (482–649 °C) range. A 25% Cr content provides effective oxidation resistance up to 3000 °F (1649 °C). Hot strength may be obtained by a selection of nickel or cobalt as the alloy base. The cobalt-base alloys, protected by 20 to 25% Cr and strengthened with 5 to 15% tungsten, provide the highest hot hardness above 1200 °F (649 °C). The abrasion resistance will depend on the carbon content.

Nickel, cobalt, and chromium provide corrosion resistance and promote oxidation resistance. It is evident from this brief discussion that the selection of surfacing alloys is dependent on a knowledge of both composition and structure. Shown in Table 5.1 is the AWS classification of surfacing alloys, starting with the hardest and most abrasive resistant and proceeding to the tougher, more impact resistant types.

WELDING PROCESS SELECTION FOR HARD SURFACING

Oxyacetylene Process.   Before a hard surfacing alloy is applied, it is necessary to prepare the surface. The surface must be free of all grease, dirt, and other foreign matter. All sharp edges that may be subjected to shock should be rounded off if possible. If a sharp edge is required, as in the case of a punch, shear blade, or die, a recess should be machined for the hard surfacing, as shown in Fig. 5–9a. If severe impact will not be encountered, the corners may be machined as shown in Fig. 5–9b.

The oxyacetylene flame is generally used where a smooth hard surface is to be achieved with a minimum of finishing. It is especially useful in applying a crack-free deposit of tungsten carbide particles. The small pieces of tungsten are held

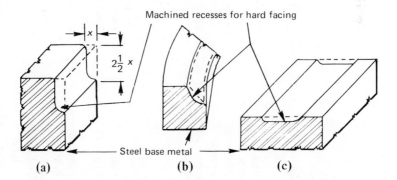

FIGURE 5–9.   *The machined recess at (a) when filled with hardsurfacing material and ground will be well supported to act as a cutting edge. If severe impact will not be encountered, the recess may be made as shown at (b). If the alloy is to be applied to a flat area, it may be placed in a groove as shown at (c) to create a wear strip or the entire surface may have a padding type layer. (Courtesy of American Welding Society.)*

TABLE 5.1. *Classification of surfacing alloys* (Courtesy of the American Welding Society, *Welding Handbook*, 5th ed., sec 3.).

Classification by Basic Types	Important Features	Successful Applications
Tungsten carbide deposits	Maximum abrasion resistance	Oil well rock drill bits and tool joints
Granules or inserts		A wide range of severely abrasive conditions
Coarse granule tube rods	Worn surfaces become rough	
Fine granule tube rods	Best performance when gas welded	
High chromium irons	Excellent erosion resistance	
Multiple alloy type	Hot hardness from 800–1200 °F (427–649 °C) with W & Mo	Abrasion by hot coke
Martensitic type	Can be annealed and rehardened	Erosion by (1000 °F) (538 °C) catalysts in refineries
Austenitic type	Oxidation resistant	Agricultural equipment in sandy soil
Martensitic alloy irons	Excellent abrasion resistance	General abrasive conditions with light impact
Chromium-tungsten type	High compressive strength	Machine parts subject to repetitive metal-to-metal wear and impact
Chromium-molybdenum type	Good for light impact	
Nickel-chromium type		
Austenitic alloy irons	More crack-resistant than martensitic irons	General erosion conditions with light impact
Chromium-molybdenum type		
Nickel-chromium-types		
Chromium-cobalt-tungsten alloys	Hot strength and creep resistance	
High carbon (2.5%) type	Brittle and abrasion-resistant	Hot wear and abrasion above 1200 °F (649 °C)
Medium carbon (1.4%) type	Tough and oxidation-resistant	Exhaust valves of gasoline engines; valve trim of steam turbines
Low carbon (1.0%) type		

Material	Properties	Applications
Nickel base alloys		
Nickel-chromium-boron type	Good hot hardness and erosion resistance	Oil well slush pumps
Nickel-chromium-molybdenum tungsten type	Corrosion resistance	
Nickel-chromium-molybdenum type	Resistant to exhaust gas erosion	Exhaust valves of trucks, buses and aircraft
Nickel-chromium type	Oxidation resistant	
Copper base alloys	Anti-seizing; resistant to frictional wear	Bearing surfaces
Martensitic steels		General abrasive conditions with medium impact
High carbon (0.65–1.7%) type	Fair abrasion resistance	
Medium carbon (0.30–0.65%) type	Good resistance to medium impact	Hot working dies
Low carbon (below 0.30%) type	Tough, economical	General low-cost hard facing
Semi-austenitic steels	Tough, crack resistant	Base for surfacing or a build-up to restore dimensions
Pearlitic steels	Crack resistant and low in cost	
Low alloy steel	Suitable for build-up of worn areas	
Simple carbon steel	A good base for hard facing	
Austenitic steels	Tough; excellent for heavy impact	General metal-to-metal wear under heavy impact
13% manganese—1% molybdenum type	Fair abrasion and erosion resistance	Railway trackwork
13% manganese—3% nickel type	Lower yield strength	
13% manganese-nickel-chromium type	High yield strength for austenitic types	
High carbon nickel-chromium stainless type	Oxidation and hot wear resistant	Frictional wear at red heat; furnace parts
Low carbon nickel-chromium stainless type	Oxidation and corrosion resistance	Corrosion resistant surfacing of large tanks

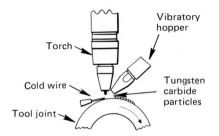

FIGURE 5-10. *A schematic view of the inert gas tungsten-arc hardsurfacing process.*

in a mild steel tube; as the tube is melted, the tungsten carbide particles become fused to the surface in a matrix of mild steel.

Manual Metal Arc Process. The stick electrode is the most adaptable as to position and location and is, therefore, the most used for short runs in metal arc hard surfacing. It is also available in a wide range of alloys.

GTA Hardsurfacing. The use of the inert gas, tungsten-arc process for hardsurfacing is recommended where a flawless deposit is required on new construction rather than on building up a worn surface for repair. Shown in Fig. 5-10 is a schematic view of how tungsten carbide particles are fed down from a vibratory hopper. The particles feed down to the base metal where they are fused on by the addition of molten metal that is produced as the wire is fed into the arc. This composite surface is very abrasive resistant, yet not brittle. Beads may be made in stringer fashion or by oscillation. Oscillation is preferred because it provides a better tungsten–carbide distribution throughout the deposit. The auxillary or "cold" wire can be added to obtain any desired properties in the matrix of the deposit. It is termed a cold wire because it is not involved in producing the arc energy. The use of the added matrix tends to reduce cracking on those applications where the base material is a high carbon alloy.

GMA Hardsurfacing. The inert gas metal arc was at first considered slow for hard surfacing. The process has since been modified so that, in addition to the wire coming down through the arc, an additional cold wire is fed automatically into the weld pool. The addition of the cold wire increases the deposition rate by a factor of 1.5 to 2.5. A deposition rate of 21 to 35 lb/h (9.52–15.87 kg) is possible. This process is equally as applicable for depositing an overlay of copper or stainless steel.

The Submerged Arc Process. The submerged arc process is of most value for hard facing when it is done in large runs of similar work. Multiple electrodes and oscillation techniques have made this method fast. The welding head can be made to oscillate back and forth up to 3 in. (7.62 cm). A forward speed of 3 to 5 in./min (7.62–12.7 cm/min) can be maintained with an oscillation of

FIGURE 5–11. *Pellet rolls hardsurfaced by the submerged-arc oscillating technique. (Courtesy Linde Division, Union Carbide Corp.)*

2 in. (5.08 cm). The pellet rolls shown in Fig. 5–11 have been hardsurfaced using the oscillating technique. A mild steel wire was used in conjunction with the submerged-arc hardsurfacing composition, which produced a deposit hardness of RC 58. The main disadvantage of the high heat input is possible cracking of the hardened overlays.

FLAME SPRAYING

One of the easiest ways to improve the surface of a metal is to apply a thermal spray coating, for example, ceramics are sprayed on rocket nozzles to improve heat resistance. Hard surfacing alloys are sprayed on machine components to resist wear. Aluminum is sprayed on ship hulls to resist corrosion. The principle of flame spraying may be briefly described as feeding a suitable wire or powder through a gun, where it is heated and atomized by an oxyacetylene, oxypropane flame, or an electric arc and then propelled by compressed air or other means to become embedded on the surface of the workpiece. Adhesion of the coating to the substrate may range from simple mechanical interlocking to complete metallurgical bonding, depending upon the process used.

Surface Preparation. The surfaces to be sprayed must be absolutely clean and sufficiently roughened to offer the maximum amount of mechanical anchorage for the coating. The alternatives employed are blasting, machining, or the use of a bond coat.

Blasting. The use of clean, sharp, crushed steel grit or aluminum oxide blasted against the surface by compressed air will provide reentrant angles for mechanical bonding.

Machining. Surfaces that will be machined after thermal spraying need an exceptionally strong bond. If a heavy coating or built-up surface is required, an

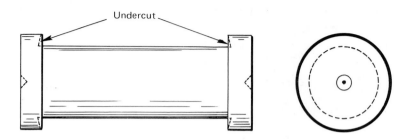

FIGURE 5–12. *Undercuts are made to provide additional anchorage for the flame sprayed metals.*

undercut will provide anchorage at the edges, as shown in Fig. 5–12. Additional anchorage is provided by machining grooves in the surface. A standard 1/8 in. (3.17 mm) wide cut-off or parting tool may be ground down to 0.045 or 0.050 in. (1.143–1.270 mm) wide and rounded on the end. The grooves are cut about 0.025 in. (0.635 mm) deep and about 0.015 in. (0.381 mm) apart. The holding power of the grooved surface is greatly increased by rolling down the ridges with a knurling tool, Fig. 5–13.

Preparation of Internal Surfaces. When the outside surfaces of shafts are metallized, each successive layer cools and contracts, making it fit more tightly, similar to a shrink-fit sleeve. Greater care must be exercised in applying coatings to internal surfaces where the shrinkage may work to a disadvantage and tend to

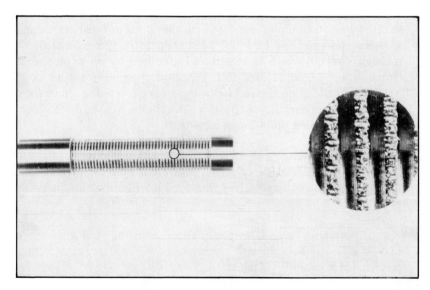

FIGURE 5–13. *The holding power of the grooved surface can be enhanced by knurling. (Courtesy of Metco Inc.)*

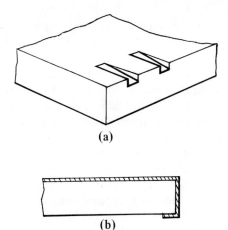

(a)

FIGURE 5-14. *Methods of treating flat stock to keep the sprayed coating from lifting as it shrinks: (a) cutting tapered grooves into the edge or (b) spraying over the edge.*

(b)

pull the metallizing away from the base metal as it cools. To overcome this difficulty, it is recommended that, after the internal surface has been prepared and prior to depositing the coating, the part be heated up to 350 °F (177 °C). As the metal cools, the stresses will tend to be compressive rather than tensile, aiding the bonding process. The inside diameter preparation may be accomplished by going over the surface with a boring tool, using a medium coarse feed so the bore will not be too rough.

Preparation of Flat Surfaces. Shrinkage stresses that tend to lift the coating away from flat surfaces may be overcome either by spraying over the edge, to secure a clamping action, or by cutting short slots, about 1/16 in. (1.587 mm) deep at the outer edge, which taper inward about 2 in. (5.08 cm), Fig. 5-14. The base metal may be preheated to 350 °F (177 °C) to equalize the coating stresses.

It is important that the parts to be metallized not be handled after cleaning, as oil of any kind will impair the bond. If this is not possible, the part should be wrapped in a clean cloth or handled with clean gloves. Any oil or grease that may come in contact with the work surface must be removed by vapor degreasing or other chemical cleaning method prior to spraying. Solvent cleaning may also be used if it is applied generously enough to carry the contamination off the surface. The choice of solvent will be dependent on the degree of cleanliness required. For work that will subsequently be prepared by blasting, almost any solvent will serve if properly used.

Heating is sometimes the only way to eliminate oil completely. Parts taken from service may be oil soaked. Cleaning can be accomplished by heating the casting until all the oil boils out of it. A 500 to 600 °F (260–316 °C) temperature should be maintained until the oil ceases to come to the surface and all smoking stops. Aluminum and zinc die castings that have been exposed to oil may have to be boiled in trichlorethylene before they can be successfully metallized.

Bond Coat. The first layer of metal sprayed on the prepared metal surface is referred to as the *bond coat.* It is essentially molybdenum and is used because it adheres to a smooth surface. The surface to be metallized must be undercut (reduced in size) an amount equal to the minimum coating thickness plus an allowance for wear. The minimum coating thickness is determined by first deciding what the minimum wear allowance should be and then adding to this the minimum permissible coating thickness. On a one-inch diameter shaft, this allowance will usually be 0.010 in. (0.025 mm) on the radius. For shafts larger than one inch, an additional 0.005 in. (0.127 mm) may be added for each additional inch of diameter until about 0.040 in. (1.016 mm) is reached. The thickness of the coating may be reduced if the spray is well atomized and even, as opposed to a coarse uneven layer. The coating may also be reduced if the finish process consists of grinding rather than other machining methods.

Before the bonding coat is sprayed on the workpiece, surfaces that are not to be coated must be masked or oiled. Care must be taken, if oil is used, that it does not touch or run onto the prepared surfaces. The oil may be boiled off by running the flame of the gun over the area.

Thermal Spraying Processes. There are four main methods of applying thermal-sprayed metallic and ceramic coatings. These are the rod-and-wire method, the powder method, the plasma method, and the flame plating method.

Rod-and-Wire Method. The rod-and-wire method of flame spraying is based on the theory that coating materials become fully molten as they pass through the application flame. The process utilizes compressed air to atomize the molten metal or oxides and to project them against a prepared surface where they are embedded, assuring good mechanical adhesion, Fig. 5–15. Most common metals vaporize to some extent when they pass through the intense heat of the oxyacetylene

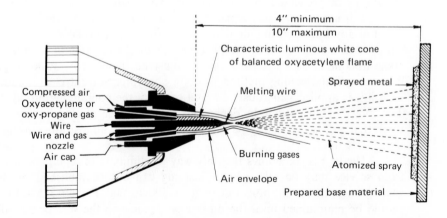

FIGURE 5–15. *This type of equipment is used for wire type metallizing or rod type application of ceramic coatings.* (*Courtesy of Metco Inc.*)

flame. This is of small consequence, however, if the wire is fed too slowly, there will be a loss of deposit efficiency and an increase in oxides in the deposit. As a rule of thumb, the relative surface speed between the gun and the work ranges from 35 to 100 ft/min (10.66–30.48 m/min). The traverse speed should be from about 1/16 to $\frac{1}{4}$ in. (1.58 to 6.35 mm) per revolution.

The compressed air used in the process also helps cool the work surface so that coatings may be successfully applied not only to metals but also to glass, wood, asbestos, and certain plastics. New electronic controls have contributed to make the metallizing process fully automatic, Fig. 5–16. The system is made not only to start and stop the metallizing guns when desired but to supervise the process. A heat sensitive "eye" watches the flame and turns on an alarm system as well as shuts off the machine if it does not perform properly.

Powder Method. The powder-spray method utilizes a welding torch with a modified tip that permits the powdered metal to be sprayed through the flame. A carrier gas—argon, helium, nitrogen, carbon dioxide, or compressed air—conveys the powdered metal to the torch tip. The fuel gas may be acetylene or hydrogen, Fig. 5–17.

The material may not be molten as it leaves the gun. The powder bonds mechanically, but can be fused to the substrate either by the torch flame (which contacts the work surface) or by the kinetic energy released on impact.

Plasma Method. The plasma flame is created by passing gas such as argon, helium, hydrogen, or nitrogen through a high amperage arc. The ionization and subsequent recombination of the gas produces temperatures up to 30,000 °F (16,490 °C). Thus, even such high melting point materials as tungsten, thorium dioxide, and hafnium carbide can be sprayed. The coating is fed to the spray gun as a powder, Fig. 5–18.

Since the plasma is inert, materials can be sprayed with very little contamination. The process is more costly than flame spraying, but it produces coatings with greater density and better adhesion. Typical applications include rocket nozzles and components for jet engines that require high heat resistance. Although the process can be used for depositing the most refractory of materials, it can also be used to advantage on the more common materials such as stainless steel, Nichrome, aluminum, and molybdenum, because contamination and residual oxides are minimized.

Flame Plating. The flame plating process has a combustion chamber where oxyacetylene gas is detonated by a spark plug at the rate of 260 detonations/min. Powder fed to the chamber is driven by the explosions into the substrate at velocities on the order of 2500 ft/sec (762 m), Fig. 5–19.

Equipment for the process is expensive and cumbersome, and a specially constructed application room is required to contain the 150 dB noise created. Job facilities are maintained by Linde Div. of Union Carbide Corp., who have the proprietary rights to the process.

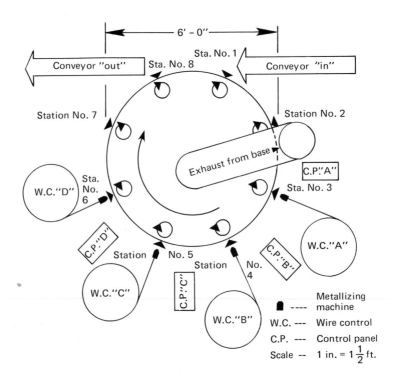

FIGURE 5–16. *A schematic plan layout of an automated system for thermal spraying. (Courtesy of Metco Inc.)*

Station No. 1—*loading.*
Station No. 2—*preheat to approximately 250°F.*
Station No. 3—*apply SPRABOND .0015" to .002" thick. Spray time 15 to 20 seconds.*
Station No. 4—*SPRABRONZE AA, first coating.*
Station No. 5—*SPRABRONZE AA, second coating.*
Station No. 6—*SPRABRONZE AA, third coating.*
Station No. 7—*inspection, or spare for increased production.*
Station No. 8—*unloading.*

This machine indexes once every 60 seconds and 10 seconds are allowed for this operation. An automatic timing device shuts down the metallizing machines B, C, and D for this intermediate period. Metallizing machine A operates for 15 to 20 seconds and the wire feed is shut off for 40 to 45 seconds.

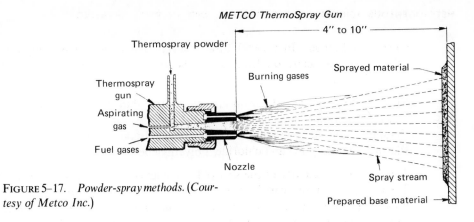

FIGURE 5–17. *Powder-spray methods. (Courtesy of Metco Inc.)*

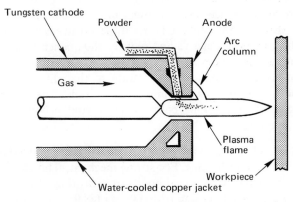

FIGURE 5–18. *A high-current electric arc is struck in the controlled atmosphere of a special nozzle. A gas (usually argon) is passed through the arc, where it ionizes to form a stream of plasma. As the gas continues through the nozzle, it recombines to create temperatures as high as 30,000 °F (16 490 °C). Powder is fed to the stream, where it is melted and ejected from the gun at velocities of approximately 300 mph (482.7 km/h). Bonding is usually mechanical. (Courtesy of Machine Design.)*

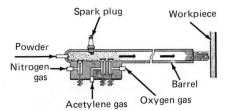

FIGURE 5–19. *Gases are fed to a combustion chamber at one end of a cannon-like torch, where they are ignited by a spark plug at the rate of 260 firings per minute. Powder is also fed to the chamber, so that the explosions drive the powders down the barrel and out of the torch at velocities of approximately 1,700 mph (2735 kph). The powders embed in the substrate upon impact, and bonding is further aided by the release of kinetic energy. (Courtesy of Machine Design.)*

157

Coating Selection. In general, any material that does not decompose, vaporize, sublime, or dissociate on heating can be thermal sprayed. Examples of sprayed materials are:

Metals—tungsten, nickel, chromium, tantalum, aluminum, zirconium.
Alloys—Nichrome, stainless steel, babbit, bronze.
Ceramics—alumina, zirconia, beryllia, spinel, zircon, glass.
Plastics—nylon, polyvinyl chloride, polyethylene.

The chemical reactivity of the sprayed particles is important to coating adherence. The potential reactivity with the substrate is a major factor in establishing a strong bond. Conversely, reactivity with the thermal source can contaminate the particle surfaces and greatly impair bonding. Also, selective oxidation or vaporization on the sprayed coat produces gross changes in particle chemistry.

Most coatings adhere better if the substrate is preheated. Preheating makes the substrate more reactive and reduces the quench rate of the impacting particles. Preheating also serves to drive off surface moisture, which can inhibit bonding. It also serves to expand the substrate so that differential contraction between it and the hot particles is minimized as it cools. This reduces residual stress. Preheating must be carefully controlled to avoid producing an oxide film that will inhibit bonding.

Types of Bonds. Many thermal spray coatings are bonded to the substrate primarily by mechanical interlocking. For strong mechanical bonds, the substrates should be as rough as possible. Mechanical bond strengths range from below 1000 psi (6.89 MPa) for a flame sprayed coating to 10,000 psi (68.95 MPa) for a flame plated coating.

Alloy bonds, which are usually stronger than mechanical bonds, can be produced by spraying metals with high melting points, such as tungsten on stainless steel and molybdenum on mild steel. Typical bond strengths for sprayed coatings on cold-rolled steel are shown in Table 5.2.

TABLE 5.2. *Typical Test Results for Sprayed Coatings on Cold-Rolled Steel*

Sprayed Material	Test Type	Bond Strength, psi (MPa)		
		Grit Blasted Surface	Molybdenum Undercoat	24 Pitch Thread Groove & Mo Undercoat
Low carbon steel	Shear	6,970 (48.02)	8,850 (60.97)	16,000 (110.3)
(Spraysteel 10)	tension	2,660 (18.32)	2,150 (14.81)	3,050 (21.01)
Nickel-chromium steel	Shear	8,330 (57.39)	9,500 (65.45)	20,500 (141.25)
(Metcoloy #1)	tension	3,040 (20.94)	2,250 (15.50)	3,050 (21.01)
High purity aluminum	Shear	2,830 (19.49)	4,400 (30.32)	5,550 (37.96)
(Metco aluminum)	tension	1,030 (7.09)	1,640 (11.30)	2,400 (16.54)
Aluminum—6% silicon	Shear	4,500 (31.00)	5,100 (35.14)	8,150 (56.15)
(Metco SF aluminum)	tension	1,700 (11.71)	2,050 (14.12)	3,100 (21.35)

Courtesy of Metco Inc., *Metallizing Handbook.*

Typical Applications for Thermal Spray Coatings. Thermal spray coatings are used to improve surface properties of metals and for rebuilding. Examples of improved surface properties include: electrical contacts and oxide insulation sprayed on motors and induction coils; aluminum and zinc sprayed on ships, docks, piers, bridges, and factory structures; plastics sprayed on steel; tin sprayed on steel tanks in a brewery or food processing plant; chromium carbide and hard surfacing alloys applied to machinery, knife blades, turbine blades, and gripping surfaces; zirconium dioxide applied to extrusion dies; oxide coatings sprayed on high temperature parts to reduce heat loss; and aluminum oxide coatings on satellite surfaces to control heat gain from solar radiation.

Examples of parts rebuilding are: shafts and other machine parts rebuilt with alloy steels, molybdenum, and other metals; and cracks, holes, and other surface imperfections repaired in castings.

EXERCISES

Questions and Problems

5–1. (a) Compare the tensile strength of brazed aluminum joints made as shown in Fig. P5–1. Each of these joints develops 30 ksi (206.8 MPa) of strength in both shear and tension.
(b) Is $3t$ the usual overlap for a scarf joint?
(c) Why is a scarf joint seldom used?
(d) What is the main advantage of a scarf joint?

5–2. (a) Compare the tensile strength of a brazed butt joint (as made in Fig. P5–1) with a properly made lap joint. The joint develops 45 ksi (310.3 MPa) of strength in both shear and tension.
(b) What is the objection to the use of a lap joint for some installations?
(c) How can the objection be partly overcome?

5–3. Two pieces of aluminum tubing are to be joined as a brazed lap joint. The larger tube is 2 in. (5.08 cm) in diam with a wall thickness of 1/8 in. (3.175 mm). The nearest standard size 1/8 in. (3.175 mm) wall tube to fit inside is 1.750 in.

(4.445 cm) in diam. (a) What do you recommend be done to make a proper fit for brazing?
(b) What size should the smaller tube be?
(c) Both tubes are made out of 2024-T3 aluminum with a tensile strength of 64,000 psi (441.3 MPa), and a shear strength of 15,000 psi (103.4 MPa). How much overlap should be used? The shear strength of the brazing alloy is approximately 20,000 psi (137.9 MPa). The desired factor of safety is 3.

5–4. A stainless steel pipe 2 in. (5.08 cm) in diam is to be joined with a copper pipe. The inside diameter of the stainless steel pipe is 1.750 in. (4.445 cm). The copper pipe will have to be machined slightly on the outside diameter (o.d.) to obtain the desired size for a brazed joint. The coefficient of expansion for stainless steel is 8.5×10^{-6} in./°F (8.5×10^{-6} °C^{-1}), and for copper it is 9.8×10^{-6} (9.8×10^{-6} °C^{-1}). What size should the o.d. of the copper tube be? A nickel alloy filler material will be used with a brazing temperature of 2100 °F (1149 °C).

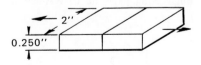
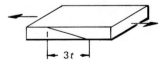

FIGURE P5–1.

5–5. An 8 in. (20.32 cm) diam aluminum tank is to be fabricated. The dome type ends will be fastened to the cylinder as a brazed lap joint. The material used is alloy 3003–H14, 1/8 in. (3.175 mm) thick, which has a yield strength of 25,000 psi (172.4 MPa). The tank will be required to hold pressures of 8,000 psi (55.16 MPa). How much overlap will be required if the shear strength of the brazed joint is 20,000 psi (137.9 MPa)?

5–6. A 1/2 in. (12.70 mm) diam mild steel bar has been grit blasted and thermal sprayed with a molybdenum undercoat and finally with 0.020 in. (.51 mm) nickel-chromium steel. The bar is used to support a tensile load of 5,000 lb (34.47 MPa). Would this much load have any affect on the metal-sprayed coating?

5–7. (a) A 10 in. (25.4 cm) diam pipe is being used as a conveyor tube for crushed rock. There is excessive wear on the inside of each of the 30° elbows. How would you propose to increase the wear resistance on the inside surface of the elbow?
(b) Select a hardsurfacing type alloy for this application and tell why you chose it.

5–8. Discuss the accuracy of this statement:

"Brazing is not well suited to mass production techniques."

5–9. (a) What is meant by a eutectic solder?
(b) Give an example of two eutectic solders.

5–10. Two lap joints are made in 16 gauge (1.5875 mm) aluminum sheet stock, one by brazing and the other by soldering. Make a chart as shown and rate each joint as best, good, or fair.

Joining Method	Strength	Appearance	Qualities of Visual Inspectability	Ease of Field Repair
Brazing				
Soldering				

5–11. A 3 in. (7.62 cm) diam steel shaft has been removed from a pump for repair. One bearing surface is badly worn, 0.015 in. (.38 mm) in places. You decide to repair it by flame spraying and grinding.
(a) How should the surface by prepared?
(b) How thick a bond coat should be used?
(c) If the final spray coating is a nickel-chromium steel, what would be the shear strength?

BIBLIOGRAPHY

Books and Pamphlets

A Complete Guide to Successful Silver Brazing, Silver Brazing Division, The American Platinum Works, Newark, N.J.

Easy-Flo and Sil-Fos Low Temperature Brazing, Handy and Harman, New York, N.Y.

INGHAM, H. S., AND SHEPARD, A. P., *Flame Spray Handbooks*, vols. 1, 2, and 3. Metco Inc., Long Island, N.Y., 1964.

LINDBERG, R. A., *Processes and Materials of Manufacture*, Allyn & Bacon, Inc., Boston, Mass., 1964.

Welding Handbook, American Welding Society, 5th ed., sec. 3, 1965.

Periodicals

ADKINS, H. E., AND RIDEOUT, R. A., "Techniques for Torch Brazing Aluminum," *Welding Engineer*, August 1959.

BARNETT, O. T., and editor, "Brazing, the Process, the fluxes, the Alloys," *Welding Engineer*, August 1958.

GRISAFFE, J. S., "Thermal-Spray Coatings," *Machine Design*, July 20, 1967.

PEASLEE, R. L., "The Brazement—Design and Application," American Society of Mechanical Engineering, paper no. 61-WA-259.

6

Adhesive Bonding

INTRODUCTION

Manufacturers are constantly searching for easier, faster methods of joining product components. Although adhesives or "glues" have long been used as a standard joining method in the woodworking industry, their introduction to the metal working industry is comparatively recent. The older protein glues made from animal hides, hoofs, soybeans, etc., have been largely replaced by a variety of epoxy and polyester resins and natural/synthetic rubbers. The phenominal growth of adhesives as a bonding agent (20 % per year) has been part of the synthetic polymer explosion. It is truly an interdisciplinary science involving physicists, physical chemists, organic chemists, metallurgists, and technologists.

The use of adhesives probably started when early man noticed matted hair in dried blood or the solidification of tacky pitch between two pieces of wood. Archaeological evidence derived from artifacts shows natural resins were used by early man to fasten arrows and spearheads to shafts. The Egyptians used glues to fasten veneers to coffins. Sir Isaac Newton was far ahead of his time when he wrote in his *Opticks* two-and-one-half centuries ago, "There are agents in nature able to make the particles of joints stick together by very strong attractions, and it is the business of experimental philosophy to find them out." It is only within the past few years that Newton's adhesion phenomenon has been delved into and some progress made in understanding the forces operating in adhesion.

The Nature of Adhesion. Many theories have been advanced as to the forces at work in an adhesive. Some of the more common explanations of bonding can be broadly classified as:

1. Chemical or molecular.
2. Diffusion.
3. Mechanical.
4. Electrostatic.

CHEMICAL OR MOLECULAR THEORY OF ADHESION

Of the three different theories offered, the adsorption theory, which deals with molecular forces, is considered the most valid as well as the most basic. These forces may be classified as long or short range molecular interactions, or primary or secondary forces. The principal types of primary chemical bonds are *ionic* and *covalent*.

Ionic Bonds. The ionic bond is the force that results when a positive ion and a negative ion attract each other. Each ion acts as a nucleus surrounded by a rigid spherical distribution of electrons. One example of ionic bonding is the sodium crystal. The extra electron in its valence shell can be released, the sodium atom then becomes a positively charged ion. An ion of chlorine, on the other hand, can easily add an electron to its outer shell, making it eight, and completing the stable configuration for an atom, Fig. 6–1. Ionic type forces are found in nearly all bonds between metallic and nonmetallic elements. That is not to say "pure ionic bonding" is taking place, but probably 90 percent is ionic with the remainder of other types.

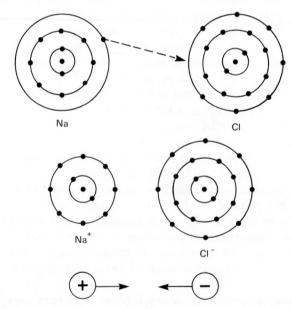

FIGURE 6–1. *Ionic bonding occurs due to electron transfer to produce stable outer shells, which is the result of positive and negative ions that are mutually attracted by covalent forces.*

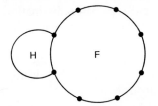

FIGURE 6-2. *Polarization occurs at an electrical imbalance in asymmetrical molecules shown here as H and F. The imbalance produces an electrical dipole with one end* (+) *and the other* (−). *The negative and positive ends, in turn, provide for secondary attraction between molecules as in unlike charges.*

Ionic bonds are strong within the range of $2\,\text{Å}$ (Å = angström, an atomic measure of size equal to $10^{-8}\,\text{cm}$) in a crystal lattice and are extremely resistant to heat (stable), yet they are quite easily dissociated by polar solvents such as water.

Covalent Bonds. Covalent forces are highly directional bonds of 1 to $2\,\text{Å}$ in length. They may be considered the extreme opposite of pure ionic bonding. There is no excess net charge on one atom over the other. Covalent bonds are based on *sharing* of electron pairs of the valency orbits between similar atoms. They may be said to occupy overlapping atomic orbitals or even molecular bonding orbitals. If the bonds are formed from dissimilar atoms, they have polar characteristics, which is due to unequal sharing of the electron pair, Fig. 6-2. The sharing may be single, double, or triple pairs of electrons. Covalent bonds are by nature extremely strong and thus very important as a bonding force in adhesives, when they can be made to react with the adherent. This is dependent on whether or not the electronic requirements in the outer shells of the two materials have already been met.

Some bonds are partially ionic and partially covalent in character. When two atoms have different degrees of electronegativity, the bond between them will be partially ionic in character. If the atomic orbitals of the two atoms are such that they overlap and if the electrons are available to occupy the resulting molecular orbitals, the bond will be partially covalent in character.

Diffusion Theory. Voyutskii and Vakula* have proposed that adhesion between dissimilar polymers is best explained on the basis of diffusion or the interdiffusion of the molecules of similar phases, or interpenetration of one phase by the molecules of the other. If, for example, in a group of molecules a vacancy exists, there is a probability of a molecule of the second phase moving into it. This theory may be applicable to some polymers, but by its nature it requires interfacial forces sufficiently great to cause polymer segments to diffuse across the interface. It also assumes ideal wetting conditions between the adherent and the substrate.

Electrostatic Theory. Electrostatic charges are observed when adhesive bonds are ruptured. Derjaguin** and his co-workers have calculated the energy required to peel polymer films from metal and glass substrates. They assume that, over the small distances involved, the stripping action is the same as separating charged plates of an infinite capacitor. The force required for separation is constant until the charge is dissipated by conduction, electron emission, or dielectric breakdown in the

---

* Voyutskii, S. S., and Vakula, V. L., *Journal of Applied Polymer Science*, 1963.

** Derjaguin, B. V., and Smilga, V. P., *Proceedings of the International Congress*, 1960.

atmosphere. It is assumed that with rapid separation dissipation by conduction is negligible, and that the dielectric breakdown in the air gap is limited.

The electrostatic theory seems to be somewhat in error. It has been verified experimentally that adhesives applied to widely varying substrates have negligible change in performance. This would not be true if they were dependent on the electrostatic energy levels of various substrates.

Mechanical Theory. The mechanical theory of adhesion is based on the formation of a joint in which the adhesive flows into the pores of the substrate so that after curing it is hooked or locked into position. This theory may apply to porous substrates such as wood or concrete, however, the same adhesive performs equally well on nonporous substrates. The mechanical theory has now given way to the theory that the polarity of the adhesive is a much more important factor.

ADHESION AND WETTING

If an adhesive is going to be effective, it must thoroughly wet the surface, or another way of expressing it, is that there must be an affinity of the adhesive for the adherent. The affinity can be predicted by knowing the cohesive energy densities. A laboratory method of predicting the affinity of an adherent for the substrate is to measure the contact angle θ for incompletely wetting liquids. If a drop of liquid is placed on the surface of the substrate and sufficient time is allowed for the liquid to spread, an equilibrium angle θ is formed with the surface, Fig. 6–3. When the drop spreads into an immeasurably thin film, the contact angle is zero. When the drop does not spread, the contact angle is high. This may be caused by a number of factors: The substrate or adherent surface is oily or dusty; or the adhesive lacks affinity for the surface, a condition that results when the adhesive has too high a surface tension or is too viscous to flow readily.

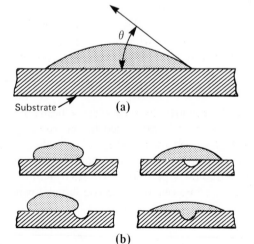

Substrate   **(a)**

(b)

FIGURE 6–3. *The wetting characteristics of an adhesive can be determined by the contact angle θ (a). A low or zero angle θ indicates excellent wetting characteristics. Poor and good wetting action is shown at (b).*

From a practical aspect, it can be seen that poor adhesion may be the direct result of incomplete wetting at the interface. This means, generally, when bonded structures are formed by applying a fluid adhesive on a rigid substrate, the wetting process requires the adhesive to flow around asperities and into interfacial interstices. Diffusion and spreading are enhanced by strong interactions between the phases. The interactions are stronger when the intermolecular attraction of cohesion within the liquid and diffusion and spreading are rapid. When capillary forces assist in wetting, the maximum rates are obtained.

The following preparations may be taken to promote a low adhesive contact angle: (1) Remove all dust, oil, and liquid that could trap air or prevent the adhesive from contacting the surface. (2) Etch or roughen the surface to give the adhesive more area over which to interlock with the substrate—conversely, a rough surface may have to be made smooth so that the adhesive will flow over it. (3) Apply a thin film of adhesive to displace air from deep pores or cracks. (4) Treat the surface chemically. (5) Electroplate the surface, thereby coating it with a metal that has a greater attraction for the adhesive.

Good wetting is merely an indication of low interfacial tension and does not by itself predicate adequate total adhesion. Resins of low polarity will rarely give adequate physical adhesion to metal surfaces. Metals require polymers whose polarities are reasonably high but not excessive.

POLYMERS OR RESINS

A brief introduction to the basic structure of plastics follows for the reader who is unfamiliar with plastic materials (resins). Plastics, or polymers, whether they are made from agricultural products or mineral products have one thing in common, their structure is made up of giant molecules. Each molecule has hundreds or thousands of atomic groups or repeat units. The backbone of these repeat units is the carbon chain. Attached to it are one or more of seven other elements: hydrogen, oxygen, nitrogen, chlorine, fluorine, sulfur, and silicon.

When other elements join the carbon chain, it usually acquires four other atoms. These smaller units are called *mers* from the Greek *meros*, meaning part. Thus when single *mers* or *monomers* join together they become polymers. The way these monomers add into a chain is dependent on their relative reactiveness. If there are free radical groups, molecules having unsatisfied bonds, they join together to complete their structure—a single large molecule, Fig. 6–4. Mers may add in randomly, alternately, or in blocks. Graft copolymers are formed by attaching a side chain to an already polymerized main chain.

Cross-linking. Some polymers form a network of cross-linked structures, Fig. 6–5. When a considerable amount of cross-linking takes place so that there is a transition from the reasonably fluid state to a "gel" state, the polymer passes a critical point

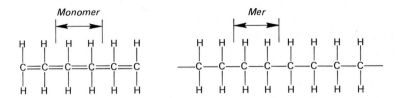

FIGURE 6–4. *The polymerization of ethylene. The original double bond of the ethylene monomer is broken to form two single bonds that connect two adjacent mers and create a polymer.*

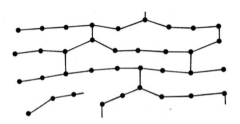

FIGURE 6–5. *A schematic representation of cross-linking between molecular chains thereby restricting movement.*

known as the thermosetting point. The process is not reversible. Common examples of thermosetting adhesives are: phenolic, epoxy, polyester, urea, and melamine formaldehyde. The cross-linking is accomplished by means of a catalyst or the application of heat to form hard, nonflexible, waterproof bonds.

CLASSIFICATION OF ADHESIVES

There are several approaches to the classification of adhesives. One method is based on the bond forming process, or how the adhesive passes from the liquid to the solid state. Adhesive bonds are usually formed by one or a combination of three processes:

1. Hardening or solidifying from the melted to the solid state upon cooling, e.g., polyamide hot melts.
2. Loss of a solvent (which may be water), by evaporation, e.g., polyvinyl acetate or contact cement.
3. Curing, which is the chemical joining together of small molecules (cross-linking) to form large ones, e.g., epoxy adhesives.

In the hot melt and solvent type adhesives, the large linear or slightly branched molecules form bonds as shown schematically in Fig. 6–6.

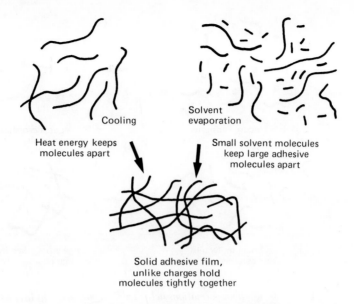

Cooling

Solvent evaporation

Heat energy keeps molecules apart

Small solvent molecules keep large adhesive molecules apart

Solid adhesive film, unlike charges hold molecules tightly together

FIGURE 6–6.   *Bond formation in hot melt and solvent adhesive systems.*

Adhesives that form bonds by curing, i.e., by forming polymer molecules after they are applied to the surfaces, are more chemically complex than the hot melt or solvent types. The molecules are very small when applied to the surfaces. The curing process involves getting these small molecules to join together to form large polymer molecules. This is accomplished by curing agents or catalysts. Curing agents, also called accelerators, are used up in the curing process while catalysts are not, Fig. 6–7. Most catalyst type adhesives will cure without the catalyst but the process is so slow that normal production rates could not be maintained. The catalyst is not used up during the large molecule or polymer formation, but is free to continue the building process. The cure type bond with its cross-linked polymer structure has better heat and mechanical resistance than hot melt or solvent type adhesive systems.

Another method often used in engineering of classifying adhesives is *structural* and *nonstructural*. The structural adhesives have as their main characteristic the ability to carry heavy loads. Chemical welding is another suitable term, since in most cases the structural adhesives are thermosetting prior to being exposed to physical stress. The nonstructural adhesives are, in general, comprised of animal, vegetable, and related proteins. However, some synthetic resins are also in the non-structural group, but since they are not generally used in metal fabrication, they will not be discussed here.

The structural adhesives can be divided into two groups, based on the properties of the cured bond line, as *brittle* or *nonbrittle*. The brittle type consists of the thermosetting resins such as epoxies and phenolics. They are noted for

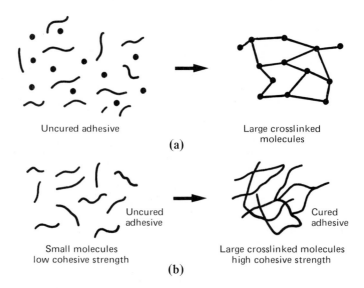

Uncured adhesive

Large crosslinked
molecules

(a)

Uncured
adhesive

Cured
adhesive

Small molecules
low cohesive strength

Large crosslinked molecules
high cohesive strength

(b)

FIGURE 6–7.   *Bond formations activated by a curing agent (a) and by a catalyst (b).*

having high shear strengths but low peel strengths. (Peel strength was illustrated in Fig. 5–4.)

The nonbrittle type is made up of combinations of thermoplastic, thermosetting, and elastometric (rubber like) resins. This group gives added bond line flexibility and impact resistance. They also have good shear, tensile, and peel strengths.

Brittle Adhesives. Of the "brittle" adhesives, the epoxies are the most widely used for structural joining. The basic chemical structure is a condensation product of phenol and acetone known as Biphenol A, Fig. 6–8. Epoxy adhesives are usually used as a solventless fluid or a thixotropic paste with a separate curing agent. Such two-part systems normally require only contact pressure and curing at room temperature. Depending upon the type of curing agent, temperatures up to 350 °F (177 °C) are used to obtain optimum properties.

Epoxies. One-part epoxy systems are also produced, these are cured with heat and are very useful in bonding dissimilar metals. One-part systems, heat cured, are generally stronger and more heat resistant than two-part adhesives.

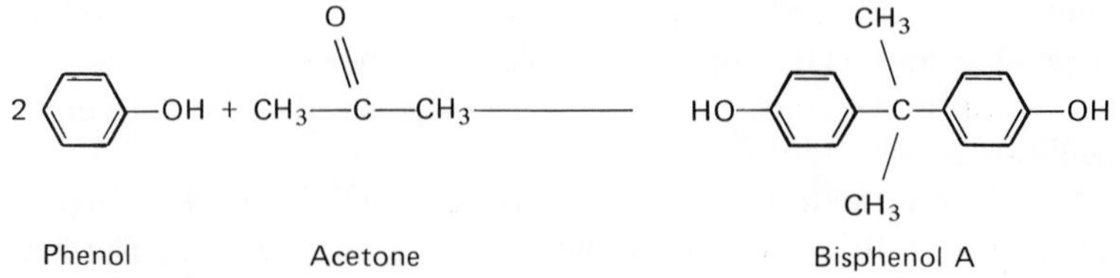

Phenol   Acetone   Bisphenol A

FIGURE 6–8.   *The basic chemical structure of epoxy resin.*

Cure, in either type of epoxy, is accomplished by cross-linking, activated by heat or a catalyst. Room temperature cures are satisfactory for many applications. However, heat can reduce the time required from days in some cases to hours and minutes.

Although epoxy adhesives develop high bond strengths [about 3,000 psi (20.68 MPa)], being brittle they have low flexibility, poor impact strength, and limited temperature resistance, 300 °F (149 °C).

Hot Melt Epoxies. Single-component epoxy adhesives are sometimes used in the "hot melt" form. These are made by partially reacting the resin and hardener to a brittle solid. The solids are available in either rod or powder form and must be melted prior to or during application. Parts to be bonded are usually heated; the solid adhesive is melted when it comes into contact with the parts. Solids in stick form may also be fed into an electrically heated gun and applied directly to the joint area. Parts are joined and clamped while the adhesive is still fluid. Alternately, the adhesive can be allowed to solidify, then be reheated, and the parts joined at some future time. The adhesive may be stored for a year or more without adverse effects on properties. Polyethylene is also a hot melt adhesive when made as a copolymer with one of the following: polypropylene, nylon, or polyester.

Phenolics. Phenolics are used to describe a wide range of synthetic resins formed by a reaction of phenol with an aldehyde. Phenols are characterized by a hydroxyl group attached to an aromatic nucleus and the aldehydes by the presence of the aldehydic carbonyl groups. The most common of these compounds is phenol-formaldehyde, Fig. 6–9.

Phenolic resins have been used extensively in bonding plywood since WWII. Now, with improvements in technology and the increased availability of substituted phenols and more complex aldehydes, the physical characteristics of phenolic resins have been modified. These modifications have been made by addition of other reactive components during polymerization or by blending with natural and synthetic polymers.

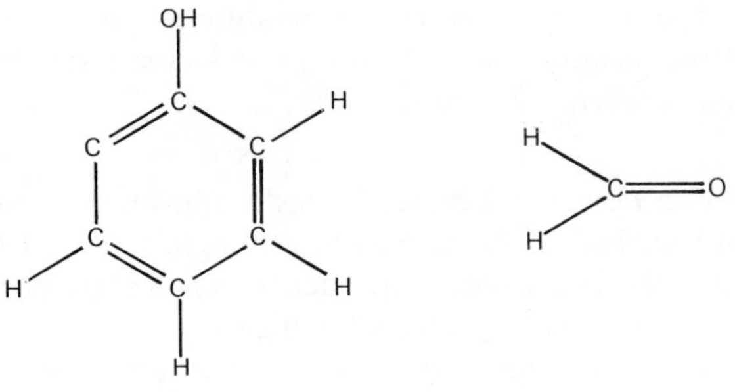

FIGURE 6–9. *The formulation of phenol-formaldehyde.*

Higher temperature resistance is obtained with an epoxy-phenolic compound, which combines excellent thermal characteristics of the phenolic resin with the adhesion properties of the epoxy. This system is capable of short-term operation at 700 °F (371 °C) and continuous use at 350 °F (177 °C). Retention of 1300 psi (89.63 MPa) shear strength on an aluminum substrate after a 200 hour exposure at 550 °F (288 °C) has been obtained with epoxy-phenolic adhesives.

Phenolic adhesives, available as liquids, melts, films, and reinforced tapes, produce water and formaldehyde during cure [at approximately 300 °F (149 °C)]; thus they must be cured under clamping pressure [usually 100 psi (7 kg/cm²)] to prevent the joint from being forced apart.

Nonbrittle Adhesives. "Nonbrittle" adhesives are commonly called "alloys" since they are formulations of thermosetting resins with thermoplastic *elastomers* (synthetic rubbers), or other resins. This blend is noted for its excellent *peel* strength, good flexibility, and ability to absorb vibration. The principal types of structural nonbrittle adhesives are modified phenolics and epoxies and polyurethane. Phenolic modifiers include nylon or vinyl compounds and elastomers such as neoprene or nitrile rubber.

Neoprene Phenolics. Neoprene phenolics have good adhesion to most metals and plastics and are particularly suited to joining thin sheet metal parts. Most of the early neoprene-phenolics required a solvent (a liquid that would wet the surfaces to be bonded), however, this is no longer true. In fact these adhesives have progressed to where they are available in tapes and films in addition to liquid. All types must be cured with heat and pressure.

Nitrile Phenolics. This blend of phenolic and acrylo-nitrile butadine rubber is the type that is commonly used to bond friction materials to steel for brakes and transmission parts in automobiles. They combine high strength, about 4000 psi (27.57 MPa) shear strength, with high temperature performance and have excellent resistance to water, salt spray, oils, and hydrocarbon fuels. Many formulations can be subjected to temperatures as high as 350 °F (177 °C) continuously and will withstand service temperatures as high as 500 (260) to 600 °F (316 °C) for brief periods. The minimum service temperature is about −60 °F (−51 °C). In recent times nitrile phenolics in both solvent and tape forms have become more widely used than neoprene phenolics.

Nylon Phenolics. Nylon in this combination is used to give a flexibility to the bond line. The room temperature strength is over 4000 psi (27.57 MPa). Some applications for these adhesives include: bonding of the backup saddle to electrotype shells in forming printing plates for high speed rotary presses; the lamination of magnets in particle accelerators; and the formation of flexible bonds between metals, nylon and metals, and nylon and nylon. High nuclear resistance, as well as strength, is an important factor in their use in making magnets for accelerators.

Modified Epoxies. Modified epoxies are made by using polymeric curing agents such as polyamides, isocyonates, and phenolics or by alloying with polymeric film formers such as polysulfides, nylon, and alkyds. The modified epoxies may be either heat curing or cured at room temperature by the addition of a chemical activator. The heat curing types give the highest shear strengths and maintain their strength over a wide temperature range.

The modified epoxies are excellent for metal-to-metal bonding and bonding to concrete. One of their advantages derives from being 100 percent solids; there is no problem of solvent evaporation, which makes for minimum shrinkage when joining impervious surfaces. They are able to wet metal, glass, and concrete surfaces easily. Although they are more flexible than the rigid adhesives, they still lack flexibility in the cured bond line. Epoxies modified by elastomeric components are best in overcoming this problem. One of the most used epoxies is epoxy-nylon.

Epoxy-Nylon. Generally, nylon is considered incompatible with epoxy resins. However, there are "alcohol-soluble nylons" that are also soluble in certain polar and nonpolar solvent mixtures. The epoxy-nylon adhesives have found applications in the fabrication of supersonic aircraft because of their high tensile [8250 psi (56.88 MPa) at room temperature] and shear strengths, and also their outstanding peel strength. They also exhibit desirable properties at very low temperatures.

Polyurethane. Polyurethanes have been used to make foams, coatings, elastomers, and, more recently, adhesives. In general, polyurethanes are a nonbrittle adhesive made by reacting a hydroxylated polyester or polyether with isocyanate. Polyurethanes provide high peel strength bonds to many plastics and to metal substrates. The isocyanates produce clean bonding surfaces by reacting with surface moisture and hydrated oxide layers on metals.

Several precautions are necessary in the use of polyurethane adhesives. First they must be stored under dry conditions, because water reacts with the isocyanate and lowers the reactivity and effectiveness of the adhesive. Secondly, isocyanates may cause dermatitis. Proper handling procedures and adequate ventilation are definitely required. Finally, polyurethane adhesives have short lives after the two components are mixed. A comparison of shear strength vs. temperature is given for several types of epoxy resins and polyurethane in Table 6.1.

ADHESIVES WITH CHEMICALLY-BLOCKED REACTANTS

Cyanoacrylate Adhesives. Chemically blocked reactants refer, in this case, to adhesives that will not become activated by themselves but can be catalyzed by atmospheric moisture. An example of this type of adhesives is the cyanoacrylates. When pressed to a very thin film, they will harden very rapidly (sometimes in 10 to 15 seconds). Cyanoacrylate is a monomer that has a very low viscosity; however, it can have a polymer added to thicken it, such as acrylate polymers or cellulose esters. The commercial name given to this adhesive is Eastman 910. It is also marketed by the Armstrong Cork Company.

TABLE 6.1. *Summary of Basic Structural Adhesives*

Adhesive	Form	Cure	Advantages	Limitations	Applications
Epoxy, single component	Viscous paste or sprayable liquid	300 to 450°F (149–232°C), little or no pressure	Long shelf life, no metering or mixing required	Poor peel strength, requires high temperature cure	Bonding aluminum tubing in air conditioners and evaporators, bonding flat metal sheets to other shapes
Nylon–epoxy	Unsupported adhesive film and a solvent solution	350 to 450°F (177–232°C), 25 psi (17.57 kg/cm²) pressure	No solvent in unsupported film, extremely high strength to 180°F (82°C)	Poor bond strength above 180°F (82°C), poor shelf life, must be refrigerated	Bonding aluminum honeycomb core and skin components and flat metal sheets
Epoxy, hot melt	Solid rod or powder	Melt at 200°F (93°C), cure between 350 and 450°F (177–232°C)	Long shelf life, can be applied and cured later, can be preformed	Brittle bond	Fastening metal tubes, bonding metal components. A combination adhesive and gap filling compound used in high volume production
Polyurethane	Two-component 100% solids or a solvent solution	Room temperature to 350°F (177°C)	Bonds well to most plastics and metals, bonds have good peel strength, large range of bond flexibility available	Susceptible to moisture absorption, poor bonds result if adhesive is contaminated with moisture	Bonding some plastics, small metal parts to similar or dissimilar metals, for elevated or cryogenic temperature [down to −400°F (−240°C)]

The advantage of cyanoacrylate adhesives is the ability to cure rapidly at room temperature. This eliminates the need for complicated jigs, fixtures, and drying ovens. Activation by atmospheric moisture present on a bonding joint eliminates the need for mixing components and the concern for limited pot life. Strongly acid surfaces will retard the cure indefinitely. Polymerization is postulated to be ionic, induced by the strong electromeric effects of both the nitrile and alkoxycarbonyl groups attached to the same carbon atom. Bond strengths are highest for polar substrates such as glass, porcelain, metals, polar plastics, and wood.

The disadvantages of cyanoacrylate adhesives are high cost, relatively poor heat resistance, poor shock resistance, failure to bond with acidic surfaces, and a thin glue line requirement. Each of the disadvantages listed is being subjected to research. It is believed that within a few years scientists will discover more systems that will expand the use and reduce the cost of adhesives of this type.

Anaerobic Adhesives. This is another chemically blocked adhesive, usually marketed under the trade name Loctite.™ This group of adhesives is known as acrylate acid diesters. These are essentially monomeric thin liquids that polymerize to form a tough plastic bond when confined between close fitting metal parts. The adhesive, when in contact with the air, remains as a liquid, but when contact is made with a metal and the air is excluded, anaerobic polymerization occurs. All common metals can be bonded, also glass, ceramics, and phenolic plastics. Some plated metals require a primer such as ferric chloride. Anaerobic adhesives are stable for a year or more when kept in contact with the air or oxygen. This is usually accomplished by packaging in small polyethylene bottles with adequate air space. The cure action is essentially a freeradical type addition polymerization.

Because of their excellent penetrating qualities, anaerobic adhesives are frequently used as liquid lock nuts for screws and bolts applied either before or after assembly. It is also used as a retaining compound for slip-fitted joints as a replacement to press fitting. Thus, much of the high cost in close tolerance manufacturing, with its attendant reduced fatigue life, can be eliminated. Also, anaerobic adhesives seal the joints against leakage and protect the mating surfaces from corrosion.

ADHESIVES WITH MECHANICALLY-BLOCKED REACTANTS

Molecular Sieves. A mechanically blocked system for adhesives was pioneered by the Linde Division of Union Carbide Corporation. The Linde molecular sieves are synthetic metal alumino-silicates having three-dimensional crystals. The crystals contain water that can be driven off by heating without collapsing the crystal lattice. The sieves are white, free-flowing powders ranging from 1 to 3 micro-inches in diameter. Each particle of a molecular sieve powder contains billions of

™ Trademark of the Loctite Corporation.

tiny cavities or cages connected by channels of unvarying diameter. The cavities can function as retainers for chemical compounds that can be driven off by heating or displacing with water.

As an example a molecular sieve epoxy resin adhesive can now be packaged as a one-part system and still have a pot life of several months. In this case the molecular sieves are loaded with reactive polyamine. The polyamine can be released when needed by heat or a releasing agent. The release agents are preferentially absorbed at elevated temperatures.

Microencapsulation. Reactants can be separated from each other by a relatively new technique, microencapsulation. The solvent or other reactive material, whether liquid or solid, is kept in microscopic capsules. When the adhesive is to be activated, the capsules can be broken by either heat or pressure. The tiny capsules vary in structure, but in most solvent activated adhesives, the solvent makes up 85–94 percent of the capsule by weight. The ratio of adhesive to capsules in a dry film is about 1 to 3. Small quantities of plasticizers and tackifiers may also be contained in the capsules. An added advantage of this type of adhesive is it can be used to precoat the surfaces to be joined.

SURFACE PREPARATION

The first step in preparing surfaces for adhesive bonding is similar to that of soldering, there must be good fit-up. The next consideration is the surface condition. It must be free of dust, oxide films, and liquids. If any liquid has been absorbed into the surface of the substrate, it will hinder the penetration of the adhesive.

If high strengths are not required the surface preparation can be quite simple. In the case of aluminum, the oxide film may be removed by wiping it with methyl ethyl ketone (MEK) or trichlorethylene. The surface must be *rubbed* so that all loosely adhering oxide film is removed. The bond will then form with the remaining oxide layer, which is tenaciously held to the metal surface. When no oxide appears on either a clean application cloth or a clean wiping cloth, the surface is considered to be free of loose oxides.

If the surfaces are roughened by sandblasting, wire brushing, or sanding, a greater surface area will be available for the adhesive. This, however, is not the main factor; the change in chemical properties of the surface layer, making it highly reactive, causes better adhesion.

For some metals, particularly aluminum, chemical cleaning will give maximum strength. This consists of a vapor degrease with trichlorethylene followed by: a rinse, a sulfuric acid-chromic etch, rinse, and forced air drying. If this type of cleaning is not suitable to field conditions, a tungsten carbide disc may be used to abrade the surface followed by a degrease with methyl ethyl ketone or acetone. After chemical cleaning a warm, 170 °F (77 °C) water rinse is used. Carbon steels offer maximum bonding strength after treatment with a dry abrasive blast of 80 grit aluminum oxide followed by a flushing with clean trichlorethylene for degreasing followed by a spray rinse of clean water. The parts should be dried immediately by placing in an oven at 140 °F (60 °C).

Because of the likelihood of contamination and oxide formation, it is desirable to use the prepared materials within hours after treatment. If storage is necessary, the metal should be kept in airtight containers. Care must be taken to see that etched or cleaned surfaces are not touched; handlers must wear clean cotton gloves.

ADHESIVE APPLICATION AND CURE

Adhesives can be applied in a variety of ways. Epoxies, for example, can be formulated for spraying, brushing, dipping, roll-coating, dusting, extruding, and trowelling, Fig. 6–10. Hot melts are often applied with an adhesive gun. Tape type adhesives have become quite popular because they eliminate the need for mixing, and the application will be a known uniform thickness.

Glue line thickness refers to the amount of adhesive that remains after pressure has been applied and curing has been completed. For example, to achieve an ultimate glue line thickness of 0.001 to 0.003 in. (0.03 to 0.13 mm) anywhere from 0.005 to 0.015 in. (0.13 to 0.38 mm) of 20 percent solid, wet type adhesive (*lay-down*) must be applied. The actual application of the adhesive to the part can be done either with one thick coat on one of the components or with two thin coats, one on each component. Of the two methods, one thin coat on each of the two surfaces is generally better.

If the adhesive is applied in solution, time must be allowed for evaporation of the solvent before the two members can be brought together. Solvent reduces the viscosity of the adhesive and lowers the contact angle and surface tension. The solvent also allows the adhesive to penetrate the pores of a porous substrate. In industrial applications, the adhesive is often applied as a polymer melt or as a solvent-free mass having high viscosity. The contact angle may be in excess of 60° with incomplete wetting. In this case, the degree of penetration into a porous substrate will be dependent on the applied pressure, since little will be obtained by capillary action. The amount of penetration into the substrate is not important as pointed out in the section on the nature of adhesion. It may even be detrimental since it can cause starvation at the bond line.

Rough surfaces must have enough adhesive applied to fill in small depressions, plus enough to achieve the desired glue line thickness. The gap between the two surfaces should not exceed a few thousandths of an inch.

Curing Time. Adhesives may be cured by solvent release, cross-linking, or pressure. Some adhesives, such as the one-part epoxies, contain latent catalysts that are activated by heat; cross-linking results with only contact pressure. There are no volatile solvents to be driven off. Phenolics, on the other hand, do have volatile solvents and need high pressure during cure to ensure against solvent entrapment. The time required varies depending on the heat, pressure, and catalyst used. The adhesive manufacturers' specifications should contain information on optimum

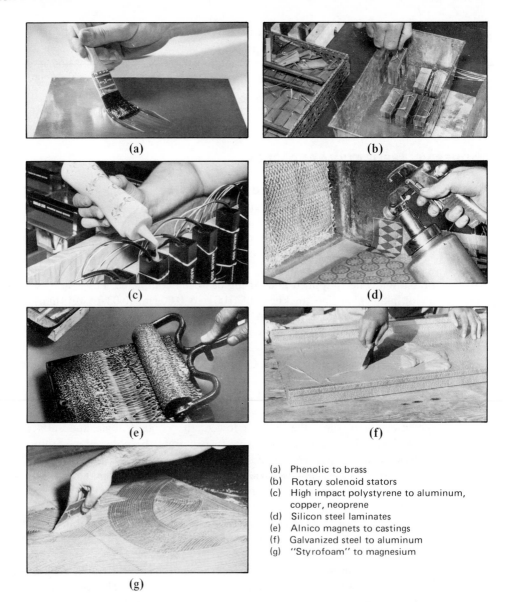

(a) Phenolic to brass
(b) Rotary solenoid stators
(c) High impact polystyrene to aluminum,
    copper, neoprene
(d) Silicon steel laminates
(e) Alnico magnets to castings
(f) Galvanized steel to aluminum
(g) "Styrofoam" to magnesium

FIGURE 6–10. *Methods that can be used to apply epoxy adhesives include brushing, dipping, roller coating, dusting, extruding, trowelling, and spraying. (a) Phenolic to brass. (b) Rotary solenoid stators. (c) High impact polystyrene to aluminum, copper, neoprene. (d) Silicon steel laminates. (e) Alnico magnets to castings. (f) Galvanized steel to aluminum. (g) "Styrofoam" to magnesium. (Courtesy of Epoxy Technology, Inc.)*

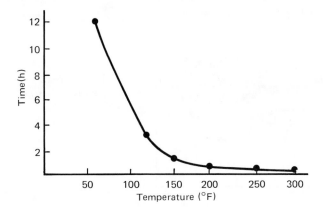

FIGURE 6–11. *The time and temperature relationship for curing a typical two-part epoxy adhesive.*

conditions. It is important that solvents be driven off or allowed to evaporate at the prescribed rate in order to develop full strength.

Adhesives that cure by cross-linking are often heated to speed up the reaction. The cure temperature may reach 300 to 400 °F (149–204 °C). A cure time vs. temperature plot for a typical two-part epoxy adhesive is shown in Fig. 6–11. Too high a temperature can result in "overcure." Overcured bonds have too many cross-links formed in their structure, making the joint interface brittle. Too short a cure or too low a temperature will result in an insufficient number of bonds and a soft, weak interface.

The one-part systems have a shorter shelf life than the two-part because cross-linking proceeds slowly even at storage temperatures.

Assembly by Reactivation. Certain adhesives are known as the *dry-film* type. The adhesive may be applied several days in advance of assembly. The surfaces are then reactivated by one of several methods:

1. The surface is rubbed lightly with a sponge filled with solvent or filled with a liquid adhesive of the same type.
2. The adhesive is heated with infrared lamps, or by other means to the reactivation temperature. The parts should be positioned prior to reactivation since the tack life is short.

Shown in Fig. 6–12 are the parts of a honeycomb assembly that utilizes a dry-film adhesive, which can be reactivated by either solvent or heat. It is used to cut labor costs through minimum handling, yet it insures a sound bond.

FIGURE 6–12. *Dry film adhesives are used for ease of assembly.*

PHYSICAL PROPERTIES OF ADHESIVE JOINTS

A knowledge of adhesive theory is necessary to an understanding of the process; however, it may be even more practical to see what relationships exist between the molecular forces mentioned and the physical properties of adhesive joints in shear and tensile strength.

As with solid materials, the calculated theoretical strength is always higher than the actual physical strength obtained in practice. This has been explained at least in part by the occurrence of microscopic voids and cracks in both the adhesive and the substrate, these result in localized fracture under stress. Another explanation of the less than theoretical strength is that the idealized surface for the adherend does not exist. The ideal surface should be atomically planar with a perfect lattice structure beneath it. The surface should also be free of physically or chemisorbed film. Outside of strict laboratory conditions, these idealized conditions do not exist. Metals always have a layer of oxide many ions thick.

MECHANICAL PROPERTIES OF ADHESIVE JOINTS

The mechanical properties of an adhesive joint are the usual factors inspected or tested in all metal joints, namely, tensile strength, shear, peel, cleavage, fatigue, creep, and bend. The most common stresses are shown in the sketches, Fig. 6–13.

Tensile Strength. The tensile strength of an adhesive joint is determined by the amount of load per unit area required to break the bond when the acting forces are applied perpendicular to the bond. It is expressed in psi or kg/sq mm. Standard tests are carried out on two solid metal cylinders bonded at the faces. After being pulled apart, the surfaces are examined to see if the break was due to cohesive failure of the adherend or a partial adhesive–adherend interface break.

Similar tests are carried out to determine bond tensile strength between dissimilar materials. Materials that have low tensile strengths are tested by placing

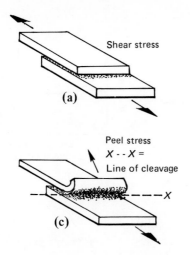

Shear stress

(a)

Peel stress
X - - X =
Line of cleavage

— — — — — — X

(c)

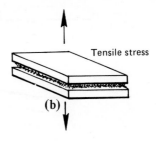

Tensile stress

(b)

FIGURE 6–13. *Common stresses encountered in adhesive joints.*

them between the adhesive coated faces of the standard test rods. This saves machining these materials.

Shear Strength. Shear stress is a force that tends to slide adherends in opposite directions, acting in the plane of the adhesive layer. In other words, it is the total stress exerted along the adhesive plane by the adherends. Lap shear bond strengths are directly proportional to the width of overlap, but the unit strength decreases with length. The optimum shear strength of bonded joints is dependent, among other things, upon the shear modulus of the adherend and its optimum thickness. The thickness may vary from 0.002 in. (0.05 mm) for high modulus to 0.006 in. (0.15 mm) for low modulus material.

The average tension-shear values of a lap joint are influenced by the geometry of the joint as well as the elastic properties of the adhesive and adherent. Figure 6–14 shows a simple lap joint loaded in tension. You will note the tendency of the joint to shift position so that the tensile load is located near the center of the joint, as shown on the corresponding stress distribution curve. High peel stresses develop at the edges of the joint. Shown at Fig. 6–14b is the same joint but with tapered edges and a more even stress distribution. If bending occurs with overstressing, tears will develop at the points of maximum stress.

A schematic view of how the stress is transmitted from the substrate through the viscoelastic material is shown in Fig. 6–15. The atoms are displaced from their thermal equilibrium positions. Stress can only be transferred by distortion. Distortion results in a change of volume or shape, or both. Failure occurs when the stresses overcome the cohesive forces of the joint. As shear occurs, an irreversible distortion takes place caused by a stretching and flow of molecules, as shown schematically in Fig. 6–16.

Compressive loading may also be used to determine the adhesive shear of a lap joint. The results are usually higher than for tensile shear due to more uniform distribution of shear stresses in the film.

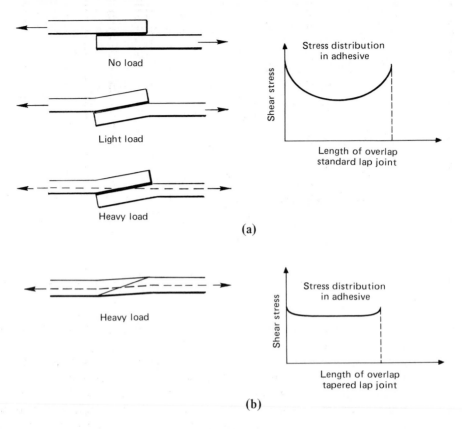

(a)

(b)

FIGURE 6–14. (a) *The action of an adhesive lap joint under various load conditions and the resulting stress distribution.* (b) *The same joint with tapered edges has a more even stress distribution.*

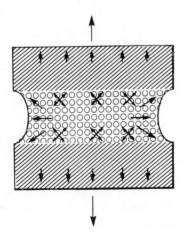

FIGURE 6–15. *A schematic view of how stress is transmitted from the substrate through the viscoelastic material.*

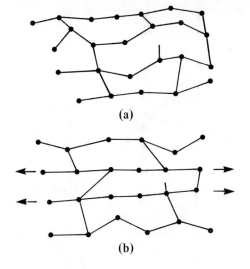

(a)

(b)

FIGURE 6–16.   (a) *Schematic of an unstressed cross-linked polymer chain.* (b) *A stressed cross-linked polymer chain.*

Peel strength.   Peel strength is the force in pounds per inch or kg/cm^2 that is required to strip a flexible member from another member, which may be flexible or rigid, Fig. 6–17. Adhesive lap joints subjected to peel tests on thin aluminum strips have shown that failure need not be in the adhesive. Nylon-epoxy, for example, has sufficiently high peel strength to cause failure in the substrate rather than in the adhesive compound, Fig. 6–18.

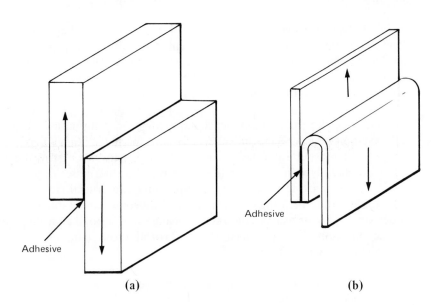

Adhesive

Adhesive

(a)   (b)

FIGURE 6–17.   *Adhesive joints being tested.* (a) *Shear test.* (b) *Peel test.*

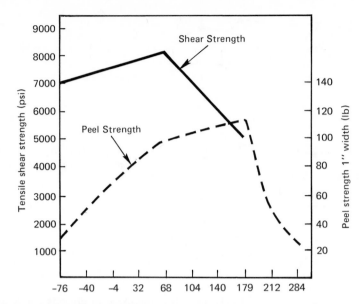

FIGURE 6–18. *The strength/temperature relationship for epoxy–nylon adhesive using an aluminum alloy lap joint.*

Cleavage Test. The American Society for Testing and Materials method, D-1062, describes a cleavage test in which a prying force is introduced at one end of a bonded specimen splitting the bond in two. The cleavage specimen is shown in Fig. 6–19. Strength is expressed in lb/in^2 of width.

Fatigue Testing. Fatigue testing consists of repeated application of a given load or deformation on an adhesive system. Fatigue life is dependent on frequency, amplitude, temperature, and mode of stress application. These variables are controlled and specified. The fatigue strength is defined as the minimum stress at which bond failure occurs after 10 million cycles of known frequency.

Creep. Creep is the dimensional change or deformation occurring in an adhesive-bonded specimen under stress over a period of time. This phenomenon occurs in thermosetting adhesives modified with natural or synthetic rubbers. It may also take place in thermosetting adhesives, but to a lesser degree.

No standard tests are given. Temperature is a crucial factor, and therefore, tests are generally performed at a number of different temperatures. The creep rate is determined by measuring the rate of change in the bond in a direction parallel to the direction of the applied stress. Typical creep data for a nitrile-rubber phenolic is shown in Fig. 6–20.

Bend Test. The flexibility of an adhesive may be tested by applying it to a strip of metal. After the adhesive has dried or cured, it is bent over a mandrel to determine its resistance to rupture.

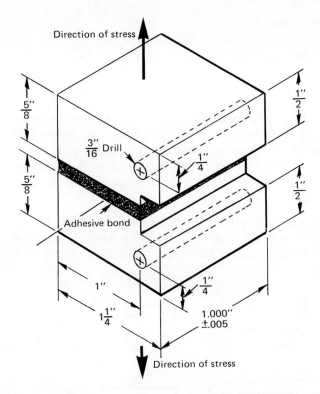

FIGURE 6–19. *A cleavage test specimen as designated by ASTM method D-1062. (Reprinted by permission, copyright American Society for Testing and Materials.)*

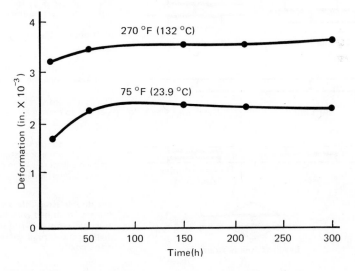

FIGURE 6–20. *Typical creep data showing the effect of temperature on a lap joint of nitrile rubber phenolic adhesive. The bonds are subjected to the same continuous stress.*

JOINT DESIGN

Adhesive joints should be designed so that the bond area is stressed in the direction of its maximum strength. Maximum strength is, of course, a function of the bond area. Since lap joints afford the greatest bond area, they are preferred. Some variations of lap and butt joints are shown in Fig. 6–21. Good and bad aspects are also given. The plain butt joint is quite unsatisfactory since the adhesive area is relatively small and stress concentrations are high.

Four practical approaches for producing lap joints with more uniform stress distribution are:

1. Reduce bending of the joint for a given load, by stiffening the adherends.
2. Permit the joint to bend more easily for a given load, by decreasing the thickness of the adherends, particularly toward the edges of the joint.

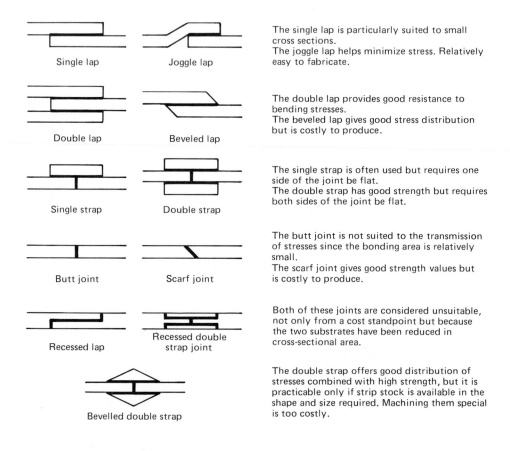

Single lap   Joggle lap

The single lap is particularly suited to small cross sections.
The joggle lap helps minimize stress. Relatively easy to fabricate.

Double lap   Beveled lap

The double lap provides good resistance to bending stresses.
The beveled lap gives good stress distribution but is costly to produce.

Single strap   Double strap

The single strap is often used but requires one side of the joint be flat.
The double strap has good strength but requires both sides of the joint be flat.

Butt joint   Scarf joint

The butt joint is not suited to the transmission of stresses since the bonding area is relatively small.
The scarf joint gives good strength values but is costly to produce.

Recessed lap   Recessed double strap joint

Both of these joints are considered unsuitable, not only from a cost standpoint but because the two substrates have been reduced in cross-sectional area.

Bevelled double strap

The double strap offers good distribution of stresses combined with high strength, but it is practicable only if strip stock is available in the shape and size required. Machining them special is too costly.

FIGURE 6–21. *Lap, strap, and butt type adhesive joints.*

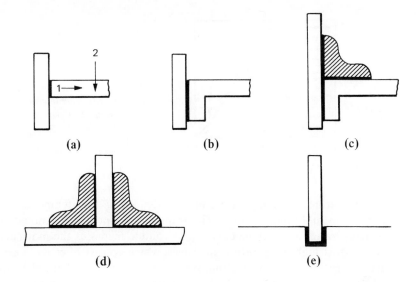

FIGURE 6–22. *Angle joint (a) stressed in direction **1** has good strength; however, in direction **2** cleavage develops. Angle joint (b) is only suitable for light stresses, considerably higher strength can be achieved by reinforcement as shown at (c). If two sheets must be joined at right angles, the recommended joint design is shown at (d) and (e).*

3. Increase the toughness of the adhesive.
4. Decrease the length of overlap.

Angle or intersection joints observe the same principle of increasing the bond area, as shown in Fig. 6–22.

ADHESIVE BONDING AND TEMPERATURE

Adhesives are constantly being subjected to increasingly hostile environments. Consequently, formulations have appeared that maintain good strength at elevated temperatures. An approach has been to put poor heat conductors in the adhesive such as carbon fibers, cork composites, and cast silicon elastomers.

Most adhesives perform well at room temperature, and many at temperatures of a hundred degrees or higher, but there are relatively few that perform well beyond 350 °F (177 °C). A high softening temperature is not the only requirement for a successful high temperature adhesive. The material must also be able to withstand, for a reasonable amount of time, attack by agents that oxidize and degrade the resin.

Once the temperature approaches the glass transition point of a low melting point adhesive, plastic flow results in deformation of the bond and a loss of strength.

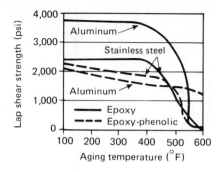

FIGURE 6–23. *A comparison of the strength of epoxy and epoxy-phenolic adhesive bonding, aluminum to aluminum and stainless steel to stainless steel. (Courtesy of Machine Design.)*

The glass transition temperature may be defined as that temperature below which there is no molecular motion. The transition point may be noted by a break in the temperature vs. volume curve since volume increases the molecular action.

Most high temperature adhesives, which consist of cross-linked networks of large molecules, have no melting points. Their limiting thermal factor is strength reduction due to oxidation. Oxidation starts a progressive breaking (scission) of the molecular bonds of the long chain molecules. This scission results in loss of strength, elongation, and toughness—leading to failure of the joint.

Oxidative degradation of an adhesive is affected by any catalytic reaction between the adhesive and the substrates. As an example, under identical conditions an adhesive used for bonding stainless steel may oxidize more than when bonding aluminum, but an adhesive with a different chemical make-up may show exactly the opposite effect, Fig. 6–23.

High temperature curing agents can extend the service temperature of epoxy resins to as high as 500 °F (260 °C). These curing agents increase the number of cross-links between the epoxy molecules.

Polyimides are specifically designed for high temperature service. For continuous exposure at 500 °F (260 °C), they are unmatched by any other commercially available adhesive, Fig. 6–24. Although polyimides are not cross-linked (they are thermoplastics), they are both rigid and thermally stable up to 700 °F (371 °C).

Polyimide polymers are available as resinous enamel solutions, cured films, and impregnated glass fabric tapes. The conventional method of applying the adhesive

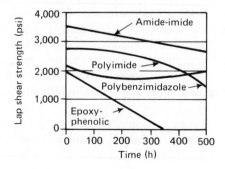

FIGURE 6–24. *Strength vs. thermal-aging time up to 500 °F (260 °C) for glass-fabric supported adhesives on stainless steel. (Courtesy of Machine Design.)*

is by a glass fabric carrier; however, unsupported film or liquid resin may also be used. Good adhesion is obtained on both perforated or smooth substrates. A cure temperature of 500 °F (260 °C) for 90 minutes and 15 to 200 psi pressure assure a uniform bond line. Optimum properties are obtained with a postcure of 1 hour at 600 °F (316 °C).

Silicone adhesives are pressure sensitive materials that adhere well to a variety of substrates, including metals, glass, plastics, and other silicone materials. Application can be by brush, spray, or dip. These adhesives are mixed with a catalizer and cured for five minutes at 300 °F (149 °C). They are flexible and can be used in long-term applications at temperatures to 500 °F (260 °C).

Fluorosilicone adhesives are available as solventless pastes that cure when exposed to moisture in air. The cure time is slow, about 120 hours for a 1/8 in. (3.17 mm) thick layer. This slow cure is preferred for bonding fluorosilicone parts to almost any substrate.

Table 6–2 summarizes the properties and recommendations for the use of high temperature adhesives.

Advantages and Disadvantages of Adhesive Bonding. There are many advantages of adhesive bonding that have made it attractive to the metal working world. The fact that industry is now using more than 100 million dollars per year of adhesives attests to the benefits they bring to manufacturing. Some of the main advantages may be listed briefly:

1. The ability to join unlike materials.
2. The ability to join like or unlike materials without the stress concentrations usually associated with other methods, Fig. 6–25.
3. Adhesives do not conduct electricity, therefore, electrolytic corrosion is not set up in joints between dissimilar metals.
4. Flexible adhesives can absorb shock and vibration, increasing the fatigue life of metal parts.
5. Lighter materials can more easily be used and stiffening members can often be eliminated.
6. Bonded joints present a smooth, often improved, appearance.
7. Materials can be attached to very thin metal parts, often impractical by any other method.
8. Less after-finishing is required.
9. Adhesives often permit extensive design simplifications.
10. Large areas can be bonded in a relatively short time.
11. Adhesives provide sealing action in addition to bonding.

Disadvantages.

1. Adhesives are more subject to deterioration by environmental conditions than metal bonding.

TABLE 6.2. *Typical Properties of Resinous High-Temperature Adhesives (Courtesy of Machine Design)*

Adhesive	Cure Condition						Maximum Service Temperature (°F) (°C)		Lap-Shear Tensile Strength† (psi) (MPa)	Environmental Recommendations
	Temperature (°F) (°C)		Time (min)	Clamping Pressure (psi) (kg/m²)			Short Term* (°C)	Long Term (°C)		
Epoxy	75–350	23.9–177	120–10	0	(0)		500 (260)	400 (204)	2,500 (17.24)	Resists moisture, brine, petroleum fuels, and oils; degrades slowly in air at maximum service temperature
Epoxy-phenolic	350	(177)	...	15–50	(0.10–0.34)		1,200 (649)	500 (260)	2,000 (13.79)	Similar to epoxy, but degrades faster in air at maximum service temperature
Polymide	500	(260)	60	15–100	(0.10–0.689)		1,000 (538)	600 (316)	2,200 (15.17)	Resists oxidation at >600°F (316°C) better than other adhesives
Polybenzimidazole	600	(316)	60	...			1,000 (558)	500 (260)	1,900 (13.10)	Oxidation resistance at high temperature almost equal to that of polyimide
Phenolic	300	(177)	30	100	(0.689)		...	600 (316)	2,000 (13.79)	Degrades rapidly at high temperatures
Cyanoacrylate	75	(23.9)	2	0	(0)		325 (163)	285 (140)	1,160 (7.80)	Degraded by UV, moisture, and alkali materials
Polysulfone	700	(371)	5	80	(0.55)		...	300 (149)	2,200 (15.17)	Resists acids, alkalis; dissolved by polar organic solvents and aromatic hydrocarbons; degraded by moisture
Silicone	300	(177)	5	0	(0)		700 (371)	500 (260)	38 (0.026)	Resists most oils, acids, and bases; non-oxidizing; dissolved by solvents such as xylene and toluene
Fluorosilicone	70	(21.1)	120 h	0	(0)		700 (371)	500 (260)	...	Particularly resistant to fuels and oils

* Less than 2 min.
† At maximum long-term service temperature.

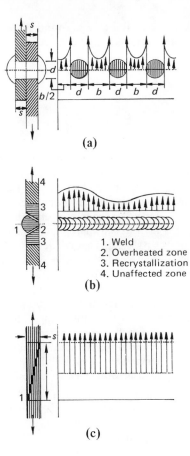

(a)

(b)

1. Weld
2. Overheated zone
3. Recrystallization
4. Unaffected zone

(c)

FIGURE 6-25. *Stress concentrations in various kinds of joints. Joint (a) shows stress concentrations at the edges of the rivet. Joint (b) shows superimposed welding stresses. Joint (c) shows the even stress distribution of a stepped adhesive butt joint.*

2. Adhesive joints are difficult to inspect once assembled.
3. Some rather elaborate jigs and fixtures may be needed to apply heat and pressure.
4. Adhesives have less strength than some other joining methods. The poor resistance to peel may require the usage of additional mechanical fasteners at stress points.
5. The release of toxic or harmful chemicals during cure must be provided for with some systems.

Applications.   Present day applications of adhesives in industry are so extensive that only a few examples will be given. The automobile industry is probably the largest single user of bonding adhesives. Current estimates are that each automobile uses 18 different applications of adhesive bonding. A modern military plane may use up to 1000 lbs. Northrop Aircraft Company has compared the cost of conventional trailing wing sections using the conventional riveted construction to those with adhesive bonding. They found the conventional method cost $1,312.00 as compared to $149.00 by adhesive bonding. A similar savings was found in the construction of ailerons. There are numerous instances where the cost of the rivets alone was more than the total cost of the adhesive and the man-hours of assembly.

EXERCISES

Problems

6–1. A stainless steel tank 3 ft (0.912 m) in diameter is fabricated out of 14 gauge [0.080 in. (2.03 mm)] stainless steel. The two domed ends are fitted onto the rolled and welded cylinder with a $\frac{1}{2}$ in. (12.70 mm) lap joint. The tank will be used to contain a hot brine solution at 250 °F (121 °C) and a pressure up to 500 psi (3.45 MPa). (a) What type of adhesive do you recommend? (b) Give reasons why you chose this adhesive and why you believe it will be adequate for the requirements.

6–2. An aluminum fuel gas pipeline connection is shown in Fig. P6–2. The pipe will be buried in the ground and will be required to remain leak-proof at 500 to 700 psi (3.45–4.82 MPa). State the following:
(a) Recommended surface preparation.
(b) Recommended adhesive.
(c) Any special considerations that may affect the joint that are not mentioned in the problem. If a straight epoxy adhesive is used, would this exceed the basic strength of the material?
(d) What type of preliminary test program should be used before actually installing the pipeline for service?

6–3. An adhesive lap joint is made between two strips of 0.060 in. (1.524 mm) thick aluminum 2 in. (5.08 cm) wide. The recommended overlap length is $10t$ (1.52 mm). Compare this joint in tensile shear strength with the same joint when four solid rivets are substituted. The rivets used are 0.175 in. (4.445 mm) in diameter and are made from 5056-0 aluminum. The environmental conditions of the test range from room temperature to −60 °F (−51 °C). (a) Select an appropriate adhesive and compare the lap shear and tensile strengths with that of a riveted joint. (b) Look up, in an aluminum handbook or other reference, the tensile and shear strengths of 2024-T3 and 5056-0 aluminum. What is the basic tensile strength for one thickness of this material? (c) Will the joint strength in any case, as given, exceed the strength of the material? (d) If cryogenic temperatures are not encountered but mainly room temperature, what will the lap shear strength of the joint be with an epoxy adhesive?

6–4. Sometimes it may be feasible to reinforce an adhesive joint with a mechanical fastener as shown in Fig. P6–4. If the joint is intended to be flexible, the added fastener must not make it rigid. The breaking strength of a combination joint subjected to static shear loading is determined by the strength of the metal fasteners. Under dynamic stress, the adhesive bond is the determining factor. This is a lap joint made with 2–3/16 in. (4.76 mm) steel plates with an overlap of 1 in. (25.40 mm). The joint width is 2 in. (5.08 cm). The bolt clearance hole is $\frac{1}{4}$ in. (6.35 mm) and the washers used are 3/4 in. (19.05 mm) in diameter. The environment will be at room temperature and an epoxy-nylon is chosen.

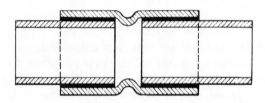

Swaged from each end toward center

FIGURE P6–2.

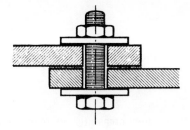

FIGURE P6–4.

(a) Determine the static and dynamic loads this joint will be able to take with a safety factor of 2. (b) Other than being the main holding element for static loads, how else do the bolts and washers make a better joint? (c) Would it be better to have a tight fit between the bolt and the plates? Why or why not?

6–5. Shown in Fig. P6–5 is an adhesive lap joint reinforced with a tubular rivet. You will note this is a tight fit whereas the bolt shown in the previous problem was a clearance fit. (a) Why can this joint be made tight and still allow some flexibility? (b) The lap joint is made 2 in. (5.08 cm) wide with 1 in. (2.54 cm) of overlap. Two hollow rivets with 15/32 in. (23.8125 mm) o.d. and 7/32 in. (.21875 mm) i.d. are used. The rivets and joint are made out of 2024–T4 aluminum with a shear strength of 41,000 psi (282.7 MPa). The lap joint is made out of 20 gauge [0.0375 in. (0.934 mm)] aluminum of the same designation. What static load can the joint stand in tensile shear with a safety factor of 2? The rivet o.d. = .468 in. (11.90 mm) and the i.d. = .218 in. (5.556 mm).

6–6. Complete the following table to show the properties of some of the adhesives discussed in this chapter.

Adhesive	Type* TS or TP	Advantages	Limitations
Straight epoxy			
Phenolic			
Polysulfide			
Cyanoacrylate (Eastman 910)			

* Thermosetting or thermoplastic.

6–7. (a) How much creep could you expect in a 2 × 2 in. (5.08 × 5.08 cm) square properly made nitrile rubber lap joint when subjected to a temperature of 270 °F (132 °C) for 200 hours, as compared to 100 hours at 75 °F (23.9 °C)? (b) At the 75 °F (23.9 °C) temperature, how long would it take for 1/8 in. (3.175 mm) of creep?

6–8. Compare the bond strength of aluminum to aluminum and stainless steel to stainless steel when aged at 300 °F (149 °C) if the joints are 2 × 2 in. (5.08 × 5.08 cm) square and made with epoxy.

Questions

6–1. What are some conditions that would make an adhesive bonded joint the preferred type?

6–2. Is there any difference between a "glue" and an adhesive?

6–3. Why are the theories of mechanical or electrostatic bonding discredited?

6–4. What are the forces that make an adhesive cohesive and adhesive?

6–5. Why is it that some adhesives can develop strong bonds at room temperature without a catalyst?

6–6. What happens in the molecular structure of a thermosetting adhesive when it cures?

6–7. How can a thermosetting adhesive be formulated to be flexible?

6–8. What are the main principles in adhesive joint design?

6–9. What is one of the main weaknesses of adhesive bonding and how is it being approached?

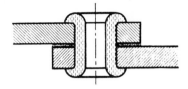

FIGURE P6–5.

6–10. How do metallic bonds operate in an adhesive system?

6–11. What is the general relationship between cure temperature and strength?

6–12. Is the tensile–shear strength of a lap joint directly proportional to the area of overlap? Explain.

6–13. Why is joint preparation more important on some nonferrous metals than on ferrous metals?

6–14. How may the wetting effectiveness of adhesives for a given substrate be evaluated?

6–15. Why are adhesive bond strengths less than theoretical?

BIBLIOGRAPHY

Books

Adhesive Bonding Aluminum, Reynolds Metals Company, Richmond, Virginia, 1966.

Advances in Chemistry, "Contact Angle, Wetability and Adhesion" (Kendall Award Symposium), Am. Chem. Soc., ser. no. 43, Washington, D.C., 1964.

ALNER, D. J., Ed., *Aspects of Adhesion*, vols. 1, 2, and 3, University of London Press, 1963, 1964, and 1965.

BARRETT, C. S., *Structure of Metals*, McGraw-Hill, New York, N.Y., 1952.

BATTISTA, O. A., *Fundamentals of High Polymers*, Reinhold Publishing Corp., New York, N.Y., 1958.

BIKERMAN, J. J., *The Science of Adhesive Joints*, Academic Press, New York, N.Y., 1961.

BILLMEYER, F. J., Jr., *Textbook of Polymer Science*, Interscience Publishers (Div., John Wiley and Sons), New York, N.Y., 1962.

BODNER, M. J., Ed., *Symposium on Adhesives for Structural Applications*, Interscience Publishers (Div., John Wiley and Sons), New York, N.Y., 1962 and 1966.

DELMONTE, J., *The Technology of Adhesives*, Hafner Press, New York, N.Y., 1965.

GUTTMAN, W. H., *Concise Guide to Structural Adhesives*, Reinhold Publishing Corp., New York, N.Y., 1961.

HAVWINK, R., AND SALMON, G., Eds., *Adhesion and Adhesives*, Elsevier Publishing Company, New York, N.Y., 1965.

HURD, J., *Adhesives Guide*, British Scientific Instrument Research Association Research Report M.39 (1959).

KATZ, I., *Adhesive Materials, Their Property and Usage*, Foster Publishing Company, Long Beach, Calif., 1971.

KOEHN, G. W., Ed., *Industrial Adhesives*, Armstrong Cork Company, Lancaster, Pa., 1959.

MARTIN, A. F., *Structural Adhesives*, McGraw-Hill Encyclopedia of Science and Technology, vol. 1, p. 66, New York, N.Y., 1960.

PARKER, R. S. R., AND TAYLOR, P., *Adhesion and Adhesives*, Pergamon Press, New York, N.Y., 1966.

PATRICK, R. L., Ed., *Treatise on Adhesion and Adhesives*, Marcel Dekker, Inc., New York, N.Y., 1967.

SKEIST, I., Ed., *Handbook of Adhesives*, Reinhold Publishing Corp., New York, N.Y., 1962.

VAN VLACK, L. H., *Elements of Material Science*, Addison-Wesley Publishing Co., Inc., Reading, Mass., 1964.

Periodicals

BIKERMAN, J. J., "Making Polyethylenes Adhesionable." *Adhesives Age*, Feb. 2, 1959.

BLACK, J. M., AND BLOMQUIST, R. F., "Polymer Structure and the Thermal Deterioration of Adhesives in Metal Joints." *Adhesives Age*, May 5, 1962.

BRYSON, F. E., "Heat-Resistant Adhesives." *Machine Design*, June 15, 1972.

CASSIDY, P. E., JOHNSON, J., AND LOCKE, C., "The Relationship of Glass Transition Temperature to Adhesive Strength." *The Journal of Adhesion*, July 1972.

CLARKE, J. A., "Methods and Materials for Achieving Adhesive Joints." *SME Engineering Conference Paper*, AD69-104.

DERJAGUIN, B. V., AND SMILGA, V. P., *Surface Activity*, Proc. Interm. Congr. 3rd, Cologne, **II** sec. B, 349 (1960).

IRVING, R. R., "Adhesive Bonding: Its Day Is Coming." *Iron Age*, April 1968.

LEVINE, M., ILKKA, G., AND WEISS, P., "Wetability of Surface-Treated Metals and the Effect on Lap Shear Adhesion." *Adhesives Age*, July 6, 1964.

LONDON, F., *Physik*, **63:** 245 (1930).

PETRIE, E. M., "High Temperature Structural Adhesives." *Machine Design*, May 15, 1969.

ROSELAND, L. M., "Structural Adhesives and Composite Materials." *Machine Design*, March 17, 1966.

————, "Structural Adhesives and Composite Materials at Cryogenic Temperatures." *Machine Design*, March 17, 1966.

SCHENEBERGER, G. L., "Chemical Aspects of Adhesive Bonding." *Adhesives Age*, April 1970.

SHARPE, L. H., "Assembling with Adhesives." *Machine Design*, August 18, 1966.

VOYUTSKII, S. S., AND VAKULA, V. L., *J. Applied Polymer Sci.*, **7:** 475 (1963).

TABLE 7.0.   *A Partial List of Screw Thread Sizes Showing Both Metric and Decimal Sizes. (The asterisked areas are not official ISO designations but are given to show how they would match the unified series.)*

New Thread Code No.	ISO Designation Number	O.D. in mm	Pitch in mm	Decimal o.d.	Threads per Inch	Present American Unified Thread Sizes	Best Metric Replacement for American Thread Sizes
00548–01058		5.486	1.058	.2160	24	# 12–14 UNC	
00548–00907		5.486	0.907	.2160	28	# 12–28 UNF	
00548–00794		5.486	0.794	.2160	32	# 12–32 UNEF	
00550–01000*	M5.5*	5.5*	1*	.2165*	25.4*		# 12–24 UNC*
00550–00900*	M5.5*	5.5*	0.9	.2165*	28.2*		# 12–28 UNF*
00600–00750	M6 × 0.75	6	0.75	.2362	33.8		
00600–01000	M6	6	1	.2362	25.4		
00635–01270		6.35	1.27	.2500	20	1/4–20 UNC	
00635–01058		6.35	1.058	.2500	24	1/4–24 NS	
00635–00940		6.35	0.941	.2500	27	1/4–27 NS	
00635–00907		6.35	0.907	.2500	28	1/4–28 UNF	
00635–00794		6.35	0.794	.2500	32	1/4–32 UNEF	
00635–00635		6.35	0.635	.2500	40	1/4–40 NS	
00650–01250*	M6.5*	6.5*	1.25*	.2558*	20.3*		1/4–20 UNC*
00650–00900*	M6.5*	6.5	0.9*	2.558*	28.2*		1/4–28 UNF*
00700–00750	M7	7	0.75	.2756	33.8		
00700–01000	M7	7	1	.2756	25.4		
00793–01411		7.937	1.411	.3125	18	5/16–18 UNC	
00793–01270		7.937	1.27	.3125	20	5/16–20 NS	
00793–01058		7.937	1.058	.3125	24	5/16–24 UNF	
00793–00907		7.937	0.907	.3125	28	5/16–28 NS	
00793–00794		7.937	0.794	.3125	32	5/16–32 UNEF	
00800–01000*	M8 × 1*	8*	1*	.3149*	25.4*		5/16–24 UNF*
00800–01250	M8	8	1.25	.3149	20.3		
00800–01500*	M8*	8*	1.5*	.3149*	16.9*		5/16–18 UNC*
00900–01000	M9 × 1	9	1	.3543	25.4		
00900–01250	M9	9	1.25	.3543	20.3		
00950–01500*	M9.5*	9.5*	1.5*	.3739*	16.9*		3/8–16 UNC*
00950–01000*	M9.5*	9.5*	1*	.3739*	25.4*		3/8–24 UNF*
00952–01588		9.525	1.588	.3750	16	3/8–16 UNC	
00952–01270		9.525	1.27	.3750	20	3/8–20 NS	
00952–01058		9.525	1.058	.3750	24	3/8–24 UNF	
00952–00907		9.525	0.907	.3750	28	3/8–28 NS	
00952–00794		9.525	0.794	.3750	32	3/8–32 UNEF	

7

Mechanical Fasteners

Industrial type mechanical fasteners no longer refer to just nuts, bolts, and rivets, but to an almost infinite variety of ingenious devices used to hold two or more parts together. The art and science of fastener engineering has had an astounding growth in the past decade. The diversity of fasteners is so great that some large manufacturers have had to re-examine their assembly methods and reduce their inventory of fasteners through standardization.

Systems engineering can be applied to fastener selection in the following ways: (1) careful selection of the right mating surfaces to realize maximum performance, (2) selection of fasteners to provide a trouble-free assembly line, and (3) choice of fasteners that will minimize maintenance and repair.

A fastener is often a highly engineered product which has frequently undergone long critical periods of material and process selection. The joint to which it is applied may be complex and may even be the most highly stressed point of the assembly. Fasteners that loosen or wear excessively in service may cause catastrophic failure of the whole assembly.

Fasteners may be classified into five main types:

1. Threaded.
2. Pins.
3. Washers and retaining rings.
4. Rivets.
5. Quick-operating fasteners.

THREADED FASTENERS

Origin of Threads. Leonardo da Vinci foresaw the social implications of the screw thread and wrote on the subject at considerable length. He was the first to conceive of the use of a master screw for the repetitive cutting of threads. He

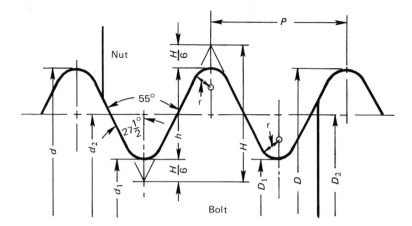

FIGURE 7-1. *The British Standard Whitworth thread. The letters shown refer to the following thread description: d = thread diameter or outside diameter (o.d.), d_2 = pitch diameter, d_1 = root or minor diameter, h = thread depth, and P = thread pitch.*

was, of course, far ahead of his time as the need for screw threads had not yet developed. Henry Maudslay, an English inventor, is credited with making the first screw cutting lathe in about 1797. It is said that he worked ten years before completing a master screw that he considered precise enough to serve in his lathe. Many machines were soon developed embodying Maudslay's basic principles to fulfill the demand for screws. Another problem soon developed in that inter-changeability was nonexistent. It was rare that a screw made on any one machine was the same size and form as that made on another machine. In 1841, after a long struggle Sir Joseph Whitworth, an engineer who worked with Maudslay, offered a series of standard pitches and diameters based on a 55 degree thread profile, Fig. 7-1. This thread was adopted as a British standard in 1905.

Screw thread development was taking place in other countries as well. In 1864, William Sellers of Philadelphia proposed a standard screw thread with a 60 degree included-angle profile. It gradually became known as the American National thread, a name it still retains.

During the Second World War, a Unified standard between Great Britain, Canada, and the United States was worked out. The form, or thread profile, is essentially that of the American National thread. Shown in Fig. 7-2 is the basic screw thread nomenclature, (a); and in (b), is shown a comparison of the American National thread and the Unified thread. The crest of the Unified thread was left optional, either flat or rounded.

The International Standards Organization (ISO) has been working to standardize screw threads throughout the world. They have reached agreement in their recommendation to use ISO metric threads for screw fasteners. This thread, Fig. 7-3, closely resembles the German Industrial Standard (Deutsche Industriale Normen), or DIN.

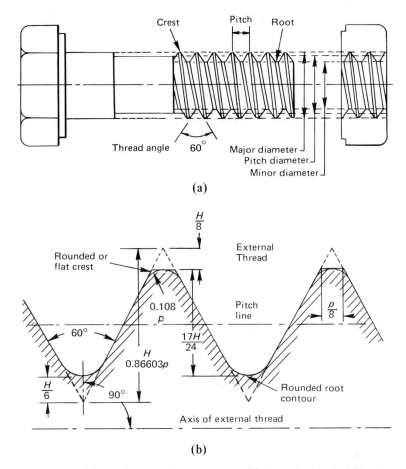

FIGURE 7–2. (a) *Standard thread nomenclature.* (b) *A standard unified thread form.*

Metric Threads. Considerable progress has been made in the international standardization of screw threads. The ISO thread form is essentially the same as the Unified thread form. Parallel series of Unified inch and ISO metric screw threads have been set forth, see chapter opening page. For countries on the metric system, a series of recommended diameter/pitch combinations have been selected from the ISO standards and are shown in Table 7.1. You will note that our chapter opening table contains additional sizes of metric threads other than the ISO. Thus, a 6.35 thread o.d. with a 1.27 mm pitch is equivalent to a $\frac{1}{4}$–20 UNC thread; and a 6.35 with a 0.907 mm pitch is equivalent to a $\frac{1}{4}$–28 UNF thread.

Thread Fits. Thread fit is a term used to designate the tightness that results between mating thread members for a given combination of allowances and tolerances. It is not a criterion of quality. It is related to three factors: pitch diameter, lead, and flank (or half) angle. The "yardstick" for specifying fit is the pitch diameter.

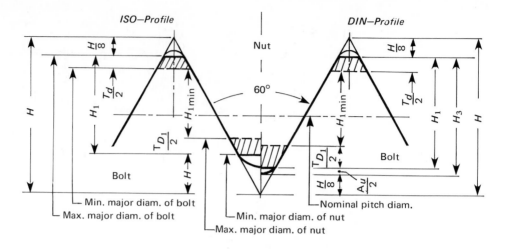

FIGURE 7-3. *A comparison of the International Standards Organization (ISO) thread profile and the Deutsche Industrial Normen (DIN) that is extensively used and is a simplified industrial equivalent to American National (AN), Military Standards (MS), and National Aircraft Standards (NAS).*

TABLE 7.1. *ISO Metric Screw Threads for Screws, Bolts, and Nuts*

Diameters				Pitches			
Primary		Secondary		Coarse		Fine	
mm	in.	mm	in.	mm	in.	mm	in.
6	0.2362			1	0.0394		
		7	0.2756	1	0.0394		
8	0.3150			1.25	0.0492	1	0.0394
10	0.3937			1.5	0.0590	1.25	0.0492
12	0.4724			1.75	0.0689	1.25	0.0492
		14	0.5512	2	0.0787	1.5	0.0590
16	0.6299			2	0.0787	1.5	0.0590
		18	0.7087	2.5	0.0934	1.5	0.0590
20	0.7874			2.5	0.0984	1.5	0.0590
		22	0.8661	2.5	0.0984	1.5	0.0590
24	0.9449			3	0.1181	2	0.0787
		27	1.0630	3	0.1181	2	0.0787
30	1.1811			3.5	0.1378	2	0.0787
		33	1.2992	3.5	0.1378	2	0.0787
36	1.4173			4	0.1575	3	0.1181
		39	1.5354	4	0.1575	3	0.1181

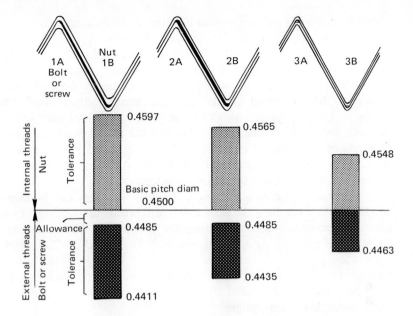

FIGURE 7-4. *A comparison of pitch diameter tolerances and mating thread allowances for three classes of fit. The thread represented is a $\frac{1}{2}$-13 UNC. (Courtesy of Machine Design.)*

The three classes of thread fit are shown in Fig. 7-4. The tolerances and allowances are shown for a half-inch nominal diameter bolt with 13 threads per inch for the Unified National Coarse series. This can be written $\frac{1}{2}$-13UNC. The same general relationships exist for other threads of the Unified series. The three classes of fit are: Class 1A/1B (loose), Class 2A/2B (medium), and Class 3A/3B (tight).

Class 1A/1B is seldom specified as it has a somewhat limited application. Class 2A/2B is the most commonly used thread fit; its tolerances and allowances are best suited to the production of bolts, screws, nuts, and other threaded fasteners. The Class 3A/3B fit does not have allowances; at maximum external-thread and minimum internal-thread pitch diameters, there is complete contact of the mating threads.

Design Considerations. The strength of mating threads depends on adequate thread engagement or overlap of the threads in a transverse direction and the length of that engagement. Strength is determined by mechanical testing. It would appear that the closer fit and tighter tolerance of a Class 3A/3B thread would be the strongest, but this is not necessarily true. Tests have shown that the stressing of a bolt and the transfer of loads through the threads is an extremely complex problem in elasticity and plasticity of materials. Ample evidence exists to show the desirability of having plenty of "breathing room" between mating threads—to accommodate local yielding, thread bending, and adjustment due to varying elastic deformations throughout the length of engagement.

In general, the Class 2A/2B thread fit is used for low and medium strength materials [up to 150,000 psi (1 034 MPa)]. However, with higher tensile strength materials of lower ductility where there is practically no plasticity, a closer fit (Class 3A/3B) is required. Brittle materials cannot "adjust" to distribute the load on the length of engagement, and these must have added clearances. In addition, the need for rounded fillets at the roots of the thread becomes more acute as the plasticity of the material decreases. Also, for elevated temperatures, it is usually desirable to provide a clearance fit. An allowance minimizes the possibility of thread seizure and also provides space for a lubricant.

Unless great care is exercised, threaded fasteners can be exposed to harsh handling before reaching the assembly line. Threads can easily be nicked. A thread allowance can usually accommodate this slight damage. Also, assembly by high-cycle wrenching is more easily accomplished with 1A/1B and 2A/2B thread fits.

Fastener Loading. The materials joining engineer must consider the principal types of stresses to be encountered by the fastener in service, i.e., tensile loads, shear loads, shock loads, and fatigue.

Tensile Loads. The correct size fastener for a tensile load may be determined by the formula:

$$S_y = \frac{P}{A}$$

where S_y = tensile yield strength in psi.
P = total load in lb.
A = net working area in in^2.

Normally the area at the root of the thread is considered the net working area. Tests have shown, however, that thread rolling has a strengthening effect, the amount of which is dependent on the alloy of the bolt and the treatment. Indications are that the net working area can be based on the mean pitch/root diameter rather than the root diameter, Table 7.2 (see also Table 7.4).

TABLE 7.2. *Ultimate Strength of Fastener Materials* (Courtesy of *Machine Design*)

Material	Tensile Strength, s_t (psi)	Shear Strength, s_s (psi)
1010 Carbon steel	50,000	30,000
303 Stainless steel	90,000	65,000
316 Stainless steel	90,000	60,000
Brass (yellow)	60,000	35,000
Silicon bronze	70,000	40,000
2024-T4 Aluminum	68,000	41,000

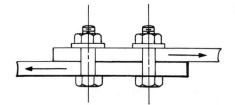

FIGURE 7-5. *Two bolts are used to hold two steel straps under a shear load of 25,000 lb (172.4 MPa).*

Shear Loads. Fastener shear loads may be determined as:

$$P_s = S_s A n$$

where P_s = fastener shear load in lb.
S_s = fastener shear stress in psi.
A = area in shear of each fastener in in.^{2}.
n = number of shear planes.

To determine the shear stress on the fastener, use the following form of the shear load formula:

$$S_s = \frac{P_s}{A n}$$

Example: Two bolts are used to join two $\frac{1}{2}$ × 2 in. metal straps, as shown in Fig. 7-5. The load supported in shear by the bolts is 25,000 lb. Assume the bolt material required to meet environmental conditions is 303 stainless steel, which has an allowable shear strength of approximately 65,000 psi, Table 7.2. Calculate the exact size of the bolts required if the threaded section is entirely out of shear zone. (NOTE: Neglect bending forces.)

Solution:

$$A = \frac{P}{S_s} = \frac{25,000}{65,000}$$

$$= 0.385 \text{ in.}^2 = \frac{\pi d^2}{4}$$

$$d = \sqrt{\frac{.385 \times 4}{3.14}} = 0.700$$

The nearest standard size of the bolts required will be two $\frac{3}{8}$-16UNC, or two $\frac{3}{8}$-24UNF (see Table 7.0). The equivalent ISO thread is M9.5.

Bearing Load. Bearing load is a compressive force on the bolted plates. Failure may occur when the diameter or area of the bolt or rivet head is not large enough. The bearing stress is expressed as:

$$Sb = \frac{P}{dt}$$

TABLE 7.3. Maximum Wrenching Torque and Axial Load for Typical Fasteners (Courtesy of Machine Design)

Bolt Size (2A)	1010 Carbon Steel		303 Stainless Steel		316 Stainless Steel		Brass (yellow)		2024-T4 Aluminum	
	Torque (lb-in.)	Axial Load, P (lb)	Torque (lb-in.)	Axial Load, P (lb)	Torque (lb-in.)	Axial Load, P (lb)	Torque (lb-in.)	Axial Load, P (lb)	Torque (lb-in.)	Axial Load, P (lb)
2–56	2.2	121.0	2.5	137	2.6	143	2.0	110	1.4	77
2–64	2.7	151.2	3.0	168	3.2	179	2.5	140	1.7	95
3–48	3.5	175.0	3.9	195	4.0	200	3.2	160	2.1	105
3–56	4.0	204.0	4.4	224	4.6	244	3.6	183	2.4	122
4–40	4.7	197.0	5.2	218	5.5	231	4.3	180	2.9	122
4–48	5.9	253.0	6.6	284	6.9	296	5.4	232	3.6	155
5–40	6.9	248.0	7.7	277	8.1	292	6.3	226	4.2	122
5–44	8.5	314.5	9.4	348	9.8	362	7.7	286	5.1	189
6–32	8.7	295.8	9.6	326	10.1	343	7.9	268	5.3	180
6–40	10.9	381.5	12.1	425	12.7	445	9.9	346	6.6	230
8–32	17.8	534.0	19.8	595	20.7	620	16.2	485	10.8	324
8–36	19.8	594.0	22.0	660	23.0	700	18.0	540	12.0	360
10–24	20.8	540.8	22.8	590	23.8	645	18.6	482	13.8	372
10–32	29.7	801.9	31.7	855	33.1	895	25.9	700	19.2	517
¼–20	65.0	1350	75.2	1550	78.8	163	61.5	1280	45.6	855

1/4 –28	90.0	1930	94.0	2050	99.0	213	77.0	1660	57.0	1230
5/16 –18	129	2140	132	2200	138	229	107	1770	80	1330
5/16 –24	139	2390	142	2420	147	252	116	1980	86	1470
3/8 –16	212	3040	236	3370	247	253	192	2740	143	2040
3/8 –24	232	3460	259	3860	271	405	212	3160	157	2340
7/16 –14	338	4130	376	4580	393	480	317	3880	228	2780
7/16 –20	361	4560	400	5050	418	527	327	4130	242	3050
1/2 –13	465	5050	517	5650	542	590	422	4600	313	3400
1/2 –20	487	5550	541	6170	565	645	443	5050	328	3740
9/16 –12	613	6050	682	6750	713	706	558	5500	413	4080
9/16 –18	668	6900	752	7750	787	810	615	6330	456	4700
5/8 –11	1000	8720	1110	9700	1160	1010	907	7900	715	6230
5/8 –18	1140	10,500	1244	11,500	1301	1200	1016	9350	798	7400
3/4 –10	1259	9200	1530	11,200	1582	1160	1249	9150	980	7180
3/4 –16	1230	9400	1490	11,400	1558	1190	1220	9350	958	7300
7/8 –9	1919	12,500	2328	15,200	2430	1580	1905	12,400	1495	9700
7/8 –14	1911	12,800	2318	15,600	2420	1640	1895	12,800	1490	10,100
1 –8	2332	16,200	3440	19,700	3595	2060	2815	16,100	2205	11,750

where d = the diameter of the fastener head.

           t is the thickness of the material.

Preload and Vibration. Increasing a fastener preload increases the friction forces in the joint and thereby increases its vibration resistance. If the dynamic environment is not severe, the increased friction forces may be sufficient to resist motion and loosening will be prevented. If, on the other hand, the dynamic environment is severe enough to overcome the increased preload or the resulting added friction forces, internal torque will cause loosening. Thus, the net result of an increased preload is a slightly increased vibration life under more severe dynamic conditions.

Thread Pitch and Loosening Torque. The theoretical internal loosening torque is directly proportional to the helix angle of the thread. A coarse pitch (large helix angle) will generate a greater internal loosening torque than a fine pitch thread. As an example, a $\frac{3}{8}$–16 thread produces 50% greater internal loosening torque than a $\frac{3}{8}$–24 thread for an equal preload condition.

Shock Loads. When a shock load is applied to a bolt in service, the resultant stress in the bolt is little more than tightening stress until the total shock load exceeds the tightening load. This is because bolt elongation produced by the shock load relieves an equal amount of compression in bolted members. The corresponding reduction of compressive force substantially counterbalances the shock load. Therefore, the initial tension is sufficient to hold the clamped surfaces together.

The maximum shock resistance of a mounted piece of equipment or structure is given by:

$$G = \frac{PN}{2W}$$

where G = shock resistance (g-factor).

           P = the tightening load based on bolt tension factors as shown in Table 7.3.

           N = number of fasteners.

           W = the weight in pounds of the part fastened in place. A sudden load is considered to be twice as severe as the same load applied statically, therefore, the factor 2 is used.

Example: Find the shock resistance (g-factor) required to prevent separation in tension of four 6–32 UNC-2A, type 303 stainless steel bolts, which are used to mount a 10 lb transformer. Assume that the g-force must not exceed the internal load on the bolts due to tightening.

Solution: From Table 7.3, P = 326 lb. Then by substitution:

$$G = \frac{326\,(4)}{2\,(10)} = 65.2\,\text{lb}$$

TABLE 7.4. *Stress Area for Fasteners* (Courtesy of *Machine Design*)

Bolt Size	Stress Area	Bolt Size	Stress Area
2–56	.0034	$\frac{5}{16}$–24	.0560
2–64	.0037	$\frac{3}{8}$–16	.0747
3–48	.0045	$\frac{3}{8}$–24	.0853
3–56	.0049	$\frac{7}{16}$–14	.1028
4–40	.0056	$\frac{7}{16}$–20	.1154
4–48	.0062	$\frac{1}{2}$–13	.1376
5–40	.0074	$\frac{1}{2}$–20	.1560
5–44	.0078	$\frac{9}{16}$–12	.1769
6–32	.0085	$\frac{9}{16}$–18	.1980
6–40	.0090	$\frac{5}{8}$–11	.2201
8–32	.0132	$\frac{5}{8}$–18	.2505
8–36	.0140	$\frac{3}{4}$–10	.3266
10–24	.0166	$\frac{3}{4}$–16	.3660
10–32	.0190	$\frac{7}{8}$–9	.4518
$\frac{1}{4}$–20	.0303	$\frac{7}{8}$–14	.5006
$\frac{1}{4}$–28	.0349	1–8	.5937
$\frac{5}{16}$–18	.0503	1–12	.6520

This value does not represent failure but rather that point at which separation of the rigid, clamped surfaces is likely to occur.

To cause actual rupture of a joint for a particular application, the external, variable dynamic forces must exceed the load due to tightening. This load is now based on the ultimate working strength of the bolt.

Ultimate shock load may be calculated based upon the use of the mean equivalent area or "stress area" as given in Table 7.4. Proper values may be substituted in the following formulas:

$$\text{tensile} \quad G = \frac{S_t A N}{2W} \qquad \text{shear} \quad G = \frac{S_s A N}{2W}$$

Values of ultimate tensile and shear strengths for six typical fasteners are listed in Table 7.3.

Example: Find the size of stainless steel bolts (4 bolts, type 303 stainless steel) required to support a 40 lb frame in tension when subjected to a 35 g force.

Solution: From Table 7.3, $S_t = 90,000$ psi. Then by solving the tensile stress equation for A and substituting:

$$A = \frac{2 G W}{S_t N}$$

$$= \frac{2(35)(40)}{90,000(4)} = 0.0078 \text{ in.}^2$$

A safety factor must be selected, based upon the application. A safety factor of 2 for a noncritical area, for example, would change the minimum area to 0.0078 × 2 − 0.0156 in^2. The nearest standard bolt size from Table 7.4 would be 10–24. An alternative might be to use six 6–32 bolts. For critical applications, the safety factor must be revised upward.

Torque. Fatigue failure of highly stressed, bolted joints is dependent upon the bolt design as well as the tightening technique. Because of its shape, a bolt is subjected to a notch effect where the shank joins the head, at the thread root and at the thread runout on the shank. This notch effect is minimized in bolts that have a fillet at the intersection of the shank and head, a rounded thread root, and an undercut at the thread runout.

A properly tightened nut is one that applies a tensile load to the bolt that is less than the elastic limit but greater than the service load, thus holding the parts together. However, working loads are added to residual stress in the bolt shank. To prevent fatigue failure, the sum of these two stresses must be held below the elastic limit of the bolt. That is the reason it is so important to apply only the correct tension when setting up a nut.

Two types of torque wrenches are available. On one type a pointer on a scale indicates the amount of force being applied; the other type employs a dial to show the torque. In some cases the dial (second type) can be set so a light flashes when the desired amount of torque is applied. The amount of torque to be applied can be determined by the formula:

$$T = C D P$$

where T = torque (in.–lb).
C = coefficient of sliding friction.
D = bolt diam (in.).
P = total load (lb).

The coefficient of friction will, of course, vary with the condition of the bearing surfaces. Lubricated nuts appear to give more accurate and consistent readings on the wrench. The clearance between the hole and the bolt should not be more than 3 per cent of the bolt diameter, and preferrably less for highly stressed applications.

THREAD TAPPING

The process of cutting or forming threads on the surface of a hole (in a nut, pipe, block of metal, etc.) with a tap is referred to as tapping a thread. Common types of taps are taper, plug, and bottoming, shown in Fig. 7–6. On a through hole, as shown in Fig. 7–7a, it is only necessary to use a taper tap. On a blind hole, (b) and (c), it is customary to start with a taper or plug tap and

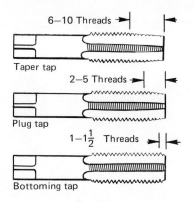

Taper tap

Plug tap

Bottoming tap

FIGURE 7–6. *A hand tap set consists of taper, plug, and bottoming taps. (Courtesy of DoAll Company.)*

finish with a bottoming tap. For adequate fastening strength, a hole need only be tapped at a depth up to $1\frac{1}{2}$ times the tap diam. This leaves a large margin of safety.

The drill size used before tapping the hole (often referred to as the tap drill size) may be determined from the formula:

$$D_t = D - \frac{1}{N}$$

where D_t = tap drill size.
           D = screw diam.
           N = number of threads/in.

Table 7.5 may also be used to select the proper drill size with the desired percentage of thread. A 75 percent thread (one with the top 25% of the thread removed due to an oversize drill being used) is recommended for commercial practice.

Threaded Fastener Types. The wide variety of threaded fasteners has evolved from many specialized applications that now are standard. Common types of screws include: machine screws, capscrews, setscrews, self-tapping screws, self-locking screws, pre-assembled screws, and special head screws, Fig. 7–8.

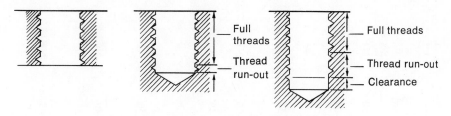

FIGURE 7–7. *Tapped holes:* (a) *through,* (b) *blind,* (c) *blind with clearance. (Courtesy of American Machinist.)*

TABLE 7.5. *Tap Drill Sizes*

Thread Size	Pitch Series	Tap Drill Size	Decimal Equivalent of Tap Drill	% of Thread (Approx.)
0–80	NF	56	.0465	83
		$\frac{3}{64}$	.0469	81
1–64	NC	54	.0550	89
		53	.0595	67
1–72	NF	53	.0595	75
		$\frac{1}{16}$	.0625	58
2–56	NC	51	.0670	82
		50	.0700	69
		49	.0730	56
2–64	NF	50	.0700	79
		49	.0730	64
3–48	NC	48	.0760	85
		$\frac{5}{64}$	.0781	77
		47	.0785	76
		46	.0810	67
		45	.0820	63
3–56	NF	46	.0810	78
		45	.0820	73
		44	.0860	56
4–40	NC	44	.0860	80
		43	.0890	71
		42	.0935	57
		$\frac{3}{32}$	.0938	56
4–48	NF	43	.0890	85
		42	.0935	68
		$\frac{3}{32}$	.0938	67
		41	.0960	59
5–40	NC	40	.0980	83
		39	.0995	79
		38	.1015	72
		37	.1040	65
		14	.1820	73
		13	.1850	67
$\frac{1}{4}$–20	UNC	9	.1960	83
		8	.1990	79
		7	.2010	75
		$\frac{13}{64}$	.2031	72
		6	.2040	71
		5	.2055	69
$\frac{1}{4}$–28	UNF	3	.2130	80
		$\frac{7}{32}$	.2188	67
$\frac{5}{16}$–18	UNC	F	.2570	77
		G	.2610	71
$\frac{5}{16}$–24	UNF	H	.2660	86
		I	.2720	75
		J	.2770	66
$\frac{3}{8}$–16	UNC	$\frac{5}{16}$	.3125	77
		O	.3160	73
$\frac{3}{8}$–24	UNF	Q	.3320	79
		R	.3390	67
$\frac{7}{16}$–14	UNC	T	.3580	86
		$\frac{23}{64}$	.3594	84
$\frac{7}{16}$–20	UNF	W	.3660	79
		$\frac{25}{64}$	.3906	72
$\frac{1}{2}$–13	UNC	$\frac{27}{64}$	.4219	78
$\frac{1}{2}$–20	UNF	$\frac{29}{64}$	.4531	72
$\frac{9}{16}$–12	UNC	$\frac{15}{32}$	.4688	87
		$\frac{31}{64}$	.4844	72
$\frac{9}{16}$–18	UNF	$\frac{1}{2}$	.5000	87
		0.5062	.5062	78
$\frac{5}{8}$–11	UNC	$\frac{17}{32}$	.5312	79
$\frac{5}{8}$–18	UNF	$\frac{9}{16}$	.5625	87

Thread	Series	Drill	Decimal	No.
5–44	NF	38	.1015	80
		37	.1040	71
		36	.1065	63
6–32	NC	37	.1040	84
		36	.1065	78
		$\frac{7}{64}$	.1094	70
		35	.1100	69
		34	.1110	67
		33	.1130	62
6–40	NF	34	.1110	83
		33	.1130	77
		32	.1160	68
8–32	NC	29	.1360	69
8–36	NF	29	.1360	78
		28	.1405	65
10–24	NC	$\frac{9}{64}$	.1406	65
		27	.1440	85
		26	.1470	79
		25	.1495	75
		24	.1520	70
		23	.1540	66
10–32	NF	$\frac{5}{32}$	.1562	83
		22	.1570	81
		21	.1590	76
		20	.1610	71
12–24	NC	$\frac{11}{64}$	.1719	82
		17	.1730	79
		16	.1770	72
		15	.1800	67
12–28	NF	16	.1770	84
		15	.1800	78

Thread	Series	Fraction	Decimal	No.
$\frac{3}{4}$–10	UNC	0.5687	.5687	78
		$\frac{41}{64}$	.6406	84
		$\frac{21}{32}$	.6562	72
$\frac{3}{4}$–16	UNF	$\frac{11}{16}$	.6875	77
$\frac{7}{8}$–9	UNC	$\frac{49}{64}$	.7656	76
$\frac{7}{8}$–14	UNF	$\frac{51}{64}$	.7969	84
1–8	UNC	0.8024	.8024	78
		$\frac{13}{16}$	.8125	67
		$\frac{55}{64}$	.8594	87
1–12	UNF	$\frac{7}{8}$	.8750	77
		$\frac{29}{32}$	.9062	87
$1\frac{1}{8}$–7	UNC	$\frac{59}{64}$	.9219	72
		$\frac{31}{32}$	.9688	84
		$\frac{63}{64}$	.9844	76
$1\frac{1}{8}$–12	UNF	$1\frac{1}{32}$	1.0312	87
$1\frac{1}{4}$–7	UNC	$1\frac{3}{64}$	1.0469	72
		$1\frac{3}{32}$	1.0938	84
$1\frac{1}{4}$–12	UNF	$1\frac{5}{32}$	1.1562	87
		$1\frac{11}{64}$	1.1719	72
$1\frac{3}{8}$–6	UNC	$1\frac{3}{16}$	1.1875	87
		$1\frac{13}{64}$	1.2031	79
		$1\frac{7}{32}$	1.2188	72
$1\frac{3}{8}$–12	UNF	$1\frac{9}{32}$	1.2812	87
		$1\frac{19}{32}$	1.2969	72
$1\frac{1}{2}$–6	UNC	$1\frac{5}{16}$	1.3125	87
		$1\frac{21}{64}$	1.3281	79
$1\frac{1}{2}$–12	UNF	$1\frac{13}{32}$	1.4062	87
		$1\frac{27}{64}$	1.4219	72
$1\frac{3}{4}$–5	UNC	$1\frac{17}{32}$	1.5312	84
		$1\frac{35}{64}$	1.5469	78
2–$4\frac{1}{2}$	UNC	$1\frac{25}{32}$	1.7812	76

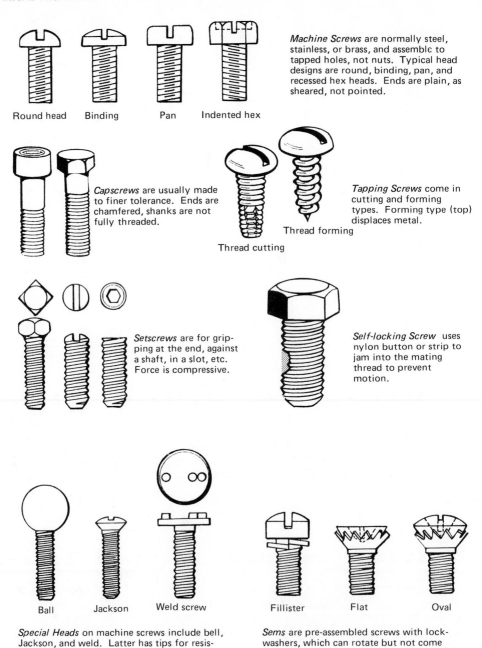

Machine Screws are normally steel, stainless, or brass, and assemble to tapped holes, not nuts. Typical head designs are round, binding, pan, and recessed hex heads. Ends are plain, as sheared, not pointed.

Round head Binding Pan Indented hex

Capscrews are usually made to finer tolerance. Ends are chamfered, shanks are not fully threaded.

Tapping Screws come in cutting and forming types. Forming type (top) displaces metal.

Thread forming

Thread cutting

Setscrews are for gripping at the end, against a shaft, in a slot, etc. Force is compressive.

Self-locking Screw uses nylon button or strip to jam into the mating thread to prevent motion.

Ball Jackson Weld screw

Fillister Flat Oval

Special Heads on machine screws include bell, Jackson, and weld. Latter has tips for resistance welding, to concentrate current.

Sems are pre-assembled screws with lockwashers, which can rotate but not come off. They are widely used in mass assembly industries such as automotive, appliance, electrical, radio, and TV.

FIGURE 7–8. *Common types of screws. (Courtesy of American Machinist.)*

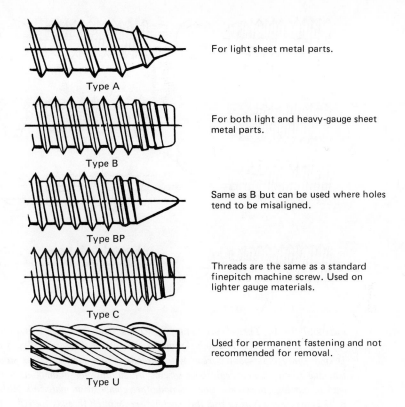

For light sheet metal parts.

Type A

For both light and heavy-gauge sheet metal parts.

Type B

Same as B but can be used where holes tend to be misaligned.

Type BP

Threads are the same as a standard finepitch machine screw. Used on lighter gauge materials.

Type C

Used for permanent fastening and not recommended for removal.

Type U

FIGURE 7–9. (a) Thread-forming tapping screws as listed in American Standard ASA B18.6.4–1958.

Tapping Screws. Shown in Fig. 7–9a and Fig. 7–9b are two types of tapping screws, thread forming and thread cutting. The thread forming, Fig. 7–9a, produce a secure joint by displacing and forming the material adjacent to the screw. The pilot hole size is made so that no further material need be removed and a maximum of metal to metal contact is made as the thread is formed by the screw. The thread cutting screws, Fig. 7–9b, are made with chip cavities and produce a mating thread by removing material from the engaged section. This type is used especially in thicker materials.

Tapping screws should always be driven in place to provide proper thread action and holding power. The ideal tightening torque is between 2/3 and 3/4 of the stripping torque. Tapping screws may be used in almost all materials including steel, cast iron, aluminum, zinc, brass, plastic, glass fibers, asbestos, and resin impregnated plywood. Self-drilling tapping screws, Fig. 7–10, have a milled drill point on the end of a hardened screw. This ensures the right size pilot hole to eliminate one operation. General specifications and dimensions of standard tapping screws are listed in ASA B18.6.4–1958.

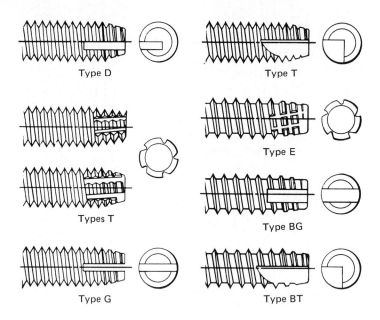

Type D   Type T

Types T   Type E

Type BG

Type G   Type BT

FIGURE 7–9. (b) *Thread cutting tapping screws. In each screw the first threads cut chips that produce a mating thread for the rest of the screw. Types F, G, D, and T are blunt-point screws with threads of the same pitch as standard machine screws. In general, these screws are suitable for aluminum, zinc and lead die castings, plywoods, asbestos, and other composition materials. Although there is overlapping as to use, the machine screw pitch threads were developed for brittle or granular material and the second, B types, were developed for very friable plastics such as urea compositions. (American Standard ASA B18.6.4–1958.)*

Bolts. Bolts are threaded fasteners that are designed for insertion through holes in parts to be assembled, they are normally tightened in place by torquing the nut. Common types are machine bolts, stove bolts, carriage bolts, lag screws or lag bolts, T-bolts, and hanger bolts, Fig. 7–11.

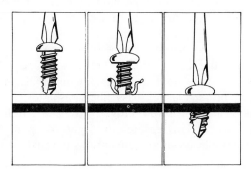

FIGURE 7–10. *Self-drilling, tapping screws are able to provide their own hole and threads in one operation. (Courtesy of Shakeproof Division of Illinois Tool Works, Inc.)*

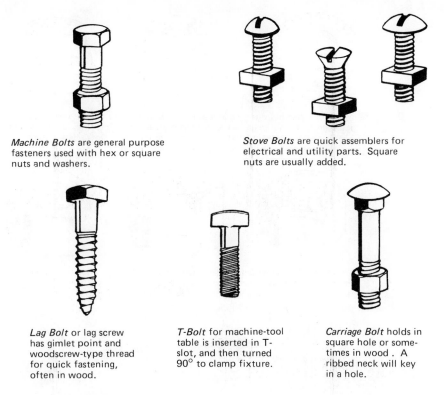

Machine Bolts are general purpose
fasteners used with hex or square
nuts and washers.

Stove Bolts are quick assemblers for
electrical and utility parts. Square
nuts are usually added.

Lag Bolt or lag screw
has gimlet point and
woodscrew-type thread
for quick fastening,
often in wood.

T-Bolt for machine-tool
table is inserted in T-
slot, and then turned
90° to clamp fixture.

Carriage Bolt holds in
square hole or some-
times in wood . A
ribbed neck will key
in a hole.

FIGURE 7–11.   *Common types of bolts. (Courtesy of American Machinist.)*

Nuts.   Common nuts are classified as finished (close tolerances) and heavy (looser
fit for large clearance holes and high loads). Jam nuts, usually thinner nuts, are used
in pairs to lock each other onto the threads of the bolt.

Lock Nuts.   Wherever there is any possibility of vibration that may cause bolted
joints to loosen, some type of locking device should be considered. Many types
have been developed and can be classified as free spinning, prevailing torque,
and spring action, Fig. 7–12.

Free Spinning.   Free spinning lock nuts are usually solid stock types with
machined special features such as integral tooth washers or concave bottoms,
which flatten out slightly when tightened. (The thread is distorted slightly, which
causes the nut to lock.)

Prevailing Torque.   Prevailing torque nuts include deformed threads, plastic
inserts, or other devices that tighten up on the bolt thread immediately as the
nuts are loaded; these maintain a constant load against loosening. Some of them
have holes that are not round. Prevailing torque refers to the extra resistance to
turning that is always being exerted.

 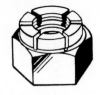 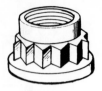

Spin-Lock Nut has outer-edge teeth on a thinned web. As nut bears down, teeth grip, and nut edge flexes to keep pressure constant. Flexloc nut (middle) has threads with distorted pitch that are tighter inside slots. Self-locking nuts (right) have non-circular threads that squeeze on bolt.

Prevailing Torque Nuts include security nut (left), which has a threaded elliptical insert inside a standard body, *Nylok Nut* (middle) has a nylon insert, *Lokuts Nut* (right) designed for fragile assemblies, surface contact.

Spring-Action Speednut spins on easily, but pressure of arched base against work holds it.

Palnut is a similar spring-action fastener, which uses the whole circle to exert pressure.

FIGURE 7–12. *Self-locking nuts may be classified as free spinning, prevailing torque, and spring action. (Courtesy of American Machinist.)*

Spring Action. Spring action lock nuts provide a locking action when they are pulled up against a surface. Many are stamped from sheet metal and are heat-treated to provide a constant tension.

Inserts. Threaded inserts provide re-usable threads in a material that may not tap well, or in which tapping a thread may be costly. Outside of repair work, they are often used on original equipment as an alternative to threads in soft, low strength materials such as plastics or aircraft sandwich structures, Fig. 7–13.

PINS, WASHERS, AND RETAINING RINGS

Pins. Pins are sometimes used to replace keys, splines, rivets, or bolts. One prime advantage is that they usually work in simple drilled holes. The two main types are machine pins and radial locking pins. Machine pins are commercial straight

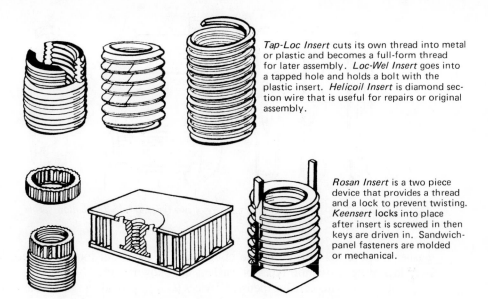

Tap-Loc Insert cuts its own thread into metal or plastic and becomes a full-form thread for later assembly. *Loc-Wel Insert* goes into a tapped hole and holds a bolt with the plastic insert. *Helicoil Insert* is diamond section wire that is useful for repairs or original assembly.

Rosan Insert is a two piece device that provides a thread and a lock to prevent twisting. *Keensert* locks into place after insert is screwed in then keys are driven in. Sandwich-panel fasteners are molded or mechanical.

FIGURE 7–13. *Various types of threaded inserts. (Courtesy of American Machinist.)*

pins, hardened and ground dowel pins, clevis pins, tapered pins, and standard cotter pins, Fig. 7–14. Radial locking pins are of two basic types; solid with grooved surfaces and hollow spring pins, which can be either slotted or spiral wrapped, Fig. 7–15. In assembly, radial forces produced by the elastic action at the pin surface develop a secure frictional grip against the hole wall. These pins are reusable and can be removed and reassembled several times before there is an appreciable loss of holding power.

Washers and Retaining Rings. Washers are used to distribute the compressive stress over a wider area than the bolt, screw head, or nut; and they may keep the surface from being damaged by the fastener. The most important use of the

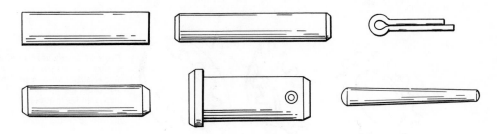

FIGURE 7–14. *Machine pins are used for locating and for locking. Shown here are commercial straight, dowel, clevis, taper, and cotter pins.*

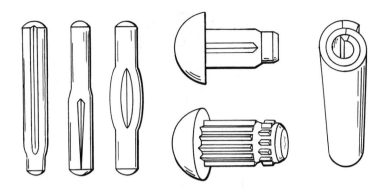

FIGURE 7–15. *Radial locking pins are of two types, grooved or hollow spring.*

washer, however, is to prevent the bolt, screw, or nut from loosening once it has been locked up. Common types of washers are shown in Fig. 7–16.

Retaining rings are inexpensive devices used to provide a removable shoulder and to accurately locate, retain, or lock components on shafts or in boxes and housings, Fig. 7–17. Since they are usually made from spring steel, retaining rings have a high shear strength and impact capacity. Some rings are designed to take up end play caused by accumulated tolerances and wear in the parts being retained. Most types slip or snap into grooves. Some are flat stock and others are formed from wire. In most cases, they permit the assembly of parts without the use of bolts and washers.

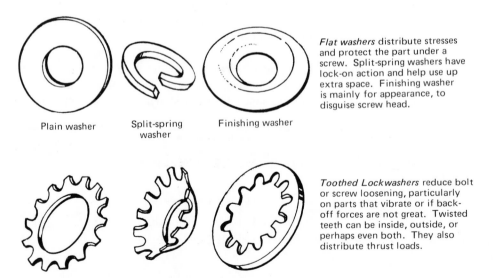

Plain washer   Split-spring washer   Finishing washer

Flat washers distribute stresses and protect the part under a screw. Split-spring washers have lock-on action and help use up extra space. Finishing washer is mainly for appearance, to disguise screw head.

Toothed Lockwashers reduce bolt or screw loosening, particularly on parts that vibrate or if back-off forces are not great. Twisted teeth can be inside, outside, or perhaps even both. They also distribute thrust loads.

FIGURE 7–16. *Common types of washers.*

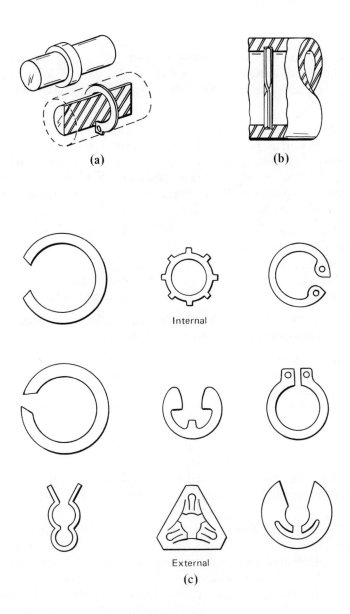

(a)

(b)

Internal

External

(c)

FIGURE 7–17. *(a) Retaining rings are used to form removable shoulders on pins or shafts, and (b) inside of holes. Retaining rings are made to fit into narrow grooves and can be quite easily removed with simple tools. Various internal and external retaining rings are shown in (c).*

Resistance Welded Fasteners.   Another important method of securing fasteners in place is by means of resistance welds, both spot and projection. This form of fastener was discussed in Chapter 4 under resistance welding.

Special Purpose Fasteners.   Many fasteners do not fit into the more common types already discussed and these may be classified as follows: plastic fasteners, spring clips, self-sealing fasteners, and quick operating fasteners.

Plastic Fasteners.   Plastic fasteners are mechanical devices, bolts, nuts, screws, rivets, and specially engineered fasteners made from plastics designed to hold parts together. The main difference between these fasteners and those already discussed is that they are made out of plastic. They are made to the same dimensions and standards as metal fasteners. Some of the outstanding properties of plastic fasteners are:

Corrosion resistance.	Self-sealing capacity.
Nonelectrical conductance.	Nontoxicity.
Self-locking capacity.	Light weight.
Chemical resistance.	Antimagnetic properties.
Self lubrication.	Dent and scratch resistant.

Although plastics, like metals, include many different families of materials, only a few are predominant for fasteners, these are:

High impact ABS.	Polycarbonate, unfilled.
Acetal homopolymer.	Polyethylene, high density.
Fluorocarbons, TFE.	Polypropylene, general purpose.
Nylon, type 6/6.	Rigid PVC, normal impact.

Most of these materials can have glass or metallic filler added to the base resin to improve strength, stiffness, useful temperature range, and specific gravity. This information can be obtained from the supplier.

The pilot hole size used for thread forming or tapping screws in plastic materials is usually taken as the pitch diameter of the screw.

Spring Clips.   Spring clips, some of which are shown in Fig. 7–18 are generally considered to be one-piece, self-sufficient fasteners. The spring tension fastening principle eliminates vibration loosening, allows for design flexibility, compensates for tolerance buildup and misalignment, and minimizes assembly damage.

Spring clips are normally made from steel with .5 to .8% carbon. Generally they are formed in the annealed stage and then hardened to Rockwell C45–50. Spring tensions are controlled by varying the width and thickness of steel used. Plastics are being used also and have in some cases replaced metal spring clips. Plastic designs have near infinite possibilities and are limited only by their holding capabilities. Neoprene and vinyl clips provide a soft cushion to protect easily damaged parts.

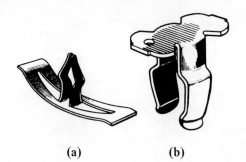

FIGURE 7–18. *Dart type spring clips. Clip (a) has moderate holding power and will not damage delicate components. Clip (b) has more strength. (Courtesy of Machine Design.)*

(a)   (b)

Dart type molding clips, Fig. 7–19, are commonly used for securing two panel surfaces together or in securing molding strips.

Self-sealing Fasteners. An auxiliary function of a fastener may be to seal the joint against leakage of gas or liquids. A number of fasteners have been developed with built-in sealing elements, Fig. 7–20. To make the whole joint pressure tight, the fasteners must be placed as close as possible with a minimum edge distance. Fasteners with fairly large heads or with washers are best. Usually a gasket is used in the joint or a coat of caulking compound or sealant is applied to the contact surfaces. The insulating properties of some sealing fasteners make them useful in preventing electrolysis, as well as reducing vibration and noise.

Factors to consider in choosing a sealant fastener are:

1. Cost vs. guaranteed exposure time.
2. Material vs. chemical attack.

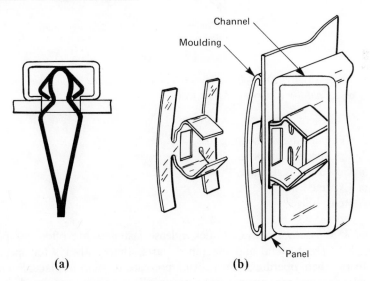

(a)   (b)

FIGURE 7–19. *Dart type molding clips. Double dart (a) holds the molding with an auxiliary short dart. Arched arms (b) support a wide molding. (Courtesy of Machine Design.)*

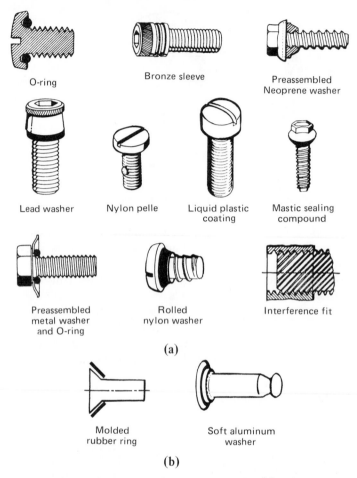

O-ring

Bronze sleeve

Preassembled
Neoprene washer

Lead washer

Nylon pelle

Liquid plastic
coating

Mastic sealing
compound

Preassembled
metal washer
and O-ring

Rolled
nylon washer

Interference fit

(a)

Molded
rubber ring

Soft aluminum
washer

(b)

FIGURE 7-20. *Self-sealing screws are shown at the left in (a) and rivets on the right (b). (Courtesy of American Machinist.)*

3. Temperature vs. sealing loss.
4. Pressure vs. support.
5. Corrosion resistance.
6. Expected life.
7. Reusability.

Quick Release Fasteners. Quick release fasteners are often complex mechanisms made of sheet metal and machined parts, almost always for special applications. Most of them operate against spring pressure to snap in place. Sometimes a special tool is required for removal. They are usually used where repeated access is necessary. Quick release fasteners may be classified into four types; lever-actuated, turn operated, slide action, and push-pull, Fig. 7-21.

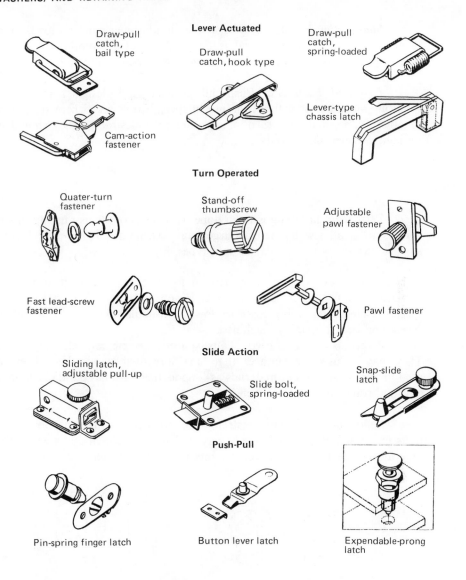

FIGURE 7–21. *Quick release fasteners are classified into four categories as shown.* (*Courtesy of Machine Design.*)

Lever Actuated Fasteners. The draw-pull catch is designed to secure boxes or chests with coplaner surfaces at the parting line of the lid. The draw-pull fastener is also excellent for gasket compression on container lids, hinged covers, and for use on electrical control boxes.

Turn Operated Fasteners. The quarter turn fastener is used on access panels, hinged doors, etc., or wherever frequent access is necessary.

221

Slide Action Fasteners.   The slide action fastener converts loads on a panel to shear within the fastener and is intended to restrict motion in at least two directions. It is usually used to restrict motion perpendicular to it.

Push-Pull Fasteners.   Push-pull fasteners are designed for applications requiring quick access without the use of special tools and where space limitation or other factors make lift, turn, or slide action undesirable. They should be used on lightly loaded panels only.

RIVETS

Permanent fastening is often done with rivets on items ranging in size from bridges to small jewelry items. Large heavy structures require solid rivets, but smaller items may use tubular or split types, Fig. 7–22.

Advantages.   The primary reason for riveting is that it provides a low in-place cost. The initial cost of rivets is substantially below that of screws or special fasteners. Assembly costs are low: semitubular rivets are clinched on high speed, hopper-fed, riveting machines.

Rivets also serve as pivot shafts, spacers, electric contacts, stops, or inserts. They may be used on surfaces that have already been finished. They may be used to join most types of materials as long as there are flat parallel surfaces available and room to upset or clinch it.

Limitations.   The fatigue strength of rivets is less than that for comparable bolts or screws. High tensile loads may pull out the clinch and severe vibrations may loosen the fastening. Riveted joints are neither water nor airtight, although

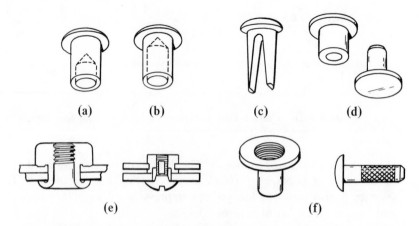

(a)   (b)   (c)   (d)

(e)   (f)

FIGURE 7–22.   *Rivet types are numerous. Shown here are (a) semi-tubular, (b) tubular, (c) split, (d) compression, (e) threaded, and (f) drilled.*

they could be made so by sealing compounds. If they must be disassembled for maintenance or repair, the rivets will have to be drilled out and replaced. They should not be used where the dimensional variation must be maintained as low as ± 0.001 in. (0.03 mm).

TYPES OF RIVETS

Semitubular. Semitubular rivets are the most widely used type for small assemblies. When properly specified and set, this rivet becomes essentially a solid member since the hole depth is just enough to form the clinch.

Tubular. Tubular rivets can be used to punch their own holes in fabric, plastic sheet, and other soft materials.

Bifurcated or Split. The bifurcated rivet is split to produce prongs that make their own holes through fiber, wood, plastic, or metal.

Compression. Compression rivets consist of a solid (male) and a tubular (female) member. Together they form an interference fit when pressed in place. Because the heads can be controlled from both sides, they are often used on cutlery where uniform appearance is important. The flush heads present a neat appearance and prevent the accumulation of dirt.

Rivet Strength. The effective cross-sectional area in shear for a rivet is usually considered to be the hole size, except in structural work where it is based on the rivet diameter.

Blind Rivets. Blind rivets derive their name from the characteristic of being headed from one side of the assembly, an essential factor in operations on enclosed products. Most blind rivets have a central shank that is pulled, either by mechanical or explosive action, to expand the shank securing the parts to be joined, Fig. 7–23. Blind rivet use is increasing even in applications where both sides are accessible, because they simplify assembly. They provide for easy portability of tooling and are installed with the minimum of time and effort.

EXERCISES

Problems and Questions

7–1. Two $\frac{1}{2}$ in. thick steel plates are to be bolted together to form a 4 in. lap joint, 8 in. wide. Four $\frac{1}{2}$–13 UNC carbon steel bolts are used. The joint will be subject to shear stresses up to 100,000 psi. Are these bolts adequate? Explain.

7–2. (a) You have a $\frac{3}{8}$–16 UNC thread and want to find the nearest metric size thread to match it. What would it be?

(b) Is there an ISO thread equivalent to a $\frac{3}{8}$–16 UNC thread?

(c) If no metric thread chart was available to you, how could you approximate the o.d. of an M6 thread in decimal inches?

7–3. A weight of 2000 lb is to be attached to a mild steel screw eyebolt. The bolt is threaded into a $\frac{3}{4}$ in. thick steel plate. (a) What size bolt would you recommend and why?

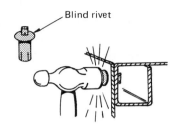

Setting a blind rivet.

The rivet can be removed, if necessary by driving out the center pin with a nail punch and prying out the remainder.

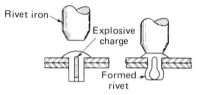

Explosive rivet, ignited by an electrical charge, blows its bottom to head the shank.

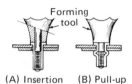

(A) Insertion   (B) Pull-up

Rivnuts are headed by a pull bolt that can be unthreaded out of the fastener.

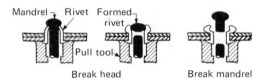

Break head   Break mandrel

Pop rivets are also for inaccessible holes. The central pull pin has special contouring that allows deforming the rivet shank with a straight pull, and then lets the pin separate when the rivet is fully formed.

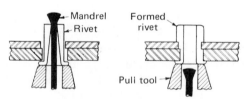

Chobert rivets are also one-side fasteners, upset by drawing a wire tipped with a wedge shape through the central hole. These will take care of slight hole misalignment. They also come in self-plugging types.

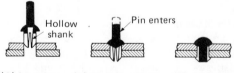

(A) Insertion   (B) Partially driven   (C) Assembly

Plasti-rivets are one-piece units for any kind of assembly, but their strength is nowhere near that of a metal fastener. They are easy to install, and not impossible to remove and use over again.

FIGURE 7–23. *Examples of blind rivets.* (*Courtesy of American Machinist.*)

(b) To what depth should the hole be tapped (with full threads) if a 75% thread is used?

(c) What drill size should be used?

(d) If hand tapping is done, what types of taps should be used?

(e) Approximately how deep should the hole be, including the clearance at the bottom? Assume the thread runout and the clearance will each equal about 2 threads.

7–4. Shown in Fig. P7–4 is a beam bolted to a column. This joint is subjected to shock loads of 25 g. The normal weight on the joint is 1,900 lb. The bolts used are carbon steel. What size must the bolts be for safe construction in a noncritical area?

7–5. (a) If $\frac{3}{4}$–10 bolts are used in the joint shown in Fig. P7–5 and the joint is subjected to fatigue loading conditions, what size clearance holes should be used?

(b) Can this hole size be obtained with a standard drill? (Assume steps of 1/64 in.)

(c) Should the holes be reamed? Why?

7–6. (a) Shown in Fig. P7–6 is a partial view of a bolted assembly. There are eight, 6–32 UNC brass bolts used. The bolts are used to mount a bearing housing that will support a total load of 40 lb. The g-force must not exceed the internal load on the bolts due to tightening. What shock resistance can the assembly tolerate and not have the bolts show any signs of separation?

(b) What shock load would cause separation of the bolts?

7–7. (a) What should the torque reading be on a 10 in. torque wrench for a $\frac{3}{4}$–10 UNC 303 stainless steel bolt if the coefficient of friction is 0.12? Four bolts are used to support a static weight of 40,000 lb.

(b) What would the maximum wrenching torque be for each bolt?

(c) What is the maximum load that these four bolts can support?

7–8. (a) What size hole should be used for $\frac{1}{4}$–20 UNC threads if self-tapping screws are used in thermoplastic materials?

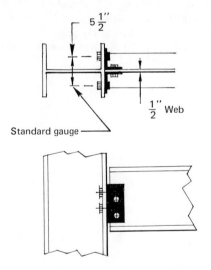

FIGURE P7–4.

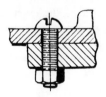

FIGURE P7–5.

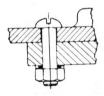

FIGURE P7–6.

(b) What type of self-tapping screw thread is recommended?

7–9. (a) Determine the maximum safe load for a lap joint shown in Fig. P7–9. The rivets are $\frac{1}{2}$ in. (12.700 mm) in diam and the allowable stresses are: tension, 18,000 psi (124.1 MPa); shear, 14,000 psi (9653 MPa); and bearing, 26,000 psi (262.0 MPa). Assume each rivet takes an equal share of the load.
(b) Determine the bearing strength for this joint.

7–10. (a) Four $\frac{5}{16}$–18 UNC bolts are used to support a 3/4 hp motor weighing 40 lb (18.14 k) on a tension ceiling mount. If this is a simple static load, will these bolts be adequate? Assume the yield strength is 3/4 of the tensile strength.
(b) What would the shock resistance be if a safety factor of 2 is used?

7–11. (a) A double lap joint as shown in Fig. P7–11 is held together by a bolt. It is subjected to a shear stress of 20,000 lb (9072 k). Assume the bolt has an allowable shear stress value of 30 ksi (206.8 MPa). Calculate the exact size of the bolt body required, assuming there are no threads in the shear plane.
(b) Would a $\frac{7}{8}$–9 UNC bolt be satisfactory if the threads are in shear?

7–12. Preload is an important factor if bolts are to provide optimum performance. Preload, or tightening, that is insufficient will allow stress fluctuations in the joint and lead to fatigue failure. Too much preload, on the other hand, may cause fracture of the bolt or plastic elongation with a consequent loss of clamping force.
(a) Suggest a couple of different methods that can be used to check for proper preload.

7–13. Assuming a $\frac{1}{2}$ in. (12.70 mm) diam stud welded bolt is to be used, what thickness should the base plate be to develop full fastener strength?

7–14. (a) What would be the approximate tensile strength of a low carbon steel $\frac{1}{2}$ in. (12.70 mm) diam stud? (b) The yield strength?

7–15. Compare the tap drill size of $\frac{1}{4}$–20 tap by formula and by Table 7.5. Use a 75% thread.

7–16. Two 3/4 in. (19.05 mm) thick steel plates are bolted together with four $\frac{3}{4}$–10 bolts. The bolts will be threaded into the bottom plate. Determine the following:

(a) Tap drill size for 75% thread.
(b) Type of tap or taps.
(c) Drill size for hole in top plate.
(d) What would be the proper torque reading on these bolts?

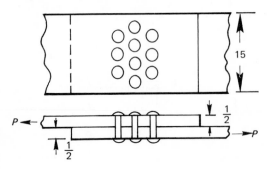

FIGURE P7–9.

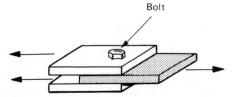

FIGURE P7–11.

BIBLIOGRAPHY

Books

Fastener Standards, Industrial Fasteners Institute, Cleveland, Ohio, 1970.

LINDBERG, R. A., *Processes and Materials of Manufacture*, Allyn and Bacon, Inc., Boston, Mass., 1964.

MacKENZIE, R. V., *Screw Threads, Design Selection, and Specification*, The Industrial Press, New York, N.Y., 1961.

Standards and Dimensions for Taps and Dies, Metal Cutting Tool Institute, New York, N.Y., 1955.

Periodicals

BELFORD, R. B., "Screw Thread Fits." *Machine Design*, November 7, 1963.

D'AGSTINO, R., "Selecting Threaded Fasteners for Shock Loads." *Machine Design*, February 14, 1963.

FINKELSTON, R. J., "How Much Shake Can Bolted Joints Take?" *Machine Design*, Oct. 5, 1972.

"Fastening and Joining." *Machine Design*, Reference Issue, 1971.

SCHREMMER, G., "How To Keep Bolted Joints Tight." *Machine Design*, Oct. 5, 1972.

SPROAT, R. L., " The Future of Fasteners," *Machine Design*, Dec. 22, 1966.

"U.S.A. Goes Metric." Pamphlet, Beloit Tool Corporation, South Beloit, Illinois.

Basic Safety Rules

7. Light acetylene before opening oxygen valve on torch.

8. NEVER use oil on regulators, torches, fittings or other equipment in contact with oxygen.

HERE LIES
WILLIE DOYLE
HE THOUGHT
EVERYTHING
WORKED
BETTER
WITH OIL.

9. Do not use OXYGEN as a substitute for compressed air.

10. Keep heat, flames and sparks away from combustibles.

4. Open cylinder valve SLOWLY.

5. Do not use or compress acetylene (in free state) at pressures higher than 15 PSI.

6. Purge oxygen and fuel gas passages (individually) before lighting torch.

1. Blow out cylinder valves before attaching regulators.

2. Release adjusting screw on regulators before opening cylinder valves.

3. Stand to one side of regulator when opening cylinder valve.

15 psi

Fundamentals of Combustion

1. KINDLING POINT
How hot must the fuel be to burn?

2. BURNING RATIO
How much oxygen is required to burn a quantity of fuel?

3. BURNING RATE
How fast will the fuel burn?

4. B.T.U. OUTPUT
How MUCH heat is produced by the burning fuel?

5. TEMPERATURE OUTPUT
How HOT will the fuel burn?

1 Coal   Paper   Steel wool

2 Atmosphere   Low oxy   Pure oxy

4 Volume

5   6000 °F
5000 °F
4000 °F
3000 °F
2000 °F
1000 °F

Purpose

To develop a better understanding of safety practices in the use and application of gas welding, cutting, heating and regulating equipment.

Triangle of Combustion

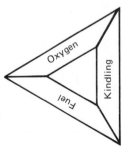

Oxygen

Fuel

Kindling

All these factors must be present to support combustion.

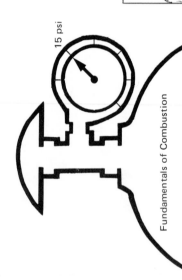

FIGURE 8–0. *Safety practices in the use and applications of gas welding, cutting, heating, and regulating equipment. (Courtesy of Tescom Corporation.)*

8

Filler Materials, Fluxes, and Gases

FILLER MATERIALS AND FLUXES

A materials joining engineer is expected to match filler material to the base material, a decision that is not easy and if incorrect could lead to joint failure. To as many as one thousand kinds of base metal alloys, there are perhaps less than one hundred kinds of filler metals. Generally it is a simple matter to learn the chemistry of the filler materials; however, it is not an uncommon situation when the chemistry of the base metal is unknown and the materials joining engineer can do little more than venture an educated guess. This chapter introduces the American Welding Society (AWS) classification system for filler materials and suggests applications for each.

By AWS definition, filler materials are any materials added in making a welded, brazed, or soldered joint and in overlaying a surface. The base materials are intended to include both metal and plastics. This discussion is limited to only those materials listed by the AWS, except for filler materials for plastics, which can be obtained from the Modern Plastics Encyclopedia. An AWS designation guarantees these materials meet certain minimum specifications as outlined and tested by the AWS. Copies of these specifications and testing procedures may be obtained from the AWS headquarters; a fairly complete list of AWS specifications follows in Table 8.1.

AWS Specifications for Filler Materials. Specification code numbers are given for each group of filler materials; for example, all mild steel covered electrodes are listed as A5.1XX. The A5 is used for all filler materials. The 1 designates mild steel covered electrodes and the XX, when used, is for the year the specifications were last updated. Each of the AWS filler materials will be discussed briefly.

TABLE 8.1. *AWS Filler Metal Specifications*
(These Specifications are available from the American Welding Society, Inc., 2501 N.W. 7th Street, Miami, Florida 33125.)

Specification for:	AWS Designation
Mild steel covered arc welding electrodes..........................	A5.1
Iron and steel gas welding rods	A5.2
Aluminum and aluminum-alloy arc welding electrodes	A5.3
Corrosion resisting chromium and chromium-nickel, steel covered welding electrodes	A5.4
Low-alloy steel covered arc welding electrodes	A5.5
Copper and copper-alloy arc welding electrodes	A5.6
Copper and copper-alloy welding rods	A5.7
Brazing filler metal...	A5.8
Corrosion-resisting chromium and chromium-nickel, steel welding rods and bare electrodes	A5.9
Aluminum and aluminum-alloy welding rods and bare electrodes	A5.10
Nickel and nickel-alloy covered welding electrodes...................	A5.11
Tungsten arc welding electrodes....................................	A5.12
Surfacing welding rods and electrodes..............................	A5.13
Nickel and nickel-alloy bare welding rods and electrodes	A5.14
Welding rods and covered electrodes for welding cast iron	A5.15
Titanium and titanium-alloy bare welding rods and electrodes.........	A5.16
Bare mild steel electrodes and fluxes for submerged arc welding........	A5.17
Mild steel electrodes for gas metal arc welding	A5.18
Magnesium alloy welding rods and bare electrodes	A5.19
Mild steel electrodes for flux-cored arc welding.....................	A5.20
Composite surfacing welding rods and electrodes	A5.21
Flux-cored corrosion resisting chromium and chromium-nickel steel electrodes...	A5.22

Mild Steel Covered Electrodes. Mild steel covered electrodes comprise one of the larger groups of filler materials. The AWS classification is printed on the coating near the bare end of each electrode. The classification has a letter followed by four numerals, for example, E6010.

E ... electrode.

60 ... minimum tensile strength, in 1000 psi (6.89 MPa), of deposited weld metal in the as-welded condition.

1 ... recommended positions for specific electrodes to make a satisfactory weld:

"1" all positions (flat, horizontal, vertical, and overhead).

"2" limited to flat position or horizontal fillets only.

"3" flat or downhand position only.

"4" The 4th digit has meaning only in terms of the 3rd. Together they indicate the following:

Designation	Current*	Covering Type
EXX10	DCRP only	Organic
EXX11	AC or DCRP	Organic
EXX12	AC or DCSP	Rutile
EXX13	AC or DCSP/RP	Rutile
EXX14	AC or DCSP/RP	Rutile, iron powder
EXX15	DCRP only	Low hydrogen
EXX16	AC or DCRP	Low hydrogen
EXX18	AC or DCRP	Low hydrogen, iron powder
EXX20	AC or DCRP/SP	High iron oxide
EXX24	AC or DCSP/RP	Rutile, iron powder
EXX27	AC or DCSP/RP	Mineral, iron powder
EXX28	AC or DCRP	Low hydrogen, iron powder

* DCRP means direct current reverse polarity.
 DCSP means direct current straight polarity.
 AC means alternating current.

ELECTRODE AND ROD DESCRIPTIONS

Most mild steel electrodes are made from SAE 1010 rimmed steel, which has a low carbon content without being deoxidized. The main types of mild steel electrodes, coverings, weld positions, recommended currents, typical applications, advantages, and disadvantages are summarized in Table 8.2.

Iron and Steel Welding Rods. Iron and steel gas welding rods are bare steel rods. They are intended for use in all positions, limited only by the skill of the operator. Only three different rods comprise this group. They are the RG 45, RG 60, and RG 65. The "R" identifies a welding rod rather than an electrode and the "G" indicates gas as the source of welding energy. The two digits give the tensile strength in thousands of psi for the weld metal in the as-welded condition. All rods must contain less than 0.040% phosphorus and 0.02% aluminum.

Mild Steel Electrodes for Gas Metal Arc Welding. This group of mild steel electrodes is designed for the welding of mild and low alloy steels in a gas atmosphere. The AWS classification system for E705–1 mild steel electrodes used in gas metal arc welding is:

E ... designates an electrode.
70 ... designates the minimum as-welded tensile strength in 1000 psi.
S ... designates a bare, solid electrode.

TABLE 8.2. *Classification and Selection of Mild Steel Covered Electrodes*

AWS Classification	Coating	Weld Position	Type of Current*	Characteristics	Applications
E6010	High cellulose	All positions	DCRP	Deep penetrating spray type arc, thin friable slag	Shipbuilding, bridges, buildings, and pressure vessels
E6011	High cellulose potassium	All positions	AC/ DCRP	Similar to those of E6010	Same as those for E6010
E6012	High titania	All positions	DCSP DCRP	Medium penetration, quiet type arc, single pass, high speed	Where poor fitup exists
E6013	Rutile and other easily ionized materials	All positions	AC	Shallow penetration, excellent radiographic quality	Sheet metal
E7014	Same as E6012 and E6013, plus iron powder	All positions	AC/DC	Medium rate of deposition, medium penetration similar to the E6012	Mild and low alloy steels
E7015	Limestone	All positions	DCRP	Low hydrogen, moderate penetration, heavy friable slag	High strength/high carbon alloy steels, high sulphur steels, malleable iron, spring steels, steels to be enameled, and selenium steels
E7016	Same as E7015, plus potassium silicate or other potassium salts	All positions	AC	Same as E7015	Same as E7015
E7018	Same as E7016, plus iron powder (25% to 40% by weight)	All positions	AC/ DCRP	High lineal speeds, low penetration, low spatter, smooth quiet arc, and globular type transfer	High strength, high carbon or alloy steels

TABLE 8.2. *Classification and Selection of Mild Steel Covered Electrodes* (Cont.)

AWS Classification	Coating	Weld Position	Type of Current*	Characteristics	Applications
E6020	High iron oxide, sodium type	Horizontal filler	AC/DC	Spray type arc, heavy slag, medium to high penetration, high deposition rates, excellent radiographic properties	Pressure vessels, heavy machine bases, structural parts, specialized procedures
E7024	Same as E6012 and E6013 plus 50% of coating weight of iron powder	Horizontal and flat position fillets	AC/DC	Smooth, quiet arc, low spatter, low penetration, used at high lineal speed	Fillet welds on low, medium, and high carbon steels, and low alloy steels.
E7027	Same as E6020 plus 50% by weight of iron powder in coating	Fillet and groove welds in flat and horizontal positions	AC/DC	Spray type metal transfer, high deposition rates at high lineal speeds, medium penetration, low spatter losses, and heavy friable slag	Heavy sections
E7028	Similar to that of E7018 plus 50% by weight of iron powder in coating	Horizontal and flat position	AC/ DCRP	Spray type transfer	High strength high carbon or alloy steels

* DCRP means direct current reverse polarity.
 DCSP means direct current straight polarity.
 AC means alternating current.

233

U ... if used instead of "S," designates an emissive coated solid electrode.

(Emissive means that the electrode is coated with an arc stabilizing coating.)

1 ... indicates the electrode's manufactured chemical composition.

When these electrodes are used with an argon-oxygen shielding gas, their chemical composition remains fairly intact during the change to weld metal. However, when the same electrodes are used with straight CO_2 shielding, the content of carbon, manganese, silicon, and other deoxidizers becomes significantly reduced in the main alloying elements. Wires with high deoxidizers can be used to successfully weld steels with rusty and dirty surfaces.

Mild Steel Electrodes for Flux-cored Arc Welding. Mild steel flux-cored electrodes are used with or without a CO_2 gas shielding for the welding of mild and low alloy steels. These electrodes are used in the gas metal-arc process. Their classification is based on the factors of: whether CO_2 gas is required as a separate shield, the type of current, their usefulness for either single or multiple pass applications, chemical composition, and the mechanical properties of the deposited weld metal. The "T" in an E70T–1 electrode class designates a composite or powder cored electrode. The other digits and letters are conventional. The composite or powder is contained within the wire by one of the several different designs shown in Fig. 8–1. The major alloying elements in flux-cored wires are manganese, silicon, nickel, chromium, molybdenum, vanadium, and aluminum; their use is for carbon and low alloy steels. A flux-cored wire can be expected to give a high rate of deposition.

Bare Mild Steel Electrodes. The AWS classification system for bare mild steel electrodes and fluxes for submerged-arc welding is more complex than the general practices followed for filler materials. For example, a classification of EL8K for an electrode means:

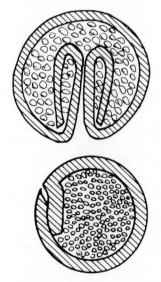

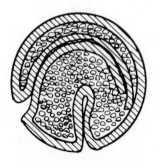

FIGURE 8–1. *Cross-sectional views of flux-cored wires.*

E ... that it is an electrode.

L ... that the electrode contains a low manganese content, 0.60% or less. If this letter is "M" instead of "L," it would indicate a medium amount of manganese—or less than 1.25%. If the letter is "H," this would indicate a high manganese content or less than 2.25%.

8 ... the nominal carbon content of the electrode in points.

K ... that this electrode is made from silicon killed steel.

Fluxes. The flux used in submerged-arc welding greatly influences the chemistry of the deposited metal and, therefore, requires discussion along with the electrode. Fluxes are classified according to the mechanical properties of the weld metal. For example, a flux classification of F71 means:

F ⋯ the material is a flux.

7 ... the guaranteed minimum tensile strength in 10,000 psi (68.95 MPa) for the filler material when used with this flux.

1 ... the minimum impact properties of weld metal that is produced from the combination of flux and filler material such as:

"1" ... 20 ft lb (27.1 joules) at 0 °F (-18 °C)
"2" ... 20 ft lb at -20 °F (-29 °C)
"3" ... 20 ft lb at -40 °F (-40 °C)
"4" ... 20 ft lb at -60 °F (-51 °C)

A commonly used combination of flux and wire for submerged-arc welding is "F71–EM12K."

The electrode compositions classified in this specification differ in carbon, manganese, and silicon content. The sulfur and phosphorus content may also vary within the maximums permitted. Furthermore, the choice of raw materials by the steel producer will introduce varying amounts of tramp elements such as nickel, copper, or chromium. Each of these elements will affect the properties of the weld metal and the welding operation. The effect of different compositions with any given flux and combination of welding conditions may be generalized.

Increased carbon, manganese, and silicon content in the electrode results in an increase of the elements in the weld metal. Increased manganese only in the electrode increases the manganese content in the weld metal without affecting the silicon content, thus increasing the manganese–silicon ratio. Increased carbon in the electrode may accelerate the manganese–silicon reactions in the melt during welding. This will result in less manganese recovery and greater silicon recovery in the weld metal.

Increased carbon and manganese content in the electrode will increase the yield strength and the ultimate tensile strength of the weld metal. A lesser effect is obtained from increased silicon. Similarly, tramp elements, such as copper, nickel, and chromium, may increase the tensile strength and reduce the ductility of the weld metal in measurable amounts.

Notch toughness is inversely related to the phosphorus content of the weld metal. The phosphorus may originate in the base metal, flux, or electrode. When category "2" fused fluxes are used, the optimum notch toughness of the weld metal will be obtained with electrodes containing low carbon, manganese, and silicon content. Notch toughness is reduced by the use of electrodes containing either greater manganese or silicon content, or both, even with the same category "2" fused fluxes.

Electrodes containing high manganese, such as EH14, and especially when used with category "1" fused fluxes, promote weld metal soundness by reducing cracking and porosity. Electrodes containing large amounts of silicon increase the fluidity of the melt and its wet-ability. This results in improved reinforcement shape, more regular weld edges, and sounder welds at maximum welding speeds.

Submerged arc welding fluxes are granular, fusible mineral compounds in various proportions and quantities. Manufactured by several different methods, the general types are named for their method of manufacture and are known as fused, bonded, and mechanically mixed fluxes. Each type appears to have certain unique characteristics that affect the mechanical and chemical properties of the weld metal, the operating performance, and the handling of environmental variables.

Low Alloy Steel Covered Arc Welding Electrodes. Low alloy steel covered arc welding electrodes use the same AWS classification system as that for the mild steel covered arc-welding electrodes. The added suffixes A, B, C, D, G, or M identify the chemistry of the deposited metal, which in turn implies its properties and uses, such as:

E7010-A1 ... a carbon molybdenum steel electrode.
E9015-B3 ... a chromium molybdenum steel electrode.
E8016-C2 ... a nickel steel electrode.
E10016-D2 ... a manganese molybdenum steel electrode.
E7020-G ... other low alloy steel electrodes.
E12018-M ... an electrode conforming to military specifications.

Steels welded with low alloy steel electrodes are generally used for specific purposes. To successfully weld low alloy steels, one should become familiar with the weldability of each steel being welded. The electrodes should chemically match the base material.

Aluminum and Aluminum Alloy Arc Welding Electrodes. Only two aluminum and aluminum alloy arc-welding electrodes meet AWS specifications, namely, A1–2 and A1–43 electrodes. The A1–2 electrode is intended for use with pure aluminum, whereas the A1–43 electrode is intended for use with high strength aluminum alloys. The greatest difference between the two groups is silicon content; A1–43 electrodes contain from 4% to 6% silicon whereas the A1–2 must contain less than 1% silicon.

Aluminum and Aluminum Alloy Welding Rods and Bare Electrodes. This group of welding rods was developed for use with the oxyacetylene torch, carbon arc, atomic-hydrogen, and GTA/GMA welding processes. Bare electrodes are intended for use with GMA welding processes in an inert gas atmosphere. The AWS classification system is the same as that for similar filler materials. An electrode (E) classification that meets the A5.10-XX prescribed tests can be used either as an electrode or welding rod; however, the reverse is not necessarily true.

Control of surface contamination for aluminum filler metals and the base metal is critical. Oils, shop dirt, oxide films, and condensation are all contributors to inferior weldments. The welding engineer must be aware of this and take care to keep all aluminum filler metals clean and dry.

As mentioned at the beginning of this chapter, there are many more varieties of base metals to be welded than there are filler metals; aluminum is no exception. To have a filler metal with the same properties and chemistry as the base metal is not always possible; therefore, a substitution must be made. Only about thirteen aluminum filler rods and electrodes are available for the welding of approximately twenty-nine aluminum base alloys. The optimum matching of filler metal to base metal is shown in Table 8.3. The total of alloying elements used in aluminum filler metal is generally less than 6 %. The metals most often used as alloys with aluminum are silicon, copper, manganese, magnesium, chromium, and zinc. The exact amounts of each in a particular electrode can be found in AWS specification A5.10XX.

Magnesium Alloy Welding Rods and Bare Electrodes. Magnesium alloy welding rods are designed for use with GTA welding processes. Bare electrodes are for use with the gas metal arc (GMA) welding processes. This American Welding Society classification is based on the earlier American Society for Testing Material recommended practices, B275, a codification of light metals and alloys, cast and wrought. The magnesium alloy being welded determines the filler metal to be used; although a worthy generalization for many cases, it becomes very critical when welding magnesium alloys. Some filler materials will cause undesirable galvanic effects between the weld deposits and the base metal. Others do not give good color matches between the base metals and weld metals. The AWS magnesium filler metals and the correct base metal are shown in Table 8.4. The major chemical elements found in magnesium filler metals are: aluminum, beryllium, manganese, zinc, zirconium, rare earth, copper, iron, nickel, and silicon.

Copper and Copper Alloy Arc Welding Electrodes. Copper and copper alloy arc welding electrodes include solid, stranded bare and covered copper, and copper alloyed arc welding electrodes. These electrodes should be used with the manual shielded metal arc, GMA, and submerged arc welding processes.

The AWS classification system uses the "E" to indicate an electrode rather than a welding rod or brazing filler material. The chemical symbol is included throughout to identify the base alloys and principal alloying elements, such as ECuSn. If there is more than one classification in any one group, a letter A, B, etc., is used to identify the particular class. In some instances where further subdividing is required, a number, 1, 2, and etc., is used. For example, a copper-aluminum-iron alloy electrode would be identified as ECuAl–A2.

TABLE 8.3. Guide to the Choice of Filler Metal for General Purpose Welding of Aluminum (Courtesy of American Welding Society)

Base Metal	319, 333 354, 355 C355	13, 43, 344 356, A356 A357, 359	214, A214 B214, F214	7039 A612, C612 D612, 7005^k	6070	6061, 6063 6101, 6151 6201, 6951	5456	5454
1060, EC	ER4145^{c,i}	ER4043^{i,f}	ER4043^{e,i}	ER4043^i	ER4043^i	ER4043^i	ER5356^c	ER4043^{e,i}
1100, 3003 Alclad 3003	ER4145^{c,i}	ER4043^{i,f}	ER4043^{e,i}	ER4043^i	ER4043^i	ER4043^i	ER5356^c	ER4043^{e,i}
2014, 2024	ER4145^g	ER4145	…	…	ER4145	ER4145	…	…
2219	ER4145^{g,c,i}	ER4145^{c,i}	ER4043^i	ER4043^i	ER4043^{f,i}	ER4043^{f,i}	ER4043	ER4043^i
3004 Alclad 3004	ER4043^i	ER4043^i	ER5654^b	ER5356^e	ER4043^e	ER4043^b	ER5356^e	ER5654^b
5005, 5050	ER4043^i	ER4043^i	ER5654^b	ER5356^e	ER4043^e	ER4043^b	ER5356^e	ER5654^b
5052, 5652^a	ER4043^i	ER4043^{b,i}	ER5654^b	ER5356^{e,h}	ER5356^{b,c}	ER5356^{b,c}	ER5356^e	ER5654^b
5083	…	ER5356^{c,e,i}	ER5356^e	ER5183^{c,h}	ER5356^e	ER5356^e	ER5183^e	ER5356^e
5086	…	ER5356^{c,e,i}	ER5356^e	ER5356^{e,h}	ER5356^e	ER5356^e	ER5356^e	ER5356^e
5154, 5254^a	…	ER4043^{b,i}	ER5654^b	ER5356^{b,h}	ER5356^{b,c}	ER5356^{b,c}	ER5356^b	ER5654^b
5454	ER4043^i	ER4043^{b,i}	ER5654^b	ER5356^{b,h}	ER5356^{b,c}	ER5356^{b,c}	ER5356^b	ER5554^{c,e}
5456	…	ER5356^{c,e,i}	ER5356^e	ER5556^{e,h}	ER5356^e	ER5356^e	ER5556^e	
6061, 6063, 6101 6201, 6151, 6951	ER4145^{c,i}	ER4043^{b,i}	ER5356^{b,c,e}	ER5356^{b,c,h,i}	ER4043^{b,i}	ER4043^{b,i}		
6070	ER4145^{c,i}	ER4043^{e,i}	ER5356^{c,e}	ER5356^{c,e,h,i}	ER4043^{e,i}			
7039 A612, C612 D612, 7005^k	ER4043^i	ER4043^{b,h,i}	ER5356^{b,h}	ER5039^e				
214, A214 B214, F214	…	ER4043^{b,i}	ER5654^{b,d}					
13, 43, 344 356, A356 A357, 359	ER4145^{c,i}	ER4043^{d,i}						
319, 333 354, 355, C355	ER4145^{d,c,i}							

Base Metal	5154 5254[a]	5086	5083	5052 5652[a]	5005 5050	3004 Alc. 3004	2219	2014 2024	1100 3003 Ac. 3003	1060 EC
1060, EC	ER4043[e,i]	ER5356[c]	ER5356[c]	ER4043[i]	ER1100[c]	ER4043	ER4145	ER4145	ER1100[c]	ER1260[c,j]
1100, 3003 Alclad 3003	ER4043[e,i]	ER5356[c]	ER5356[c]	ER4043[e,i]	ER4043[e]	ER4043[e]	ER4145	ER4145	ER1100[c]	
2014, 2024	...	...	...	...	...	...	ER4145[g]	ER4145[g]		
2219	ER4043[i]	ER4043	ER4043	ER4043[i]	ER4043	ER4043	ER2319[c,f,i]			
3004 Alclad 3004	ER5654[b]	ER5356[e]	ER5356[e]	ER4043[e,i]	ER4043[e]	ER4043[e]				
5005, 5050	ER5654[b]	ER5356[e]	ER5356[e]	ER4043[e,i]	ER4043[d,e]					
5052, 5652[a]	ER5654[b]	ER5356[e]	ER5356[e]	ER5654[a,b,c]						
5083	ER5356[e]	ER5356[e]	ER5183[e]							
5086	ER5356[b]	ER5356[e]								
5154, 5254[a]	ER5654[a,b]									

NOTE 1 Service conditions such as immersion in fresh or salt water, exposure to specific chemicals, or a sustained high temperature [over 150°F (66°C)] may limit the choice of filler metals.

NOTE 2 Recommendations in this table apply to gas shielded-arc welding processes. For gas welding, only R1100, R1260, and R4043 filler metals are ordinarily used.

NOTE 3 Filler metals designated with ER prefix are listed in AWS specification A5.10.

[a] Base metal alloys 5652 and 5254 are used for hydrogen peroxide service. ER5654 filler metal is used for welding both alloys for low-temperature service [150°F (66°C) and below].

[b] ER5183, ER5356, ER5554, ER5556, and ER5654 may be used. In some cases they provide: (1) improved color match after anodizing treatment, (2) highest weld ductility, and (3) higher weld strength. ER5554 is suitable for elevated temperature service.

[c] ER4043 may be used for some applications.

[d] Filler metal with the same analysis as the base metal is sometimes used.

[e] ER5183, ER5356, or ER5556 may be used.

[f] ER4145 may be used for some applications.

[g] ER2319 may be used for some applications.

[h] ER5039 may be used for some applications.

[i] ER4047 may be used for some applications.

[j] ER1100 may be used for some applications.

[k] This refers to 7005 extrusions only.

NOTE 4 Where no filler metal is listed, the base metal combination is not recommended for welding.

TABLE 8.4. *Magnesium Alloy Welding Rods and Bare Electrodes* (Courtesy of American Welding Society)

Base Alloy	Base Alloy — Filler Alloy[a]										
	AM100A	AZ10A	AZ31B&C	AZ61A	AZ63A	AZ80A	AZ81A	AZ91C	AZ92A	EK41A	EZ33A
AM100A	AZ101A[a] / AZ92A										
AZ10A	AZ92A	AZ61A / AZ92A									
AZ31B&C	AZ92A	AZ61A / AZ92A	AZ61A / AZ29A								
AZ61A	AZ92A	AZ61A / AZ92A	AZ61A / AZ92A	AZ61A / AZ92A							
AZ63A	b	b	b	b	AZ101A[a] / AZ92A						
AZ80A	AZ92A	AZ61A / AZ92A	AZ61A / AZ92A	AZ61A / AZ92A	b	AZ61A / AZ92A					
AZ81A	AZ92A	AZ92A	AZ92A	AZ92A	b	AZ92A	AZ101A[a] / AZ92A				
AZ91C	AZ92A	AZ92A	AZ92A	AZ92A	b	AZ92A	AZ92A	AZ101A[a] / AZ92A			
AZ92A	AZ92A	AZ92A	AZ92A	AZ92A	b	AZ92A	AZ92A	AZ92A	AZ101A[a] / AZ92A		
EK41A	AZ92A	AZ92A	AZ92A	AZ92A	b	AZ92A	AZ92A	AZ92A	AZ92A	EZ33A[a]	
EZ33A	AZ92A	AZ92A	AZ92A	AZ92A	b	AZ92A	AZ92A	AZ92A	AZ92A	EZ33A	EZ33A
HK31A	AZ92A	AZ92A	AZ92A	AZ92A	b	AZ92A	AZ92A	AZ92A	AZ92A	EZ33A	EZ33A
HM21A	AZ92A	AZ92A	AZ92A	AZ92A	b	AZ92A	AZ92A	AZ92A	AZ92A	EZ33A	EZ33A
HM31A	AZ92A	AZ92A	AZ92A	AZ92A	b	AZ92A	AZ92A	AZ92A	AZ92A	EZ33A	EZ33A
HZ32A	AZ92A	AZ92A	AZ92A	AZ92A	b	AZ92A	AZ92A	AZ92A	AZ92A	EZ33A	EZ33A
K1A	AZ92A	AZ92A	AZ92A	AZ92A	b	AZ92A	AZ92A	AZ92A	AZ92A	EZ33A	EZ33A
LA141A	c	c	EZ33A	b	b	b	b	b	b	c	c
M1A	AZ92A	AZ61A / AZ92A	AZ61A / AZ92A	AZ61A / AZ92A	b	AZ61A / AZ92A	AZ92A	AZ92A	AZ92A	AZ92A	AZ92A
MG1	c	c	AZ92A	c	b	c	c	c	c	AZ92A	EZ33A
QE22A	c	c	c	c	b	c	c	c	c	EZ33A / AZ92A	EZ33A
ZE10A	AZ92A	AZ61A / AZ92A	AZ61A / AZ92A	AZ61A / AZ92A	b	AZ61A / AZ92A	AZ92A	AZ92A	AZ92A	EZ33A	AZ92A
ZE41A	c	c	c	c	b	c	c	c	c	EZ33A	EZ33A
ZK21A	AZ92A	AZ61A / AZ92A	AZ61A / AZ92A	AZ61A / AZ92A	b	AZ61A / AZ92A	AZ92A	AZ92A	AZ92A	AZ92A	AZ92A
ZH62A / ZK51A / ZK60A / ZK61A	b	b	b	b	b	b	b	b	b	b	b

Base Alloy

Filler Alloy

Base Alloy (row) \ Base Alloy (column)	HK31A	HM21A	HM31A	HZ32A	K1A	LA141A	M1A MG1	QE22A	ZE10A	ZE41A	ZK21A	ZH62A ZK51A ZK60A ZK61A
HK31A	EZ33A[a]											
HM21A	EZ33A	EZ33A										
HM31A	EZ33A	EZ33A	EZ33A									
HZ32A	EZ33A	EZ33A	EZ33A	EZ33A[a]								
K1A	EZ33A	EZ33A	EZ33A	EZ33A	EZ33A[a]							
LA141A	c	EZ33A	c	c	c	LA141A[a] EZ33A						
M1A MG1	AZ92A	AZ92A	AZ92A	AZ92A	AZ92A	c	AZ61A AZ92A					
QE22A	EZ33A	EZ33A	EZ33A	EZ33A	EZ33A	EZ33A	c	EZ33A[a]				
ZE10A	EZ33A AZ92A	EZ33A AZ92A	EZ33A AZ92A	EZ33A AZ92A	EZ33A AZ92A	EZ33A	AZ61A AZ92A	EZ33A AZ92A	AZ61A AZ92A			
ZE41A	EZ33A	EZ33A	EZ33A	EZ33A	EZ33A	c	c	EZ33A	c	EZ33A[a]		
ZK21A	AZ92A	AZ92A	AZ92A	AZ92A	AZ92A	c	AZ61A AZ92A	AZ61A AZ92A	AZ61A AZ92A	AZ92A	AZ61A AZ92A	
ZH62A ZH51A ZK60A ZK61A	b	b	b	b	b	b	b	b	b	b	b	EZ33A[a]

NOTE 1 When more than one filler metal is listed, they are listed in order of preference.

NOTE 2 In an emergency most alloys may be welded with strips cut from the base metal.

[a] Cast alloys are generally welded with filler metals having the same or similar composition as the base metal in order to achieve maximum strength and proper response to postweld heat treat schedules. Lacking the availability of suitable rods of such alloys, the commercially available filler metals listed will provide equivalent weldability but with the possibility of some reduction in strength. Alloy LA141A is a non-standard filler metal but is available upon special inquiry.

[b] Welding not recommended.

[c] No data available for welding this combination.

[a] Filler metals A261A, AZ92A, EZ33A and AZ101A are classified both as electrodes (E) and as welding rods (R).

TABLE 8.5. *Classification, Characteristics, and Uses of Copper Electrodes*

Classification	Characteristics and Uses
ECuSi	Used primarily for welding copper silicon alloys, copper zinc alloys, and copper. The shielding gas for GMA welding is argon or helium or a mixture of both. Covered electrodes are often used for welding silicon bronzes when the GMA process is not available or is uneconomical.
ECuSn	Used for joining phosphor bronzes of similar compositions, brasses, cast iron, and mild steels. Preheat and interpass temperatures of 400 °F (204 °C) are often necessary. Electrodes should be baked at 250 °F (121 °C) to 300 °F (149 °C) before use.
ECuSn–A	Used for joining materials of similar compositions.
ECuSn–C	The "C" indicates a higher tin content than the "A." This, in turn, provides a higher tensile and yield strength than the lower tin content ECuSn–A electrodes.
ECuNi	Used for the joining of copper nickel alloys. GMA bore filler wires are used for joining copper nickel alloys to nickel copper alloys or to steel.
ECuAl–A1	(Iron free) available only as bore wire. Used primarily for the fabrication of annealed aluminum bronze plate, sheet, and strip. Also used for the repair of castings having similar compositions and for corrosion resistant surfaces.
ECuAl–A2	(Iron-bearing) available both as a coated stick electrode and a bare wire for GMA welding. Used for joining aluminum bronzes of similar composition, high strength copper zinc alloys, silicon bronzes, manganese bronzes, some nickel alloys, some ferrous metals and alloys, and for a combination of dissimilar metals. This filler metal is also used for wear and corrosion resistant surfaces.
ECuAl–B	Used for joining aluminum bronze sheet and plate and for repairing castings. Also used for the surfacing of bearings, wear, and corrosion resistant surfaces. Preheat and interpass temperatures should be 200 to 300 °F (93 to 149 °C) for iron base metals, 300 to 400 °F (149 to 204 °C) for bronzes, and 500 to 600 °F (260 to 316 °C) for brasses. This filler material produces a deposit of higher tensile strength, yield strength, and hardness with a lower ductility than the ECuAl–A2 electrode.

Copper is of three types; oxygen free, deoxidized, and electrolytic tough pitch. The deoxidized coppers are readily weldable, producing joints of maximum strength. The ECu electrodes are available on coils, spools, or rims for GMA welding. These electrodes are deoxidized and strengthened with silicon. Other elements are frequently added to improve the quality of the weld. For joining copper in gauges thinner than 3/16 in. (0.476 cm), the GTA process is usually recommended. GMA processes are generally recommended for greater thicknesses. When welding copper materials 1/4 in. (6.35 mm) thick or heavier, a 400 to 1000 °F (204 to 538 °C) preheat is necessary to produce high quality welds. The electrode classification and characteristics for copper are shown in Table 8.5.

Copper and Copper Alloy Welding Rods. These copper and copper alloy welding rods are designed for use with oxyacetylene and GTA welding processes. The AWS classification is identical to that discussed previously in this chapter except for the letter "B," when used, as RBCuZn–A. The letter "B" immediately following "R" indicates that the rod can also be used for brazing applications. A back-up gas is often desirable. If the copper being joined is thick, a preheat may be desirable. The available copper rods and their chemistry, characteristics, and uses are listed in Table 8.6.

TABLE 8.6. *Classification, Characteristics, and Uses of Copper Rods*

Classification	Characteristics and Uses
RCu	RCu rods containing 0.15 % maximum phosphorous produce a fluid weld metal and are considered to be self-fluxing. Filler material of this type is often porous resulting in approximately half the strength to that of the base metal. The addition of silicon, tin, and manganese is small amounts can yield greater soundness. A basic acid-borax flux is often required. Used with deoxidized coppers.
RCuSi–A	RCuSi–A rods contain approximately 3 % silicon plus small amounts of manganese, tin, or zinc. The energy used when welding is either the oxyacetylene flame or the GTA process. Used to weld copper, copper silicon, and copper zinc base metals to themselves or to steels.
RCuSn–A	RCuSn–A rods contain approximately 5 % tin and up to 0.30 % phosphorus as a deoxidizer. The tin broadens the temperature range between the liquidus and solidus points, which slows up the solidification of the weld metal. Used only with the GTA process to weld copper and copper tin base metals. Fluxing is required.
RCuNi	RCuNi rods provide a high strength and a corrosion resistant joint. Welds with this kind of rod can be made with either a slightly reducing oxyacetylene flame or the GTA process. Fluxing is necessary when using the oxyacetylene flame. Used on nickels or copper nickel alloy base metals.
RCuZn	RCuZn rods are made from 60–40 copper zinc alloy, with small amounts of tin, iron, nickel manganese, silicon, and other elements added. This rod is used only with the oxyacetylene process.
RBCuZn–A	RCuZn–B is a low fuming bronze nickel welding rod. The addition of 0.15 % silicon helps to control the vaporization of the zinc. The addition of iron and manganese improves hardness and strength. Used primarily oxyacetylene welding of brass and braze welding of copper, bronze, and nickel alloys.
RCuZn–B	RCuZn–B is a low fuming bronze nickel welding rod. The addition of 0.15 % silicon helps to control the vaporization of the zinc. The addition of iron and manganese improves hardness and strength. Used primarily for braze welding steel and cast iron and for building up worn surfaces. It is also used to replace gear teeth, to weld brass, bronze, and nickel alloys.

(continued)

243

TABLE 8.6. *Classification, Characteristics, and Uses of Copper Rods (Continued)*

Classification	Characteristics and Uses
RCuZn–C	RCuZn–C is a low fuming bronze welding rod. This rod produces excellent mechanical properties and is considered to be a good all-purpose copper zinc welding rod.
RBCuZn–B	RBCuZn–B is also known as a nickel bronze welding rod and is different primarily in color from other copper zinc rods. It is used where the yellow color of brass is objectionable.
RCuZn	RCuZn is high in zinc and can be used only with a neutral or slightly oxidizing oxyacetylene flame. Fluxing is necessary. The flux is usually in powder form; however, it can be obtained as a coating on the rod or even as a liquid.
RCuAl–A2	The copper-aluminum-iron type rods are characterized by their relatively high tensile strength, yield strength, and hardness. These rods are used to join aluminum bronzes or similar composites, high strength copper zinc alloys, silicon bronzes, some copper nickel alloys, ferrous metals, and dissimilar metals. The rods are also used for repairing castings and for building up.
RCuAl–B	RCuAl–B welding rods are used to weld annealed aluminum bronze plate, sheet, and strip. Due to the formation of aluminum oxide, only the GTA process is used.

Nickel and Nickel Alloy Covered Welding Electrodes. The AWS classification system for nickel and nickel alloy covered welding electrodes is typical of the AWS classification system used for other filler materials and discussed earlier in this chapter. Cleanliness is critical when welding with nickel. Any foreign material containing sulfur, such as oil, threading compounds, marking pencils, and temperature indicating materials may cause embrittlement.

AWS nickel and nickel alloy covered welding electrodes are divided into five groups; namely, ENi, ENiCu, ENiCr, ENiCrFe, and ENiMo. The ENi electrode is almost pure nickel and is used primarily to join pure nickel base metals to themselves and to steel. The ENiCu electrode may contain up to 15 % copper and is used to weld nickel copper alloys. It may also be used for welding the clad side of nickel copper clad steels. The ENiCr electrode may contain up to 17.5 % chromium. This electrode is used to weld nickel base alloys where high chromium content must be retained in the weld metal, to weld nickel base alloys to steel, to surface steel with a nickel-base alloy, and to weld the clad side of nickel base alloy clad steel. This electrode is also used to surface carbon steels or low alloy steels. The ENiCrFe electrode contains up to 12 % iron. This electrode is used for welding nickel-chromium-iron alloy to itself and for welding dissimilar metals, such as carbon steel, stainless steels, pure nickel, and nickel base alloys to themselves or to each other. In some cases, this electrode is also used for surfacing steel with nickel-chromium-iron alloys. The ENiMo electrode contains up to 30 % molybdenum. It is used for welding nickel molybdenum alloy to itself, for welding nickel molybdenum alloy plate or sheet cladding to a steel backing, and for welding nickel molybdenum alloy to dissimilar alloys such as nickel base, cobalt base, and iron base.

Nickel and Nickel Alloy Bare Welding Rods and Electrodes. Nickel and nickel alloy bare welding rods and electrodes are for use with the oxyacetylene torch, atomic hydrogen, GMA, GTA, and submerged arc welding processes. There are four basic groups of nickel filler metals; they are Ni, NiCu, NiCr, and NiCrFe. Each of these groups contains several subgroups. Each electrode and rod has a different chemistry and is recommended for a specific application.

Corrosion Resistant Chromium and Chromium Nickel Steel Covered Welding Electrodes. Chromium and chromium nickel steel electrodes must yield weld metal in which the chromium exceeds 4 % and the nickel does not exceed 50 %. The number 15 or 16 indicates the intended current for the particular electrode. For example, the coating of the E308–15 contains calcium or alkaline earth minerals and is used with DCRP only; whereas, the E308–16 electrode, coated with titanium and potassium, is designed to be used with either ac or dc reverse polarity. Straight polarity is not recommended for this kind of electrode.

The AWS classification system for chromium and chromium nickel steel electrodes is similar to that of other filler metals where the "E" indicates electrode. The three digits designate the electrode chemistry. The letters occasionally following the three digits indicate a modification of the standard composition as shown in Table 8.7.

Corrosion-Resistant Chromium and Chromium Nickel Steel Welding Rods and Bare Electrodes. This group of bare rods is used with atomic hydrogen and GTA welding processes. The bare electrodes are for use with submerged arc and GMA welding processes. The rods and electrodes in this group are classified as being corrosion resistant and therefore must contain in excess of 4 % chromium and less than 50 % nickel.

The AWS classification system is identical to that discussed in previous sections for rods and electrodes. However, this group can be used as either a filler rod or bare electrode, so both the "E" and "R" are used preceding the three digit number (ER 308). The AWS corrosion resistant chromium and chromium nickel steel welding rods and bare electrodes are classified and described in Table 8.8.

Titanium and Titanium Alloy Bare Welding Rods and Electrodes. Titanium and titanium alloy bare welding rods are used with the GTA welding process, and the electrodes are used with the GMA welding process. The AWS classification system follows the same pattern as previously discussed. For example, an ERTi–8Al–2Cb–1Ta would average 8% aluminum, 2% columbium, and 1% tantalum. Titanium is a reactive metal; therefore, special attention is given to the nitrogen, hydrogen, oxygen, and carbon content in the electrode.

TABLE 8.7. *Classification and Selection of Corrosion Resisting Chromium and Chromium Nickel Steel Covered Welding Electrodes*

AWS Classification	Main Chemical Ingredient	Applications
E308	18–21.90 % chromium, 9–11 % nickel, and 0.08 % carbon.	Base metal of similar composition.
E308L	Same as for 308, except maximum of 0.04 % carbon.	Where resistance to intergranular corrosion is important.
E309	22–25 % chromium, 12–14 % nickel, and 0.15 % carbon.	Similar base metals in wrought or cast form. Dissimilar metals such as 18–8 to mild steels.
E309cb	Same as 309, except for the addition of columbium and a reduction in carbon limit to 0.127.	For 347 clad steels. For dissimilar metals such as columbium stabilized 18–8 stainless steel to mild steel.
E309MO	Same as 309, except for the addition of molybdenum and a reduction in carbon limit to 0.12%.	For 316 clad steels and for dissimilar metals such as molybdenum containing austenitic steel to carbon steel.
E310	25–28 % chromium, 20–22 % nickel, and 0.20 % carbon.	To weld hardable steels such as armor plate, clad steels, and for certain dissimilar high strength steels.
E310Cb	Same as E310, except for addition of max. 1% columbium and reduction of carbon to 0.12 %.	For welding type 347 clad steels or for welding dissimilar metals such as molybdenum bearing stainless to carbon steels.
E310MO	Same as for E310Cb except that 2–3 % molybdenum replaces the columbium.	For welding type 316 clad steels or for welding dissimilar metals such as molybdenum bearing stainless steels to carbon steels.
E312	28–32 % chromium, 8–10.5 % nickel, and 0.15 % carbon.	To weld cast alloys of similar composition or dissimilar metals where one of the pieces is high in nickel.
E16–8–2	14.5–16.5 % chromium, 7.5–9.5 % nickel, 1.0–2.0 % molybdenum, and 0.10% carbon.	For welding types 316, 317, and 347 stainless steels when used in high pressure, high temperature piping systems.
E316	17–20 % chromium, 11–14 % nickel, 2.0–2.5 % molybdenum, and 0.08 % carbon.	For welding similar alloys containing 2 to 3% molybdenum. For certain high temperature service applications.
E316L	Same as E316, except for the low carbon of 0.04 %.	For welding extra-low carbon, molybdenum bearing austentic alloys, whose usual specification for carbon is 0.03 maximum.

AWS Classification	Main Chemical Ingredient	Applications
E317	18–21 % chromium, 12–14 % nickel, 3–4 % molybdenum, and 0.08 % carbon.	Usually used only where severe corrosion is expected from sulfuric and sulfurous acids and their salts.
E318	17–20 % chromium, 11–14 % nickel, 2.0–2.5 % molybdenum, 0.08 % carbon, and 6 × C min. to 1% max. of columbium.	Recommended where intergranular corrosion due to carbide precipitation must be prevented.
E330	14–17 % chromium, 33–37 % nickel, and 0.25 % carbon.	For the repair of defects in alloy castings and the welding of castings and wrought alloys of similar composition, where heat and scale resisting properties above 1800 °F (895 °C) is important.
E347	18–21 % chromium, 9–11 % nickel, 0.08% min. to 1% max. of columbium, and 0.08 % carbon.	Where high temperature strengths are important. Are generally used for the welding of chromium nickel alloys of similar composition.
E349	18–21 % chromium, 8–10 % nickel, 0.35–0.65 % molybdenum, and 0.13 % carbon.	Usually used to weld such steels as 19–9W mo., 19–9DL, or 19–9DX— giving high temperature rupture strengths.
E410	11–13.5% chromium and 0.12 % carbon.	Used where preheat and postheat treatments are necessary and applicable. Also used for surfacing of carbon steels where corrosion/erosion or abrasion resistance is required, such as in valve seats.
E430	15–18% chromium and 0.10 % carbon.	Where preheating and postheating of base material is required.
E502	4–6% chromium and 0.10 % carbon	For welding of pipe and tubing with similar compositions. Must be preheated and postheated.
E505	8–10.5 % chromium	For welding tubes and pipe castings of similar base metals; preheat and postheat treatment necessary.
E7Cr	6–8 % chromium	For welding tubes, pipe, and castings of similar base materials; must be preheated and stress relieved.

TABLE 8.8. *Classification and Description of Corrosion Resistant Chromium and Chromium Nickel Steel Welding Rods and Electrodes*

Classification	Chromium	Nickel	Uses	Comments
ER308	19	9	To weld base metal of similar composition.	Carbon content is 0.20% unless suffix "S" is used, then carbon content drops to 0.08%.
ER308L	19	9	To weld base metal of similar composition.	Carbon content 0.03% max., not as strong at elevated temperature as the columbium stabilized alloys.
ER309	25	12	To weld similar alloys in wrought and cast. Also used where high corrosion resistance is required and to join dissimilar metals such as 18–8 to mild steels.	
ER310	25	20	To weld similar base metals and hardenable steels.	
ER312	29	9	To weld cast alloys of similar composition and to weld dissimilar metals where one is high in nickel.	
ER316	18	12	To weld similar alloys containing 2% molybdenum.	Contains about 2% of molybdenum.
ER316L	18	12	Same as 316.	0.03% max. of carbon, not as strong at elevated temperature as stabilized alloys.
ER317	Above that of ER316.	Above that of ER316.	Used where there are severe corrosion applications involving sulfuric and sulfurous acids and their salts.	Higher content of molybdenum than found in ER316.
ER318	Same as ER316	Same as ER316	To weld metals of similar composition.	Columbium has been added.

Designation				
ER321	19	9	To weld chromium nickel base metals of similar composition.	Titanium added.
ER330	15	35	To weld base metals that have heat and scale resistant properties at temperatures above 1800 °F (982 °C). Also to repair casting and wrought alloys of similar composition.	
ER347	19	9	To weld chromium–nickel base metals of similar composition, stabilized either with columbium or titanium.	Columbium added as a stabilizer.
ER348	19	9	Same as ER347.	Tantalum max. 0.10 %.
ER349	19	9	To weld base metals with high temperature rupture strength.	Columbium 1%, molybdenum 0.5%, and tungsten 1.4%. The filler metal is commonly referred to as 19–9W mo.
ER410	12	0	To weld alloys of similar composition. Overlays on carbon steels for corrosion/erosion or abrasive resistance.	An air hardening steel, therefore requires preheat and postheat treatment.
ER420	12	0	For increased resistance to abrasion.	A higher carbon content than for ER410.
ER430	15–17	0	For good corrosion resistance.	Optimum mechanical properties and corrosion resistance when heat treated after welding.
ER502	4–6	0	For welding material of similar composition, usually in the form of pipe and tubing.	0.50 % molybdenum, heat treatment recommended.

Reactive metals are those with extreme sensitivity to certain elements, such as nitrogen and oxygen, when heated to high temperatures. If the titanium electrodes exceed the permissible carbon, oxygen, or nitrogen content, then certain equivalents are permitted. The equivalents are detailed as follows.

1. Carbon ... If the carbon content exceeds the limit of 0.07 %, the rod or electrode will be acceptable if otherwise within the following limits: oxygen $\leq 0.10\%$, hydrogen ≤ 100 ppm, nitrogen $\leq 0.03\%$. Under no conditions shall the carbon exceed 0.10% unless agreed to by the purchaser.
2. Oxygen ... If the oxygen content exceeds the limit of 0.15 %, the rod or electrode will be acceptable if otherwise within the following limits: carbon $\leq 0.05\%$, hydrogen ≤ 100 ppm, nitrogen $\leq 0.03\%$. Under no conditions shall the oxygen exceed 0.20% unless agreed to by the purchaser.
3. Nitrogen . . . If the nitrogen content exceeds the limit of 0.05 %, the rod or electrode will be acceptable if otherwise within the following limits: carbon $\leq 0.05\%$, oxygen $\leq 0.10\%$, hydrogen ≤ 100 ppm. Under no conditions shall the nitrogen exceed 0.07 % unless agreed to by the purchaser.

All titanium filler metals are classified as both electrodes and rods. Their main alloying elements are aluminum, vanadium, tin, chromium, molybdenum, columbium, and tantalum. This is the only group of filler metals where the hydrogen, oxygen, and nitrogen percentages are stressed. These filler metals are used to weld base metals of the same chemistry.

WELDING RODS AND COVERED ELECTRODES FOR WELDING CAST IRON

Cast iron welding rods are used with the oxyacetylene torch and the carbon arc torch. The covered electrodes are for use with the shielded metal-arc welding process. This group of filler metals is used for welding gray cast iron, malleable cast iron, and ductile cast irons. The AWS classification system for cast iron filler material closely follows the standard pattern. One exception is the suffix used in electrode identifications such as ENiFeCI, where the CI is used to indicate that these electrodes are designed to be used specifically for welding cast iron. This eliminates confusion with other nickel base electrodes used exclusively with nickel base, base metals. In addition to nickel alloy electrodes for welding cast iron, there are also mild steel electrodes; these carry the letters "St" to indicate steel instead of nickel. The AWS welding rods and electrodes for welding cast iron are classified according to their chemistry under the following system:

RCI and ECI . . . the major alloying elements are carbon and silicon.
RCuXX and ECuXX . . . the major alloying elements are copper, nickel, tin, zinc, and aluminum.
ENiCI . . . the major alloying elements are carbon, silicon, manganese, iron, nickel, and copper.
EST . . . the major alloying elements are manganese and iron.

Excellent joint properties and color match can be achieved when the welding is done with the oxyacetylene process and the proper filler metal. A flux is always used with the oxyacetylene process. The purpose of the flux is to increase the fluidity of the iron silicate slag that forms on the molten puddle. Preheat and postheat treatments are important when fusion welding cast iron. When arc welding cast iron either ac or dc reverse polarity current is used. Welding electrodes are used for making repairs and for the fabrication of new parts.

SURFACING RODS AND ELECTRODES

Surfacing materials are available in both rods and covered electrodes. The AWS classification system for surfacing welding rods and electrodes is similar to that used for the other filler metals. Applying surfacing rods changes the properties at the surfaces of an existing unit; for instance, it hardens the teeth on gears or the metal ways on a lathe. Toughness, wear resistance, and impact are some other properties improved by surfacing. Some typical applications include cutting tools, forming dies, rock crushers, agriculture machinery, exhaust valves, and hundreds of other such items. Surfacing materials are highly alloyed with several metals such as copper, cobalt, molybdenum, tungsten, and others. Surfacing filler metals are grouped under five divisions as shown in Table 8.9. For specific chemistry and uses, consult the AWS specification.

TABLE 8.9. *Classification and Description for Surfacing Rods and Electrodes*

Classification	Characteristics and Uses
RFe5–A EFe5	High in carbon, retain a high number hardness property at elevated temperatures. Used on cutting tools, shearing dies, and ingot tongs.
RFeCr EFeCr	High in carbon, have a high hardness property. Used on plowshares, steel mill grinders, and sand blasting equipment.
RCoCr ECoCr	Have excellent heat, corrosion, and oxidation resistance properties. Used for valve trim in steam engines and pump shafts.
RCuZn ECuAl ECuSi RCuSi ECuAl RCuSn ECuSn	Have excellent surface bearing, and wear-resistant and corrosion-resistant surface properties. Used on gear teeth, cams, dies, and wear plates.
RNiCr	Have excellent metal-to-metal wear resistance, abrasion resistance, and retention of hardness at elevated temperature. Used on screw conveyors and cams, and cement pump screws.
EFeMn	Have excellent metal-to-metal wear and impact properties. Are work hardening alloys. Used on railroad frogs and rock crushers.

Composite Surfacing Welding Rods and Electrodes.   The composite surfacing welding rods and electrodes, including tungsten carbide surfacing rods and electrodes, are designed for use with the oxyacetylene, GTA, and atomic hydrogen welding processes. Other straight length bare and covered electrodes are designed for use with the shielded metal-arc welding process. The AWS classification system for composite surfacing rods and electrodes is similar to the general specifications except for those containing tungsten carbide. An example of this would be "RWC–2." The "WC" indicates that the filler metal consists of a mild steel tube filled with granules of fused tungsten carbide. The numeral gives the mesh size limits for the tungsten carbide granules. The AWS rods and electrodes available in this group are:

RFe5 and EFe5 . . . high speed filler metals.
EFeMn . . . austenitic manganese steel electrodes.
RFeCr–A and EFeCr–A . . . austenitic high-chromium iron filler metals.

Tungsten Carbide Rods and Electrodes.   Are selected for a specific quality such as abrasion resistance, impact resistance, or a combination of the two. Composite surfacing is used on the wearing surfaces of earth-moving equipment, brick-making equipment, steel-mill equipment, and agricultural equipment.

Surfacing with Powdered Filler Metal.   Several varieties of surfacing powders are available on the open market; generally they are either metallic or ceramic in nature. The metallic powders consist mainly of tungsten-carbide, nickel, aluminum, chromium, cobalt, molybdenum, copper, and iron. The ceramic powders are made from mixtures of aluminum oxide, zirconium oxide, and a number of other oxides in smaller quantities. These powders provide a large variety of desirable surface properties, such as corrosion resistance, hardness, abrasive resistance, impact resistance, and heat resistance, just to mention a few. Most surfacing powders are passed through a gas stream, such as oxygen and acetylene, and exit from the torch with the gas. The kind of bond achieved when surfacing with powdered metals is either mechanical or fused. A mechanical bond occurs when the powder is sprayed onto a cold base, whereas the fused bond results from depositing the powder onto a "sweating" surface of the base metal.

PLASTIC FILLER MATERIALS

There is no AWS classification system for plastic filler materials. Only thermoplastic plastics are weldable. Although several welding processes are used to join plastics, only the "hot-air" processes permit the use of filler materials. The composition of plastic filler material is usually the same as that of the material being joined.

For special cases, plasticizers and other ingredients are added to improve the weldability and physical properties of the weld joint. Plastic filler materials are usually flat, round, or triangular in shape. The triangular shapes usually fit the prepared joint better and give better joint properties and, therefore, are used more often than the others. Fluxes are never required when welding thermoplastics; however,

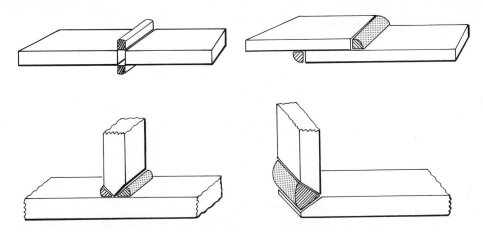

FIGURE 8–2. *Plastic filler materials are available in several different shapes and sizes specifically designed to match the prepared joint.*

in some cases, an inert gas atmosphere is desirable. Edge preparation for weld joints in plastics is similar to that for metals (see Fig. 8–2). When welding thermoplastics, neither the base material nor the filler material becomes molten; they only soften.

BRAZING FILLER METAL

Brazing filler metal is an alloy of metals used in the brazing process discussed in Chapter 5. Brazing filler metals must melt at a temperature lower than the melting temperature of the base metal being joined and must be able to flow between closely fitted joints by capillary attraction.

The AWS classification for brazing filler materials is comprised of seven groups as follows: aluminum silicon, copper phosphorus, silver, precious metals, copper and copper zinc, magnesium, and nickel. The classification system for brazing materials is identical to those discussed earlier for AWS-specifications. In welding, a knowledge of melting temperatures of metals is important; however, for brazing the terms *solidus* and *liquidus* are more often used. Solidus refers to the highest temperature at which a metal is totally solid, just prior to melting. Liquidus is the lowest temperature at which the metal is completely liquid. Between the solidus and liquidus states, the braze filler metal is in plastic form.

An acceptable method of marking brazing filler metals has not been developed. Filler metal cleanliness is critical for brazing materials, and most of the tried methods of marking such as inking, stamping, dyeing, notching, and painting are sources of contamination. Codes permit braze welding filler metals of 1/8 in. and larger to be stamped; however, the smaller sizes and all braze filler metals must be tagged. The seven basic groups of brazing filler materials, their chemistry, characteristics, and uses are found in Table 8.10.

TABLE 8.10. *Characteristics and Uses of Brazing Filler Materials*

Classification	Characteristics and Uses
BAlSi	Aluminum silicon brazing filler metals are used to join the listed grades of aluminum and aluminum alloys: 1060, EC, 1100, 3003, 3004, 5005, 5050, 6053, 6061, 6062, 6063, 6951, and cast alloys A612 and C612. The suffix numbers, such as BAlSi–3, indicate the specific uses, specified process, limitation, and precautions for each filler. Joint clearances of 0.006 to 0.010 in. are desirable. Fluxing and cleaning are essential to quality brazing.
BCuP	Copper phosphorus filler metals are used to join copper and copper alloys. These filler materials are self fluxing when used with copper base materials.
BAg	Silver brazing filler metals are used for joining most ferrous and nonferrous metals except aluminum and magnesium. A joint clearance for silver brazing of 0.002 to 0.005 in. is necessary for good capillary action.
BAu	Precious metals brazing filler metals are used for the joining of iron, nickel, and cobalt base metals where resistance to oxidation or corrosion is required. These filler metals have a low rate of interaction, therefore, are limited to thin gauge metals. The processes used are induction, furnace, or resistance heating in a reducing atmosphere or in a vacuum with no flux.
BCu and RBCuZn	Copper brazing filler metals are used for joining ferrous and nonferrous metals. Copper zinc brazing filler metals are used for the same materials and under the same circumstances as the BCu filler metals. The corrosion resistance of these filler metals is generally inadequate for joining copper, silicon, bronze, copper nickel, or stainless steels.
BMg	Magnesium brazing filler metals are used for joining A210A, K1A, M1A, AZ31B, and ZE10A magnesium base metals and compositions. Filler metals are used to help prevent ignition of the base metal. Joint clearance must be maintained at 0.004 to 0.010 in. Corrosion resistance is good if the flux is completely removed.
BNi	Nickel brazing filler metals have good corrosion and heat resistant properties. The AIST 300 and SISI400 series of stainless steels and nickel and cobalt base alloys are the most commonly used base metals with BNi brazing metals. This brazing filler metal retains good properties over a range of temperatures from the cryogenics [−400 °F (−240 °C)] to 1800 °F (982 °C). Best results are attained when the brazing operation takes place in a vacuum, a pure dry hydrogen atmosphere, or a pure dry argon atmosphere. If a vacuum brazing device is not available, then BNi filler metal can be used with a torch, furnace, or induction furnace provided a suitable flux is used. This brazing filler metal is used for joining turbine blades, jet engine parts, and honeycomb structures, and for nuclear applications.

TABLE 8.11. *Relative Solderability of Metals, Alloys, and Coatings* (Courtesy of American Welding Society)

Base Metal, Alloy, or Applied Finish	Flux Requirements			Soldering Not Recommended
	Non-corrosive	Corrosive	Special Flux and/or Solder	
Aluminum			X	
Aluminum bronze			X	
Beryllium				X
Beryllium copper		X		
Brass	X	X		
Cadmium	X	X		
Cast iron			X	
Chromium				X
Copper	X	X		
Copper chromium		X		
Copper nickel		X		
Copper silicon		X		
Gold	X			
Inconel			X	

TABLE 8.11. (*Continued*)

Base Metal, Alloy, or Applied Finish	Flux Requirements			
	Non-corrosive	Corrosive	Special Flux and/or Solder	Soldering Not Recommended
Lead	X	X		
Magnesium			X	
Manganese bronze (high tensile)				X
Monel		X		
Nickel		X		
Nichrome			X	
Palladium	X			
Platinum	X			
Rhodium		X		
Silver	X	X		
Stainless steel			X	
Steel		X		
Tin	X	X		
Tin bronze	X	X		
Tin lead	X	X		
Tin nickel	X	X		
Tin zinc	X	X		
Titanium				X
Zinc		X		
Zinc die castings			X	

SOLDER

Solder is classified into several groups of metal alloys designed to surface, seal, or join materials. It is used with a wide variety of base metals, see Table 8.11. Soldering is often a heterogeneous process when the solder has an entirely different chemical composition than the base metal with which it is being used. Solder is almost always used with a flux. The melting temperature of a solder is always lower than the melting temperature of the metal with which it is used.

Tin Lead Solders. Tin lead solders make up the largest group and are used for joining most metals. A commonly used tin lead classification is 30/70. The numerals to the left of the diagonal line represent the percentage of tin in the solder. The numerals to the right give the lead content. The eutectic or mixture of tin lead with the lowest liquidus temperature is 63/37. The solidus–liquids temperature of this alloy is 361 °F (182 °C). The 40/60 and 50/50 are the most commonly used, general purpose solders. The 30/70 is referred to as a body solder and is used in the manufacture of new automobile bodies, and for filling joints and surface irregularities. The 60/40 solder is considered a high quality solder and is often used for electrical connections. The ASTM classification for tin lead solders is given in Table 8.12.

Tin-Antimony-Lead Solders. The use of tin-antimony-lead solders is limited to metals containing no zinc. Antimony is occasionally added to tin lead solders to conserve tin. The mechanical properties of tin lead solder can be improved by the addition of up to 6 % antimony, however, not without impairment of soldering characteristics.

Tin Antimony Solders. Tin antimony solders retain better strength than tin lead solders at temperatures above 300 °F (150 °C). This alloy also has higher electrical conductivity than the tin lead solders. Tin antimony solders are often used for food handling equipment where lead is not permitted.

Tin Silver Solders. Tin silver solders are reserved for use in the joining and sealing of precision instruments. As the silver content goes up, so does the cost.

TABLE 8.12. *Tin Lead Solders** (Courtesy of American Welding Society)

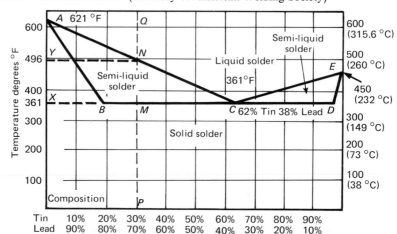

ASTM Solder Classification	Composition (weight %)		Temperature °F (°C)		
	Tin	Lead	Solidus	Liquidus	Pasty Range
5A	5	95	572 (300)	596 (313)	24 (13)
10A	10	90	514 (268)	573 (301)	59 (33)
15A	15	85	437 (225)	553 (289)	116 (64)
20A	20	80	361 (183)	535 (279)	174 (96)
25A	25	75	361 (183)	511 (266)	150 (83)
30A	30	70	361 (183)	491 (255)	130 (72)
35A	35	65	361 (183)	477 (247)	116 (64)
40A	40	60	361 (183)	455 (235)	94 (52)
45A	45	55	361 (183)	441 (227)	80 (44)
50A	50	50	361 (183)	421 (216)	60 (33)
60A	60	40	361 (183)	374 (190)	13 (7)
70A	70	30	361 (183)	378 (192)	17 (9)

* NOTE: The constitutional diagram of the melting characteristics of tin-lead solders shows the melting characteristics of tin-lead solders as determined by composition.

Tin Zinc Solders. Solders containing zinc are used with aluminum. The solidus temperature is constant at 390 °F (200 °C), but the liquidus temperature rises sharply when the zinc content goes above 20 %. The higher the zinc content, the better the corrosion resistance. But, the higher zinc content makes the solder more difficult to use.

Lead Silver Solders. Lead silver solders have the property of high strength at moderately high temperatures. Lead silver solders are susceptible to atmospheric corrosion due to humidity when in storage and may become unusable. However, by adding 1 % tin to the alloy, the corrosive action can be reduced, with an additional improvement in wetting and flow properties.

Cadmium Silver Solders. Cadmium silver alloys provide high temperature, high strength solders. A butt joint of copper to copper using cadmium silver solders can produce a tensile strength of 25,000 psi (172.4 MPa).

TABLE 8.13. *Common Forms of Solder Bars* (Courtesy of American Welding Society)

Type	Characteristics
Pig	Available in 50 and 100 lb (22.7 and 45.4 kg) pigs.
Slabs	Weight 15 to 35 lbs (6.8 to 15.9 kg). Common sizes are 36 × 3 × 1 in. (91.4 × 7.6 × 2.5 cm), 24 × 3 × $\frac{3}{4}$ in. (61.0 × 7.6 × 1.97 cm), and 18 × 3 × $\frac{3}{4}$ in. (45.7 × 7.6 × 1.9 cm).
Cakes or ingots	Rectangular or circular in shape, weighing 3, 5, and 10 lbs (1.4, 2.3, and 4.5 kg).
Bars	Available in numerous cross sections, weights, and lengths.
Paste	Available as a mixture of powdered solder and suitable flux in paste form in quantities of 1 lb. or more.
Ribbon or tape	Thicknesses of 1/16 to 3/16 in. (0.159 to 0.476 cm), 1/8 to 1 in. (0.318 to 2.54 cm) in width, on reels or spools and in 12 in. (30.5 cm) lengths.
Segment or drop	Triangular bar or wire cut into any desired number of pieces or lengths up to 500 pieces per pound.
Pulverized or powdered	Screen sizes from 50 to 200 mesh.
Foil	Thicknesses of 0.00125 in. (0.03175 mm) or more, on rolls 5 to 18 in. (12.7 to 45.72 cm) wide, weighing 20 to 30 lbs (9.1 to 13.6 kg) each.
Sheet	Thicknesses from 0.010 to 0.100 in. (0.0254 to 0.254 cm) in sizes not larger than 24 × 36 in. (61.0 × 91.4 cm).
Wire, solid	Diameters of 0.010 to 0.30 in. (0.0254 to 0.762 cm) on spools weighing 1, 5, 10, 20, 25, 50, or 65 lbs (0.045, 2.3, 4.5, 9.1, 11.3, 22.7, or 29.5 kg).
Wire, flux-cored	Either acid or rosin cored, 1/64 (0.0397), 1/16 (0.159), 5/64 (0.198), 5/32 (0.397), and 1/8 in. (0.318 cm) in diameter, on spools weighing 1, 5, 10, 15, 20, 25, and 50 lbs (0.45, 2.3, 4.5, 6.8, 9.1, 11.3 and 22.7 kg). The flux may be incorporated with the solder in single or multiple hollows or in external parallel grooves.
Preforms	Unlimited range of sizes and shapes to meet special requirements.

Zinc Aluminum Solders. Aluminum alloys give good strength and high corrosion resistance at the joint. Their high solidus and liquidus temperatures may give some heat and flux problems.

Fusible Alloy Solders. The fusible alloy solders all have a bismuth base. They have a melting temperature lower than the tin lead group and are used for step soldering, for soldering near heat sensitive material, and for soldering on heat treated surfaces.

Indium Solders. Indium, when added to lead silver solders, improves the wetting properties. Tin lead solders, when alloyed with 25 % or more indium, have excellent corrosion resistance and melt at low temperatures. A 50 % indium and 50 % tin alloy has the property of wetting to glass. This characteristic makes possible the soldering of glass to glass and glass to metal. The low vapor pressure of this alloy makes it useful for sealing in vacuum systems.

Solder comes in many different forms and sizes. Several of the more common sizes and shapes are shown in Table 8.13.

Cadmium Zinc Solders. The presence of zinc indicates that this solder is used with aluminum. It gives average strength and corrosion resistant joints. The 40 % cadmium, 60 % zinc solder is used for spot soldering aluminum.

GASES USED IN MATERIAL JOINING

Initially, the only gases used in material joining were oxygen, acetylene, and hydrogen. The gas or combination of gases produced the energy necessary for melting the material. Today, a number of gases serve one of three purposes: they serve as a shield to protect the materials being joined from atmospheric contamination; they create energy by chemical reaction for melting the metal; and they resist the flow of electrons between the cathode and anode to produce heat. A complete list of gases used in welding and their properties are given in Table 8.14.

Shielding Gases. Shielding gases are used primarily to protect the heated material from atmospheric contamination. They may or may not be inert. Most frequently used shielding gases include helium, argon, nitrogen, and carbon dioxide. Helium, argon, and xenon are inert or chemically inactive; their outer shells of electrons are complete. Helium, for example, is complete with only two electrons in its outer shell (see Fig. 8–3). Argon and xenon each have eight electrons in their stable outer shells. Since these gases have no tendency to give up or add electrons, they are considered stable and desirable shielding gases.

All metals have a tendency to combine with oxygen and nitrogen, especially at elevated temperatures, where they form oxides and nitrides or binary compounds that produce inferior welds. Some metals when molten absorb hydrogen, which upon solidification causes porosity and embrittlement. Therefore, it is important that metals being welded are shielded from the atmosphere.

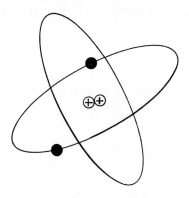

FIGURE 8–3. *Helium is an inert gas because it is balanced with two electrons in its outer shell.*

Shielding gases are either used singly or in combination with each other or with other gases for shielding and wetting purposes, Table 8.15. Oxygen and hydrogen are often used with a shielding gas to give some specific desirable characteristic; however, their primary purpose is not shielding. Gas mixtures and their volumes influence the penetration patterns in welded joints, Fig. 8–1b.

Helium (He).   Natural gas is one percent helium. To separate the helium, a combination of liquification and filtering techniques is used. Liquification is lowering the temperature of a gas under pressure until it changes to a liquid. This is done by pumping the gas to be liquified through a series of compressors. At a temperature of $-310\,°F$ ($-185\,°C$), every gas except helium and nitrogen liquifies. The remaining gas is then filtered through an activated charcoal that has been chilled with liquid nitrogen. The end product is 99.995% pure helium.

Helium weighs less than any other element except hydrogen. It is odorless, colorless, tasteless, and nonpoisonous. It does not burn or explode, and it does not chemically combine with any other element or compound. Helium has a high ionization potential, 24 electron volts (eV), which results in a hotter arc than argon gas. The higher the ionization potential, the more difficult it is to initiate the arc. Like other gases helium can be stored as a liquid, which saves space and reduces shipping costs.

The properties of helium make it a desirable gas for several material joining techniques. Because the weight of helium is one-eighth that of air, it is generally used in confined areas. Air movements of any kind will remove the helium from the area to be shielded from the atmosphere. Helium is most often used as a shield when welding nonferrous metals; however, He or a mixture with other gases can be desirable for welding some ferrous metals and their alloys. The joining processes most adaptable to the use of helium gases are GTA and GMA, Table 8.16.

Argon (Ar).   Argon is extracted from the atmosphere by the liquification technique; it boils off as a byproduct when making oxygen at a temperature of $-303\,°F$ ($191\,°C$). Argon is stored in cylinders as a pressurized gas at ambient temperatures or as a liquid at temperatures below $-300\,°F$ ($185\,°C$). It is refined to a 99.95% purity with an average of 50 ppm of impurities.

259

TABLE 8-14. *The Identity and Applications of Gases Used In Welding and Related Processes.**

Gas	Sym	Heat Value Btu/ft³ (MJ/m³)	Flame propagation fps (m/s)	Function	Flame/Arc Temperature in O	Applications
Acetylene	C_2H_2	1480 (55.10)	17.7 (5.39)	F*	5,600 °F 3,206 °C	Used with O_2/air to generate heat for welding, brazing, and soldering.
Air	NA	NA	NA	SC*-I*	NA	Supports combustion with certain soft flames and as an ionized gas with plasma arcs.
Argon	A	NA	NA	S*-I	Est. 25,000 °F up to 13,671 °C	Used as an inert shield with several processes and as an ionized gas in the GTA, GMA, and plasma processes.
Butane	NA	3132 (11.66)	Fast	F	5,252 °F 2,900 °C	Used as a fuel gas for soft flames and in the cutting process of carbon steels.
Carbon dioxide	CO_2	NA	NA	S-I	10,000 °F Est. 5,538 °C	Used as a shielding gas with carbon steels and also ionized with the arc.
Flamex	NA	2510 (93.46)	14.5 (4.42)	F	6,000 °F 3,315 °C	Used as a fuel gas for soft flames and with the cutting process for carbon steels.
Helium	He	NA	NA	S-I	Est. 40,000 °F up to 22,204 °C	Used as an inert shield with several processes and as an ionized gas with the GTA, GMA, and plasma processes.

Hydrogen	H_2	274 (10.20)	Slow	4,622 °F 2,550 °C	F	Used as a fuel gas for underwater cutting and welding, and as a chemical source of energy in the atomic hydrogen welding process.
Mapp	NA	2450 (91.22)	7.9 (2.41)	5,301 °F 2,927 °C	F	Used as a fuel gas for soft flames and in the cutting operation of carbon steels.
Natural gas	NA	1046 (38.95)	8.2 (2.50)	4,600 °F 2,538 °C	F	Used as a fuel gas for soft flames and in the cutting operation of carbon steels.
Nitrogen	N_2	NA	NA	NA	S-I	Reserved for special applications, such as shielding back-up gas, can also be ionized in an arc. CAUTION: Process creates nitrous oxide.
Oxygen	O	NA	NA	NA	S & SC	Supports combustion and is used in small quantities in shielding gases to promote wetting.
Propane	C_2H_3	2509 (93.42)	5.9 (1.80)	4,579 °C 2,526 °C	F	Used as a fuel gas for soft flames and in cutting operations of carbon steels.
Xenon	Xe	NA	NA	NA	E	Used with the pulsed laser welder to excite the ruby atoms to a higher energy level.

* The following symbols and meanings are used in the table:

F—Fuel gas.
SC—Supporter of combustion.
S—Shielding gas.
I—Ionizer.
E—Exciter.
NA—Not applicable.

TABLE 8.15. *Shielding Gases Matched to Metal and Process* (Courtesy of *Welding Engineer*)

Metal	GMA	GTA
Mild steel	Carbon dioxide: high quality, low-current out-of-position welding.	Argon preferred. Helium gives greater penetration, but more difficult to handle.
Low alloy steel	Argon plus 2% oxygen: eliminates undercutting tendency, removes oxidation. Also acceptable, 80% argon, 20% carbon dioxide.	Argon for manual: easier to handle. Helium for automatic: provides higher speeds.
Stainless steel	Argon plus 5% oxygen: improves arc stability when using dc straight polarity.	Argon for thin gage: controlled penetration. Helium for heavier gage: greater penetration.
Nickel, Monel, Inconel	Argon: good wetting, decreases weld metal fluidity.	Argon for manual: easier to handle. Helium for automatic: provides greater speeds.
Aluminum alloys	Argon: with dc reverse polarity, removes surface oxides. For Mg-Al, use 75% helium, 25% argon: high heat input reduces porosity tendency, cleans surface oxide.	Argon: preferred for ac, has arc stability, good cleaning action. For dc straight polarity use helium: stable arc, higher welding speeds.
Magnesium	Argon: with dc reverse polarity removes surface oxides.	Argon: good cleaning action, use ac, dc rarely used.
Titanium	Argon: reduces heat-affected zone, improves metal tranfer.	Argon for manual: easier to handle. Helium for automatic: provides higher speeds.
Deoxidized copper	75% helium, 25% argon preferred: good wetting, higher heat input to counter high thermal conductivity. Argon for light gages.	Helium preferred: high heat input to counter high thermal conductivity. 75% helium, 25% argon: stable arc, somewhat reduced heat input.
Aluminum bronze	Argon: reduced penetration, generally used for surfacing.	Argon: reduced penetration, generally used for surfacing.
Silicon bronze	Argon: reduces crack sensitivity on hot short material.	Argon: reduces hot short tendencies of material.

TABLE 8.16. *Shielding Gases Matched to Applications* (Courtesy of *Iron Age*)

Shielding Gas	Process*	Common Applications
Pure Argon	GMA–DCRP	Al, Ti, Mo, Zr, Ni-base alloys
	GTA–AC	Al, Mg
	GTA–DCSP	Stainless steel
Pure Helium	GMA–DCRP	Al
	GTA–DCSP	Cu, mild and stainless steel
Pure CO_2	GMA–DCRP	Mild steel (solid wire with buried arc and dip transfer, flux-cored wire)
Pure N_2	GMA–DCRP	Cu
A–0.05 to 0.15% O_2	GMA–DCRP	Al
A–0.5 to 2% O_2	GMA–DCRP	Mild and stainless steel, low alloy steel
	GMA–DCSP	Mild and stainless steel, Ni-base alloy (overlay)
A–5% O_2	GMA–DCSP	Mild steel
A–15 to 25% CO_2	GMA–DCRP	Mild steel, low alloy (spray and dip transfer)
	GMA–DCSP	Mild steel
25% A–75% He	GTA–AC	Al, stainless steel
	GTA–DCSP	Cu
	GMA–DCRP	Cu
20% A–70% He–10% CO_2	GMA–DCRP	Mild and stainless steel (dip transfer)
80% A–15% CO_2–5% O_2	GMA–DCRP	Mild and stainless steel (spray and dip transfer)

* DCSP—direct current straight polarity.
DCRP—direct current reversed polarity.
AC—alternating current.
GMA—gas metal arc.
GTA—gas tungsten arc.
(Reprinted with Permission from *Iron Age,* Aug. 6, 1964. Copyrighted 1964 by Chilton Co.)

Argon is $1\frac{1}{3}$ times heavier than air; therefore, it is not affected as much by drafts and other air movements and provides a more stable shield than helium. It has a low ionization potential (15.8 eV) and a relatively long ion recombining time, which makes it a desirable gas for GTA welding. The low ionization potential provides easier arc starting and makes it possible to have limited control over penetration patterns.

Argon can be used alone for the welding of nonferrous metals; however, when a specific penetration geometry is desired, a gas mixture may be required (Fig. 8–4). When welding low carbon steels with the GMA process, a 75% Ar/25% CO_2 gas mixture is often used to reduce spatter. Argon and argon mixtures are used with both the GTA and GMA welding processes.

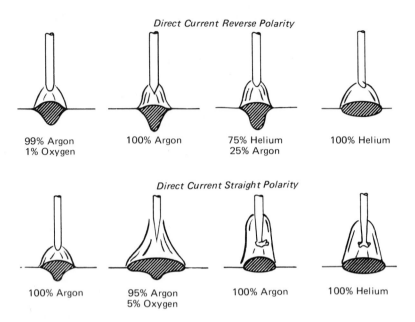

Direct Current Reverse Polarity

99% Argon
1% Oxygen

100% Argon

75% Helium
25% Argon

100% Helium

Direct Current Straight Polarity

100% Argon

95% Argon
5% Oxygen

100% Argon

100% Helium

FIGURE 8–4. *Gas mixtures and their ratios influence the penetration patterns in welded joints. (Courtesy of American Technical Society.)*

Carbon Dioxide (CO_2). Carbon dioxide is produced by burning fuels containing carbon. These fuels are coke, oil, or natural gas. For welding purposes, CO_2 is stored in cylinders at approximately 850 psi pressure and at a temperature of 70 °F (21 °C). Due to its low boiling point, a full cylinder is two-thirds liquid with the top one-third a gas at the above pressure. As the gas is drawn off from the top third of the cylinder, the CO_2 liquid automatically changes to a gaseous state; however, if the rate is faster than 30 cfh (9.144 m/h = 0.00254 m/sec) per cylinder, icing (dry ice) will form in the orifices of the regulator. If this occurs, heaters must be installed or extra tanks of CO_2 added by a manifold system. A heater is simply an electrical element placed around the gas line to keep the CO_2 at the proper temperature as it is being drawn from the cylinder. A manifold system makes it possible to maintain an adequate supply of CO_2 by dividing the total gas requirements between two or more cylinders.

Gaseous CO_2 is 53% heavier than air, which makes it an ideal shielding gas for drafty shops and outdoor welding. At room temperature, CO_2 is stable and relatively inert. However, at elevated temperatures such as those in the welding arc process, it becomes active and will disassociate into CO and O. To prevent the free O from combining with the base metal, a deoxidizing filler wire is used. A deoxidizing wire contains elements with a high affinity for oxygen; they combine readily with it. The deoxidizers used in welding wires are aluminum, titanium, manganese, and silicon. The CO_2 provides an extremely hot arc that results in deep penetration and high torch travel speeds.

The use of CO_2 gas for shielding is limited to carbon and stainless steels where it may be used either pure or as a mixture. Mixtures give desirable characteristics for special applications (Table 8.16). Typical CO_2 mixtures are: 15–25% CO_2 with the remainder argon; 20% Ar, 70% He, and 10% CO_2, or 80% Ar, 15% CO_2, and 5% O. CO_2 and CO_2 gas mixtures are used only with the GMA welding process. CO_2 is also used with continuous gas laser welders providing improved efficiency over the pulsed type.

Nitrogen (N_2). Nitrogen comprises 80% of the atmosphere; it, too, is separated by the liquification process. Nitrogen boils off at $-320\,°F$ ($195\,°C$). Storage is in cylinders, as a gas under high pressure, or as a liquid at reduced temperatures.

If N_2 is permitted to combine with iron it forms nitrides that cause brittleness and decreased ductility in weld metal. It enters the weld metal through destructive distillation or disassociation of the air caused by the arc. As it passes through the arc, both nitrous (N_2O) and nitric (NO) oxides form. N_2O is an anesthetic gas (laughing gas), whereas NO is lightly toxic; both have a sweetish odor.

The use of N_2 for material joining applications is very limited. It is the recommended shielding gas when welding deoxidized copper and is an additive to argon for welding some of the austenitic stainless steels. N_2 is also used as an inexpensive back-up gas when welding stainless steel and certain alloy steels. A back-up gas is a controlled shield on the back side of the joint area, used to protect the metal from the atmosphere and prevent the formation of oxides. A 95% N_2 and 5% H_2 mixture produces a bright shiny root pass when used with stainless steel. N_2 has been used as the ionized gas in plasma cutting of stainless steels, but the formation of N_2O demands extremely effective ventilation systems for health reasons.

Fuel Gases. The fuel gases most often used for joining materials are acetylene, propane, hydrogen, natural gas, butane, Mapp, and Flamex. A fuel gas may be defined as a gas used to provide heat by combustion. With the exception of H_2, fuel gases are classified as hydrocarbons and have those characteristics shown in Table 8.14. For metal joining, a fuel gas when burned with oxygen should have the following characteristics:

1. A high temperature flame.
2. A high rate of flame propagation.
3. Adequate heat content and a minimum chemical reaction of the flame with the base and filler metal.

Acetylene (C_2H_2). C_2H_2 is produced by a chemical reaction between calcium carbide (CaC_2) and water (H_2O), as shown previously (Fig. 8–5). It is often generated as needed in the fabricating plant, or it may be purchased in cylinders. The cylinder for acetylene storage is more complex than other gas cylinders. This is necessary because acetylene, being an unstable gas, cannot be safely compressed

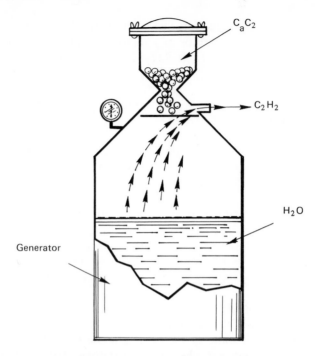

FIGURE 8–5. C_2H_2 *is produced by a chemical reaction between calcium carbide* (CaC$_2$) *and water* (H$_2$O). (*Courtesy of Welding Engineer.*)

above 15 psi (0.103 MPa) in the pure state. By filling the cylinder with a porous material saturated with acetone, acetylene can be safely compressed and stored at 275 psi (1.896 MPa).

C_2H_2 when mixed with $2\frac{1}{2}$ parts of oxygen produces the highest flame temperature (6300 °F/3482 °C) of any fuel gas (see Table 3.2). It is also a fast burning gas, 330 ft/sec (100.6 m/sec), second only to hydrogen, which burns at 800 ft/sec (243.8 m/sec). C_2H_2 is generally regarded as a source of energy for fusion welding. However, due to its low efficiency rate, Tables 8.17 and 8.18, it is seldom used for fusion welding, except for metals with low melting temperatures.

TABLE 8.17.   *A Comparison of the Melt-Off Efficiency Rates for the Commonly Used Welding Processes*

Energy Source	Melt-Off Efficiency for Aluminum	Melt-Off Efficiency for Steel	Intensity of Heat Source in Kw/In.^{2}
Oxyacetylene	<2%	10%	2
GTA	10%	40%	100
GMA	20%	60%	200
Submerged arc	—	80%	400
Plasma	60%	100%	1000
Electron beam	100%	100%	>10,000
Laser	100%	100%	>10,000

TABLE 8.18. *Fuel Gas Data Sheet*

Characteristics		Acetylene	Propane	Methane	Mapp	Propylene
Maximum flame temperature						
	°F	5720	5130	5040	5340	5240
	(°C)	(3160)	(2832)	(2782)	(2949)	(2893)
Oxygen to fuel gas ratio for maximum temperature		1.5 to 1	4.5 to 1	1.9 to 1	3.5 to 1	3.6 to 1
Oxygen to fuel ratio for neutral flame or lower efficiency usage		1.1 to 1	2.9 to 1	1.3 to 1	2.5 to 1	2.6 to 1
Gross heat of combustion						
	Btu per cu ft	1470	2563	1000	2406*	2371
	(MJ/m^3)	(54.73)	(95.43)	(31.23)	(89.59)	(88.28)
Fraction of available Btu's (Joules) released in primary flame						
	%	35	11	4	22**	16
(inner cone)						
	Btu	507	280	40	517	379
	(kJoules)	(535)	(295)	(42)	(545)	(400)
Heat intensity [heat transfer to 3/8 in. (0.9525 cm) diam probe at 10,000 Btu/hr (2928.75 Watt) exposure]						
	Btu/in.^{2}/min.	220	150	130	170	
	(Mwatt/m^2)	(5.992)	(4.086)	(3.541)	(4.630)	
Cubic feet of gas per pound (cu m of gas per kg) at 70°F (21°C)		14.7	8.65	24.1	8.85	9.03
	(m^3/kg)	(0.918)	(0.540)	(1.505)	(0.552)	(0.564)
Pounds of gas per 100 cubic feet [multiply × price/lb to get price per 100 cu ft (2.832 cm^3)]		6.80	11.56	4.15	11.30	11.05
Specify gravity of gas (Air = 1) @ 60°F (16°C)		0.91	1.55	0.62	1.48	1.45

* From suppliers' literature claims.
** American Welding Society Journal, August 1969, p. 641.

When comparing the volumes of different fuel gases used by ındustry, acetylene ranks first as a fuel gas for oxyacetylene cutting of carbon steels, and its use is equally common for soldering, brazing, and braze welding.

Hydrogen (H). Gaseous H is produced by the electrolysis of water or by the catalytic steam-methane reforming process. It is stored as a gas in cylinders at

2000 psi (13.79 MPa), or as a liquid at $-423\,°F$ ($-251\,°C$). Hydrogen has a heat value of 275 Btu's per cubic foot ($10.24 \times 10^7 \, MJ/m^3$) compared to that of acetylene, 1475 ($54.92 \times 10^7 \, MJ/m^3$). When used as a fuel gas with oxygen, hydrogen produces a flame temperature of $4622\,°F$ ($2550\,°C$), which is approximately $1650\,°F$ ($917\,°C$) less than the oxyacetylene flame.

Due to oxy-hydrogen's low temperature flame, its use is limited to fusion welding of low melting point metals and thin sections of aluminum, magnesium, and their alloys. The low temperature flame and low heat input make for good control of the molten pool. It is also used as a heat source for the fusion welding of lead up to 3/8 in. (0.9525 mm) thick. Hydrogen is generally substituted for acetylene as a fuel gas for underwater cutting at depths exceeding twenty-five feet (7.62 m). At this depth, water pressure makes acetylene an unsafe fuel gas. Because of acetylene's unstable atomic structure, it should never be compressed beyond 15 psi (0.1034 MPa). Due to its soft flame and low heat output, the hydrogen flame is often used for brazing, soldering, and braze welding applications. Another use for hydrogen in metal joining is in the atomic hydrogen process explained earlier. In furnace brazing operations, hydrogen is also used to provide a dry, oxygen free atmosphere that prevents metal discoloration during joining.

Propane (C_3H_8). This C_3H_8 gas is abundant in North America. It is a heavy gaseous hydrocarbon in the methane series, occurring naturally dissolved in crude petroleum. One-hundred pounds of liquid propane will vaporize into 850 cu ft (24.07 m³) of gas.

One cubic foot of C_3H_8 produces 2520 Btu's of heat, compared to 1475 Btu's of heat for C_2H_2. The maximum flame temperature for oxy-propane is $5300\,°F$ ($2927\,°C$), compared to $6300\,°F$ ($3462\,°C$) for oxyacetylene. To achieve this temperature, five volumes of O_2 for each volume of C_3H_8 are required. Therefore, three times as much oxygen is needed to produce the same volume of heat when using propane instead of acetylene. This high ratio of O_2 to C_3H_8 produces a highly oxidizing flame, which, when used for fusion welding of steel, oxidizes rather than melts the metal. For comparative purposes, in cylinders of approximately equal dimensions (12 in. × 38 in. for acetylene and $14\frac{1}{2}$ in. × 42 in. for propane), the acetylene cylinder contains only 405,075 Btu's, whereas the propane cylinder will produce 2,146,250 Btu's. The average tare weight of the acetylene cylinder is 230 lbs, compared to an average weight of 85 lbs for the propane cylinder. Weight becomes an important factor when considering transportation and handling costs. Liquid petroleum (LP) fuel gases are slow burning when compared to the flame propagation rate for acetylene or hydrogen. Due to this slow burning feature, flashbacks, backfiring, and pre-ignition are an uncommon occurrence with LP gases.

The LP fuel gases, due to their oxidizing characteristics, are not suitable for the fusion welding of ferrous materials. These gases are used extensively for brazing and soldering applications. In addition to use as a fuel gas for metal joining, propane is used as a fuel gas for the preheating flame in oxygen cutting of carbon steels.

Natural Gas: Methane (CH_4) *and Ethane* (CH_3CH_3). Natural gas is collected in oil fields and is normally transmitted by pipeline to the consumer at 400 psi pressure. It can also be compressed to a liquid (LNG) for storage. Natural gas, like propane, has low flame propagation rates, and it produces an extremely oxidizing flame at high temperatures. One cubic foot of natural gas produces an average heat value of 1050 Btu's. To liberate the same volume of heat as one cubic foot of propane, $2\frac{1}{2}$ cubic feet of natural gas is required. The principal use for natural gas in fabrication is for soldering, brazing, and for the preheat flame when cutting carbon steels.

Mapp.[TM] Mapp is a stabilized methylacetylene gas manufactured by the Dow Chemical Company. It is a compound composed of methylacetylene, propadiene, propylene, and other related compounds. Mapp gas produces a highly oxidizing flame similar to that of LPG and LNG. Oxy-mapp has a flame temperature of 5300 °F (2927 °C), which is less than that of an oxyacetylene flame. Mapp gas is less sensitive to shock than C_2H_2 and storage cylinders weigh considerably less. Mapp gas, like LPG and LNG gases, is limited in its application: for preheat flames when cutting carbon steels, powder cutting stainless steels, soldering, brazing, and metal spraying.

Flamex.[TM] Flamex is one of the newer fuel gases on the market. It is manufactured by the Flamex Company. Little descriptive information is available; however, it is a hydrocarbon and basically a propane gas with additives. Flamex is a stable gas as compared to the instability of C_2H_2. Although the claims of manufacturers differ, it is perhaps the hottest of all fuel gases when used with O_2. The Btu value/ft^3 (MJ/m^3) of Flamex is nearly twice that of acetylene; however, its flame is less concentrated and therefore has a lower efficiency rate. Flamex, like other fuel gases except acetylene, has a soft flame and its use is limited to a preheat for cutting carbon steels and for brazing, soldering, and metalizing.

Other Gases. In addition to the gases that have been discussed, there are two gases, oxygen and xenon, that are important to materials joining and are not classified as either shielding or fuel gases.

Oxygen (O_2). Oxygen is produced in large quantities by electrolysis of water or extraction from the atmosphere through the liquification process (Fig. 8–6). The latter is the more economical process and hence is the one most often used. The electrolysis technique is used primarily for the production of hydrogen, with oxygen being collected only as a byproduct. Oxygen is stored as a gas in cylinders at pressures of 2200 psi or as a liquid in insulated containers at −293 °F (−178 °C). Some large consumers of oxygen, such as steel mills that use basic oxygen furnaces, pipe liquid oxygen directly from the liquifier to the industrial site.

[TM] Trademark of Dow Chemical Co.
[TM] Trademark of the Flamex Co.

Principles of Oxygen Production Cycle

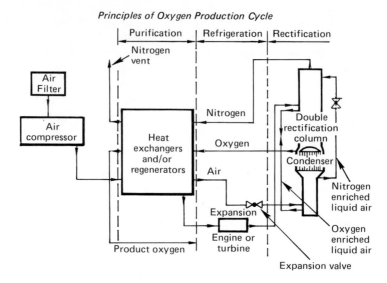

FIGURE 8-6. *Oxygen* (O) *is produced from the air in large quantities by the liquefaction process.* (*Courtesy of Welding Engineer.*)

Oxygen is a colorless, odorless, and tasteless gaseous element. Oxygen provides rapid oxidation, as observed when oxygen cutting carbon steels. When mixed with a fuel gas, it increases the combustion rate and produces a flame hot enough for melting many metals or for applying preheat and postheat treatments. O_2 is in great demand for several manufacturing applications, but is best known for its part in producing the oxyacetylene flame used for flame cutting carbon steels, fusion welding, brazing, braze welding, and soldering. However, it is also used for metal spraying and flame cleaning. Oxygen is also used in mixture with shielding gases to improve wetting and arc stability (Table 8.16). Wetting refers to the ease of flow for filler metal, and to its bonding ability with the base metal. The greatest use of oxygen in manufacturing is in steel mill furnaces.

Xenon (Xe). Xenon is a rare inert gas. Laser welding is the reason for mentioning it. The blue-green light produced by a xenon gas filled flashtube is used to excite the chromium atoms of a ruby crystal to a higher energy state. This increased activity is responsible for the impulse and intensity of the laser beam.

SAFETY APPLICATIONS WHEN USING GASES

To be safety conscious, one must be aware of the potential hazards. In material joining applications where gases are being used, many precautions must be taken. Some shielding gases produce poisonous gases as they pass through the arc. The more potentially dangerous are ozone, oxides of nitrogen, carbon monoxide, and trichloraethylene and perchloraethylene decompositions. There is also the danger from poisonous fumes being discharged from the metal being welded. Metals such as beryllium should be welded only under optimum ventilation conditions. A better understanding of the safe use of gases can be gained by referring to the AWS publication titled *Safety in Welding and Cutting*, ANSI Z49.1-1973, which is the basis for the requirements of the Occupational Safety and Health Act (OSHA) for welding. Safety practices in the use and applications for gas welding, cutting, heating, and regulating equipment are further detailed in the chapter frontispiece (Fig. 8–0).

EXERCISES

Problems

8–1. Metals usually change characteristics and properties when they are alloyed with one another. A typical example of this is the Sn/Pb solders. Table 8.12 shows a typical tin/lead diagram. From this diagram determine the following:

(a) What is the significance of a 63/37 Sn/Pb mixture?

(b) Why does a 30/70 Sn/Pb mixture make a better autobody solder than a 60/40 Sn/Pb mixture?

(c) Determine which has a higher cost, 25/75 or a 75/25. State the reasons for your answers.

(d) Select several commonly used Sn/Pb solders and suggest typical applications based on chemistry.

8–2. Over the next few years, US industries will be actively engaged in converting from the English system of measurements to the metric system. Throughout this period of conversion, many welding procedures and drawings will arrive in the shop with one or the other. In view of the fact that temperatures are of utmost importance to the welding engineer, he must be in a position to readily convert °C to °F, and vice versa.

(a) Convert the eutectic temperature of Sn/Pb solder from °F to °C.

(b) Convert the melting temperature of SAE 1010 steel (1510 °C) to the °F melting temperature. (Show your work.)

8–3. Your shop is welding $\frac{1}{4}$ in. (6.35 mm) T aluminum with the GTA process. A square butt joint is prescribed, and you wish to get as deep a penetration as possible. Which of the gases will you use and why?

8–4. Schematically show the penetration geometry for each of the following gases or gas mixtures:

(a) 95% Ar and 5% O_2 with DCSP.

(b) 100% H with DCRP.

(c) 75% H and 25% Ar with DCRP.

8–5. Given a supply CaC_2 and H_2O, what end product could you expect? Schematically show the principle for generation.

8–6. You are the production engineer for the "X" Company. To date you have farmed out all your production flame cutting jobs. However, because of difficulties with your subcontractors in meeting production schedules, your company is considering purchasing equipment and setting up a cutting department within your facilities. Prior to making this decision, you must answer a number of questions. State the questions and the logical discussion that would most likely take place prior to making the final decision.

Questions

8–1. Name the technical organization that approves the standards for all filler materials.

8–2. What is the AWS specification identification for filler materials?

8–3. Identify the following: A5.13–70.

8–4. Where is the AWS classification number placed on each type of electrode?

8–5. Identify the following: E7018.

8–6. Define a rimmed steel.

8–7. Identify the following: RG–60.

8–8. Identify a E10016–D2 coated electrode.

8–9. Identify a ECuAl–A2 coated electrode.

8–10. Name the three basic types of copper.

8–11. Name the seven basic groups for brazing filler materials.

8–12. Define the terms, liquidus and solidus.

8–13. Why isn't the identification of brazing filler materials by inking permissible?

8–14. Identify the following filler material: ERTi–6Al–2Cb–1Ta–1Mo.

8–15. Identify the following when used with submerged arc welding: F71–EM12K.

8–16. Gases are used for one or more of three specific purposes in materials joining. Name the three purposes.

8–17. Define an inert gas.

8–18. Name the inert gases used in materials joining applications.

8–19. Define a fuel gas.

8–20. Name the seven fuel gases used for joining and cutting metals.

8–21. In addition to shielding the metals being joined from the atmosphere, what other effects may these gases have on the joint?

8–22. With what welding processes would you always use an inert gas? Why?

8–23. Explain why propane and natural gas do not make good fuel gases for the welding of carbon steels.

8–24. When will hydrogen generally be used as a fuel gas?

8–25. Where is xenon gas used in materials joining? Why?

8–26. The fuel gases acetylene and natural gas are both lighter than air. What safety feature does this suggest?

BIBLIOGRAPHY

"Acetylene." Pamphlet G-1, 6th ed., Compressed Gas Association, New York, N.Y., 1966.

Aluminum and Aluminum-Alloy Arc-Welding Electrodes, AWS (American Welding Society), A5.3-69, Miami, Fla.

Aluminum and Aluminum-Alloy Welding Rods and Bare Electrodes, AWS, A5.17-69, Miami, Fla.

Bare Mild Steel Electrodes and Fluxes for Submerged Arc Welding, AWS, A5.10-69.

Braze Safety, AWS (1971 ed.).

Brazing Filler Metal, AWS, A5.8-69.

Brazing Manual, 2nd ed., AWS, 1963.

Composite Surfacing Welding Rods and Electrodes, AWS, A5.21-70.

Copper and Copper-Alloy Arc-Welding Electrodes, AWS, A5.6-69.

Copper and Copper-Alloy Welding Rods, AWS, A5.7-69.

Corrosion-Resisting Chromium and Chromium-Nickel Steel Covered Welding Electrodes, AWS, A5.4-69.

Corrosion-Resisting Chromium and Chromium-Nickel Steel Welding Rods and Bare Electrodes, AWS, A5.9-69.

FELMLEY, C. R., "Gas Shielding for Arc Welding." *The Iron Age*, August 6, 1964.

"Flux-Cored Arc Welding with Carbon Dioxide (CO_2) Gas Shielding." Hobart Brothers Technical Center, 1967.

GOODWIN, HARRIS A., "The Fuel Gases." *Welding Engineer*, December 1961.

Guest Editors, "Helium Centennial Year." *Cryogenic Engineering News*, May 1968.

"Helium A Place in the Sun." *Welding Design and Fabrication*, January 1968.

HOLT, G. D., "Linde Grooms Liquid Hydrogen for Industrial Roles." Linde Division–Union Carbide, New York, November 13, 1962.

"Industrial Gases." Air Reduction Corporation, no. 4-62, 75M-8509, 1962.

Iron and Steel Gas-Welding Rods, AWS, A5.2-69.

KISIELEWSKI, R. V., "Statistical Analysis of Fuel

Gases for Flame Gases." University of Wisconsin MS Thesis, 1972.

LINDBERG, R. A., "Processes and Material of Manufacture." Allyn and Bacon, Inc., Boston, Mass., 1964.

Low-Alloy Steel Covered Arc-Welding Electrodes, AWS, A5.5-69.

Magnesium-Alloy Welding Rods and Bare Electrodes, AWS, A.19-69.

Mild Steel Covered Arc-Welding Electrodes, AWS, A5.1-69.

Mild Steel Electrodes for Flux Cored Arc Welding, AWS, A5.20-69.

Mild Steel Electrodes for Gas Metal-Arc Welding, AWS, A5.18-69.

Modern Joining Processes, AWS, 1966.

Modern Plastics Encyclopedia, Modern Plastics, Highstown, N.J.

MOTL, B. C., "Welding with the Carbon Dioxide Welding Process." Allyn and Bacon, Inc., Boston, Mass., 1964.

Nickel and Nickel-Alloy Bare Welding Rods and Electrodes, AWS, A5.14-69.

Nickel and Nickel-Alloy Covered Welding Electrodes, AWS, A5.11-69.

"O_2." *Welding Engineer*, February 1961.

PETSINGER, R. E., "Innovations in LNG Applications." *Cryogenic Engineering News*, March 1968.

"Production Equipment for Gaseous and Liquid-Oxygen, Nitrogen, Argon and Other Gases." Superior Air Products Co., Newark, N.J., 1968.

"Recommended Safe Practice for Plasma Arc Cutting," AWS, A6-3-69, 1969.

"Recommended Safe Practices for Gas Shielded Arc Welding." AWS, A6.1-66, 1966.

ROSENTHAL, C. H., "Welding of Austenitic Stainless Steels." Vol. 5, 1-2, Svetsaren, ESAB, Goteborg, Sweden, 1969.

SAACKE, F. C., "Oxygen is Neither Combustible Nor Explosive; Often Miscalled." *Welding Engineer*, November 1960.

"Safe Handling of Compressed Gases." Pamphlet P-1, 5th ed., Compressed Gas Association, New York, N.Y., 1965.

Safety in Welding and Cutting, AWS, ANSI Z49.1, 1973.

"Safety Starts Here." Air Reduction Corporation, 12-64-50M-8886, Murry Hills, N.J., 1964.

Solder, American Smelting and Refining Co. (ASARCO), 1961.

Solder, Its Fundamentals and Usages, 2nd ed., Kester Solder Company, 1961.

Soldering Manual, 1st ed., AES, 1959.

Standard Method for Evaluating the Strength of Brazed Joints, AWS, C3.2-63.

"State of Wisconsin Employee Safety Handbook." State of Wisconsin, 1967.

Surfacing Welding Rods and Electrodes, AWS, A5.13-70.

"The Impact of Cryogenics on the Welding Industry." *Welding Design and Fabrication*, August 1960.

"The Shielding Gases—Their Function in the Welding Arc." *Welding Engineer*, August 1961.

"The 60th Year of the Gas Welding Torch." *Welding Engineer*, May 1960.

Titanium and Titanium-Alloy Bare Welding Rods and Electrodes, AWS, A5.16-70.

Welding, 1st ed., Kaiser Aluminum and Chemical Sales, Inc., 1967.

Welding Environment, AWS, 1974.

Welding Handbook, 6th ed., sec 3A, AWS, Miami, Fla., 1970.

Welding Processes, British Welding Research Assoc., Houldcroft, P. T., Cambridge University Press, 1967.

Welding Rods and Covered Electrodes for Welding Cast Iron, AWS, A5.15-69.

9

Welding Metallurgy

BASIC STRUCTURE OF METALS

The optical microscope, the electron microscope, and the field-ion microscope have done much to transform metallurgy from an art to a science. As in many other areas, the art preceded the science. For centuries workers welded metals together and made steel by heating solid iron in contact with carbon without much notion of the diffusion processes involved. In 1896 Sir William Roberts-Austen, an English metallurgist, measured the diffusion of gold into lead by a simple experiment. He fused a thin disk of gold on the end of an inch long cylinder of pure lead, put the cylinder in a furnace at 392 °F (200 °C), and kept it there ten days. He then sliced up the cylinder into thin slices and measured the amount of gold that had diffused into each slice. A considerable amount of gold had traveled all the way to the opposite end of the cylinder, and some of the lead had even diffused into the gold disk. Further investigations showed that one heated metal would diffuse into another even when their surfaces were merely pressed together.

In this atomic age, diffusion in solids is no longer surprising. We realize that even the most rigid solid is, after all, a rather loose collection of atoms. In the crystals that compose a metal, the atoms are arranged in a fixed lattice from which they are hard to displace. However, no crystal lattice is perfect and fully packed. It always has some vacancies or "holes" into which diffusing atoms can jump. Having jumped into a vacancy, an atom leaves a vacancy in its former place; an adjacent atom can then move into this hole and thus, by a continual shuffling of atoms, can migrate through the crystal. This vacancy mechanism of diffusion is one of many similar mechanisms that explain the movement of atoms in the solid state. The study of space lattices and crystal structures helps explain how solid state welding can take place and how one form of alloying is possible.

Welding metallurgy, in a limited sense, is confined to the weld metal and heat affected zone, but in a broader sense, includes all aspects of metallurgy. For this reason a brief general background of the principles of metallurgy will be presented before discussing the properties of the weld and the adjacent heat affected zone.

CRYSTAL STRUCTURE OF METALS

By following the structure of metals from the liquid to the solid state, we obtain a basic concept of the crystal formation.

Solidification. The process of solidification in metals begins with nucleation and growth of crystals normally at the cooler outer edges of the metal. Solid state physicists describe the process as a "cooperative" phenomena in that large numbers of atoms acting in concert are involved. Just why the atoms of one solid substance should arrange themselves to form a cube and others, a hexagon is a matter of free energy relationships. The final structure is determined by a number of complex considerations, but for simplification the structure that results is the one with lowest energy.

After nucleation the structure grows until crystals, or as they are often called in metallurgy, "grains," are formed. The grains join with other grains to form a dendritic or "pine tree" columnar structure, as shown in Fig. 9–1. Other grains of different shape can also form, especially after heat treatment and other mechanical operations.

Space Lattices. The crystal structure that results, and in a sense makes up the dendritic structure, is based on space lattices. A space lattice refers to an orderly geometric arrangement of imaginary points. The atomic arrangements found in most metals, i.e., (1) the body-centered cubic (BCC), (2) the face-centered cubic (FCC), and (3) the hexagonal close packed (HCP), Fig. 9–2, are actual space

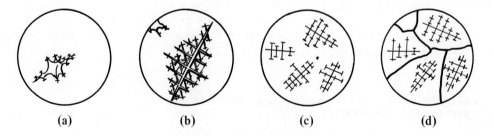

| (a) | (b) | (c) | (d) |

FIGURE 9–1. *Solidification of a pure metal. (a) Nucleation and the first dendritic structure is formed. (b) (c) The dendritic structures join together. Development is stopped by interference with adjacent structures and the container forming grain boundaries.*

275

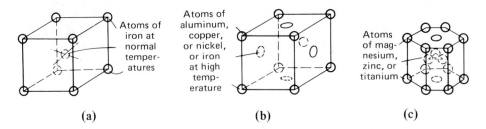

(a)   (b)   (c)

FIGURE 9–2.   *Atomic arrangements in the crystal structure of metals: (a) body-centered cubic (BCC); (b) face-centered cubic (FCC), and (c) hexagonal close-packed (HCP).*

(a) *Body centered cubic*
    crystal structure
(b) *Face centered*
    cubic structure
(c) *Close packed hexagonal*
    structure

lattices with one atom for each point in space. Other crystal structures are possible based on these lattices with more than one atom per point.

Lattice Imperfections.   The atomic concept of a "crystal" may conjure up pictures of billions upon billions of identical atoms stacked in a perfect array. Actually, orderliness is marred by a missing atom here, or a foreign particle there.

For a long time these imperfections were disregarded. Metallurgists made predictions based on perfect atomic structures. This was fine in theory, but it did not explain why pure iron, calculated to stand a stress of two million pounds per square inch, yielded at only 30,000 psi (206.8 MPa). This discrepancy, and many others, remained unsolved for many years until it was suggested that the resistance to plastic deformation depends not on the average properties of the almost perfect lattice but on the individual properties of previously unknown imperfections. This weakest link, known as a *dislocation*, was proposed independently in 1934 by G. I. Taylor in England and E. Orowan in Germany. The hypothesis, since proved by the electron microscope, is the foundation of many present day theories concerning the plastic deformation of metals.

ALTERING AND CONTROLLING MECHANICAL PROPERTIES OF METALS

The principle of strengthening metals is based on methods of impeding slip in the atomic structure. This can be accomplished by either mechanical working, heat treatment, or alloying; frequently these methods are used in combination.

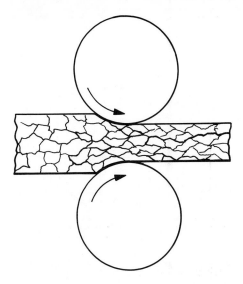

FIGURE 9-3. *The cold rolling process distorts and elongates the grain structure, thereby increasing the number of dislocations and the strength of the metal.*

Mechanical Working. Mechanical working usually refers to cold working, which includes hammering, rolling, drawing, or pressing at temperatures below that of the recrystallization temperature of the metal. Cold rolling, as shown schematically in Fig. 9-3, distorts and elongates the grain structure. As the grains become deformed, the number of dislocations increases. The dislocations tend to pile up at the grain boundaries, retarding slip, increasing the strength of the metal but making it more difficult to form.

Heat Treatment. Heat treatment is a term used to denote a process of heating and cooling materials in order to obtain certain desired properties, or often the best properties the material can offer. As an example, a shear blade must have its structure controlled to produce just the right combination of hardness, wear resistance, and toughness in order for it to successfully cut other metals. The reasons for heat treating ferrous metals may be briefly stated as:

1. To change the microstructure.
2. To relieve internal stresses.
3. To alter the surface chemistry by adding or deleting elements.

    Normalized Steel. The most common condition of steel is as a *normalized* structure. After the steel is produced in the furnace, it is poured into ingots for cooling and then rolled into sheets, plates, rods, and so forth. Thus, the steel has proceeded from an elevated temperature to room temperature in an orderly manner. This process along with other common heat treating processes, such as annealing, hardening, tempering, stress relieving, and spheroidizing, can more easily be understood by following the iron–carbon equilibrium diagram shown in Fig. 9-4.

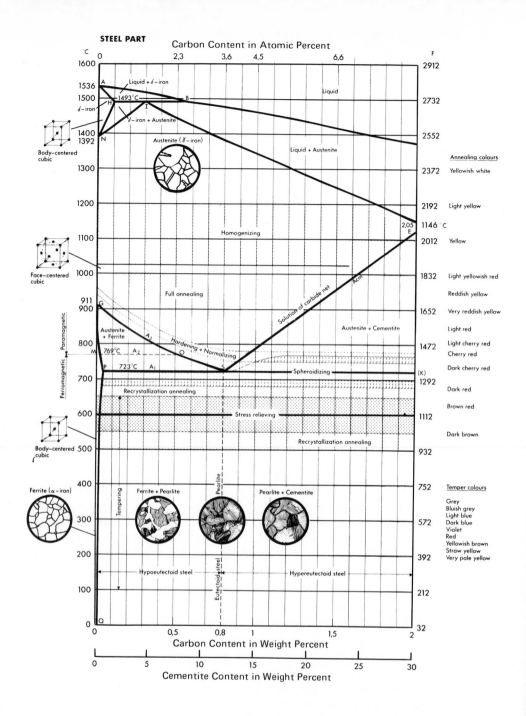

FIGURE 9–4. *Iron–carbon equilibrium diagram.* (*Courtesy of Struers Scientific Instruments.*) (*Compiled by Svend Engell-Nielsen, The Technological Institute, DK-2630 Ta°strup, Denmark.*)

As steel cools from the elevated temperature, it passes through several stages as shown on the diagram: from liquid, to liquid and austenite, and then to austenite. On the left under the A_3 line is austenite plus ferrite, and on the right, austenite plus cementite. Finally, below the A_1 line is ferrite plus pearlite, pearlite, and pearlite plus cementite. The formation of each of these structures will be discussed.

Austenite.   Austenite is shown in the area above the GSE connecting lines and is a solid solution of carbon in a face-centered cubic iron. The A_3 line represents the initial precipitation of ferrite from the austenite. The line SE indicates the primary deposition of cementite (Fe_3C) from austenite.

Ferrite.   Ferrite is an alpha (α) iron with a body-centered cubic lattice that exists in the very narrow area at the extreme left of the diagram below 1674 °F (797 °C).

Cementite.   Cementite is a very hard, brittle compound of iron and carbon, Fe_3C, containing 6.67 % carbon.

Pearlite.   Pearlite is a two phase structure consisting of thin, alternate layers of iron carbide (cementite) and ferrite, as shown in Fig. 9–5 and in the enlarged structural views on the diagram.

Normally, as the metal cools slowly from austenite, an automatic separation of ferrite and the ferrite cementite mixture (pearlite) occurs, as shown in the diagram. (The white areas represent ferrite and the lined areas, pearlite.) As the carbon content increases, it unites with greater amounts of ferrite, thus increasing the pearlite. At a point where all of the ferrite is in combination with carbon, the structure is entirely pearlite as shown by the centered microstructural view. Theoretically, this is at 0.83 percent carbon but may range from 0.75 to 0.85 percent in plain carbon steels. On this diagram it is represented at 0.80 percent carbon.

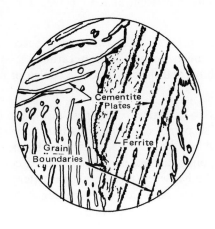

FIGURE 9–5. *Pearlite structure of steel. (Courtesy of International Nickel Co., Inc.)*

This combination of iron and carbon is known as a *eutectoid* steel. Eutectoid, taken from the Greek, means "most fusible." This combination occurs in binary alloys when a complete solid solubility does not exist. A eutectoid solidifies at a lower temperature than the melting point of either of the pure components. Steels with more than 0.80 percent carbon are called *hypereutectoid*, and those below 0.80 percent carbon are *hypoeutectoid*. During slow cooling, the Fe_3C precipitates out of the austenite until the temperature of 1333 °F (723 °C) is reached, at which time the remaining austenite changes to pearlite. The result is a pearlite structure surrounded by a network of cementite at the grain boundaries as shown in the enlargement on the iron–carbon diagram.

When steel containing more than 2.11 percent carbon is slowly cooled, the cementite separates out into iron and graphite. The fine, soft graphite forms flakes in the iron matrix, thereby producing a grey cast iron. If cooling is more rapid, a brittle structure known as white cast iron will result. Cast irons will be discussed in more detail later in the chapter.

If a specimen of hypoeutectoid steel is heated uniformly, it will be found that at approximately 1333 °F (723 °C) the temperature of the steel will stop rising even though the heat is still being applied. After a short time, the temperature will continue to rise again. Ordinarily, metals expand as they are heated, but it is found that at 1333 °F a slight contraction takes place and then, after the pause, expansion again takes place. This indicates the lower critical temperature of the steel, shown in the iron–carbon diagram as line A_1. Actually, an atomic change takes place at this point, and the structure changes from a body-centered arrangement (alpha iron) to a face-centered (beta iron) arrangement.

As the temperature increases, transformation continues until A_3, at which time the transformation of ferrite to austenite is complete. This varies with the carbon content, as shown in the diagram, and is known as the upper critical temperature. At this point most of the carbon goes into solution with the iron, so that it is now more evenly distributed. Transformation temperatures vary depending on the cooling or heating rate. As an example, in cooling, a 0.80 % carbon steel may change from austenite to pearlite at 1290 °F (699 °C); whereas, in heating, the change from pearlite to austenite may occur at 1350 °F (732 °C).

Hardening Steel. Three requirements are necessary in order to successfully harden a piece of steel: First, the steel must contain enough carbon; second, it must be heated to the correct temperature; and third, it must be cooled or quenched.

Carbon Content. In order to get extreme hardness in a piece of steel, it is necessary that the steel contain 80 points or more of carbon. Low-carbon steels (less than 25 points carbon) will not be materially affected by heat treatment. Medium-carbon steels (30 to 60 points carbon) may be toughened considerably by heat treatment, but they will not be hardened to a very great extent. High-carbon steels (60 to 150 points carbon) may be successfully hardened by simple heat treating methods (one point of carbon = 0.01% carbon).

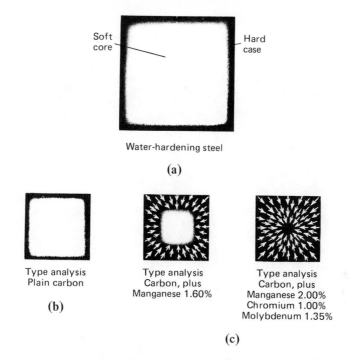

Soft core

Hard case

Water-hardening steel

(a)

Type analysis
Plain carbon

(b)

Type analysis
Carbon, plus
Manganese 1.60%

Type analysis
Carbon, plus
Manganese 2.00%
Chromium 1.00%
Molybdenum 1.35%

(c)

FIGURE 9-6. *The affect of alloys on the hardenability of steel. (a) A plain high-carbon steel gives an outer layer of hardness when heat treated. (b) If manganese is added to plain carbon steel, it can be cooled at a slower rate, usually by oil, and the hardness penetration will be deeper. (c) As more alloys are added, the steel can be cooled more slowly, as by air, and the hardness will penetrate through the section.*

Heating.   The steel must be heated above the upper critical (A_3) temperature. The actual hardening range is about 100 to 200 degrees above the upper critical temperature. Notice that this curve levels out beyond 0.80 percent carbon.

After the steel has been heated at the hardening range for a sufficient period of time to equalize the heat (about 30 min/in. of cross section), it is taken out of the furnace and quenched in a cooling media, which may be water, brine, oil, molten salts, or lead baths. The rate at which the material cools will largely determine its hardness. In order to avoid undue stress (caused by uneven cooling) during the quenching action, the part should be vigorously agitated to prevent steam pockets from forming on the surface (for water quenching).

Given the proper heat treatment, the hardness will primarily be determined by the carbon content, however the depth of hardness, known as *hardenability*, will largely be determined by the alloying elements. The effect of alloying elements on the hardenability is shown in Fig. 9-6.

When austenite is immersed in a water quench, it does not have time

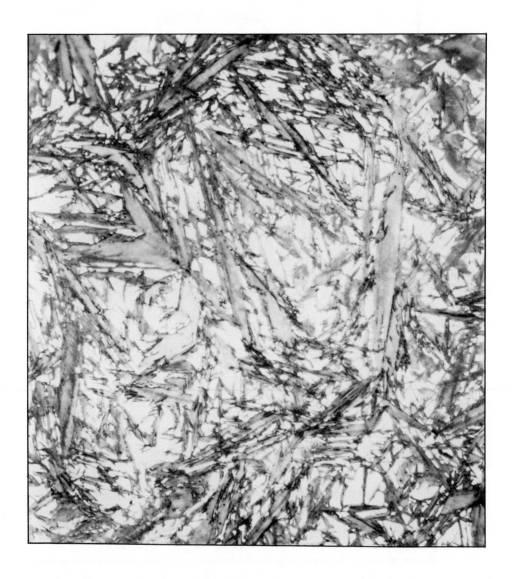

FIGURE 9–7.   *Martensite structure.* (*Courtesy of United States Steel Corporation.*)

to separate out to form pearlite. Some of the austenite transforms almost instantaneously to *martensite*, an interlaced needlelike structure as shown in Fig. 9–7 that is hard and brittle. How this change takes place can be shown by means of another diagram termed a time–temperature–transformation diagram or TTT diagram, Fig. 9–8. The diagram shows the various structures that occur as the metal is cooled and the time at which they occur. As an example, if a high-carbon

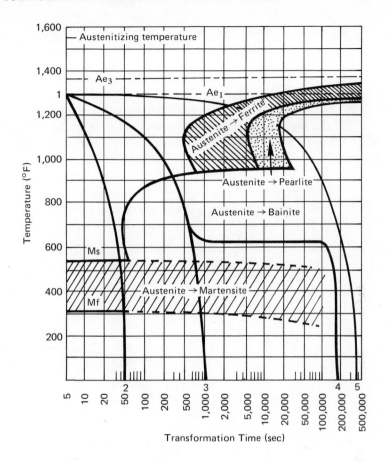

FIGURE 9–8. *A time-temperature-transformation diagram. Path 1–2 represents a rapid quench that misses the bainite curve and is transformed into martensite. Path 1–3 represents a slower quench, during which austenite begins to transform into bainite until 550 °F (288 °C); the remaining austenite then transforms to martensite. Path 1–4 shows a delayed quench. At 925 °F (496 °C), the austenite begins to transform to bainite but between 925 °F and 550 °F the quench is halted. In this case, the final cooling results in a bainite structure. In path 1–5, the austenite transforms into pearlite rather than ferrite so that at point 5 the structure is a combination of ferrite and pearlite, a soft annealed steel structure.*

steel is quenched in water such that it reaches a temperature of 550 °F (288 °C) in about five seconds or less, a martensite transformation will start (M_s). The martensite transformation will finish (M_f) at about 325 °F (163 °C). The martensite that results is a super-saturated solid solution of carbon trapped in a body-centered tetragonal structure. It involves no chemical change from the austenite stage. The unit cells of the martensite structure are highly distorted because of the entrapped carbon, which is the principle hardening mechanism. The body-centered cubic tetragonal structure is less densely packed than the face-centered cubic (FCC) lattice structure of austenite, hence there is about a 4 percent expansion causing still further localized stresses that distort the austenite matrix. The atomic space lattices that occur at the various temperatures of steel are shown at the left of the iron–carbon diagram. Normally the internal stresses will have to be relieved as soon as possible to prevent cracking by a tempering process.

If the high carbon steel had been cooled at a slightly slower rate, other structures may have formed in addition to martensite. This can be shown by a continuous cooling transformation diagram (C–T), Fig. 9–9. An end-quench

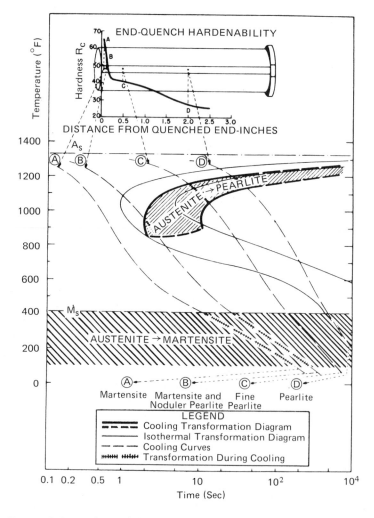

FIGURE 9–9. *End quench hardenability tests of an 8630 type eutectoid steel correlated with continuous cooling (fine line) and isothermal transformation diagrams (shaded areas). (Courtesy of United States Steel Corporation.)*

hardenability bar is superimposed on the top of the diagram. The letters A, B, C, and D represent various distances from the quenched end, the corresponding cooling curves are shown on the transformation and C–T diagrams. Thus, the hardness of the bar at A will be the highest as it has missed the "nose" of the transformation curve. At points B and C, the corresponding cooling curves are represented as being slightly slower so that they pass through the austenite–bainite and austenite–martensite transformation areas. The metal will then consist of martensite, ferrite, and bainite as shown by the micro-structures below the diagram.

Bainite. Bainite is an intermediate type structure, feathery in appearance, that forms between austenite and martensite when the cooling rate is slowed to exceed the critical rate. The critical cooling rate is that rate which is just fast enough to miss the nose of the transformation curve.

Tempering. As mentioned previously, after quenching, the steel is in a highly stressed, unstable condition. To avoid or minimize problems of cracking and distortion, the metal is reheated or tempered. Several different means may be employed to do this; conventional reheating after quenching, martempering, and austempering, as shown in Fig. 9–10.

The conventional method consists of reheating the metal immediately after quenching to a temperature less than critical. Since tempering softens the steel, the relationship of hardness to strength should be known. As an example, a steel that may have a hardness reading of $R_c 62$ (file hard) after quenching may be reheated to a temperature of 450 °F (232 °C) and then have a hardness reading of $R_c 56$. This would still be hard enough to have high strength and good wear resistance with the advantage of having a more stable structure.

In martempering, the metal is quenched in a bath of molten salts at a temperature just above the Ms line. This allows the inside of the steel to arrive at the same temperature as the outside. The steel is then removed from the bath and quenched to room temperature. If the steel is of an air hardening type, it is removed from the salt bath and allowed to cool at room temperature. This process has the advantage of reducing the stress on the steel as it changes to martensite, since the temperature has already been equalized throughout.

Austempering also pauses in the cooling curve to equalize the outside temperature of the part with that of the inside. It then is kept at that temperature to produce a bainite structure. This eliminates the additional step of reheating the metal. The hardness can be the same as that of tempered martensite depending on the transformation temperature. If 450 °F (232 °C) is used as in the example of conventional tempering, the same relative hardness of $R_c 56$ would be expected.

Annealing. Annealing in general refers to softening a metal by heating and cooling. It is generally of two types, *full annealing* and *stress relief annealing* (as was shown in the iron–carbon diagram, Fig. 9–4). Full annealing consists of heating the steel

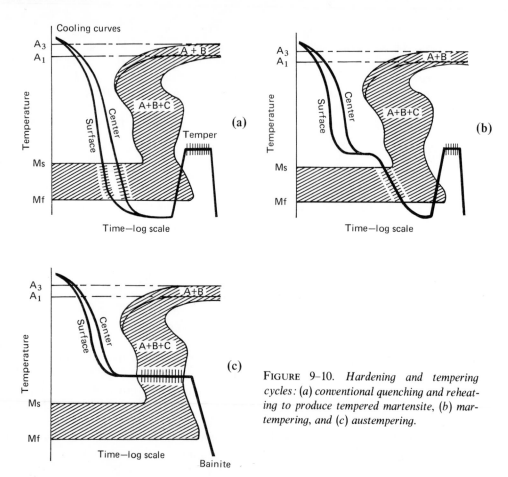

FIGURE 9-10. *Hardening and tempering cycles: (a) conventional quenching and reheating to produce tempered martensite, (b) martempering, and (c) austempering.*

about 100 °F (38 °C) above the hardening and normalizing range. It is held at that temperature for the desired length of time and then allowed to cool very slowly, usually in the furnace. This produces a soft, coarse, pearlitic structure that is stress free.

In stress-relief annealing, the metal is heated to a temperature close to the lower critical temperature, followed by any desired rate of cooling. Stress-relief annealing is usually done to soften a metal that has become strained or work-hardened during a forming operation. As shown on the iron–carbon diagram, there is a recrystallization zone. At this temperature, the grains of a cold-worked steel will recrystallize and form fine grains. There are two zones shown for recrystallization. This does not mean that recrystallization occurs only at these temperature ranges but rather that it can occur at different temperatures depending upon the amount of prior cold work. It takes energy to make the grains recrystallize and if they have a considerable amount of energy already in the structure, it requires less to make them recrystallize. Stress relieving may take place, however, without recrystallization.

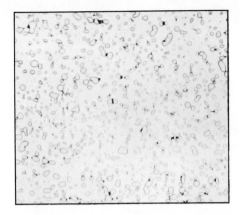

FIGURE 9-11. *Spherodized iron carbides in a matrix of annealed steel, 750X. (Courtesy of International Nickel Co., Inc.)*

Spheroidizing. When steel is tempered at a temperature just below the lower critical or A_1 line, as shown in the iron–carbon diagram, the cementite will consist of small spheroids surrounded by ferrite, Fig. 9-11. Prolonged heating (16 to 72 hours) at this temperature may be required. The process may be speeded up, particularly on smaller items, by alternately heating them to temperatures slightly below and then slightly above the lower critical. The length of time and the number of cycles will be dependent on the original structure of the steel. A fine pearlite is preferred.

Spheroidizing results in greater ductility, which improves the forming qualities as well as machinability.

Surface Hardening of Steels. The heat treating processes discussed are generally applicable to medium and high carbon steels. The depth of hardness is based on the alloy content and the quench rate. Oftentimes it is desirable to have a hard surface accompanied by a softer, tougher core. This can be accomplished by flame hardening or induction hardening on steels of medium carbon content or higher, and cast irons of suitable composition. Carburizing and nitriding processes are used on low carbon steels.

Flame Hardening. Flame hardening consists of moving an oxyacetylene flame over a part followed by a quenching spray. The rate at which the flame is moved over the part will determine the depth to which the material is being heated to a critical temperature or higher. Quenching can be built into the burner as shown in Fig. 9-12.

In flame hardening, no line of sharp demarcation separates the hardened surface zone and the adjacent layer so there is little likelihood that it will chip out or break during the service. The flame is kept at a reasonable distance from sharp corners to prevent overheating, drilled or tapped holes are normally filled with wet asbestos for edge protection. Stress relieving at about 400 °F (204 °C) is recommended for all flame hardened articles except those made from air hardened steel. Some advantages of flame hardening are:

287

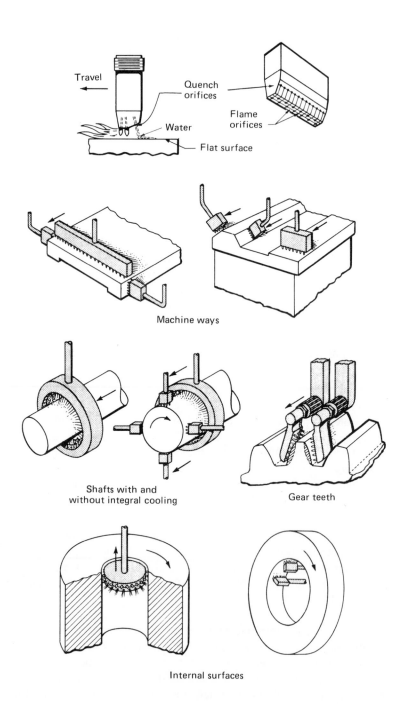

Travel

Quench orifices

Flame orifices

Water

Flat surface

Machine ways

Shafts with and
without integral cooling

Gear teeth

Internal surfaces

FIGURE 9–12. *Flame hardening with various shaped burners.*

1. Large machined surfaces can be surface hardened economically.
2. Surfaces can be selectively hardened with minimum warping and freedom from quench cracking. Examples of selective hardening are gear teeth, machine ways, cam surfaces, and engine pushrod ends.
3. Scaling of the surface is only superficial because of the relatively short heating cycle.
4. The equipment may be controlled electronically to provide precise control of the hardness.
5. The depth of case may vary to suit the part, from 1/8 to 1/4 in. (3.175–12.70 mm).

Disadvantages include: (1) To obtain optimum results, a technique will have to be developed for each design. (2) Overheating can cause cracks, or excessive distortion especially where thin sections are involved.

Induction Hardening. Induction hardening is done by heating the metal with a high frequency alternating magnetic field. Heat is quickly generated by high frequency eddy currents and hysteresis currents on the surface layer. The primary current is carried by a water-cooled copper tube, the workpiece serves as the secondary circuit. The depth of penetration decreases as the frequency of the current increases, e.g., the approximate minimum depth for 3000 Hz is 0.060 in. (1.52 mm), and for 500,000 Hz it is 0.020 in. (0.51 mm).

Induction hardening is fast even on comparatively large surfaces. As an example, a large truck crankshaft can be brought to the proper temperature and spray quenched in 5 seconds. It will have very little distortion due to the short cycle time.

Carburizing. Carburizing, as the name implies, is a method of adding carbon to the surface of low carbon steels. It may be done by pack carburizing, gas carburizing, or liquid carburizing (cyaniding).

Pack Carburizing. Pack carburizing is done by placing the low carbon steel in a heat resistant metal box and surrounding it completely with a carburizing compound. The container is then heated to the austenitic temperature. The length of time is dependent on the depth of hardness desired. Three hours at 1650 °F (899 °C) will produce a case depth of about 0.060 in. (1.52 mm). Since this process is rather slow and dirty, it has been largely replaced by gas and liquid carburizing. After carburizing, the parts are quenched and tempered just as high carbon steels.

Gas Carburizing. Gas carburizing is done by placing the low carbon steel in a heated retort in which carburizing gas (propane, natural gas, or methane) is admitted. Continuous type furnaces are available in which the parts are placed on a conveyor and are carburized, quenched, and tempered in a set sequential process.

Liquid Carburizing. Liquid carburizing is done by immersing the parts in a molten potassium cyanide salt bath, which is kept at a temperature of 1550–1570 °F

(843–854 °C). The process is often referred to as *cyaniding*. A case depth of about 0.025 in. (0.63 mm) can be obtained in two hours. Both carbon and nitrogen are added to the surface of the steel, producing a harder case than by gas carburizing. The parts are usually quenched directly from the cyanide bath and then tempered to the desired toughness.

Nitriding. Nitriding is a patented commercial process that is similar to gas carburizing. It has the advantage of requiring a relatively low temperature, 950 °F (510 °C). No scaling occurs so the parts can be finished before hardening. The case formed is very thin, ranging from about 0.001 to 0.005 in. (0.03–0.013 mm) thick, but very hard. Carburizing may range from $R_c 65$ to $R_c 67$, whereas nitriding may be in excess of $R_c 72$. Since the case is very thin, it is measured on a Rockwell Superficial hardness tester using either the 15N or 30N scale. The reading may be changed from the N scale to the C scale by the use of the hardness conversion table.

A newer method of nitriding has been developed known as *glow-discharge* or *ion-nitriding*. Instead of placing the steel workpieces in an externally heated furnace, they are made the negative electrode of a low pressure glow discharge, in a mixture of nitrogen and hydrogen gases. Under the action of an applied voltage, positive ions bombard the surface of the steel, accelerating the nitrogen to form hard alloy nitrides and delivering sufficient energy to heat the steel without requiring any external heating elements.

In addition to lower operating costs per unit of time, the time required for the process is shorter. Conventional nitriding requires about 30 hours in the furnace for a 0.010 in. (0.25 mm) case depth. A depth of 0.025 in. (0.38 mm) can be achieved in the same time by ion-nitriding. In addition, selective hardening is simplified. Conventionally, surfaces that are to be kept soft must be copper plated and later ground. In ion-nitriding, masking is accomplished by placing a sheet of mild steel in front of the area that is to be masked.

Evaluation of Surface Hardening Treatments. If the prime consideration of surface hardening is wear resistance, then it is best to choose processes such as nitriding and cyaniding where sufficient carbides are developed at the surface. If impact or torsion is involved, excess carbides will be detrimental since cracks, chipping, and spalling will develop. As the case depth increases, the ability to withstand this type of loading also increases.

THE WELD STRUCTURE AND THE ADJACENT HEAT AFFECTED ZONE (HAZ)

Welding is often described as being a small casting made in a metal mold. The properties of this casting and the adjacent metal are directly related to the thermal conditions and alloying elements involved.

Most welds are made in mild or low carbon steels. However, increasing amounts of alloy steels, particularly the high strength low alloy steels (HSLA), are being used in welded construction. To use steels in this category requires a knowledge of the metallurgical changes brought about by welding. The following discussion will first examine a single pass low carbon steel weld followed by the effect of a multiple pass weld in the same material.

A Single Pass Low Carbon Steel Weld. The weld deposit and the adjacent metal (HAZ) can be divided as shown in Fig. 9–13. The weld deposit and the six zones are classified according to the intensity and amount of heat, measured in kilojoules of energy. The zones are as follows:

1. Weld deposit.
2. Fusion zone.
3. Fully austenitized zone.
4. Transformation zone.
5. Austenite plus ferrite zone.
6. Less than critical temperature zone.

The Weld Deposit. The weld deposit on annealed steel is similar in structure to cast steel. As the metal cools from the liquid state, at about 2700 °F (1482 °C) it forms delta or austenitic crystals, depending upon the composition, in a dendritic pattern. The delta structure is primarily an interstitial solid solution of carbon in a body-centered cubic iron, while the austenite structure is an interstitial solid solution of carbon in gamma (γ) iron. As can be seen from the combination diagram (Fig. 9–13), no structural change occurs until about 1500 °F (816 °C). At that point,

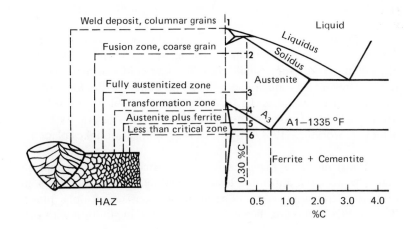

FIGURE 9–13. *A weld deposit in 0.30% carbon steel showing grain structure in the HAZ and relationship to the iron-carbide diagram. For arc welding the HAZ is quite narrow, one-eighth inch (3.175 mm) or less.*

the ferritic crystals begin to separate from the austenite so that for 0.30 carbon steel about one-third of the steel is austenite and the remaining two-thirds is ferrite. The geometrical pattern of the ferrite within the grain boundaries is called Widmann-statten structure. It is made up of white, interlaced masses distributed throughout the grains, resulting from the precipitation of ferrite in long continuous plates along the crystallographic planes. It is usually found in single pass welds made in mild steel. Certain alloying elements such as manganese, chromium, and molybdenum are known to promote its formation. Due to the coarse nature of the structure, the impact strength may fall to about one-fifth that obtained in a fully annealed structure.

The coarse columnar structure of the weld metal is usually refined by the addition of grain refiners in the filler metals used. Basically, all grain refiners provide certain intermetallics like oxides and nitrides, which act as nuclei. Grain refinement also promotes a finer distribution of impurities, such as FeS in steel, at the grain boundaries, which results in an increase in ductility and minimizes the possibility of weld cracking.

Fusion Zone. The fusion zone represents the highest temperature outside of the deposited weld metal. The grain size is usually large with a mixture of some columnar grains at one edge and normalized grains of pearlite and ferrite at the other. Adjacent to the weld deposit may be some Widmannstatten structure. If the base and filler materials are of the same composition, problems with alloying and with the coefficient of expansion will not arise.

Fully Austenitized Zone. The face-centered atomic structure and the high mobility of the atoms involved allows the carbon to become a solid solution with the ferrite. As shown in the schematic representation (Fig. 9–13), the grains are represented as of intermediate size. The size will vary with time and temperature.

Transformation Zone. Transformation occurs over a range of temperatures. The temperature requirement is dependent upon the relative amount of energy already in the structure. Small grain size and cold working represent higher energy levels, hence not as much energy need be added in the form of heat to cause the structure to recrystallize. In this case (see Fig. 9–13), the weld is represented as being made in a 0.30 % carbon hot rolled steel. The grain size is normal with no prior cold working—thus most of the recrystallization takes place above the A_3 line.

Austenite Plus Ferrite Zone. This zone of the weld represents a structure that is between the upper and lower critical temperatures and is sometimes referred to as the "transition zone." The austenite consists of alpha plus gamma iron or BCC + FCC structure. The carbon is partially in solution with the ferrite.

Less Than Critical Temperature Zone. This temperature will have little affect on the metal unless it has been previously cold worked. No austenite is formed. Some spheroidizing and softening of the structure may take place. The grains are composed of ferrite and ferrite with cementite. If the weld is stress relieved at about 1100 °F (593 °C), the cementite will spheroidize to a more noticeable extent.

Multiple Pass Welds in Low Carbon Steels. In a single pass weld in mild steel, there are coarse grains in the columnar as-cast structure. By depositing a second pass after the first bead has cooled below 1333 °F (727 °C), most of the coarse grain structure of the previous pass will be refined. Therefore, the two-pass weld in mild steel consists of a fine grained lower bead and a coarse grained upper bead. The result is improved mechanical properties of strength and toughness.

If the base material is subjected to preheating or if the second pass is applied before the metal has had a chance to cool below the lower critical level, the result will be no grain refinement. Allowing previous beads to cool to room temperature before depositing the second bead will result in less grain refinement than if the second bead is deposited while the first bead is still hot but below the critical temperature, Fig. 9–14. In terms of depth of grain refinement, only a small portion of the previous bead will have a refined grain structure if the second bead is deposited cold. If the second bead is deposited while the retained heat is at (about) 1000 °F (538 °C), a large part of the previous bead will have a refined grain structure.

Weld Effect on Cold-Worked Metals. Grain structure deformation has a strengthening effect on metal up to the point where it becomes brittle. The tensile and yield strengths will change in proportion to the reduction in area. AISI 1020 steel, for example, in the hot rolled condition, has a tensile and yield strength of 65 and 43 ksi

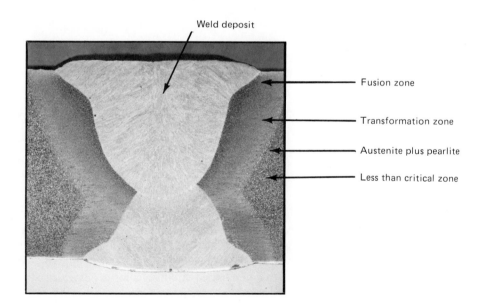

Weld deposit

Fusion zone

Transformation zone

Austenite plus pearlite

Less than critical zone

FIGURE 9–14. *In multiple pass welding, the grain structure of the underbead is refined when a bead is laid over it.*

(448.2 and 296.5 MPa), respectively. The same material in the cold rolled condition has a tensile and yield strength of 78 and 66 ksi (537.8 and 455.1 MPa), respectively. A gain of 28 % in yield strength is quite significant.

The cold worked metal is subject to recrystallization in the HAZ of the weld. The properties that can be expected after welding will be more nearly those of a hot rolled steel. The significant change in HAZ structure is, with the right time and temperature history, the possibility of having two refined zones—one due to the phase transformation phenomena and the other to the recrystallization phenomena.

FACTORS AFFECTING WELD METAL COOLING RATES

The microstructure of metals is greatly affected by cooling rates from above the upper critical temperature. Many factors affect the rate at which the weld deposit and the HAZ cool. Among these are the energy input, plate thickness, geometry, and thermal characteristics of the base metal.

Energy Input. The heat or energy input consists not only of a heat given off at the electrode or torch tip, but also the preheat temperature, if required. The term "energy input" is used to describe the amount of heat or energy used for each inch of weld, determined as follows:

$$H = \frac{60EI}{S}$$

where H is joules/in.
E is the voltage used.
I is the amperes.
S is speed in inches/min.

The units may be changed to watt seconds or joules/in. (watt second = 1 joule).

> *Example.* If a weld is made with a coated electrode at 125 amperes and 25 volts with 8 ipm travel speed, the joules per inch will be:
>
> $$\frac{125 \times 25 \times 60}{8} = 23400 \text{ J/in. or } 9212.6 \text{ J/cm.}$$

The calculation of energy input for a given weld can provide useful information in making comparisons of welds by different processes. It is also useful in being able to maintain the desired joules/in. while making changes in current, voltage, or travel speed. As, for example, the recommended conditions for a $\frac{1}{4}$ in. thick stainless steel butt weld by the gas tungsten arc or GTA process are:

$$\frac{275 \times 25 \times 60}{6} = 68750 \text{ joules/in. or } 27067 \text{ J/cm.}$$

The same weld made by the GMA process is:

$$\frac{180 \times 23 \times 60}{12} = 20700 \text{ joules/in. or } 8\,150 \text{ J/cm.}$$

The comparison readily shows the difference in heat intensity of the two processes. It also provides a useful indicator of the heating and cooling conditions in the weld. It is possible, within reasonable limits, to manipulate voltage, current, and travel independently to achieve optimum results. The recommended maximum energy input for welding various thicknesses of a quenched and tempered steel is shown in Table 9.1. The HY–130(T) steel listed is a U.S. Steel Corporation trade name. It is also very similar to ASTM A514E or A517E, which has a minimum yield strength of 100 ksi (689.5 MPa).

Shown in Fig. 9–15 is a comparison of the effect of two different energy input levels on $\frac{1}{2}$ in. steel plate. Above the center of the weld pool is shown the temperature isotherms for a weld made at 3 in. (7.62 cm)/min at 100,000 joules/in. (39370/cm). The lower half shows the temperature isotherms for a weld made at 6 in. (15.24 cm)/min at an energy input of 50,000 joules/in. (19685/cm). The preheat temperature in both

TABLE 9.1. *Recommended Procedures for Welding HY–130(T)*^{TM} *Steel* (Courtesy of American Welding Society)

Plate thickness, in. (cm)		Preheat and interpass temperature, °F (°C)		Heat input, kilojoules/in.
Gas metal arc welding procedures				
130 ksi minimum yield strength (896 MPa)				
Up to $\frac{3}{8}$	(9.52)	Not recommended*		
$\frac{3}{8}$ and including $\frac{5}{8}$	(9.52–15.87)	120 to 150	(49–66)	35
Over $\frac{5}{8}$ and including $\frac{7}{8}$	(15.87–22.22)	150 to 200	(66–93)	40
Over $\frac{7}{8}$ and including $1\frac{1}{2}$	(22.22–38.1)	200 to 275	(93–135)	45
Over $1\frac{1}{2}$	(38.1)	225 to 300	(106–149)	50
140 ksi minimum yield strength				
Up to $\frac{3}{4}$	(19.05)	Not recommended*		
$\frac{3}{4}$ and including 1	(19.05–25.4)	125 to 200	(52–93)	35
Over 1 and including $1\frac{1}{2}$	(25.4–38.0)	200 to 275	(93–135)	40
Over $1\frac{1}{2}$	(38.1)	225 to 300	(106–149)	42
Shielded metal arc welding procedures				
130 ksi minimum yield strength (896 MPa)				
Less than $\frac{3}{8}$	(9.52)	Not recommended*		
$\frac{3}{8}$ and including $\frac{5}{8}$	(9.52–15.87)	125 to 150	(52–66)	30
Over $\frac{5}{8}$ and including $\frac{7}{8}$	(15.87–22.22)	150 to 200	(66–93)	35
Over $\frac{7}{8}$ and including $1\frac{1}{4}$	(22.22–31.75)	200 to 275	(93–135)	35
Over $1\frac{1}{4}$ and including 4	(31.75–101.6)	225 to 300	(106–149)	40

^{TM} Trademark of the U.S. Steel Corporation.
* Welding of these light gauge plates by either gas metal arc or shielded metal arc processes is not recommended. For such welding, the gas tungsten arc process with appropriate controls is recommended.

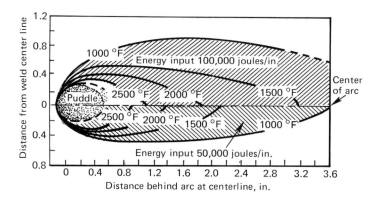

FIGURE 9-15. *A comparison of the isotherm temperature on the surface of* $\frac{1}{2}$ *in.* (12.70 mm) *plate at energy inputs of 100,000 and 50,000 joules/in.*

cases was 80 °F (26.7 °C). You will note the HAZ is reduced by decreasing the energy input and that the volume of metal heated to a temperature of approximately 1500 °F (816 °C) is considerably less at any instant of time. This effect is much more pronounced as the welding speed goes up beyond 12 in. (30.48 cm)/min. In fact, the temperature gradients normal to the direction of welding increase at an exponential rate with travel speeds above 12 in. (30.48 cm)/min.

Plate Thickness, Geometry, and Cooling Rates. Shown in Fig. 9-16 is the heat flow pattern in thin [less than $\frac{1}{2}$ in. (12.70 mm)] and thick (over $\frac{1}{2}$ in.) plates. The heat flow is considered to be two- and three-dimensional, respectively. The time a plate will remain at an elevated temperature tends to decrease with an increase in thickness. Typical cooling rates for a $\frac{1}{2}$ in. (12.70 mm) butt weld and fillet weld are shown in Table 9.2. Note that the cooling rate for fillet welds averages about three or four times that of butt welds in $\frac{1}{2}$ in. plate. This difference becomes less pronounced as the plate thickness increases and as successive beads build up the fillet weld.

Thermal Characteristics of the Base Metal. The thermal characteristics of

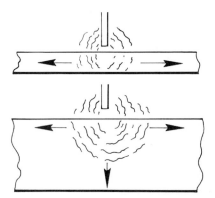

FIGURE 9-16. *Heat flow is two-dimensional in thin plates and three-dimensional in thick plates.*

TABLE 9.2.   *Typical Cooling Rates* (Courtesy of American Welding Society)

Welding Conditions Energy Input (joules/in.)	Preheat Temperature		Cooling Rate at 1200 °F (650 °C)			
			For Butt Welds in $\frac{1}{2}$ in. Plate		For Fillet Welds	
	(°F)	(°C)	(°F/sec)	(°C/sec)	(°F/sec)	(°C/sec)
50,000	72	20	20	11	80	44
50,000	250	120	13	7	62	34
50,000	400	205	9	5	36	20
100,000	72	20	approx. 7	4	18	10
100,000	400	205	approx. 3	1.7	9	5

a material are referred to as its diffusivity, or the thermal conductivity. In general, the higher the thermal conductivity is, the faster the cooling rate and the shorter the time at an elevated temperature for a given thermal cycle and peak temperature. The lower the thermal conductivity of a material is, the steeper will be the distribution of peak temperatures. Shown in Table 9.3. are common metals with their thermal diffusivity at room temperature. Also listed are the melting temperatures.

You will note, for example, that aluminum has a diffusivity of 0.912, whereas that of iron is 0.208. The peak temperature for aluminum is 1220 °F (660 °C) and that of iron is 2802 °F (1540 °C). Thus, by comparison, the HAZ of a fusion weld in iron would have a much higher peak temperature, but the distance from the edge of weld to where the metal remains at room temperature would be much less (about 23 %) than that of the aluminum.

TABLE 9.3.   *Peak Temperatures in Heat-Affected (HAZ) Zone* (Courtesy of American Welding Society)

Metal	Melting Temperature		Thermal Diffusivity at 68 °F (20 °C) (cm²/sec)
	(°F)	(°C)	
Aluminum	1220	660	0.912
Chromium	3430	1890	0.202
Cobalt	2723	1500	0.187
Copper	1981	1080	1.14
Iron	2802	1540	0.208
Lead	621	330	0.236
Magnesium	1202	650	0.873
Molybdenum	4760	2630	0.562
Nickel	2651	1455	0.236
Silver	1761	960	1.70
Tin	449	230	0.406
Titanium	3300	1820	0.063
Uranium	2065	1130	0.122
Zinc	787	420	0.414

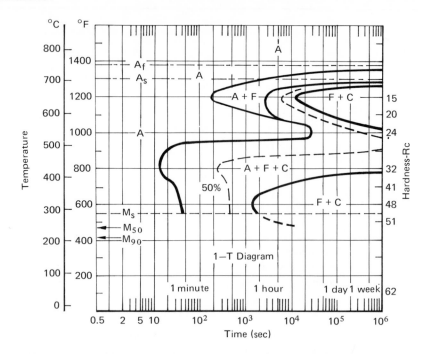

FIGURE 9–17.   *The TTT curve for 4340 steel.* (*Courtesy of United States Steel Corporation.*)

Control of HAZ Width.   The width of the HAZ can be controlled to some extent by the choice of process. For example, electron beam welding is capable of producing a very narrow HAZ due to the high concentration of energy. In contrast, an oxyacetylene flame has a relatively flat distribution of peak temperatures with a wide HAZ. A narrow HAZ is not always advantageous to have because of its higher cooling rate. This is especially true in alloy steels or in carbon steels where the carbon content is more than 30 %. As shown in the transformation curve for 4340 steel, Fig. 9–17, a fast cooling rate would produce some martensite.

CLASSIFICATION AND WELDABILITY OF STEELS

Today's competitive market requires that the welding engineer be familiar with an ever increasing array of materials and their properties in order to take full advantage of them in producing efficient, economical designs.

The greatest tonnage of any metal used in modern engineering structures and equipment is some form of iron and steel. In order to use these materials for the maximum benefit, the engineer must know the basic types, their classification, properties, and weldability.

Classification of Steels. Steel is obtained from a process of reducing iron ore in a furnace that contains principally coke and limestone. The coke (essentially pure carbon) combines with oxygen in the ore at high temperatures to provide carbon monoxide, the reducing agent. The limestone combines with other impurities in the ore to form a molten slag that can easily be removed. Pig iron has relatively large amounts of carbon (also from the coke), manganese, and silicon and must be further refined into one of three principal products: wrought iron, steel, or cast iron.

Wrought Iron. Wrought iron is nearly pure iron with approximately 4 % slag by volume. The slag, which is largely iron silicate, imparts a fibrous structure to the metal. Since the metal is of a low alloy content, it is readily welded. It is used for chains, anchors, railway couplings, crane hooks, and ornamental porch railings.

Carbon Steels.. Carbon steels may be classified by: (1) solidification characteristics, (2) chemistry, (3) mechanical properties, (4) standard specifications, and (5) end product specification.

Solidification Characteristics. Carbon steels are allowed to solidify under varying conditions that alter their characteristics. Under this classification the terms *killed, semikilled, rimmed, capped,* and *vacuum deoxidized* are used.

Killed. Killed steels, as the name implies, are chemically deoxidized so that they lie quietly in the mold as they cool. They are characterized by relatively uniform chemical composition and properties. Sheets, strips, and plates have excellent forming and drawing qualities. A large portion of the top of the ingot is "cropped" off before it is rolled.

Semikilled. These steels are just partly deoxidized before pouring into the mold. Thus some gas evolution takes place during solidification. Gases that do not escape before solidification cause blow holes. Semikilled steels are used for all but the most severe forming operations.

Rimmed. Rimmed steels derive their name from the fact that the outer rim of the ingot while in the cast iron ingot mold has considerable gas evolution during the initial solidification period. As the ingot begins to solidify, a layer of iron about three inches thick forms around the outer edge. The liquid in the center area retains the carbon, phosphorous, and sulfur. This peculiar pattern of segregation persists through forging, rolling, forming, and welding. Due to the combination of properties obtained, it is used in making stick electrodes and successfully manufacturing many other items. A comparison of killed, semikilled, and rimmed steels is shown in Table 9.4.

Capped. Capped steels are either mechanically or chemically capped to reduce the time of gas evolution. The characteristics are intermediate between rimmed and semikilled steels.

TABLE 9.4. *Comparison of Killed, Semikilled, and Rimmed Steels*

	Killed	Semikilled	Rimmed
% Carbon	Up to 1.5	Up to 1.0 %	Up to 0.3 %
% Silicon	Over 0.10	0.01 to 0.10 %	Less than 0.01
Segregation	None	Considerable	Considerable
Porosity	Very little	Little	Pronounced

Vacuum Deoxidized. Vacuum deoxidized steel is relatively new; and as the name implies, vacuum is used to extract the oxygen and other gases from the steel. This is normally done by adding deoxidizers to the melt. However, these form nonmetallic inclusions. In the newer process, vacuum is applied to the molten metal as it is poured. The carbon in the steel reacts with the oxygen to form carbon monoxide, which quickly escapes. The process causes the carbon content to drop about 0.05 %, but this can be provided for in the initial content. Since no solid oxides are formed by the process, the steel is considered to be quite "clean." A further benefit is that the gas content (hydrogen and nitrogen) can be kept low. After deoxidation, silicon and aluminum can be added to control grain size and other properties.

Chemistry Specifications. The chemistry of a steel may be specified in three ways: (1) by maximum limit, (2) by minimum limit, or (3) by an acceptable range. The chemistry of a steel refers to its "ladle analysis." A "check analysis" is supplementary to the ladle analysis and is taken from the semifinished or finished steel form. Some of the commonly specified elements are: carbon, manganese, phosphorus, sulfur, silicon, and copper.

Carbon. Carbon is the principal hardening element in steel. Hardness and strength increase with carbon content up to about 0.80 percent, but with a loss of weldability and ductility. The amount of carbon is kept low in rimmed steels for improved surface finish. By contrast, an increased carbon content improves the surface quality of killed steels.

Steels range, in carbon content, from 0.05 to 1.2 %. They are commonly classified as low, medium, or high carbon, as shown in Table 9.5.

Manganese. Manganese is a basic alloying element in steel and is carbon dependent for its effectiveness in controlling the depth of hardness. High manganese content with increasing carbon tends to lower both ductility and weldability.

TABLE 9.5. *Steel Classification as to Carbon Content*

Designation	Meaning
Low	0.05 to approximately 0.30 percent carbon
Medium	0.30 to approximately 0.60 percent carbon
High	0.60 to approximately 1.20 percent carbon

TABLE 9.6. *Main Affect of Alloying Elements in Annealed Steel*

Element	Influence on Ferrite	Influence through Carbides	Influence on Grain Size
Aluminum	Hardens by solid solution	Tends to graphitize carbon	Restricts grain growth by forming dispersed oxides or nitrides
Boron	Retards nucleation of ferrite	May form oxides or nitrides	
Copper	Shifts eutectoid composition to the right	Nil	
Chromium	Hardens slightly	Forms carbides	
Manganese	Hardens markedly	Forms carbides	
Molybdenum	Provides age hardening	Forms carbides	Raises grain coarsening temperature of austenite
Nickel	Strengthens and toughens	Tends to graphitize carbon	
Silicon	Hardens it with loss of plasticity	Graphitizes carbon	
Vanadium	Hardens moderately by solid solution	Strengthens carbides	Elevates grain coarsening temperature of austenite

Phosphorus. Phosphorus is normally an impurity and not an alloying element. Phosphorus increases strength and hardness but reduces ductility and impact toughness, particularly as the carbon content increases. Phosphorus improves resistance to atmospheric corrosion.

Sulfur. Sulfur is also normally an impurity, although commonly added to steels to increase machinability. As the sulfur content increases, transverse ductility, notch impact toughness, and weldability decrease. Sulfur has only a slight effect on longitudinal mechanical properties.

Silicon. Silicon is one of the principal deoxidizers used in the steel industry; it increases strength and hardness, but to a lesser degree than manganese. In amounts up to approximately .25 %, silicon increases toughness.

Copper. Copper improves atmospheric corrosion resistance when present in excess of 0.15 percent. Copper is detrimental to forge welding but does not affect arc or oxyacetylene welding. Some copper containing steels have greater resistance to degradation of properties due to radiation.

A general summary of the affect of alloying elements on ferrite and their influence in the formation of carbides and grain size is shown in Table 9.6.

Specification by Mechanical Properties. A common way for a designer to specify steels is by certain mechanical properties. This allows the steel producer to juggle the chemistry of the steel (within limits) to obtain the desired properties. The most common mechanical tests are bend tests, hardness tests, and a series of tension tests. Metallurgical tests may also be required to show the microstructure for decarburizations and inclusions.

Standard Specifications. Most carbon steel mill products are produced to standard specifications as prepared by those regulating agencies concerned with public welfare and safety. The largest and most influential of these is the American Society for Testing and Materials (ASTM). ASTM steels are listed in Table 9.9.

End Product Specifications. Fabricating operations such as deep drawing or welding can change the specifications for a steel. Thus, steel may be specified as having adequate properties for fabrication into an "identified" end product.

Steel producers often provide metallurgical and fabricating advice that will help achieve the desired end product most economically.

CLASSIFICATION OF ALLOY STEELS

The two most commonly used classification systems for identifying alloy steels are those provided by ASTM and the American Iron and Steel Institute (AISI). The ASTM classification of steels is given in Table 9.9.

TABLE 9.7. *American Iron and Steel Institute (AISI) Designation System for Alloy Steels*

Alloy Series	Approximate Alloy Content (%)
13XX	Mn 1.60–1.90
40XX	Mo 0.15–0.30
41XX	Cr 0.40–0.90; Mo 0.20–0.30
43XX	Ni 1.65–2.00; Cr 0.40–0.90; Mo 0.20–0.30
44XX	Mo 0.45–0.60
46XX	Ni 0.70–2.00; Mo 0.15–0.30
47XX	Ni 0.90–1.20; Cr 0.35–0.55; Mo 0.15–0.40
48XX	Ni 3.25–3.75; Mo 0.20–0.30
50XX	Cr 0.30–0.50
51XX	Cr 0.70–1.15
E51100	C 1.00; Cr 0.90–1.15
E52100	C 1.00; Cr 0.90–1.15
61XX	Cr 0.50–1.10; Va 0.10–0.15 (min)
86XX	Ni 0.40–0.70; Cr 0.40–0.60; Mo 0.15–0.25
87XX	Ni 0.40–0.70; Cr 0.40–0.60; Mo 0.20–0.30
88XX	Ni 0.40–0.70; Cr 0.40–0.60; Mo 0.30–0.40
92XX	Si 1.80–2.20

The AISI system generally uses four numbers. The first two digits refer to the alloy content and the last two (possibly three) to the carbon content in *points* of carbon, where one point is equal to 0.01 percent. Shown in Table 9.7 is a brief listing of the principal alloy types. An AISI 4150 steel, for example, would be a chromium molybdenum type containing the percentages of each element (as shown in Table 9.7) and approximately 0.50 % carbon.

Carbon Equivalent. Alloy steels present a challenge to the welding engineer. Although they are more expensive than plain carbon steels, the alloy steels achieve a wide variety of properties that make them particularly useful when increased strength, hardenability, response to tempering, and corrosion resistance is desired. Oftentimes two elements acting together will have a greater total effect than the sum of their individual quantities would indicate.

Alloy steels, containing judicious proportions of alloying elements, can often be welded with less difficulty than can carbon steels used to produce the same yield strength in the weldment. That is, a plain carbon steel would have to have a relatively high carbon content to achieve the strength of an alloy steel. As an example, an alloy steel with 0.10 % *carbon* can be equivalent to a carbon steel having 0.30 % carbon. The alloy steel will be easier to weld since there will be less tendency for brittleness and underbead cracking.

Several formulas have been postulated to determine the effect of adding various alloying elements in terms of carbon, or as it is referred to, as carbon equivalent (CE).

$$CE = \%\,C + \frac{\%\,Mn}{4} + \frac{\%\,Ni}{20} + \frac{\%\,Cr}{10} - \frac{\%\,Mo}{50} - \frac{\%\,V}{10} + \frac{Cu}{40}$$

You will notice that a given percentage of all elements are added, except molybdenum and vanadium. Whether these two elements are added or subtracted depends on whether they are present in solid solution, in which case they aid hardenability and should be treated as a positive factor, or whether they are present as complex carbides that are ineffective in hardenability. In the low alloy steel as used in construction and machine manufacture, they are taken as negative values.

The formula can only be used as an approximate guide, since it does not provide for grain size or prior heat treatment. Its usefulness is in determining welding conditions, as for example, if the CE exceeds 0.45 %, approved low hydrogen electrodes and preheating are required, Table 9.8. For a CE of not more than 0.45 %, approved low hydrogen electrodes are to be used, but preheating is not generally required except under conditions of high restraint or low ambient temperatures. If the CE is less than 0.41 %, any type of approved higher tensile electrode may be used, and preheating is only required under exceptional conditions.

Structural Steels. The American Institute of Steel Construction (AISC) lists twelve steels that have been approved and classified by the American Society for Testing Materials (ASTM). The main types of these steels are shown in Table 9.9. The AWS Structural Welding Code, D1.1-72, lists 27 steels that are approved. The structural steels listed may be placed in three broad categories:

TABLE 9.8. *Welding Precautions for Various Carbon Equivalents* (Courtesy of the *Welding Journal*)

Carbon Equivalent	Welding Procedure
<0.40	No precautions: Weldable with E6010 and E6012 electrodes
0.40–0.48	Weldable with: (a) ordinary electrodes and low preheat [200–400 °F (93–204 °C)], or (b) low hydrogen electrodes
0.48–0.55	Weldable with: (a) ordinary electrodes and moderate preheat [400–700 °F (204–371 °C)], or (b) austenitic electrodes, or (c) gas metal arc welding
>0.55	May be weldable with: (a) low hydrogen electrodes, and moderate or high preheat, or (b) austenitic electrodes, or (c) gas metal arc welding

Carbon steels.
High strength, low alloy steels (HSLA).
Heat treated, low alloy steels.

Carbon Steels.   Almost all structural carbon steels used today are of the A36 class. These steels provide an excellent material where unit stresses are low and rigidity is a main consideration. The 36 designation refers to the minimum yield strength of 36 ksi (248.2 MPa). Carbon steels provide excellent weldability, neither preheat nor postheat treatments are required. Class A36 replaces an earlier A7 carbon steel because of superior weld qualities.

TABLE 9.9.   *ASTM Structural Steel, Cast Steel, and Steel Forging Designations*

ASTM Designation	Description	Yield Strength in psi (MPa)	Tensile Strength in psi (MPa)	% Elongation in 8 in. (20.32 cm)	% Elongation in 2 in. (5.08 cm)
Structural Steels					
A36	Structural steel.	36,000 (248.2 MPa)	58 to 80,000 (399.9 to 551.6 MPa)	20	23
A242	High strength, low alloy.	50,000 (344.7 MPa)	70,000 (482.6 MPa)	19	22
A375	Low alloy, hot-rolled steel sheet.	47,000 (324.1 MPa)	67,000 (462.0 MPa)	19	
A440	High strength structural steel.	50,000 (344.7 MPa)	70,000 (482.6 MPa)	18	
A441	Low alloy structural manganese vanadium.	50,000 (344.7 MPa)	70,000 (482.6 MPa)	18	
A588	High yield strength, quenched and tempered alloy steel plate.	50,000 (344.7 MPa)	70,000 (482.6 MPa)	18	
A572	High strength, low alloy columbium–vanadium steels of structural quality.	42,000 to 65,000 (289.6 MPa to 448.2 MPa)	60,000 to 85,000 (413.7 to 586.1 MPa)	20 to 40	24
Cast Steels					
A27	Grade 65–35—Mild-to-medium strength carbon steel castings.				
A148	Grade 80–50—Steel castings for structural purpose.				
Carbon Steel Forgings					
A235	Class C1, F, and G—General industrial use.				
A237	Class A—General industrial use.				

High Strength, Low Alloy Steels (HSLA). All the structural steels listed in Table 9.9, with the exception of A36, are HSLA steels. The addition of alloying elements enhances their hot rolled strength. As an example, A572, the most economical steel of this classification, achieves a yield strength of 42 to 65 ksi (289.6 to 448.2 MPa) primarily by the addition of columbium and/or vanadium, or nitrogen with vanadium.

Grade A588 is especially suited for architectural and uncoated applications. Its corrosion resistance to atmospheric conditions is rated as 5 to 8 times that of carbon steel. An oxide coating develops that is considered naturally esthetic and need not be painted.

A440 was developed in the late 1950s as a high strength steel that was more economical than A242. Lower cost was achieved by increased use of carbon and manganese, however, ductility decreased—it is not always recommended for welding.

A441 was introduced along with A440 to be cheaper than A242 but with the same weldability as A242. Its use is now rapidly declining because A572 has higher strength values. As with A242 and A440, the chemistry is constant for all thicknesses, hence the yield load is lower for the thinnest sections.

The A572 specification covers six grades of nonweathering steels. These grades are 42, 45, 50, 55, 60, and 65. In this case the numbers represent the minimum yield point. Within any one grade, the yield point is constant for all thicknesses available. Grades 42 to 65 are for bolted or welded construction.

In general, appropriate strengths are obtained mainly by the use of columbium (0.005–0.05 %) and/or vanadium (0.01–0.10 %). The Bethlehem Steel Company steels V42 to 65 contain both vanadium and nitrogen. The A572 steels are now replacing A242, A440, and A441 steels where strength is the most important consideration.

Heat Treated Low Alloy Steels. Steels in this classification (AISI 588 and 514) may be heated, quenched, and tempered to obtain maximum yield strengths of 80 to 110 ksi (551.6 to 758.4 MPa). These steels are weldable when the proper procedures are used and no additional heat treatment is required except occasionally for stress relieving.

The heat treatment process used for these steels consists of a rapid quench in water to about 300–400 °F (149–204 °C) after heating to 1625–1675 °F (885–914 °C). Tempering consists of reheating to about 1100 °F (593 °C) and allowing to cool in still air at room temperature. Although tempering reduces strength and hardness, it greatly improves toughness. The reduction in strength and hardness is counteracted by a secondary or precipitation hardening that begins at about 950 °F (510 °C) and accelerates as the temperature is brought to the full tempering temperature of 1250 °F (677 °C). Columbium, titanium, and vanadium carbides precipitate out of solution at this temperature range. Moderate preheating and postheating is usually recommended for welding steels in this classification. Specific recommendations for welding structural steels are specified in AISC* 1.23.6, AWS D1.1–72, Table 4.1.7, and D2.0–72, Table 4.2.

A comparison of the stress–strain characteristics of various structural steels

_____
* American Institute of Steel Constructors.

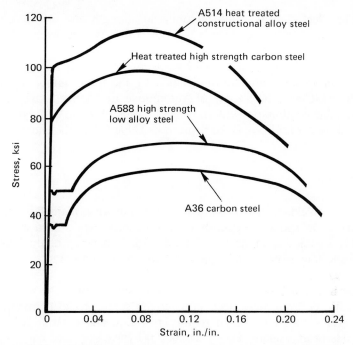

FIGURE 9–18. *Typical stress–strain curves for structural steels having specified minimum tensile properties.* (*Courtesy of United States Steel Corporation.*)

is shown in Fig. 9–18. Index numbers indicating the relative cost of HSLA steels based on unit yield strength of each material are shown in Table 9.10. The table assumes that carbon steel has a yield strength of 35,000 psi (241.3 MPa). The figures are only to be used as approximations, since cost varies with mill form, grade, added quality or chemical requirements, quantity ordered, and other market variables.

Carbon Steel and High Strength Steel Castings.   A mild-to-medium strength carbon steel casting material is listed in the AISC code as ASTM A27, class 65–35. High strength steel castings for structural purposes are designated as A148, class 80–50. The class designation refers to the tensile and yield strengths, respectively.

TABLE 9.10.   *Relative Cost of HSLA Structural Steels Based on Yield Strengths*

Steel Type	Cost Index (%)
Carbon steel	100
A–242	90
A–441	85
A–440	75
A–572	70

The effect of heat treatment on yield strength and elongation based on carbon content of various cast steels is given in ASTM specification A148–71.

Carbon and Alloy Steel Forgings for General Industrial Use. Materials in this class are commonly designated by the basic numbering system adopted by the AISC and the Society of Automotive Engineers (SAE), to show many of the possible combinations of alloying elements. A table of forging specifications, use, and mechanical properties is given in Table 9.A at the end of the chapter.

WELDABILITY

Weldability is a complex term that involves many factors, however, it may be briefly defined as the ease with which materials are welded with satisfactory results. The weldability of steels will be discussed in terms of the broad classification of steels, low carbon, alloy, quenched, and tempered types.

Low Carbon Steels. Low carbon steels can ordinarily be welded without difficulty. However, even these steels may require some special consideration when used in heavy sections. Increased rigidity, restraint, and drastic quench effect may cause underbead cracking. Heavy sections also have increased carbon content in order to maintain the same yield strength. Normally a test mill report is available for the specific analysis of the steel. From this, a decision can be made whether or not any special requirements are necessary to produce crack-free welds.

Alloy Steels. The welding of low and intermediate alloy steels such as the Cr/Mo grades is similar to welding medium carbon steels. In these steels a tendency exists for the weld to crack even though it is of the same composition as the base metal. As the weld cools, the austenite structure changes to either ferrite and carbide, bainite or martensite, or a combination of these structures. The structure that does form depends on the cooling rate, the lowest temperature reached, and the alloy content of the deposit.

The effect of alloying elements upon transformation in steel and upon the critical cooling rate is illustrated schematically in Fig. 9–19. The time, temperature, and transformation curves for plain carbon steel and an alloy steel, both containing 0.2 % carbon, are compared. Note that the addition of alloying elements such as manganese, nickel, chromium, and molybdenum produce a steel that can be cooled at a considerably slower rate and yet avoid the nose of the transformation curve.

Cooling from the austenite temperature results in a martensitic structure in the case of alloy steels, whereas the same quench rate in plain carbon steel transforms it to a relatively low strength pearlite, ferrite structure. In alloy steels, the entire weld and most of the HAZ will transform into martensite unless a preheat and postheat temperature of 800 °F (427 °C) is used.

Normally, it is considered neither necessary nor practical to attempt to prevent the formation of martensite. However, it is important to know that martensite

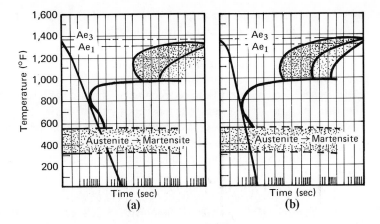

FIGURE 9–19. *A comparison of TTT curves for medium and low hardenability alloys. The medium alloy steel can have a considerably slower quench and still produce a martensite structure. A low hardenability alloy must be quenched at a severe rate (b) if complete martensite is to be obtained. However, it is likely that some softer structure will be included with the martensite. (Courtesy of Machine Design.)*

will form under cooling rates encountered during the welding of some alloy steels, and that precautionary steps should be taken during the welding process to prevent cracking. Cracks will form in Cr/Mo grades or low alloy, high strength nickel-chromium-molybdenum steels if the weldment is allowed to cool rapidly enough to be within the martensite formation range; the weldment is particularly prone to cracking if made under conditions of high restraint.

Several methods can be used to prevent cracks in alloy steel welds. One method is to preheat the weldment so that the cooling rate will be considerably reduced. The preheat temperature, usually about 500 °F (260 °C), should be maintained throughout the welding operation. Another method is to insure that no hydrogen in any form comes in contact with the molten weld pool. Low hydrogen electrodes are available, and if properly applied, will greatly reduce the likelihood of cracking; however, cracks may still occur in the HAZ of such steels as AISI 4150, 4340, and A–572.

Alternative weld processes sometimes used to produce "hydrogen free" welds are submerged arc, GMA, and GTA. A gaseous shield of argon, helium, or CO_2 does not allow significant amounts of hydrogen bearing compounds to enter the weld metal. Fluxes for submerged-arc should be stored in an oven for several hours at 250 °F (121 °C) before use to assure low hydrogen conditions.

Cracks may also be prevented in alloy steel welds by reducing the *combined effects* of high hardness and high shrinkage stresses. As an example, a crack-free weld can be made in 5 % Cr steel pipes using a relatively low preheat (200–300 °F)

(93–149 °C) with no special postheat if it is not under restraint. The weld will be martensitic with high hardness and relatively low ductility. If the same pipe is welded with the ends rigidly restrained, cracks will develop during cooling. Also, if a cellulosic coated electrode were used to make the weld, the susceptibility to cracking would be greatly increased.

In general, it is important to recognize that in welding alloy steels the cooling rate may be so fast that martensite or other brittle structures can form, thereby causing various types of cracking either in the base or weld metal. Therefore, anything that will reduce the cooling rate or embrittlement will affect the type and degree of cracking; such factors include joint design, chemistry, metal thickness, and restraint.

Broad Classification of Steels for Weldability.   Steels can be classified into four general categories as to weldability:

1. Steels that transform almost entirely to the high temperature transformation products (ferrite and cementite) under cooling rates normally encountered during welding. In this class are the low and medium carbon and plain carbon structural steels. Generally no preheat is required, particularly if low hydrogen electrodes are used. Heavy sections may require a low preheat of about 200 to 250 °F (93–121 °C). The relationship of hardness, thickness, and welding speed is shown schematically in Fig. 9–20.

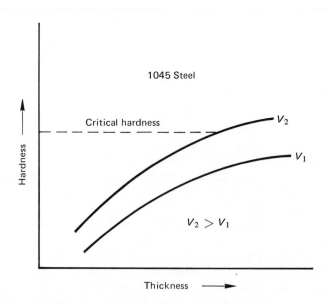

FIGURE 9–20.   *The hardness of a medium carbon steel weld is directly related to the quenching action of steel mass and the speed at which the weld is made.*

2. Steels that transform to martensite under the cooling rates normally encountered in the welding of heavy sections if a preheat is not used. Moderately high preheats (up to 400 °F, 204 °C) will retard the cooling rate sufficiently so that at least a partial transformation to ferrite and cementite occurs. Steels in this category are those containing 0.30 to 0.40 % carbon and 1.00 to 2.00 % manganese and the low carbon, low alloy steels with up to 1 % chromium in combination with lesser amounts of molybdenum and nickel.

3. Steels that normally transform almost entirely to martensite after welding. Hardenability is high and the cooling rates encountered after welding cannot be sufficiently retarded to effect a partial transformation except by preheating. In welds made in heavy sections under high restraint, controlled preheat and postheat treatments are generally required. Steels in this category are chromium molybdenum steel (2 to 9 % Cr and up to 1 % Mo), the straight chromium, martensitic stainless steels, the high carbon Ni-Cr-Mo steels, and the high strength Mn-Mo-V steels.

4. Steels of high alloy Cr-Ni, Cr-Ni-Mo, or Cr-Ni-Mn austenitic stainless steels. The cooling rates after welding these steels have little or no effect on weld quality. Cracking problems can normally be attributed to incorrect chemical analysis or excessive grain size. Included in this class are stainless steels of the 304, 316, 309, and 307 types.

STAINLESS STEELS

Since the turn of the century, "stainless steel" has been a name used to cover a large family of specialized steels. There are now over sixty different varieties, which tend to make intelligent selection difficult if not totally confusing. The usual way of classification has been three types, as shown in Table 9.11. As the name implies, martensitic stainless steel can develop martensite through heat treatment. The ferritic types are not hardenable but are tough and ductile. The austenitic alloys are the most widely used because they are extremely corrosion resistant, tough, and ductile, even at low temperatures. They cannot be hardened by heat treatment but only by cold working.

Classification. The American Iron and Steel Institute has assigned numbers to identify stainless steels as the 200, 300, 400, and 500 series. There is some over-

TABLE 9.11. *Classification of Stainless Steels*

Class	Range of Composition		
	Carbon	Chromium	Nickel
Martensitic	0.05–0.50 %	4–18 %	—
Ferritic	0.35 % max.	16–30 %	—
Austenitic	0.03–0.25 %	14–30 %	6–36%

lapping between the series. The general purpose type of each is: ferritic, 430; martensitic, 410; and austenitic, 302.

In addition there are *precipitation hardening stainless steels* such as PH 15–7 Mo, 17–4 PH, and 17–7 PH that are copywrited by Armco Steel Corporation. In this case, the main alloy is the percent chromium, which is designated by the first number. The second number is the percent nickel. There is also some molybdenum in the 15–7 Mo type. These steels can maintain high yield strengths from room temperature up to the 600–1000 °F (316–538 °C) range.

Joining Stainless Steels—Weldability. The weldability of stainless steels varies with each family group, however, three general principles apply regarding physical properties:

1. Stainless steels have a higher coefficient of thermal expansion than carbon steels, and therefore, precautions should be taken to avoid warpage. Jigs and fixtures must be designed to exert adequate restraint. Sufficient tack welds are needed to keep the parts firmly in alignment during welding.
2. Low thermal conductivity causes the heat to be conducted away from the weld at a low rate, thus the energy input can be kept low.
3. The high electrical impedance of stainless steels permits the use of smaller units for resistance type welds.

The three conditions listed are shown graphically in Fig. 9–21, by comparing them to carbon steel and aluminum.

The welding engineer should be aware of two metallurgical problems that may occur in welding stainless steels; namely, the development of a *sigma phase* (a hard iron chromium compound) and *chromium carbide precipitation.*

Sigma Phase. The hard, extremely brittle, sigma phase forms principally at the grain boundaries when the steel is heated at 950 to 1850 °F (510–1010 °C). Sigma forms slowly in the high chromium steels, somewhat faster in Cr/Ni steels such as 18–8, and most rapidly in the stainless steels that have ferrite forming elements added such as silicon, niobium, and molybdenum.

If stainless steels that are susceptible to sigma phase formation are used in high temperature service, periodic anneals of 1800 °F (982 °C) are recommended. This allows chromium to diffuse back to the impoverished areas with a consequent improvement in corrosion resistance and uniformity of structure.

Chromium Carbide Precipitation. When an austenitic stainless steel is heated to the 1000 to 1300 °F (538–704 °C) range, carbon combines with chromium and precipitates to the grain boundaries, Fig. 9–22. This does not normally affect the mechanical properties; however, for certain steels that have 0.04 % carbon or more, such as types 204, 304, and 308, there can be a decided decrease in corrosion resistance in the weld and HAZ.

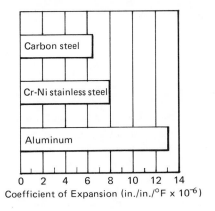

Coefficient of Expansion (in./in./°F x 10⁻⁶)

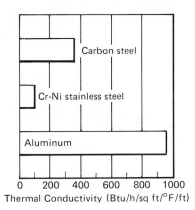

Thermal Conductivity (Btu/h/sq ft/°F/ft)

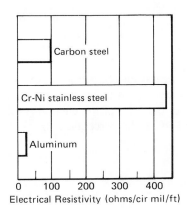

Electrical Resistivity (ohms/cir mil/ft)

FIGURE 9–21. *Stainless steel compared to carbon steel and aluminum—as to coefficient of expansion, thermal conductivity, and electrical impedance. (Courtesy of Machine Design.)*

Fortunately stainless steels are available that are stabilized, such as 347 and 321. These steels have additives (Cb/Ti) that prevent the carbides from precipitating or withdrawing from the chromium. Type 347 stainless, when welded with the same filler material, can be joined by any of the common welding processes without carbide precipitation. Type 321 stainless requires one of the inert gas processes or the loss of titanium across the arc will be too high. Carbide precipitation may also be kept to a minimum by using stainless steels designated as having extra low carbon content (ELC).

Stainless Steel Weld Deposit Composition. The weld designer needs to know if a stainless steel weld will contain a small amount of ferrite, or be mostly austenitic or martensitic. An easy method of determining the approximate composition of a stainless steel weld has been developed by Anton L. Schaeffler of the Arcos Corporation, it is known as the Schaeffler diagram, Fig. 9–23.

Chromium and the elements Si, Mo, and Cb, which act like chromium—from a microstructural viewpoint, are grouped to form the "chromium equivalent."

313

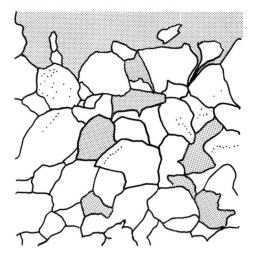

FIGURE 9–22. *Chromium carbides form at the grain boundaries, depleting certain zones of chromium and thereby making these zones susceptible to corrosion.*

Ni with the elements C and Mn, which act like nickel, are grouped to form the "nickel equivalent." The effect of each element is given on a percentage basis as shown on each axis of the diagram. Thus, in calculating the effect of each element in the composition, each element is multiplied by the percentage shown and added to the percentage of chromium or nickel.

> **Example.** Determine the approximate weld composition of a 308 stainless rod on similar base material. 308 stainless = C, .08; Mn, 2.00; Si, 1.00; Cr, 20.00; and Ni, 11.00.

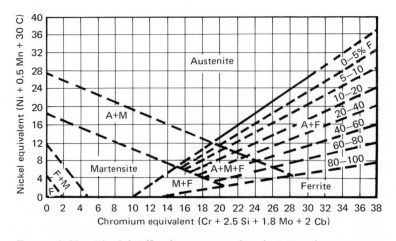

FIGURE 9–23. *The Schaeffler diagram is used to determine the composition of stainless steel weld deposits. (Courtesy of Arcos Corp.)*

Solution:

$$\text{Ni equivalent} = \text{Ni} + 0.5\,\text{Mn} + 30\,\text{C}$$
$$= 11 + 1 + 2.4$$
$$= 14.4$$

$$\text{Cr equivalent} = \text{Cr} + 2.5\,\text{Si} + 1.8\,\text{Mo} + \text{Cb}$$
$$= 20 + 2.5 + 0 + 0$$
$$= 22.5$$

By the use of the Schaeffler diagram, the composition is shown to have 5 % ferrite and the rest austenite.

A small percentage of ferrite as shown in the example will assure the designer that the weld deposit will be crack resistant. On the other hand, if the ferrite is too high, the electrode may be changed or, if martensite is to be avoided, this can also be determined by the electrode composition.

WELDABILITY OF PRECIPITATION HARDENING ALLOYS

Precipitation hardening alloys obtain their strength and hardness through a phenomenon whereby a fully dispersed second phase is precipitated from a super-saturated solid solution containing one phase. During the solution treatment, all of the β phase becomes dissolved in the α phase. This composition is only stable at or above the solidus line or the line dividing the single phase composition from the two phase composition. Quick cooling produces the supersaturated condition with the α phase retaining a greater amount of element β than is dictated by the phase diagram. In other words, if equilibrium were maintained during the cooling process, the β phase would precipitate out at preferred locations such as the grain boundaries. When the cooling rate is high, the β phase has no time to precipitate, hence the α phase retains greater amounts of element β in solution but in an unstable form.

In some alloy systems, if left at room temperature, the β phase will precipitate out and form a more stable structure. Heating to the so-called "aging" temperature range causes precipitation to occur at a faster rate. The excess β is rejected as finely divided, uniformly distributed precipitates. The effect is a distortion of the lattice structure, which retards slip, producing greater strength. The metal is said to be *precipitation hardened*. The time and temperature at which the aging takes place will affect the distribution of the β phase and the properties of the alloy.

At still higher temperatures, aging is accelerated and *overaging* may result. In overaging, the finely divided precipitate begins to agglomerate. During this process the particles grow and become fewer in number. Stress on the lattice structure is reduced. Common precipitation hardening alloys are aluminum with copper to form $CuAl_2$ upon aging; and 17–7 PH stainless steel, which contains 0.07 % C, 17 % Cr, 7 % Ni, and 1.2 % Al.

The brief review of the precipitation hardening phenomenon makes it readily apparent that the weld metal as well as the HAZ pass through temperatures that will cause modifications in the precipitated second phase. Assuming that the alloy has been properly aged prior to welding, several of the following changes can be expected in the HAZ:

1. The zone closest to the weld will undergo a resolution treatment of the second phase.
2. The next zone that just exceeds the solution temperature will be partially resolutioned.
3. The zone that falls in the aging temperature range may be further aged. The degree of overaging will vary with the welding conditions.

Generally the best results will be obtained by welding PH stainless steels in the solutionized condition with comparatively low energy input. Postweld heat treatments can be used to restore the HAZ to a stable high strength condition, but the best results are obtained on weldments that have been subjected to the solution treatment prior to welding.

MARAGING STEELS

The word "maraging" was derived from two hardening reactions: MARtensite and AGING. Strictly speaking, the steels are neither martensitic nor age hardenable. The maraging steels were first introduced in 1959 as a new type of high strength steel. Yield strengths range in the 200 to 300 ksi range. The first steels of this type contained varying amounts of nickel, up to 25 %. The most widely used, however, is an 18 % nickel type, which is commercially available in four varieties: 18 Ni 200 (1 379 MPa), 18 Ni 250 (1 724 MPa), 18 Ni 300 (2 068 MPa), and 12.2 Ni Mn 184 (1 269 MPa). The yield strength for each is designated in ksi. The fourth type has an alloy content of 12 % nickel and 2 % manganese. ASTM has designated the first three as maraging steels grades A, B, and C, respectively.

The maraging steels are essentially precipitation hardening, but aging does not take place at ambient temperatures. The martensite like structure gains its strength from an extremely large number of dislocations. The heat treating sequence for the 18 % Ni steel is accomplished by first annealing the metal by heating to 1500 °F (816 °C) to dissolve and put into solution the elements that form hardening compounds. Air cooling from the solution treatment will produce a martensitelike structure of about 30 R_c. Aging takes place when the metal is heated to 900 °F (482 °C) for three to six hours to allow precipitation of the hardening compounds. Air cooling will then produce the maraged structure at 52 R_c, as shown schematically in Fig. 9–24.

The corrosion rate of maraged steels is about half that of low alloy heat treated steels. Corrosion cracking resistance is superior to that of the HSLA steels, but protection is still required in marine and industrial environments.

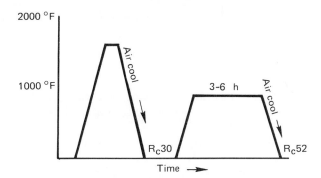

FIGURE 9–24.   *Heat treatment of 18% Ni maraging steel.*

Weldability.   The maraging steels are weldable without preheat in both the annealed and maraged conditions. Only a postweld aging treatment is required to restore properties in the HAZ and to develop good strength in the weld metal. Maraging steels are subject to hydrogen embrittlement, but they exhibit greater tolerance than other high strength steels. Postweld heating to about 900 °F (482 °C) for one to three hours usually restores *almost* full mechanical properties. Two narrow regions in the HAZ do not respond completely to postheating. One region is the grain coarsened area immediately adjacent to the weld metal and the other is the farthest extent of the HAZ, where peak temperatures during welding reach 1100 to 1200 °F (593–649 °C). This region contains stable austenite converted from the preexisting martensite. Thus, the maximum joint strength obtainable in a weldment hardened only by postweld maraging is limited by the amount of precipitation that takes place in the austenite region.

CAST IRONS

Cast irons are largely a three element alloy: iron, carbon, and silicon. These materials may be subdivided into the five following types depending on how most of the carbon occurs in the metal: gray, white, malleable, ductile (nodular), and high alloy. Within each of these types are a number of grades classified by their minimum ultimate tensile strengths.

Carbon is the single most important element in cast irons. It represents approximately 2 to 6 percent of the molten metal when poured. Less than 2 percent of the carbon can remain in solution as the metal cools. The excess carbon separates out during solidification to form free graphite, or as in the case of white cast iron, the excess carbon forms iron carbides. The excess carbon, along with a high percentage of silicon, is the key to high fluidity during the casting process and low shrinkage during the cooling process. High carbon also accounts for high damping capacity of the metal and for the good machinability.

Cast Iron Matrix Structure. The widely differing mechanical properties found in cast irons are due to the differences in matrix structures that surround the graphite. It is common to find several different matrices in one casting. The matrix can be controlled by heat treatment, however, the graphite, once formed, remains constant.

Ferrite. Ferrite grains, or free carbon grains in the matrix provide ductility. The structure is not gummy, as in low carbon steel, because of the silicon content.

Pearlite. Much of the matrix consists of pearlite, alternate layers of Fe and Fe_3C. This part of the matrix provides a strong wear resistant structure. The finer the laminations, the harder and stronger the iron.

Martensite. The martensite matrix results from a cooling rate greater than equilibrium, or faster than the carbon is able to separate out of solution with the ferrite. It is unmachinable; however, when tempered, it can be machined and made to provide a good wearing surface of optimum strength and toughness.

Acicular. Small amounts of nickel and molybdenum correlated with the cooling rate can suppress the pearlitic change point and an acicular intermediate constituent (ferrite needles in an austenitic matrix) can be produced with high mechanical properties.

Gray Cast Iron. Gray iron derives its name from the grayish appearance of the metal surface when fractured. The color is caused by the large amount of visible graphite or carbon. The composition is largely pearlite and flake graphite. Common grades are given by class, which refers to the tensile strengths. These range from class 20 to class 60.

White Cast Iron. White cast iron has a very white or bright surface when fractured. The carbon is of a combined, martensitic matrix that is very hard and brittle. It has excellent wear qualities and is used for rolling mill rolls, brake shoes, and machinery handling abrasive materials such as used in clay mixing and brick making.

Malleable Cast Iron. Malleable iron is made from white cast iron by subjecting it to a long annealing process, usually done in two stages. The first stage consists of heating the white cast iron to about 1700 °F (927 °C) so the coarse primary carbides can become graphitized into flake nodules. The temperature is then lowered to about 1325 °F (718 °C) when the iron carbide pearlitic laminations are separated and graphetized. The usual time needed for the two-stage process ranges from 30 hours in modern furnaces to 150 hours in older batch type furnaces. A fracture will reveal a white rim with a dark center, sometimes referred to as "black heart" malleable.

The two main types of malleable iron are ferritic, the original type, and pearlitic, developed later. Of the two, ferritic is more machinable and more ductile, but pearlitic has a higher strength and hardness. A five number system is used to designate the various types. A common ferritic type is 32510, which refers to a yield strength of 32,500 psi (225 MPa), and an elongation of 10 percent. The pearlitic types range in yield strength from 45 to 80 ksi (310.3–551.6 MPa) with corresponding elongation of 2 to 10 %.

Ductile Cast Iron. For generations foundrymen sought a way to transform brittle cast iron into a tough, strong, ductile material. Finally it was discovered that a small amount [about 1 lb/ton (0.4536 kg/907 kg)] of magnesium added in the melting process caused the graphite to take on a spheroidal shape. Thus, the former weakening and embrittling effects were removed. The advantages of steel, i.e., high strength, toughness, ductility, and wear resistance, were combined into an easily cast material.

Ductile irons are identified by three sets of digits that designate tensile strength, yield strength, and percent elongation. Thus a ductile iron of class 80-60-03 refers to a minimum tensile strength of 80 ksi, a minimum yield strength of 60 ksi (413.7 MPa), and an elongation of 3 % in a 2 in. (5.08 cm) test length. The microstructures of gray, malleable, and ductile cast irons are shown in Fig. 9–25.

High Alloy Irons. High alloy irons refer to appreciable alloy contents, generally more than 3 %. High alloy irons may be gray, ductile, or white. These irons have three general use classifications: wear resistance, heat and corrosion resistance, and special applications. An example of the latter is the need for an iron that will have only one-fifth the coefficient of expansion of steel to maintain high accuracies despite temperature changes. The coefficient of expansion may be varied to the desired amount within the range of 2.2 to 10.6 $\times$ 10^{-6} in./in. °F by changing the alloy content.

Weldability of Cast Irons. The purpose of producing a part by the casting process is to have completeness as far as possible with the minimum of added parts. Thus, welding is seldom used except in making repairs or in joining the cast iron to some other type of metal. Care must be exercised in welding cast irons so that the thermal conditions involved do not alter the properties of the metal.

Either of two approaches may be taken in welding gray cast iron. One involves preheating, the metal is kept uniformly heated during the welding operation to prevent localized hard spots and stresses. The oxyacetylene process is often used with a preheat of 900 to 1200 °F (482–649 °C) and slow cooling by providing postheat until the metal reaches room temperature. The usual filler rods are cast iron, containing C, 3.5; Si, 3.0; and P, 0.6 percent. Prior to welding, the surface must be cleaned. If the welding is for repair purposes, the cracks should be cut to a vee form with a chisel. Flux is necessary to keep the molten pool fluid and prevent inclusions from forming, and also blow holes.

The opposite approach, that of keeping the heat as low as possible, is done by arc welding. Short intermittent welds (one or two inches) are made and lightly peened immediately after the deposit is made. The first pass is often made with a copper nickel alloy to prevent it from becoming hard and brittle. Stainless steel electrodes of the 18–8 or 304 type are also used. A rule of thumb is not to let the

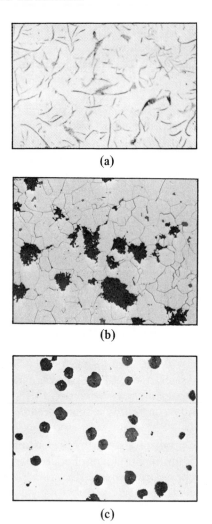

(a)

(b)

(c)

FIGURE 9-25.   *Microstructures of cast irons:* (a) *gray,* (b) *malleable, and* (c) *ductile.*

casting get any hotter than can be handled with bare hands. Cast iron repair work is often done by braze welding with a bronze or nickel copper rod and an oxyacetylene torch.

Malleable cast irons are not considered weldable, at least not in the same sense as gray cast iron. The heat required to melt the edges of the metal will completely destroy the malleable properties. A long term anneal would be required to restore the original properties. There are times, however, when stresses are low, or are in compression only, that welds can be successfully made. A commercially pure nickel rod or a 10 percent aluminum bronze rod is used.

Ductile irons can be welded in the same manner as gray iron. Most arc welding processes can be used to join it to similar materials such as carbon steel, stainless steel, or nickel. The most easily welded type of ductile iron is 60–45–10.

Cast Steels. Steels are most extensively used in the wrought form and fabricated by welding and machining. Steel castings are also used extensively especially for more intricate parts. The same AISI and SAE code systems as used on wrought steels are used on the cast materials. About 60 percent of all steel castings are in the medium carbon range.

Weldability of Steel Castings. Steel castings are welded by the same methods employed in welding wrought steel. With relatively few exceptions, all castings are stress relieved after welding, or they may be given a full heat treatment; e.g., they may be normalized, quenched, and tempered. This eliminates excessive hardness next to the weld as well as severe weld stresses. Preheating may also be used as a means of minimizing localized stresses. Preheats of 200 to 400 °F (93–204 °C) are recommended for the medium carbon steels and over 400 °F (204 °C) for castings containing higher carbon and more alloying elements.

ALUMINUM

Commercially pure aluminum has a tensile strength of about 13,000 psi (89.63 MPa). Working or strain hardening the metal can make it reach twice this figure. Much larger increases in strength can be obtained by alloying aluminum with a small percentage of other metals such as manganese, silicon, copper, magnesium, or zinc. Like pure aluminum, the alloys can also be made stronger by cold working. Some alloys are further strengthened by heat treatment, reaching nearly 100 ksi (689.5 MPa).

Aluminum Alloy Designation. Wrought aluminum alloys are identified by a four digit system. The first digit identifies the alloy group as shown in Table 9.12. The second digit indicates a modification of the original alloy or an impurity limit, and the last two digits identify the specific aluminum alloy or indicate the aluminum purity.

In the 1XXX group, the last two digits indicate the minimum aluminum percentage expressed to the nearest 0.01 percent. The second of the four digits indicates a modification of the impurity limits. If the second digit is zero, there

TABLE 9.12. *Designation of Aluminum Alloys*

Aluminum—99.00 % purity or greater	1XXX
Copper	2XXX
Manganese	3XXX
Silicon	4XXX
Magnesium	5XXX
Magnesium and silicon	6XXX
Zinc	7XXX
Other elements	8XXX

TABLE 9.13. *Common Temper Designations for Aluminum Alloys* (Courtesy of *Machine Design*)

–O–	Annealed.
–H	Strain hardened. The H is followed by a digit as –H1 indicates cold working; –H2, cold worked but partially annealed; –H3, strain hardened and stabilized. A second digit ranging from 1 to 9 indicates the degree of hardness with 9 representing extra hard.
–T1	Naturally aged.
–T2	Annealed (cast products only).
–T3	Solution heat treated and then cold worked.
–T4	Solution heat treated followed by natural aging.
–T5	Artificially aged from the as-cast condition.
–T6	Solution heat treated followed by artificial aging.

is no special control on individual impurities. Thus, an alloy containing a minimum of 99.60 percent aluminum and which requires no control of impurities would be designated as 1060.

In the 2XXX through 8XXX alloy groups, the last two digits serve only to identify the different alloys in the group. As new alloys become available, they are assigned consecutively to the list, beginning with XX01. The second digit indicates alloy modification. A zero indicates the original alloy.

The four digit number is often followed by a letter, or number and letter, to designate basic treatments used to produce tempers. Common temper designations are given in Table 9.13.

The annealed condition refers to metal that has been heated to its recrystallization temperature and cooled. The solution heat treatment consists of heating the metal to a temperature high enough to allow the alloying elements to go into solution with the base metal. This is similar to the solution heat treatment used for precipitation hardening stainless steels, described previously. As an example, 2024–T4 is a copper alloy aluminum solution, heat treated and aged naturally. During the solution treatment, the copper is finely dispersed throughout the metal. After it has cooled to room temperature, the copper begins to precipitate out of solution forming obstructions to slip in the slip planes and grain boundaries. This strengthening mechanism may take place at room temperature (natural aging), or it may be hastened by applying heat (artificial aging). The strength changes considerably. As an example, 2024–0 has a tensile strength of 27 ksi (186.2 MPa); but in 2024–T4, it goes up to 68 ksi (468.8 MPa) and in 2024–T6, 69 ksi (475.7 MPa).

Aluminum Cast Alloys. Aluminum casting alloys are generally designated by a number or a letter immediately followed by a number. However, this classification does not always follow a specific alloy grouping system. Basic temper designations are indicated by a hyphen, followed by the letter T and one or more numbers. A more significant classification of types is by process such as sand casting, permanent mold, and die casting alloys. Examples of commonly used alloys in this category are: sand casting, 319–F and 355–T6; permanent-mold casting, 43–F and

333–F; and die casting alloys, 13 and 43. The F designates as fabricated or, in this case, as cast.

JOINING ALUMINUM AND ALUMINUM ALLOYS

Weldability. As with other metals, the low alloy, non-heat treatable types are easily welded, although some of the strength gained through forming or strain hardening may be lost due to a partial anneal in the HAZ. Aluminum alloys in this classification are in the 1000, 3000, and 5000 series.

The heat treatable alloys of the 6000 series (magnesium and silicon), the 2000 series (copper), and the 7000 series (zinc with magnesium and/or copper), each have one or more materials that is best adapted to welding. In the 2000 series, 2219, which is a magnesium free alloy, is the easiest to weld. Generally, aluminum copper alloys are sensitive in the HAZ where the metal reaches temperatures between liquidus and solidus. In this area partial melting at grain boundaries forms a network containing brittle intermetallic compounds of $CuAl_2$. Farther from the weld are the solution zone, the recrystallized zone, the overaged zone, and the unaffected zone.

Because of the brittle zone adjacent to the weld, ductility in welded aluminum copper alloys is generally lower than in the aluminum magnesium alloys. The mechanical properties of welded forgings and extrusions can be severely impaired if partially melted zones are normal to the applied stress. Typical properties of aluminum welds are shown in Table 9.14.

TABLE 9.14. *Typical Properties of Aluminum Welds* (Courtesy of *Machine Design*)

Alloy	Filler Metal	Tensile Strength		Weld Elong.** (percent)
		Base Metal* (1000 psi) (6.89 MPa)	Weld (1000 psi) (6.89 MPa)	
EC†	None	9.5	10	63
1100	None	11	13.5	54
3003	None	14	16	58
Aluminum Magnesium Alloys				
5050	4043	18	20	18
5052	5356	25	28	39
5454	5554	31	35	40
Aluminum Magnesium Manganese Alloys				
5086	5356	35	39	38
5083	5183	40	43	34
5456	5556	42	46	28

NOTE: Properties listed are for GTA, ac, or GMA welds.
* Guaranteed minimum for 0 temper.
** Free-bend test.
† Electrical conductor grade, 99.60 % minimum aluminum.

TABLE 9.15. *Aluminum Alloys and Recommended Welding Methods* (Courtesy of *Machine Design*)

Welding Method	Al, Al-Mn 1100, 3003		Al-Cu 2219, 2014		Al-Mg 5000 Series		Al-Mg-Si 6000 Series		Al-Zn-Mg X7106, 7039		Al-Zn-Mg-Cu 7075, 7178	
	<1/8	>1/8	<1/8	>1/8	<1/8	>1/8	<1/8	>1/8	<1/8	>1/8	<1/8	>1/8
GTA, ac	A	B	A	B	A	A	A	B	A	B	B	C
GTA, dc	A	A	A	A	A	A	A	A	A	A	B	B
GMA	B	A	B	A	B	A	B	A	B	A	C	B
GTA, spot	B	C	B	C	B	C	B	C	B	C	C	X
GMA, spot	A	B	A	B	A	B	A	B	A	B	B	C
Resistance	A	B	A	B	A	B	B	A	A	B	A	B
Flash	B	A	B	A	B	A	B	B	B	A	B	A
Electron beam	B	B	B	B	A	A	A	B	A	A	B	B
Ultrasonic	A	X	A	X	A	X	A	X	A	X	A	X
Pressure	A	B	A	B	B	B	B	B	B	B	B	B

ALLOY TYPE

Base-Metal Thickness (in.)

NOTE: A = preferred, B = satisfactory, C = usable with care, X = not applicable.

324

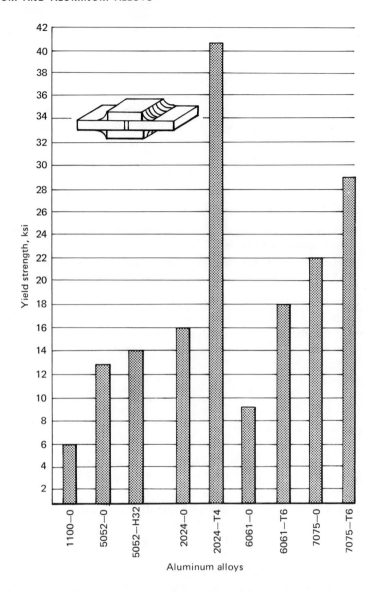

FIGURE 9–26. *A comparison of the transverse shear yield strength for various aluminum alloys made by GMA welding.*

Aluminum-zinc-magnesium-copper alloys such as 7075 and 7178 are difficult to weld, and even more difficult to repair. The weld strength is best when the heat input is limited by fast welding and good chilling methods. The welds must be postheat treated and aged. A newer class of heat treatable aluminum zinc magnesium alloys include X7106, 7039, X7005, X7002, and the Canadian 745. Most of these high strength alloys are easy to weld with 5556 and X5180, or X5039 filler rods.

Welds in these new alloys have better strength and ductility because the partially melted zones adjacent to the welds have better microstructures.

Joining Aluminum With Other Metals. Common metals to which aluminum is often joined are steel, stainless steel, copper, and copper alloys and titanium. Designing and producing good welds between aluminum and other metals involves an understanding of the metallurgical problems involved. Aluminum usually has a melting range lower than that of the metals to which it is joined. It combines with these metals to form brittle intermetallic compounds at the joint. The solution potential of aluminum often is lower than other metals and severe preferential corrosion often results unless preventive measures are taken. Three generally recognized methods avoid brittle intermetallic compounds.

1. Precoating the other metal with aluminum, silver, tin, or solder, and welding with aluminum filler metal.
2. The compounds can be squeezed out while they are molten, as in flash welding.
3. Cold welding processes, such as friction, pressure, and ultrasonic welding.

Aluminum can be welded by most of the common processes, however, the most common are GMA, GTA, spot, and seam welding. Common aluminum alloy types, thicknesses, and applicable welding methods are shown in Table 9.15. A comparison of transverse shear strengths of various aluminum alloys welded by the GTA method are shown in Fig. 9–26.

MAGNESIUM

Extreme lightness is the outstanding characteristic of magnesium. Aluminum is $1\frac{1}{2}$ times heavier and titanium is $2\frac{1}{2}$ times heavier than magnesium. Magnesium alloys have excellent machinability, hot formability, weldability, and a high strength-to-weight ratio. Another impressive quality about magnesium is its limitless supply in nature. One cubic mile of sea water contains 12 billion pounds of magnesium. Currently, almost all of the magnesium in the USA is produced from sea water.

Magnesium has insufficient mechanical strength in its pure state for structural purposes and must be alloyed with other elements. Aluminum, zinc, manganese, and zirconium are the most common alloying elements in the older room temperature alloys, while rare earth metals and thorium are the chief elements used in high temperature alloys. Beryllium and calcium are used to a lesser extent.

The ASTM classification designates the principal alloying elements as A, aluminum; Z, zinc; K, zirconium; M, magnesium; H, thorium; and E, rare earth metals. Numbers follow the alloy types representing approximate alloy percentages, e.g., AZ61A is an aluminum zinc alloy with 5.8 to 7.2 % aluminum and 0.4 to 1.5 % zinc. The letters A, B, or C may follow the number to indicate variations in treatment or composition. Temper designations are similar to those used for wrought aluminums.

Weldability of Magnesium. When heated in air to its melting point, magnesium tends to oxidize rapidly and therefore requires inert gas protection or a flux cover. The welding fumes from all commercial alloys, except those containing thorium, are neither harmful nor toxic. The welding fumes from thorium alloys are toxic if inhaled over a long period of time, but ordinarily good ventilation is all that is required. Magnesium requires approximately two-thirds the heat of aluminum for fusion welding of equal volumes. Its high thermal expansion and conductivity require good tooling for holding the parts in alignment during welding.

Although the alloying elements of aluminum and zinc effectively increase the strength of magnesium at room temperature, they also create two undesirable characteristics. The aluminum gives rise to stress and corrosion sensitivity and the zinc, to severe cracking of welds. Manganese, beryllium, and zirconium, because of their limited solubility and the formation of high mp constituents with magnesium, do not affect the welding characteristics. Rare earth metals show beneficial effects on the weldability, especially in the case of zinc alloys. None of the wrought alloys require heat treatment after welding for property improvement, but nearly all the casting alloys do. Even though postweld heat treatment may not be necessary, stress relieving is often used for dimensional stability and for avoiding stress and corrosion.

The welding processes most widely used are GTA, GMA, gas, and spot welding. Other processes that can be used include brazing, soldering, and, to a lesser extent, seam, flash, and pressure welding.

COPPER AND COPPER BASE ALLOYS

Pure copper can be classified into two main types: oxygen bearing and oxygen free. During the manufacture of copper, a process of oxidation to extract impurities from the melt is used, which results in the metal not being entirely free of oxygen. This type of copper is often referred to as "tough pitch" copper; it is not readily weldable due to the presence of copper oxides in the metal, which are susceptible to embrittlement. If the copper is further refined by reduction methods, the oxygen can be removed to produce a metal that is oxygen free and weldable by most fusion processes.

Brass. Brass is primarily an alloy of copper and zinc. A brass consisting of 70 % copper and 30 % zinc (cartridge brass) has a structure known as *alpha brass*. The zinc forms a solid solution in the copper. Under favorable conditions of annealing and cooling, the alpha brasses can contain up to 38 % zinc before other constituents begin to form.

Alpha brasses are manufactured in sheet and tube form having good strength combined with good ductility. Brasses may be work hardened and subsequently annealed or recrystallized.

When higher percentages of zinc (60 % copper and 40 % zinc) are used, a second solid solution rich in zinc forms. This structure, which has a maximum of 45 % zinc, is referred to as alpha-beta brass. The greater hardness and higher tensile strength of this alloy make it less ductile and less formable by cold working. The desirable properties are enhanced by hot working. The bulk of brass castings are made from alpha-beta alloys.

Joining Copper Base Alloys. The principle problem encountered in the welding of brass is the volitization of the zinc from the alloy. To keep the zinc loss to a minimum in gas welding, an oxidizing flame is used; and in arc welding, the shortest possible arc is recommended. A preheat is also beneficial.

Copper has a high coefficient of expansion and precautions should be taken to prevent warping during welding by the use of suitable jigging. Contraction forces will often cause cracking during the cooling time if parts are too rigidly held.

BERYLLIUM

Beryllium has very limited use in its pure form. Its major role is as an alloying element in copper alloys where it is capable of precipitation hardening. Beryllium is often referred to as a space age metal since it has a number of properties that are desirable in space vehicles, including high stiffness-to-density ratio, high strength-to-density ratio, high heat capacity, high precision elastic limit, and good mechanical properties up to 1200 °F (649 °C). It also has several characteristics that have limited its use, one of which is cost. In finished form the price may range from $1000 to $2000/lb. Also, its ductility in the short, transverse direction is low. Because of its toxic nature, beryllium metals must be handled with care.

Beryllium copper alloys are grouped as high strength or high conductivity. The high strength have 1 % beryllium and the high conductivity have less than 1 % beryllium. The latter are used where electrical and thermal conductivity are considered to be more important than strength, as in circuit breakers, switches, and welding dies. Applications of the 1 % beryllium are springs, diaphragms, aircraft engine parts, gears, bushings, and some load carrying structural elements of space craft.

Joining Beryllium. Beryllium fusion welding is limited to methods that provide 100 % protection from the atmosphere. Thus, excellent welds can be made with EB and by diffusion methods. Soldering and torch brazing require the use of an active flux to remove beryllium's tenacious and stable surface oxide. The fluxes used are acidic and generally contain chlorides and/or fluorides, therefore, must be completely removed following application. Precoating, where the excess flux can be removed and the precoat inspected, has become a standard practice. Generally, capillary flow cannot be used to develop sound joints.

A wide number of epoxy resins are used for bonding purposes. Generally any resin used in joining aluminum can be used on beryllium. Acid etching followed by thorough rinsing and drying is recommended prior to the application of the adhesive.

TITANIUM

Titanium base alloys are 60 % heavier than aluminum but are much stronger, being superior in strength to most alloy steels. It has a lower thermal conductivity and lower linear coefficient of expansion than either alloy steel or aluminum alloys. Titanium alloys are superior to all common alloys in their strength-to-weight ratio at temperatures ranging from -423 to $+1000\,°F$ (-253 to $+538\,°C$). They are also tough and have good corrosion resistance.

Titanium alloys are grouped in three classes: alpha, alpha-beta, and beta, depending on the crystallographic phase present at room temperatures. The most useful alloys of the alpha grade are Ti–5 Al–2.5 Sn and super-alpha alloy, Ti–8 Al–1 Mo–1 V, which contains a small amount of the beta phase. Alpha-beta alloys are more easily formed than the alpha alloys, as they respond to heat treatment. This class includes a wide range of compositions, from the alpha-beta general purpose type, Ti–6 Al–4 V to the highly beta-stabilized and deep-hardening alloys such as Ti–7 Al–4 Mo. The only beta alloy is Ti–13 V–11 Cr–3 Al. This material can be solution treated and aged to high strength.

Joining Titanium Alloys. Titanium alloys can be fusion welded by either the GTA or the GMA processes in the annealed, solution treated, or fully aged condition. Of the two welding processes, GTA is most often used, however, in metal thicknesses greater than 0.250 in., the advantages of having a consumable electrode make the GMA process more desirable. For maximum joint ductility, a commercially pure filler wire is recommended. It is essential to flood the weld area with an inert gas. The shielding gas may be either argon or helium, or a combination of both. Helium causes deeper penetration, but argon tends to produce a flatter, smoother weld. Argon is often used for the face shielding and helium for the back-up shielding. For thin materials (up to 0.045 in.), a copper backing bar in intimate contact with metal may be used. For heavier gauge metals, a groove from $\frac{1}{4}$ to $\frac{3}{8}$ in. (6.35 to 9.52 mm) wide and from 0.010 to 0.060 in. (0.25 to 1.52 mm) deep is cut in the copper back-up strip to make a channel for the gas.

Pressure welding may be used by heating the sections to a temperature slightly below the lower critical, or to the beta transition temperature and then forcing them together. This results in a finer grain in the weld area due to the forging action. The final assembly can be solution treated and aged. Titanium alloys may also be resistance welded without shielding due to the close proximity of the metals. Welding characteristics by this method are very similar to that of stainless steel. Friction and electron beam welding are also used on titanium.

EXERCISES

Problems and Questions

9–1. Arrange the following welding processes in order of the size of the HAZ they would produce in a $\frac{1}{2}$ in. (12.70 mm) butt welded aluminum alloy plate: stick electrode, oxyacetylene welding, electron beam, GMA, and submerged arc.

9–2. Arrange the following materials in order of their weldability: AISI steels 5100, 4140, 1018, and E51100. Discuss their differences.

9–3. In what way are some stainless steels and some aluminum alloys very much alike?

9–4. Why does an austenitic stainless steel remain austenitic even though it may be quenched rapidly from above its upper critical temperature?

9–5. (a) What is the CE of AISI 4340? Use the mid-range of alloys in determining the CE.
(b) What precautions if any need be exercised in welding this steel?

9–6. (a) Make a sketch to show the schematic representation of a plain medium carbon steel that has been heated to the upper critical and allowed to cool slowly in air.
(b) Would any precaution need to be exercised in welding this material? Explain.

9–7. What is the effect of variations in cross section on microstructure when welding alloy steels?

9–8. How do intermetallic compounds affect the strength of precipitation hardening aluminum alloys? Explain.

9–9. In what ways can a fusion weld be compared to a casting?

9–10. How does the preheating of a structure to be welded affect the microstructure of the finished weld?

9–11. Why is it that malleable cast irons are generally considered to be nonweldable?

9–12. What are the two main approaches toward fusion weld repairs in gray cast iron?

9–13. Explain the benefits of multipass welding.

9–14. How can an isothermal diagram help you determine what the microstructure of the weld will be under various conditions of temperature change?

9–15. (a) Explain what is meant by a locked-up stress in a metal.
(b) What are some methods of relieving it?

9–16. What would the structure be of a weld made in maraging steel if allowed to cool normally to room temperature? How can the normal mechanical properties be restored to the weld and HAZ?

BIBLIOGRAPHY

BLODGETT, O. W., *Weld Quality, Part I*, "Why Do Welds Crack—How Can Weld Cracks Be Prevented." *Weld Quality, Part II*, "Why Preheat—An Approach to Estimating Correct Preheating Temperature." **Bulletins G230–G231,** The Lincoln Electric Company, Cleveland, Ohio.

BRADSTREET, B. J., "Effect of Welding Conditions on Cooling Rate and Hardness in Heat-Affected Zone." *Welding Journal,* **48** Research Suppl., 499-s, November 1969.

———"Methods to Establish Procedures for Welding Low Alloy Steels," *Engineering Journal,* November 1963.

COLLINS, F. R., "Welded Aluminum Parts." *Machine Design,* March 4, 1965.

DORSCHU, K. E., "Control of Cooling Rates in Steel Weld Metal." *Welding Journal, Suppl.,* 49-s, February 1968.

Machine Design, Metals Reference Issue, February 1970.

POOLE, L. K., "Sigma—An Unwanted Constituent in Stainless Weld Metal." *Metal Progress*, June 1954.

PRIVOZNIK, L. J., "How to Prevent Weld Cracking in Steel Structures." *Metal Progress*, December 1962.

ROSENTHAL, D., "Mathematical Theory of Heat Distribution During Cutting and Welding." *Welding Journal*, 220-s, May 1941.

SAMARIN, A. M., "Some Properties of Liquid Alloys." *JISI*, February 1962, p. 95.

STOUT, P. F., MCLAUGHLIN, P. F., AND STRUNCK, S. S., "Heat Treatment Effects of Multiple-Pass Welds," *Welding Journal*, **48** Research Supplement, 155-s, April 1969.

TOYE, T. C., "Thin Metal Films and Theory of Liquid Metals." *Metallurgia*, February 1966, p. 67.

"Transparent Model Shows Inside View of How Metals Solidify." *Metallurgia*, December 1966, p. 249.

WINTERTON, K., "Weldability Prediction from Steel Composition to Avoid Heat-Affected Zone Cracking," *Welding Journal*, **40** Research Supplement, 253-s, June 1961.

TABLE 9.A. *Forging Steels, Specifications, Use, and Mechanical Properties*

ASTM Specifications	Common Use	Specified Mechanical Properties–Min.				Other Requirements
		Tensile Strength (MPa) ksi	Yield Point (MPa) ksi	Elong. in 2" %	Red. of Area	
A105	Flanges and fittings	(413.7–482.6) 60–70	(206.8–248.2) 30–36	22–25	30–38	
A181	Flanges and fittings	(413.7–482.6) 60–70	(206.8–248.2) 30–36	18–22	24–35	
A182	Flanges and fittings	(413.7–551.1) 60–80	(206.8–482.6) 30–70	20–45	35–60	201 BHN
A235	Ind. forgings	(413.7–620.5) 60–90	(206.8–379.2) 30–55	19–25	30–40	
A237	Ind. forgings	(551.6–1172) 80–170	(344.7–965.3) 50–140	11–26	38–50	
A243	Rings and discs	(413.7–1172) 60–170	(206.8–965.3) 30–140	19–26	31–52	
A266	Drums	(413.7–517.1) 60–75	(206.8–255.1) 30–37	19–26	30–42	180° × 1½ Cold bend
A288	Retaining rings	(482.6–1138) 70–165	(310.3–1034) 45–150	12–20	35–40	
A289	Non-magnetic retaining rings	(758.4–1103) 110–160	(448.2–910.1) 65–132	12–25	20–35	
A290	Gear rings	(55.16–1138) 8–165	(275.8–930.8) 40–135	11–20	27–40	163 to 401 BHN

Designation	Application	Tensile strength ksi (MPa)	Yield strength ksi (MPa)	Elongation %	Reduction %	
A291	Pinions	85–170 (586.1–1172)	40–140 (275.8–965.3)	10–20	25–45	170 to 401 BHN
A292	Generator rotors	75–115 (517.1–792.9)	40–90 (275.8–620.5)	12–24	25–40	V-Notch Charpy 10-12 ft-lb min.
A293	Turbine rotors	75–110 (517.1–758.4)	40–95 (275.8–655.0)	13–24	26–40	V-Notch Charpy 6–10 ft-lb min.
A294	Bucket wheels	90–140 (620.5–965.3)	65–125 (448.2–861.8)	15–20	40–50	V-Notch Charpy 15 ft-lb min.
A336	Drums	70–100 (482.6–689.5)	30–70 (206.8–482.6)	18–25	25–45	
A350	Flanges and fittings	60–70 (413.7–482.6)	30–40 (206.8–275.8)	25	38–50	Keyhole Charpy 15 ft-lb at 150°F.
A369	Pipe	55–60 (379.2–413.7)	30 (206.8)	14–22		
A372	Pressure vessel	60–120 (413.7–827.4)	35–70 (241.3–482.6)	15–20		
A404	Flanges and fittings	80 (551.6)	50 (344.7)	20	40	
A430	Pipe	70 (482.6)	30 (206.8)	20–45	30–50	

AMERICAN WELDING SOCIETY ⊕ STANDARD WELDING SYMBOLS

(COURTESY AMERICAN WELDING SOCIETY)

10

Weld Design

A design engineer usually does not consider himself the one to specify a manufacturing process. As a design develops, however, many detailed processes of fabrication must be selected. Although the designer may not specify the process of manufacture, the tolerances and the strength required will dictate, within narrow limits, how the fabrication is to be performed. Also, the manufacturing engineer may have numerous innovations as to how a part could be manufactured, but these are, for the most part, refinements that are limited to cost reduction. The greatest savings can be realized when the designer is familiar with the processes of manufacture or works closely with the manufacturing engineer during the planning stages.

Rarely is there a *one best* method of manufacture. Shown in Fig. 10–1 is a bearing stand that could be designed as a casting, a weldment, or a combination of casting and welding. In this case, the original design called for a casting, however, in the redesign, a composite structure was determined. The bearings were made of cast steel and welded to the steel framework. The redesign resulted in 45 percent savings. Large patterns and many core boxes were eliminated, and overall production time was reduced from 6–8 weeks to 3–4 weeks. Another advantage of a composite design is the capability to select and place the type of material where it will serve the best purpose. As in this case, high tensile strength alloy steel is used in the higher stress and wear areas, while mild steel plate is used for the main structural elements.

← FIGURE 10–0. *A complete presentation of welding symbols used by the welding industry. (Courtesy of American Welding Society.)*

335

FIGURE 10–1. *Composite design bearing stand provides 45% savings. Cast steel bearings are semi-automatically (submerged arc) welded to the fabricated mild steel frame. (Courtesy of The James F. Lincoln Arc Welding Foundation.)*

MATERIAL PROPERTIES

To insure that materials are used and placed most advantageously, the designer and the materials joining engineer must be familiar with material properties under a variety of conditions.

Tensile Properties. Tensile strength may be simply defined as a measure of the maximum stress the material can withstand while being pulled apart. More specifically, the tensile properties are measured by making a specimen as shown in Fig. 10–2.

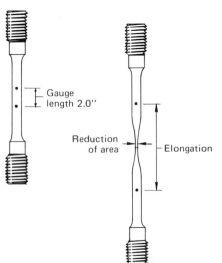

FIGURE 10–2. *A standard type tensile specimen showing elongation and reduction of area.*

This specimen has a two-inch gauge length marked on it as shown, and is placed in a tensile testing machine that is able to apply a constantly increasing load. The amount of elongation that occurs is accurately measured with an extensometer. The tensile strength is expressed as the maximum load divided by the original cross-sectional area. This strength is expressed in pounds per square inch (psi), or in metric terms as kgf/mm^2, or in the newer SI system (international system of weights and measures) as pascals (Pa). To avoid the use of prodigious numbers involved in expressing force over square metres, megapascals (MPa) are used; one thousand psi equals 6.89 MPa. The strain is the total elongation recorded by the extensometer divided by the gauge length and is expressed in in./in. The relationships are shown as:

$$\text{Stress} = \frac{\text{Load}}{A_0} \quad \text{and Strain, } \varepsilon, = \frac{\Delta l}{l_0} = \frac{l_f - l_0}{l_0}$$

where A_0 = original area.
ε = strain.
Δl = change in length.
l_0 = original length.
l_f = final length.

Note that the tensile strength is based on the original cross-sectional area. This is termed the *engineering stress* and it forms the basis upon which the engineer must make his design calculations. Notice that engineering stress is distinguished from what is termed the *true* strength, which is based on the actual or instantaneous area.

Stress–Strain Curves.   The data from the tensile test is used to plot the stress–strain curve as shown in Fig. 10–3. The curve has two main regions, elastic (*AB*)

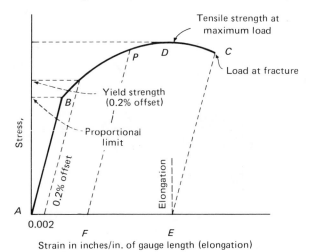

FIGURE 10–3.   *A typical tensile test plot or stress–strain curve.*

and plastic (*BC*). Point *B* represents the yield stress and *D*, the greatest stress the specimen can withstand. From *D* onwards the stress decreases since the specimen has begun to thin down locally, a process that quickly leads to fracture, *C*.

Proportional Limit. The proportional limit shown at *B* may be defined as the highest point at which the ratio of stress to strain is exactly equal i.e.,

$$\frac{\sigma_1}{\varepsilon_1} = \frac{\sigma_2}{\varepsilon_2} = \frac{\sigma_n}{\varepsilon_n}$$

Yield Point and Yield Strength. The yield point of a material is equal to the stress at which it undergoes a marked increase in strain without a corresponding increase in stress. This phenomenon is shown by the broken line in Fig. 10–4. The yield strength is usually determined either by noting the stress at the time the increase in strain occurred or by the "offset" method. The offset method is based on drawing a line parallel to the straight portion of the stress–strain curve but offset to it, usually 0.2 percent (as was shown in Fig. 10–3).

The yield strength of a material may be determined according to the type of loading: compressive, flexural, shear, tensional, or torsional. It is, however, generally assumed unless specified otherwise, that a reference is to tensile yield.

Strain Hardening. If the material represented by the curve (in Fig. 10–3) is first reduced in area by rolling, then tested, the curve obtained would be similar to that shown by *FPDC*. The yield point would be higher as shown at *P*, instead of *B*. The tensile strength would also be higher because the same maximum stress

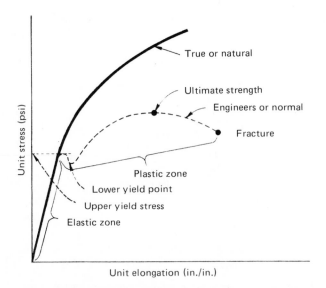

FIGURE 10–4. *A stress–strain curve for ductile material (broken line) and a true stress–strain curve based on the changing or instantaneous area.*

could be sustained by a smaller cross-sectional area. Of course, relative elongation would be smaller, FE instead of AE.

Percent Elongation. The total extension at fracture, as recorded by the extensometer or by other measurements, divided by the original length is the *elongation* and is expressed as:

$$\% \text{ elongation} = \frac{\Delta l}{l_0} \times 100$$

where Δ = the change in gauge length from the original to the final length.

Reduction of Area. The ratio of change in the cross-sectional area from the original to the final area is termed the percent reduction of area:

$$\% RA = \frac{A_0 - A_f}{A_0} \times 100$$

Modulus of Elasticity. The modulus of elasticity, or Young's modulus, is the ratio of the stress to the corresponding strain up to the proportional limit. Also known as Hooke's Law, it is expressed as:

$$E = \frac{\sigma}{\varepsilon}$$

or given units of psi $= (\text{lbs/in.}^2)/(\text{in./in.})$.

The approximate modulus of elasticity can be obtained for most materials by using this formula. Tensile strengths, modulus, and various other metal properties are given at the end of the chapter, see Table 10A.

Ductility. Ductility is defined as the ability of a material to withstand plastic deformation without rupture. The concept may also be thought of in terms of bendability and crushability. Cup drawing and twisting tests are considered to be indicators of ductility. The lack of ductility is often termed brittleness. Usually, if two materials have the same strength, the one that has the greater ductility is the more desirable.

Toughness and Resilience. The toughness of a material refers to the ability to absorb mechanical energy. Graphically, it is indicated by the total area, elastic and plastic, under the total stress–strain curve. For uniaxial tension, it is the area under the stress–strain curve out to the point of fracture. Because all parts of the tensile specimen do not deform the maximum amount, the area under the curve gives only an approximation of the metal's toughness. For resilience, the measure is how well a material absorbs mechanical energy up to its proportional limit, Fig. 10–5.

Useful tests of toughness measure the notch toughness of a material, such as the Izod and Charpy tests. The tests are quite similar in that a specimen, 0.394 in. square, is used with a 45 degree vee notch. The principle of the Izod test

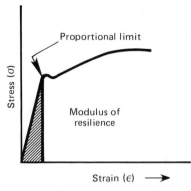

FIGURE 10–5. *The modulus of resilience, which represents how well a material will absorb energy up to its proportional limit, is shown as the shaded area under the stress–strain curve.*

is shown in Fig. 10–6. The energy required to break the bar over a range of temperatures determines the notched fracture toughness. The failure mode sometimes shifts rather abruptly from tough, at high temperatures, to brittle at low temperatures. The fracture energy for different temperatures is plotted as shown in the generalization transition curve, Fig. 10–7. The temperature where the slope is the steepest is the transition temperature, and the energy value at that point is the *ductility transition temperature*. Materials rated as tough have lower transition temperatures, and thus, require more energy for fracture.

The appearance of the fractured metal may be just as instructive as the test data. The most desirable type of break is one that shows a fibrous structure with considerable deformation near the notch. In contrast, a sharp break exposing a fully crystalline surface indicates a brittle material. Many degrees of toughness are usual and fall between these two extremes (that are revealed by the appearance of the fracture).

Notches occur in real structures when unfilleted corners are used or when

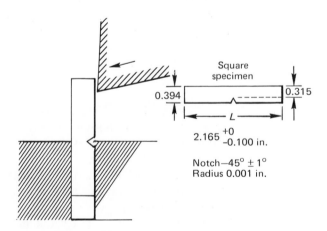

FIGURE 10–6. *The principles of the Izod test used to determine toughness during fracture of materials.*

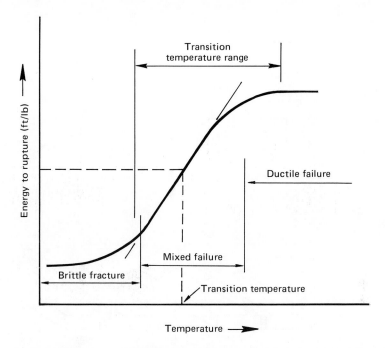

FIGURE 10–7. *A generalized transition curve showing the relationship of temperature to energy of rupture.*

welds are improperly made. These effects can be minimized by proper design and welding procedures. Abrupt changes in cross section, square corner notches or "cut-outs," undercuts, arc strikes, and scars from chipping hammers should be avoided, Fig. 10–8.

Fatigue Strength. When a material is repeatedly loaded and unloaded, it may fail even though the yield strength was never exceeded. Such a phenomena is known as *fatigue*. Fatigue loading occurs in three principal ways as shown in Fig. 10–9. The least extreme, of course, is a static load without variation, Fig. 10–9a, where the stress ratio is $R = +1$. The most extreme variation is full reversal, in which the tensile and compressive stresses of the same magnitude alternate, Fig. 10–9b. The ratio of S_{max} (tension) to S_{min} (compression) is said to be $R = -1$, which results in the lowest fatigue strength. A rotating shaft operating under a bending load is an example of this type of fatigue loading.

Unidirectional loads, Fig. 10–9c, vary from zero to maximum stress either in tension or compression. The stress ratio, R, is said to be zero under these conditions. Unidirectional with preload, Fig. 10–9d, involves loading from minimum to maximum stress without reaching zero. An example of this type of loading is a connecting rod bolt as used in internal combustion engines. The bolts are tightened to a given preload and are subjected to the reciprocating cycles of the engine.

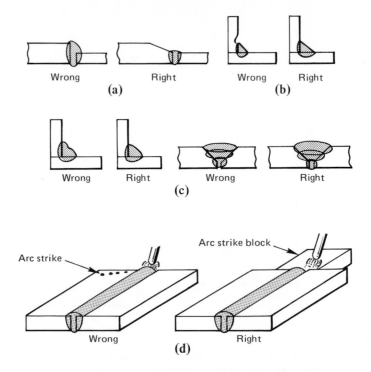

FIGURE 10–8. *Some effects to be eliminated by proper design and procedure in welding are: (a) abrupt changes in section, (b) undercuts, (c) notching effects, and (d) striking the arc on the plate to be welded but not on the weld area. The latter may appear harmless, but in medium and high carbon alloy steels the metallurgical characteristics are affected.*

When maximum stress, S_{max}, is plotted against the number of cycles to which it is subjected before failure occurs, a curve similar to that shown in Fig. 10–10 occurs. If the maximum number of cycles a member will be subjected to is known, along with the stress ratio, the fatigue strength can be determined. When the curve reaches a constant stress that is independent of the number of cycles of loading, the corresponding stress is referred to as the *fatigue limit* or *endurance limit*. This usually occurs at 2 million cycles (or more) of loading. Machine elements and highway bridges are expected to have more than 100,000 cycles of loading.

Because of the progressive nature of fatigue, characteristic "beach marks" or "clam shell" marks often form on the fractured surface, as shown in Fig. 10–11. The arrow indicates where the fatigue fracture originated. The beach marks radiated out from this point almost across the whole surface before the final separation occurred. These marks, when present, often give considerable information about the fatigue origin, relative load intensity, and the stress system.

Fatigue fractures do not always exhibit beach or clam shell marks. Sometimes fatigue failure may be indicated by a smooth bright area. A careful examination of

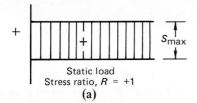

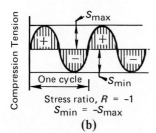

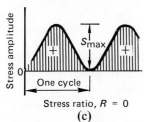

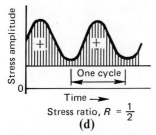

FIGURE 10–9. *Stress loading showing (a) non-fatigue condition, (b) full reversal, (c) unidirectional, and (d) unidirectional with preload.*

the fractured surface either with the unaided eye or a low magnification ($20 \times$ or less) lens can often locate the origin of the failure. Weld defects such as cracks, slag inclusions, weld undercuts, and corrosion pits are indications of fatigue origin. Severe quenching due to the sudden extinguishing of the welding arc may initiate a fatigue crack at the point of electrode change. Also, incomplete penetration of the root pass or insufficient depth of penetration in a butt weld can become sources of fatigue cracks (as were shown in Fig. 10–8). Fatigue studies of both fillet and butt welds made in low alloy structural steels show that failures are generally initiated in the region of the surface pass where the electrode is changed.

To determine the maximum allowable fatigue stress, σ_{max}, for a given type joint and/or steel, the following formula is used to reduce the allowable strength because of the fatigue condition:

$$\sigma_{max} = \frac{\sigma_{sr}}{1 - R}$$

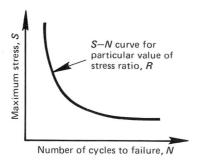

FIGURE 10–10. *A plot of maximum stress, S, and the number of fatigue cycles to failure, N.*

FIGURE 10–11. *The presence of "beach marks" usually indicates that failure was caused by fatigue. Here fracture began at a discontinuity. (Courtesy of International Harvester Co.)*

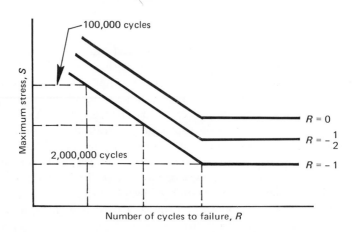

FIGURE 10–12. *Fatigue S–N curves plotted on log-log paper. Each stress ratio requires a different S–N curve.*

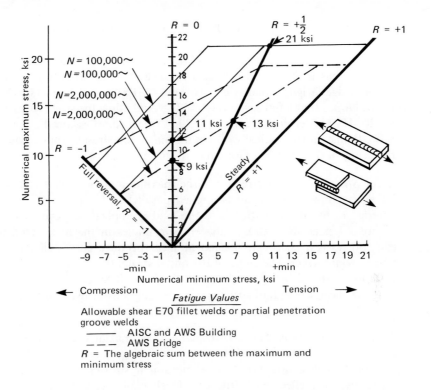

Allowable shear E70 fillet welds or partial penetration
groove welds
——————— AISC and AWS Building
— — —   AWS Bridge
R = The algebraic sum between the maximum and
minimum stress

FIGURE 10–13.   *A modified Goodman Diagram showing the 1969 AISC fatigue
values allowable for shear on the throat of fillet welds and partial penetration
groove welds. Also shown, by dotted lines, are the corresponding values from
the new American Welding Society (AWS) Structural Specification. Comparing
the two allowables at a life of 2,000,000 cycles, note that at a stress range of
R = ½, the AISC has increased the value from 13.0 ksi (89.63 MPa) to 21.0 ksi
(144.8 MPa). At a stress range of R = 0, the value has increased from 9.0 to
11.0 ksi (62.05–75.84 MPa). Previous AWS bridge fatigue values were 9.6 and
7.0 ksi (66.19 and 48.26 MPa), respectively. (Courtesy of The Lincoln Electric
Company.)*

where σ_{sr} = allowable range of stress (see AISC handbook, Tables 1, 3, and 5,
            for the various fatigue conditions).

When the maximum stress (S) and number of fatigue cycles to failure (N)
are plotted on a log–log scale, the curve changes to the straight line shown in
Fig. 10–12.

A Goodman diagram is used to summarize the results of various types of
stress cycles. A modified form of the Goodman diagram is shown in Fig. 10–13. A
comparison of the new (1969) AISC fatigue standards for shear on the throat of
fillet welds or partial penetration groove welds is shown. Also shown are the AWS
structural fatigue values.

In comparing the two allowable parameters at a life of two million cycles, it can be seen that, at a stress range of $R = +\frac{1}{2}$, the AISC has increased the value from 13.0 ksi (89.63 MPa) to 21.0 ksi (144.8 MPa). At a stress range of $R = 0$, the value has increased from 9.0 ksi (62.05 MPa) to 9.6 ksi (66.14 MPa). Previous AWS bridge fatigue values were 9.6 (66.14 MPa) and 7.0 ksi (48.26 MPa), respectively. Notice that a full reversal $(R = 1)$ would allow the stress to vary from 5.4 ksi (37.20 MPa) in tension to 5.4 ksi (37.20 MPa) in compression. For any given stress ratio, R, one may graphically obtain the maximum stress permitted during the cycle by means of the modified Goodman diagram.

Shear Strength. Shear strength is the maximum stress a material will withstand before fracture when a load is applied parallel to the plane of stress. This is in contrast to tensile strength where the load is applied perpendicular to the plane of stress. Under shear stress, adjacent slip planes in the metal tend to slide over each other. Test specimens loaded in shear produce curves similar to the tensile stress–strain curves. The slope of the initial straight line portion of the shear–stress diagram is the shearing modulus of elasticity. Results of tests indicate that the ultimate shear strength of structural steels ranges from two-thirds to three-quarters of the tensile strength.

Effect of Thickness on Stress Distribution. Multiaxial stresses may be disregarded for thin plates, since stresses transverse to the plane of the plate are negligible. For thick plates, however, three-dimensional stresses must be considered, and especially, their effect on brittleness. Thick plates have a coarser grain structure than thin plates. Thus, a higher carbon content is required to obtain the same yield strength as found in the thinner, hot rolled sections.

BENDING

Bending is characterized by the outside fibers of the part being the most severely stressed. The stress goes to zero at the neutral axis as shown in Fig. 10–14. The stress on the outer fibers depends on the section geometry, bend radius, and loading.

The deflection is dependent on the loading, section geometry, and modulus of elasticity of the material. To decrease the deflection at a given load, the section may be increased or a material with a higher modulus of elasticity may be selected.

The bending stress (σ) in a straight, round, cantilevered shaft [as shown in Fig. 10–14(b)] may be found by:

$$\sigma = \frac{My}{I}$$

where M = the bending moment.

         $y = \frac{1}{2}$ the shaft diam, or the distance from the neutral axis to the outer edge.

         I = area moment of inertia.

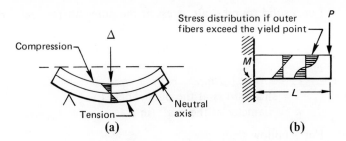

FIGURE 10–14. *Structural members subjected to transverse loads are classed according to the method of support. A simple beam supported at each end is shown at (a) and a cantilever beam is shown at (b). The moment at a section is resisted by the internal couple of compressive and tensile stresses.*

The bending moment at the supported end is equal to the force exerted on the shaft, multiplied by the distance from the unsupported end (PL). The formula can be further broken down as follows:

$$\sigma = \frac{My}{I} = \frac{(\text{PL})\left(\frac{d}{2}\right)}{\frac{\pi d^4}{64}} = \frac{(\text{PL})(32)}{\pi d^3}$$

TORSION

Torsion is the application of force to a member that causes twisting about its structural axis as shown in Fig. 10–15. The twisting force is usually referred to in terms of torsional moment or torque (T), which is basically the product of the externally applied force and the moment arm of the force. The moment arm is the distance to the centerline of rotation from the line of force and perpendicular to it. The principal deflection caused by torsion is measured by the angle of twist (θ), or by the vertical movement of one corner in framed sections.

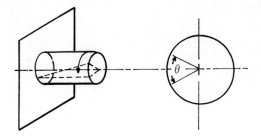

FIGURE 10–15. *A shaft in torsion. The left end is stationary; as the right end is rotated, internal shear stresses are set up in the shaft known as torsional stress.*

For a solid round shaft, stress at the outer fibers is related to torque as:

$$T = .196Td^3 \quad \text{or} \quad T = \frac{5.093T}{d^3}$$

where T = torque (lb in.).
T = shear stress at the outer fibers (psi).
d = diameter of the shaft (in.).

For a hollow shaft:

$$T = .196T \times \frac{(d_o{}^4 - d_i{}^4)}{d_i}$$

where d_o = outside diameter (in.).
d_i = inside diameter (in.).

The angle of twist (θ) may be related to torque for a solid shaft as follows:

$$\theta = 584 \frac{TL}{Gd^4}$$

where L = length of shaft (in.).
G = modulus of elasticity in shear. (NOTE: For steel, $G = 12 \times 10^6$.)

For hollow shafts:

$$\theta = 584 \frac{TL}{G(d_o{}^4 - d_i{}^4)}$$

Example. What is the minimum diameter of a steel shaft that can be used if it is 3 ft long and subjected to a torque of 50,000 lb/in. The maximum shearing stress should not exceed 8,000 psi and the angle of twist will not be greater than 2°.

Solution:

$$T = .196Td^3$$

$$50000 = .196 \times 8000 \times d^3$$

$$d = 3.17 \, \text{in.}$$

and

$$\theta = 584\left(\frac{TL}{Gd^4}\right)$$

$$L = 584\left(\frac{50000 \times 12 \times 3}{12000000 \times d^4}\right)$$

$$d = 3.17 \, \text{in.}$$

Therefore, to satisfy both conditions given in the problem, the shaft should be at least 3.17 inches in diameter.

SHEAR STRENGTH

The ultimate shear strength of a material is the maximum load it can withstand without rupture when subjected to a shearing action. One method of determining the shear strength is to use a small sheet or disk of the material which is of known thickness. It is clamped over a die, and a corresponding punch is brought down on the material. A gradually increasing load is applied until the material is completely punched through. The shear stress (S_s) may then be calculated as follows:

$$S_s = \frac{P}{\pi dt}$$

where P = the punch load in lb.
             d = punch diameter in inches.
             t = the specimen thickness in inches.

The shear strength of mild steels ranges from about 60 to 80 percent of the tensile strength.

Hardness Testing. Standardized tests have been made for determining the properties of materials. Discussed here are those used for determining hardness, tensile strength, impact strength, and fatigue. In selecting a material to withstand wear or erosion, the properties most often considered are hardness and toughness. Hardness is the property of a material that enables it to resist penetration and scratching. Hardness testing can be done by several standard methods such as Brinell, Rockwell, Vickers, and scleroscope.

Brinell. Brinell tests are based on the area of indentation a steel or carbide ball 10 mm in diameter makes in the surface of a material for a given load, Fig. 10–16. Loads used are 3,000, 1,500, or 500 kg. The load is applied for 15 seconds on ferrous metals and at least 30 seconds on nonferrous metals. When the load is released, the diameter of the spherical impression is measured with the aid of a Brinell microscope. From the diameter, a Brinell hardness number is obtained by consulting standard tables such as the sample shown, which is made by the Society of Automotive Engineers (SAE)—see Table 10.1.

Brinell hardness tests are especially good for materials that are coarse grained or nonuniform in structure, since the relatively large indentor gives a better average reading over a greater area.

Rockwell. The Rockwell hardness tester uses two types of penetrators, steel balls and a diamond cone or Brale, Fig. 10–17. The ball indentor is normally 1/16 in. diam, but larger diameters of $\frac{1}{8}$, $\frac{1}{4}$, or $\frac{1}{2}$ in. may be used for soft materials. The Brale is used for hard materials. The principle is based on measuring the difference in penetration between a minor and a major load. The minor load is 10 kg and the

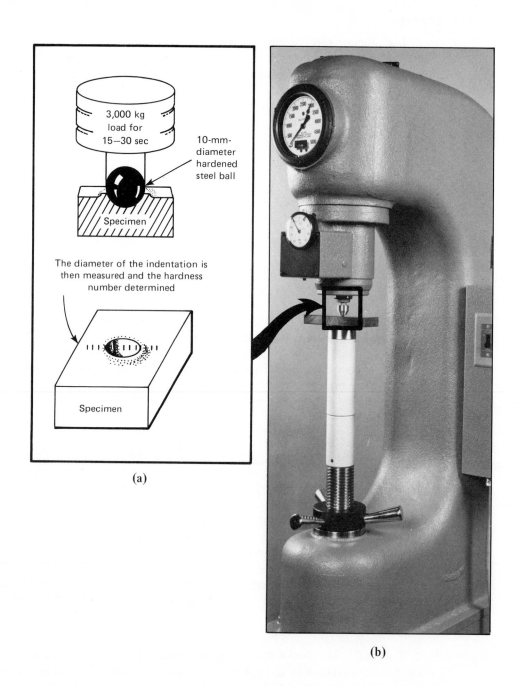

(a)

(b)

FIGURE 10–16. (a) *Principles of Brinell hardness testing.* (b) *A Brinell hardness testing machine.* (*Courtesy of Acco, Wilson Instrument Division.*)

TABLE 10.1. *Hardness Conversion Table for Hardenable Carbon and Alloy Steel (Reprinted with permission, "Copyright © Society of Automotive Engineers, Inc. 1966. All rights reserved.")*

| Brinell, 10 mm Carbide Ball, 3000 kg Load | | Diamond Pyramid Hardness Number | Rockwell | | Shore | Tensile Strength, 1000 psi |
Indentation Diam, mm	Hardness Number		C Scale 150 kg Brale	B Scale 100 kg 1/16-in. Ball		
—	—	940	68	—	97	—
—	767	880	66.5	—	93	—
2.25	745	840	65.5	—	91	—
2.30	712	—	—	—	—	—
2.35	682	737	61.5	—	84	—
2.40	653	697	60	—	81	—
2.45	627	667	58.5	—	79	—
2.50	601	640	57.5	—	77	—
2.55	578	615	56	—	75	—
2.60	555	591	54.5	—	73	298
2.65	534	569	53.5	—	71	288
2.70	514	547	52	—	70	274
2.75	495	528	51	—	68	264
2.80	477	508	49.5	—	66	252
2.85	461	491	48.5	—	65	242
2.90	444	472	47	—	63	230
2.95	429	455	45.5	—	61	219
3.00	415	440	44.5	—	59	212
3.05	401	425	43	—	58	202
3.10	388	410	42	—	56	193
3.15	375	396	40.5	—	54	184
3.20	363	383	39	—	52	177
3.25	352	372	38	(110)*	51	171
3.30	341	360	36.5	(109)	50	164
3.35	331	350	35.5	(108.5)	48	159
3.40	321	339	34.5	(108)	47	154
3.45	311	328	33	(107.5)	46	149
3.50	302	319	32	(107)	45	146
3.55	293	309	31	(106)	43	141
3.60	285	301	30	(105.5)	—	138
3.65	277	292	29	(104.5)	41	134
3.70	269	284	27.5	(104)	40	130
3.75	262	276	26.5	(103)	39	127
3.80	255	269	25.5	(102)	38	123
3.85	248	261	24	(101)	37	120
3.90	241	253	23	100	36	116
3.95	235	247	21.5	99	35	114
4.00	229	241	20.5	98	34	111
4.05	223	234	(19)	97.5	—	—
4.10	217	228	(17.5)	96.5	33	105
4.20	207	218	(15)	94.5	32	100

(Continued)

TABLE 10.1. *Hardness Conversion Table for Hardenable Carbon and Alloy Steel* (*Reprinted with permission,* "*Copyright © Society of Automotive Engineers, Inc. 1966, All rights reserved.*") (*Continued*)

Brinell, 10 mm Carbide Ball, 3000 kg Load		Diamond Pyramid Hardness Number	Rockwell		Shore	Tensile, Strength, 1000 psi
Indentation Diam, mm	Hardness Number		C Scale 150 kg Brale	B Scale 100 kg 1/16-in. Ball		
4.30	197	207	(12.5)	93	30	95
4.40	187	196	(10)	90.5	—	90
4.50	179	188	(8)	89	27	87
4.60	170	178	(5)	87	26	83
4.70	163	171	(3)	85	25	79
4.80	156	163	(1)	83	—	76
5.00	143	150	—	78.5	22	71
5.20	131	137	—	74	—	65
5.40	121	127	—	70	19	60
5.60	111	117	—	65.5	15	56

* Values in parentheses are beyond normal range and are for information only.

FIGURE 10–17. *Rockwell hardness tester with Brale penetration—used in testing hard materials.* (*Courtesy of Acco, Wilson Instrument Division.*)

major load varies with the material being tested. If the material is known to be relatively soft, the ball penetrator is used with a 100 kg load in what is known as the "B scale" (R_b). If the material is relatively hard, the diamond Brale is used with 150 kg load on what is known as the "C scale" (R_c). Other scales are available and are useful for checking extremely hard or soft surfaces. Very thin sections such as razor blades or parts that have just a thin, hard outer surface may be checked with a Rockwell superficial hardness tester. The loads range from 15 to 45 kg on what is known as the "T scale."

Microhardness Tests. Microhardness tests usually refer to tests made with loads ranging from 1 to 1,000 grams. The indenter is either a 136 degree diamond pyramid or a Knoop diamond indenter, Fig. 10–18. The Knoop indenter is a diamond ground to pyramidal form. It makes an indentation having an approximate ratio between the long and short diagonals of 7 : 1.

Prior to the advent of the microhardness tester, it had been assumed that the 136 degree diamond indenter produced a hardness number that was independent of the indenting load. In general terms, this can be accepted for loads of 1 kg and up. However, microhardness tests performed with loads of 500 g or lighter, with the Knoop indenter, and 100 g or lighter, with the diamond pyramid indenter, are a function of the test load.

The Knoop hardness number is the applied load divided by the unrecovered projected area of the indentation. Both hard and brittle materials may be tested with the Knoop indenter. The diamond pyramid hardness number (DPH) is also the applied load divided by the surface area of the indentation. The 136 degree diamond pyramid is often referred to as a "Vickers" test.

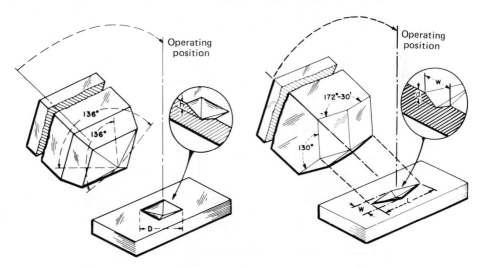

FIGURE 10–18. *The Knoop indenter (with the long impression) is shown at the left. At the right is the Diamond Pyramid or Vickers indenter. Hardness in both systems are in units of kg/mm². (Courtesy of Acco, Wilson Instrument Division.)*

Shore Scleroscope. The scleroscope presents a fast, portable means of checking hardness. The hardness number is based on the height of rebound of a diamond tipped, metallic hammer. The hammer falls free from a given height. The amount of rebound is observed on a scale in the background. The harder the material, the higher the rebound, and vice versa. Thin materials may be checked if a sufficient number of layers are packed together to prevent the hammer from penetrating the metal to an extent where the rebound is influenced by the steel anvil. This factor is known as the *anvil effect.*

DESIGNING FOR ASSEMBLY

Many factors enter into the selection of a joint design for a given application. The main considerations, however, are the desired physical and mechanical properties, the position that the joint must be in at the time it is made, the equipment available, and overall economy.

Both the American Welding Society (AWS) and the American Institute of Steel Construction (AISC) issue codes or standards to be followed in the design, fabrication, and erection of structural steel for buildings. The AWS code number for welding steel structures is listed as AWS D1.1–XX, in which buildings, bridges, and tubular structures are covered. The XX is given here to represent the year the code was issued. The AISC does not have an overall code number but is designated by the year it was issued. Code specifications are discussed in the latter part of this chapter.

From the standpoint of geometry, sheet metal and plates may be joined together in five main types of joints as shown in Fig. 10–19. Of these five joints, the two most frequently used in welded construction are butt and lap joints.

Butt Weld Joint Design. Butt welds, also known as groove welds, are made between abutting plates in the same plane and are generally described by the way the edges are prepared. In general, the least preparation required, the more

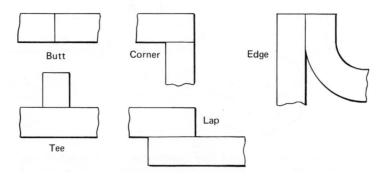

FIGURE 10–19. *The basic types of welded joints. There are many variations to each of these. Variations of the butt joints will be shown in Fig. 10–22.*

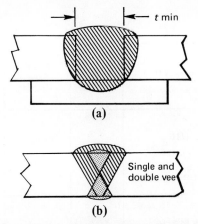

(a)

(b)

FIGURE 10–20. *The minimum preparation may lead to excessive weld deposit as in the case of a square butt joint in heavy plate (a). The weld deposit required for a single-vee compared to a double-vee butt joint (b).*

economical; however, when plates are thick, the spacing required to obtain complete penetration for manual welding may cause the amount of weld deposit to become excessive, Fig. 10–20. The type of edge preparation is largely dependent on how complete penetration can be achieved with the minimum amount of weld metal. Shown in Fig. 10–21 is a single vee butt joint, together with common terminology.

Square Edge Butt Joints. The square butt joint is often made with a *backing strip,* as shown in Fig. 10–22, along with standard types of welded butt joints. This strip may remain as part of the joint or it may be machined off later. Backing strips are commonly used when all welding must be done from one side or when the root opening is excessive.

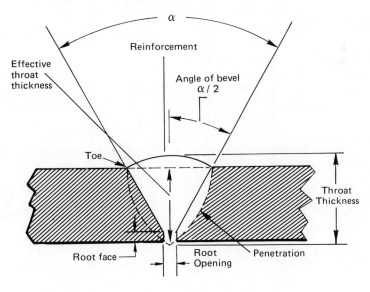

FIGURE 10–21. *A single-vee butt weld with common terms.*

WELD TYPE	Sides Welded	Thickness (in.)	Gap (in.)	Root Face (in.)	Min Included Angle (deg)
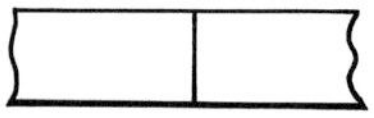 Closed Square Butt	One Both	Up to 1/16 Up to 1/8 Up to 5/8 With DP electrodes			
 Open Square Butt	Both	Up to 3/16	1/16		
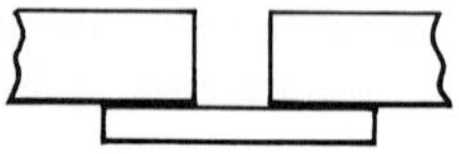 Square Butt with Backing Bar	One	Up to 3/16 Up to 1/2 Up to 1/2 With DP electrodes	3/16 5/16 1/4		
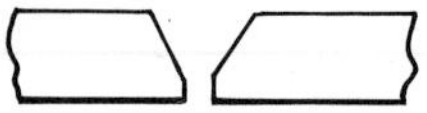 Single "V" Butt Weld	One or Both	Over 3/16 and up to 1 If made with backing strip	0 to 1/8 1/8 to 3/16	0 to 1/8	60
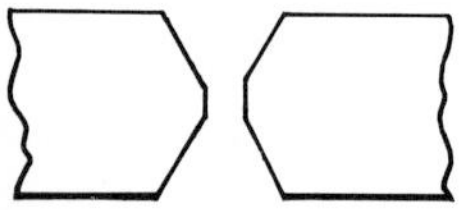 Double "V" Butt Weld	Both	Over 1/2	0 to 1/8	0 to 1/8	60
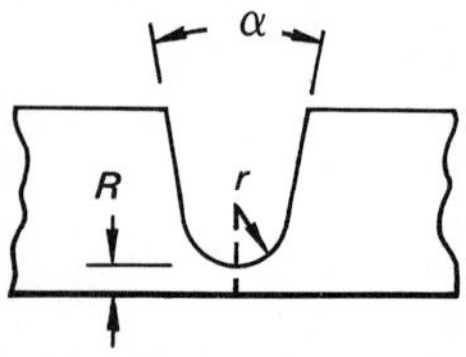 Single "U" Butt Weld	One or Both	Over 3/4	0 to 1/16	1/8 to 3/16	10 to 30

FIGURE 10–22. *Standard types of welded butt joints.*

WELD TYPE	Sides Welded	Thickness (in.)	Gap (in.)	Root Face (in.)	Min Included Angle (deg)
 Single "J" Butt Weld	Both	Over 3/4	1/8	1/8 to 3/16	20 to 30
 Single Bevel Butt Weld	Both	Up to 1 Unlimited with backing strip	0 to 1/8	0 to 1/8	45 to 50
 Double "U" Butt Joint	Both	Over 1-1/2		1/8 to 3/16	10 to 30
 Double "J" Butt Weld	Both	Over 1-1/2	1/8	1/8 to 3/16	20 to 30
 Double Bevel Butt Weld	Both	Over 1/2	0 to 1/8	0 to 1/8	45 to 50

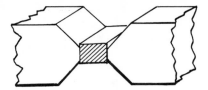

FIGURE 10–23. *Spacer strips are used between double-vee butt joints to maintain consistent root spacing for adequate penetration.*

Spacer strips as shown in Fig. 10–23 are used at intervals between double-vee joints to maintain consistent root spacing for complete penetration.

A *backing bar* is quite similar to a backing strip, however, it is only intended to be temporary and is removed after the weld is made. It consists of a heavy copper bar with a groove that provides room for full penetration of the weld metal. More recently, a tape has been developed that acts as a backing strip for closed butt joints. It consists of a flexible granular refractory layer mounted on a carrier with a wider pressure-sensitive adhesive foil protected by a paper liner, Fig. 10–24.

The root face used on all but the square butt joint provides an additional thickness of metal to minimize burn through. It should not be used if a backing strip is used. When a butt weld is made without a backing strip, *back gouging* is often necessary to eliminate fusion defects at the root face. Back gouging can be done by grinding or with an arc-air torch. The last mentioned method is more economical and leaves an ideal contour for the sealing bead. Back gouging must

(a)

(b)

FIGURE 10–24. *A tape type backing strip utilizes refractory granules on a carrier (left) and is shown in use on an aqueduct (right). (Courtesy of 3M Co.)*

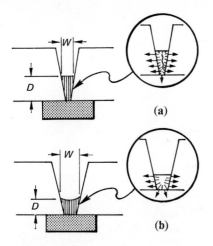

FIGURE 10–25. *The W/D ratio of weld pre-paration is related to crack-sensitivity. The wider root opening (b) provides a better W/D ratio than (a) and is not crack sensitive. (Courtesy of The Lincoln Electric Co.)*

be deep enough to expose sound weld metal and of a contour that permits complete accessibility.

Depth-to-Width Ratio for Butt Welds. Butt welds that are designed to maintain a balance between width and depth are generally less susceptible to weld cracks as shown in Fig. 10–25. Freezing starts along the weld surface adjacent to the cold base metal, and the last solidification takes place at the centerline of the weld. When the weld is deeper than it is wide, Fig. 10–25a, its surface may freeze in advance of its center. Shrinkage forces will act on the hot center core of the weld to cause a crack. The crack may or may not appear on the surface.

The critical width-to-depth ratio is almost entirely limited to the first pass. The second and subsequent passes are almost never subject to the same magnitude of stress. Thus, because of the importance of the W/D factor, the joint preparation may be inherently crack sensitive.

Butt Weld Strength. Full penetration butt welds are generally regarded as having the same load carrying capacity as the base metal. Hence, it is not necessary to calculate the strength of the weld if the weld is completed with either a backing bar or a sealing run and if the strength of the weld metal corresponds to that of the base metal.

Partial Penetration Butt and Groove Welds. The partial penetration butt weld as shown in Fig. 10–26 is only satisfactory where the joint strength requirement is low, as in tacking or positioning. As shown by the stress diagrams at the right of the welds, any bending moment on the plate will produce a high stress concentration in the weld. If the joint is tight, a load applied normal to the weld will not be too serious. The weld will be placed in tension and the unwelded side, in compression, Fig. 10–26a. The weld will also have to be able to withstand any shear forces that develop. The weakest condition of this type of weld is when the load is applied from the unwelded side, Fig. 10–26b. The weld must then carry the full

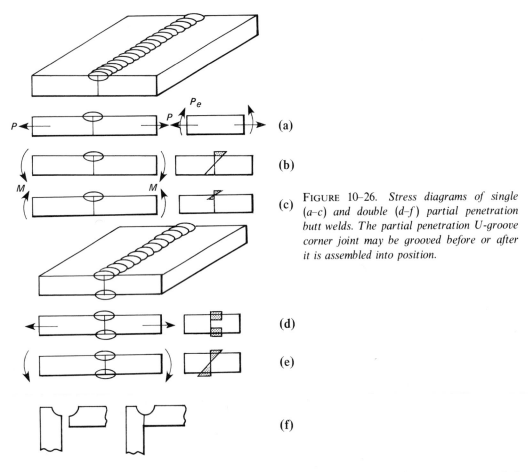

(a)

(b)

(c)

FIGURE 10–26. *Stress diagrams of single (a–c) and double (d–f) partial penetration butt welds. The partial penetration U-groove corner joint may be grooved before or after it is assembled into position.*

(d)

(e)

(f)

bending moment unaided. Loading parallel to the weld is the only type recommended for this type of joint, Fig. 10–26c.

The double welded partial penetration butt weld, Figs. 10–26d and 10–26e, offer a marked improvement. They can also be used in both tensile and bending loads when the full plate strength is not required and where fatigue is not a serious problem.

The groove welds, Fig. 10–26f, are commonly used in fabricating heavy structural members, as it results in both preparation and weld deposit savings while obtaining the required strength.

Fillet Weld Joint Design. Fillet welds may be defined as welds used to fill in a corner. They are the most common type of weld used in structural work. Variations in the use of fillet welds are shown in Fig. 10–27, along with a detailed view of a tee joint. The leg size of a fillet weld is usually equal to the height of the base material (h). The nominal weld outline, shown as dashed lines in the detail view, makes a 45 degree triangle when each leg is equal to h. The effective throat

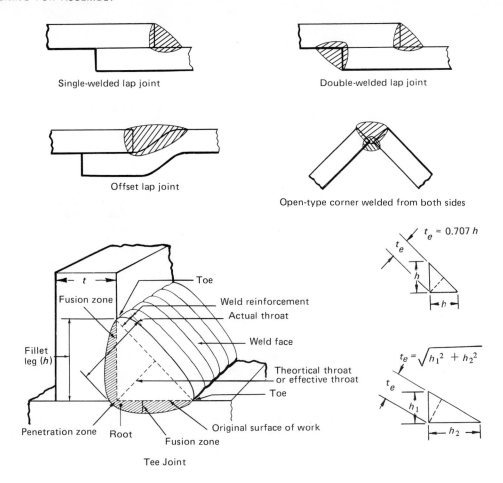

Single-welded lap joint

Double-welded lap joint

Offset lap joint

Open-type corner welded from both sides

Tee Joint

FIGURE 10–27.   *Lap joints, open corner joints, and tee joints all use fillet type welds. The fillet weld may have equal or unequal legs, but the effective throat (t_e) is taken perpendicular to the weld face from the root of the joint, as shown by the dashed line.*

(t_e) of the fillet is the distance from the root of the joint to a point perpendicular to the hypotenuse. For an equal leg (45°) fillet weld, the throat is equal to 0.707 (sine of 45°) times the leg dimension.

Shown in Fig. 10–28 is a comparison of how the theoretical and actual throat may vary between automatic and manual welding. The actual throat made by automatic welding far exceeds the requirements.

Fillet welds are not always made with leg sizes equal to h. The *maximum* size permitted by AISC–1.17.6, which states:

1. Along edges of material less than $\frac{1}{4}$ in. thick, the maximum size may be equal to the thickness of the material.

Automatic

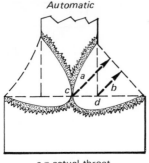

a = actual throat
b = theoretical throat

Manual

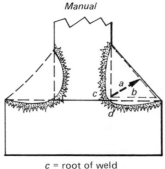

c = root of weld
d = root of joint

FIGURE 10–28. *A comparison of the actual and theoretical throat made by automatic and manual welding. (Courtesy of Lincoln Electric Co.)*

2. Along edges of material $\frac{1}{4}$ in. or more in thickness, the maximum vertical leg size shall be 1/16 in. less than the thickness of the material, unless the weld is especially designated on the drawings to be built out to obtain full throat thickness.

The *minimum* size of a fillet weld is governed by the amount of heat, and therefore, the size of the weld required to insure fusion. Most of the heat energy given off during the welding process is absorbed by the plates being joined. The thicker the plates being welded, the faster the heat is removed from the welding area. If unequal thicknesses of plates are being joined, the minimum fillet weld size will be governed by the thicker one. The minimum size fillet welds permitted by the AWS Structural Welding Code are shown in Table 10.2.

Unequal Leg Fillet Welds. Shown in Fig. 10–29 is a plot of the effect of unequal legs in a fillet weld, as leg b becomes greater than a the transverse load strength increases up to a 2b/a ratio of 2.15 in direction of the applied transverse load.

Minimum Fillet Weld Length. The ends of fillet welds, when made manually, tend to be tapered. For this reason AISC–1.17.7 limits the minimum effective length of fillet welds to four times their nominal size. If this condition cannot be met, the size

TABLE 10.2. *Minimum* Fillet Weld Size (AWS Structural Welding Code D1.1–72)*

Material Thickness of Thicker Part Joined in Inches (mm)		Fillet Weld (*h*) in Inches (mm)	
to $\frac{1}{4}$ incl.	(6.35 mm)	1/8	(3.175 mm)
over $\frac{1}{4}$	to $\frac{1}{2}$ (6.35–12.70 mm)	3/16	(4.7625 mm)
over $\frac{1}{2}$	to $\frac{3}{4}$ (12.70–19.05 mm)	1/4	(6.35 mm)
over $\frac{3}{4}$	to $1\frac{1}{2}$ (19.05–38.10 mm)	5/16	(7.9375 mm)
over $1\frac{1}{2}$	to $2\frac{1}{4}$ (38.10–57.15 mm)	3/8	(9.525 mm)
over $2\frac{1}{4}$	to 6 (57.15–152.40 mm)	1/2	(12.70 mm)
over 6	> (152.40 mm)	5/8	(15.875 mm)

* Except that the weld size need not exceed the thickness of the thinner joined part.

of the weld is considered to be one quarter of its effective length.

To reduce the stress at the ends of a fillet, it should be continued around the corner, wherever possible. The length of the end return, as it is termed, should not be less than twice the nominal size of the weld.

Fillet Weld Stress Distribution. Fillet welds may be made convex, concave, or flat, as shown in Fig. 10–30. The convex fillet makes it easier to obtain the required theoretical throat; however, if the part is subjected to dynamic loading, the stress concentration at the toes of the weld may become objectionable. The

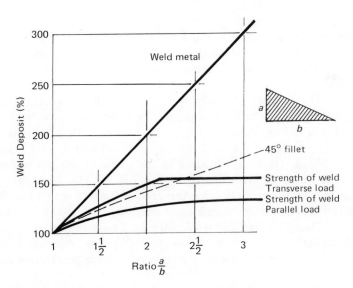

FIGURE 10–29. *As the ratio of b/a increases for transverse stress, there is an increase in load-carrying capacity up to a ratio of 2.15. (Courtesy of The Lincoln Electric Co.)*

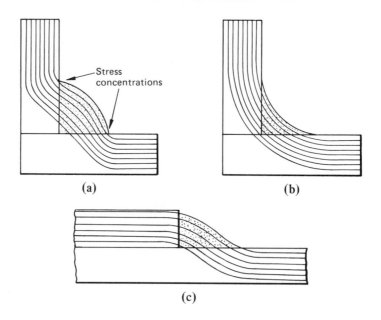

FIGURE 10–30. *Formerly the concave fillet weld (b) was favored by designers because it had smoother flow lines. Now however, the convex bead (a) is often preferred due to better crack resistance, especially in steels that require special welding procedures. The unequal leg fillet weld (c) is recommended for situations where fatigue is likely to be a factor.*

unequal legs as shown on the lap joint, Fig. 10–30c, can be used where uniform stress transfer is desired. Since it is more costly, the fillet weld should be used when the joint will be subject to dynamic loading and when there is a likelihood of severe fatigue.

The strength of a fillet weld varies with the direction of the applied load depending upon whether it is parallel or transverse to the weld. In either case, the weld will fail in shear on the plane that has the maximum shear stress. For loading parallel to the weld, the plane of rupture is at 45 degrees as shown in Fig. 10–31a. Transverse loading in the plane of maximum shear occurs at 67.5 degrees to the horizontal as shown at Fig. 10–31b. Because of this and the fact that the stress distribution for this type of loading is more uniform, the weld will carry about one-third more load in transverse shear than in parallel shear.

Two 45-degree fillet welds with leg dimensions equal to $3/4t$ will develop the full strength of the plate for either type of loading, assuming that the weld metal is made equal to that of the base metal and average penetration is obtained.

Allowable Stresses. Formerly, the allowable shearing stress recommended by AWS was 13,600 psi (93.70 MPa) for 45° fillet welds. Now, however, a shear value for weld metal in fillet or a partial penetration bevel groove weld is 0.30 of the minimum

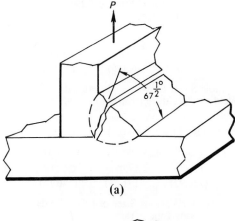

(a)

FIGURE 10–31. *The maximum shear stress occurs at an angle of 67.5 degrees for transverse loading (a), and at 45 degrees for parallel loading (b).*

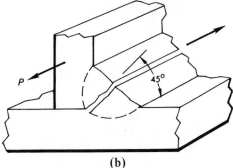

(b)

specified electrode tensile strength. Table 10.3 reflects the new values. The allowable unit force per lineal inch is also given and may be used to calculate the size of the fillet weld that is required for a given application according to the electrode used.

Example. Determine the allowable unit force (P) per inch for a 3/8 in. fillet weld made with E60 electrode.

Solution:

$$P = 0.707\ h(0.30)\ (\text{EXX})$$

$$P = 0.707 \times 3/8 \times 0.30 \times 60$$

$$= 4.77 \text{ kips/lineal inch (see Table 10.3)}$$

For E60 and E70 electrodes, the increase in allowable shear stress (T) is about 33 %. This permits a 25 % reduction in weld size while still maintaining the same allowable stress in the joint.

The new AISC code (1.14.7) gives limited credit for penetration beyond the root of a fillet weld made with the submerged arc process. Thus, for fillet welds 3/8 in. (9.525 mm) and smaller, the t_e is now equal to the leg size of the weld.

365

TABLE 10.3. *AISC and AWS Allowable Loads for Various Sizes of Fillet Welds*

	Strength Level of Weld Metal (EXX), ksi (in MPa)					
	60 (413.7)	70 (482.6)	80 (551.6)	90 (620.5)	100 (689.5)	110 (758.4)
	Allowable Shear Stress on Throat, ksi of Fillet Weld or Partial Penetration Groove Weld					
T =	18.0 (124.1)	21.0 (144.8)	24.0 (165.5)	27.0 (186.2)	30.0 (206.8)	33.0 (227.5)
	Allowable Force on Fillet Weld, Kips/Linear Inch					
f =	12.73h (87.71)	14.85h (102.32)	16.97h (116.92)	19.09h (131.53)	21.21h (146.14)	23.33h (160.74)
Leg Size (h) in	Allowable Force for Various Sizes of Fillet Welds, Kips/Linear Inch (in MPa/linear cm)					
(cm)						
1″	12.73	14.85	16.97	19.09	21.21	23.33
(25.4)	(87.71)	(102.32)	(116.92)	(131.53)	(146.14)	(160.74)
7/8″	11.14	12.99	14.85	16.70	18.57	20.41
(22.22)	(76.75)	(89.50)	(102.32)	(115.06)	(127.95)	(140.62)
3/4″	9.55	11.14	12.73	14.32	15.92	17.50
(19.05)	(65.79)	(76.75)	(87.71)	(98.66)	(109.69)	(120.58)
5/8″	7.96	9.28	10.61	11.93	13.27	14.58
(15.87)	(54.84)	(63.94)	(73.10)	(82.20)	(91.43)	(100.46)
1/2″	6.37	7.42	8.48	9.54	10.61	11.67
(12.70)	(43.89)	(51.12)	(58.43)	(65.73)	(73.10)	(80.41)
7/16″	5.57	6.50	7.42	8.35	9.28	10.21
(11.11)	(38.38)	(44.79)	(51.12)	(57.53)	(63.94)	(70.35)
3/8″	4.77	5.57	6.36	7.16	7.95	8.75
(9.52)	(32.87)	(38.38)	(43.82)	(49.33)	(54.78)	(60.29)
5/16″	3.98	4.64	5.30	5.97	6.63	7.29
(7.93)	(27.42)	(31.97)	(36.52)	(41.13)	(45.68)	(50.23)
1/4″	3.18	3.71	4.24	4.77	5.30	5.83
(6.35)	(21.91)	(25.56)	(29.21)	(32.87)	(36.52)	(40.17)
3/16″	2.39	2.78	3.18	3.58	3.98	4.38
(4.76)	(16.47)	(19.15)	(21.91)	(24.67)	(27.42)	(30.18)
1/8″	1.59	1.86	2.12	2.39	2.65	2.92
(3.17)	(10.96)	(12.82)	(14.61)	(16.47)	(18.26)	(20.12)
1/16″	0.795	0.93	1.06	1.19	1.33	1.46
(1.58)	(5.48)	(6.41)	(7.30)	(8.20)	(9.16)	(10.06)

For submerged arc fillet welds larger than 3/8 in. (9.525 mm), the t_e is obtained by adding 0.11 to 0.707h.

The higher allowable shear stress and penetration adjustment now allows the designer to obtain a given weld strength using about 35% of the weld metal deposit previously required.

Example. Two $\frac{1}{2}$ in thick plates are to be welded together with fillet welds as shown below using an E70 electrode. What will be the length of the weld required to support a 30 kips load? (See Table 10.2 for minimum weld size.)

Solution:

$$l = \frac{p}{0.707h \times \text{(allowable stress)}}$$

$$= \frac{30 \text{ kips}}{0.707 \times 0.5 \text{ in.} \times 13.6 \text{ ksi}}$$

$$= 6.25 \text{ in.} \quad (required\ with\ previous\ allowable\ loads)$$

$$l = \frac{30 \text{ kips}}{0.707 \times 0.5 \text{ in.} \times 21 \text{ kips}}$$

$$= 4.0 \text{ in.} \quad (required\ with\ present\ allowable\ loads)$$

Fillet Welds and Angle Brackets. Fillet welds encounter transverse shear in one of the most commonly used items in welding, support brackets. Shown in Fig. 10–32(a) is a bracket made from standard angle with equal legs. Some bending is encountered, but normally may be neglected if the loading arm (d) is small. End reaction (R) is assumed to be a line load. The normal and shear stresses in the weld may be calculated as follows:

$$\sigma = \frac{3\sqrt{2}\,Re}{h(H^2 = 3LH + 3Lh)}$$

$$T = \frac{\sqrt{2}\,R}{2h(H + h)}$$

where σ = normal stress.
           h = weld leg in inches.
           T = shear stress.
           H = bracket height in inches.
           L = bracket length.
           R = end reaction load.

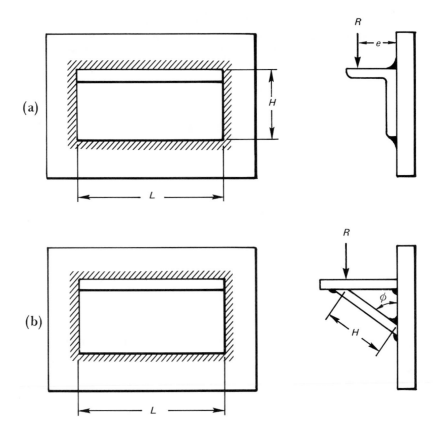

FIGURE 10–32. *The support bracket shown at* (a) *is made from a section of standard angle. The truss or gusset support shown at* (b) *is fabricated and is normally used to support heavy loads.*

The truss type support shown in Fig. 10–32(b) is normally for heavy loads. Angle ϕ should not become large since it would loose its truss-like quality. An angle of 45° is recommended. The normal and shear stress in the weld may be calculated as follows:

$$\sigma = \frac{\sqrt{2}\,R\tan\phi}{2Lh}$$

$$T = \frac{R\tan\phi}{2Lh}$$

WELDING SYMBOLS

This brief presentation of welding symbols is intended to show how easily symbols can be used to convey the exact desired information. A more complete presentation of "Standard Welding Symbols" (AWSA2.0–XX) was shown on the opening page to the chapter.

Shown in Fig. 10–33 is a welding symbol with the standard placement of information concerning the weld. Of course, in actual practice a symbol would never become this cluttered since there is not that much to say about one particular weld. You will note, for example, that the tail may be omitted when the specification of the process to be used is omitted. A further breakdown of the symbols is shown in Fig. 10–34. Here, as one example, is a chain intermittent fillet welding symbol showing the desired leg size along with the length and the increment spacing of the weld. Thus, symbols allow the designer to give rather complete information to the weldor without resorting to lengthy notes. Shown in Fig. 10–35 are some typical examples of the use of welding symbols in structural work.

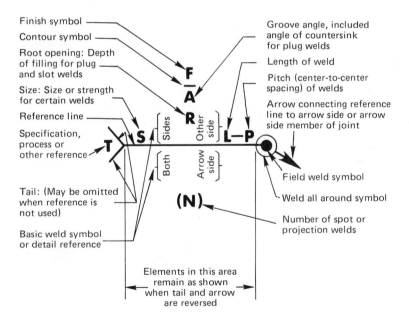

FIGURE 10–33. *The location of the elements of a welding symbol. (Courtesy of American Welding Society.)*

FIGURE 10–34. *Typical welding symbols.*
(*Courtesy of American Welding Society.*)

Back or Backing Weld Symbol

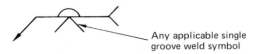

Any applicable single
groove weld symbol

Surfacing Weld Symbol Indicating Built-up Surface

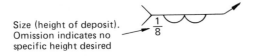

Size (height of deposit).
Omission indicates no
specific height desired

Orientation. Location
and all dimensions
other than size are
shown on the drawing

Double Fillet Welding Symbol

Size (length of leg)

Specification. Process
or other reference

Length. Omission
indicates that weld
extends between
abrupt changes in
direction or as
dimensioned

Chain Intermittent Fillet Welding Symbol

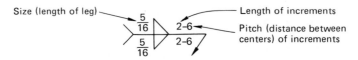

Size (length of leg)

Length of increments

Pitch (distance between
centers) of increments

Staggered Intermittent Fillet Welding Symbol

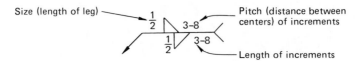

Size (length of leg)

Pitch (distance between
centers) of increments

Length of increments

Single-Vee-Groove Welding Symbol

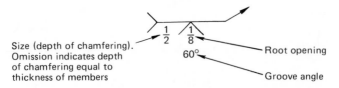

Size (depth of chamfering).
Omission indicates depth
of chamfering equal to
thickness of members

Root opening

Groove angle

WELDED CONNECTIONS

The term *welded connections* refers to the design and method of joining
standard structural members such as angles, channels, beams, and columns, some
of which are shown in Fig. 10–36. It is assumed the student will have had some
background in strength of materials; therefore, the treatment here will deal with the

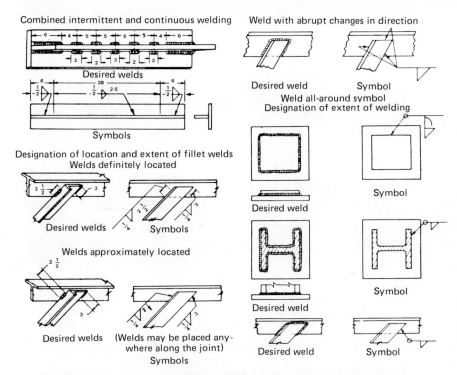

Combined intermittent and continuous welding

Desired welds

Symbols

Designation of location and extent of fillet welds
Welds definitely located

Desired welds   Symbols

Welds approximately located

Desired welds   (Welds may be placed any-
                    where along the joint)
                         Symbols

Weld with abrupt changes in direction

Desired weld   Symbol
Weld all-around symbol
Designation of extent of welding

Symbol

Desired weld

Symbol

Desired weld

Symbol

Desired weld   Symbol

FIGURE 10–35. *An example of the use of welding symbols in structural work.*
(*Courtesy of American Welding Society.*)

design of the weldments rather than the properties of the structural members. The
purpose will be to give a very brief presentation of some of the principles involved
in the design of welded structural connections. The examples chosen are of welded
beam and column type connections involving shear, bending, and torsion.

Rigid Beam and Column Connections.   Beam and column connections can be made
rigid or flexible. A rigid type connection is shown in Fig. 10–37. A first impression
of this type of connection may be that it is easy to construct and is perhaps more
than adequate for most requirements. A wide flanged (*W*) beam is used to provide
more room for the welds. As an example, an American standard six inch I-beam,
now referred to as 'S' beams, may have a three inch flange, whereas, a *W* beam
will have a six inch flange. In this case it is assumed that both bottom and top
flanges will withstand bending stresses and that the welds on each side of the web will
withstand shear. This type of construction has two distinct disadvantages:

1. The connection must withstand fairly large moments, which in turn cause large
   bending stresses.
2. It is difficult to get members to fit perfectly in the field for this type of
   connection.

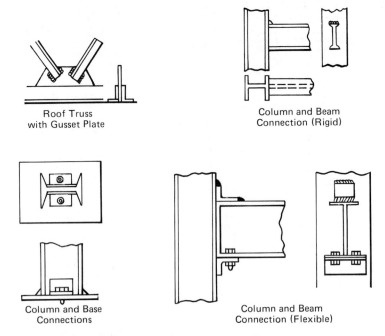

Roof Truss
with Gusset Plate

Column and Beam
Connection (Rigid)

Column and Base
Connections

Column and Beam
Connection (Flexible)

FIGURE 10–36.   *Welded type structural steel connections.*

Flexible Welded Beam and Column Connection.   Flexible beam and column connections are made so that the beam rests on an angle bracket or is held in place by a web angle connection. The simplest of these, the angle seat previously discussed in general, will be considered in more detail.

Angle Seat Connections.   The angle seat shown in Fig. 10–38 offers a short length of angle upon which the beam rests. A fillet weld is used along each side of the

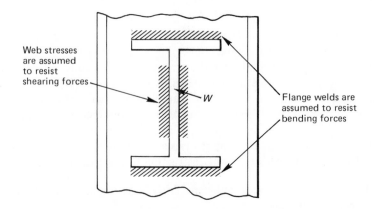

Web stresses
are assumed
to resist
shearing forces

W

Flange welds are
assumed to resist
bending forces

FIGURE 10–37.   *A rigid type connection, a wide flanged beam (W) is welded directly to a column.*

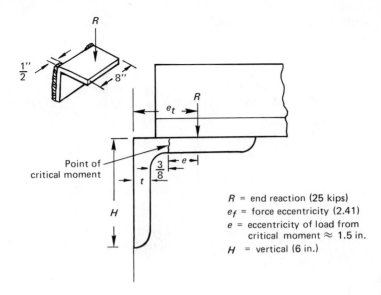

FIGURE 10-38. *Dimensions for determining the leg size of fillet welds for an angle.*

bracket with an end return at the top. Bolt holes often are provided so that both members may be locked securely into position before welding. This eliminates the need for tacking, which often causes defective welds. The beam is positioned about a half-inch from the column to allow room for rotation in a downward direction under load.

The required bracket size may be calculated in several ways depending on

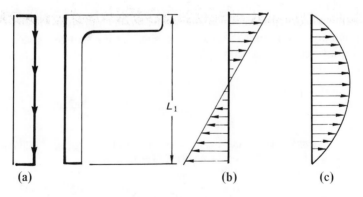

FIGURE 10-39. *The horizontal shear stress is maximum at the neutral axis of the bracket.* (a) *Vertical force.* (b) *Bending stress.* (c) *Shear stress.*

whether shear, or both moment and shear, forces are considered. The simplest case is to consider both welds as being subjected to shear only.

> *Example.* The end reaction (R) of the beam connected to two columns is 25 kips (344.7 MPa), Fig. 10–38. The allowable unit stress for a $\frac{1}{4}$ in. (6.35 mm) fillet weld is 60 ksi (413.7 MPa) and is 3.18 k/in. (21.91 MPa/25.4 mm) (Table 10.3).

> *Solution:*

$$\text{Weld length} = \frac{R}{allowable\ load\ kips/linear\ inch}$$

$$= \frac{25\,k}{3.18\ k/in.}$$

$$= 7.9\ in.$$

$$= \frac{8}{2} = 4\ in./side$$

∴   The length of weld for each side of the bracket will be 4 in. This does not include end returns.

Since the fillet size was calculated at $\frac{1}{4}$ in. (6.35 mm), the thickness of the bracket will be 5/16 in. (7.93 mm), as the AISC code requires the calculated weld strength be based on using 1/16 in. less than the thickness of the piece where the fillet weld is placed along the edge. The length of the bracket (H) must be long enough to seat the beam easily between the two $\frac{1}{2}$ in. weld returns on the bracket.

The example given assumes the shear stress will be uniform over the length of the weld, Fig. 10–39. This is the normal practice for short welds of this type. If bending is involved, the maximum tension would be at the top of the weld with maximum compression at the bottom, as shown at Fig. 10–39b. The shear stress is not uniformly distributed but will be more like that shown at Fig. 10–39c. The important point to note is that the maximum shear stress and the maximum bending stress do not occur at the same place. The combined stress is included in the following equation:

$$h = \frac{R}{2H^2.707\tau} \cdot \sqrt{H^2 + 20.25e_f{}^2}$$

> *Example.* Determine the leg size of the fillet weld required in the previous problem.
> For E60 electrode, $\tau = 18$ ksi (Table 10.3).

$$h = \frac{25}{2(6)^2.707(18)} \cdot \sqrt{6^2 + 20.25 \cdot (2.41)^2}$$

$$h = 0.488 \quad \text{or} \quad 1/2\ in.$$

A method of arriving at the thickness (t) of the angle when bending is involved is to use the following formula:

$$t = \sqrt{\frac{6Re}{SW}}$$

$$= \sqrt{\frac{6 \times 25\,\text{kips} \times 1.5}{40\,\text{kips} \times 8}}$$

$$= 0.838 \quad \text{or } 7/8 \text{ in.}$$

where   6 = constant.
           R = reaction.
           S = working stress.
          W = width of angle.
           l = load eccentricity.

Web Angle Connections. A web angle connection as shown in Fig. 10–40 is designed to transmit shear with as little moment as possible. The angle plates are usually shop welded to the beam and then welded to the column on location, after being bolted into position with high strength bolts. You will note that the beam is not made to fit flush against the column but is brought out about $\frac{1}{2}$ in. (12.70 mm) for what is known as the *setback*. The bolt is placed near the bottom as not to reduce the flexibility.

Shear and Moment in Welded Connections. Shown in Fig. 10–41 is an angle section welded to a steel plate as done ordinarily in constructing a truss. If the connection is subjected to repeated stresses, it is necessary to consider placing the welds so they will coincide with the centroid of the member. The stress in the angle shown is assumed to act along its center of gravity. If the weld is to match the

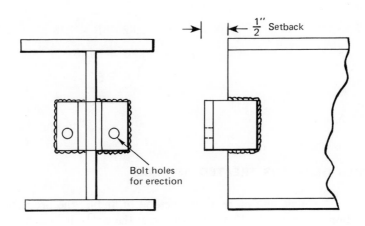

FIGURE 10–40. *A web angle connection.*

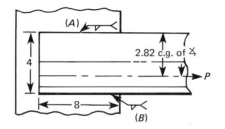

FIGURE 10–41. *An angle section welded to a steel plate. Specifications: size, 4 × 4 in. (10.16 cm); thickness, $\frac{1}{2}$ in. (12.70 mm); area, 3.75 in.² (94.61 mm); steel type A36; and electrode E60.(Courtesy of Welding Design and Fabrication.)*

angular stress, it will have to be asymmetrically placed (*B* will have to be longer than *A*).

> ***Example.*** What is the total length of weld necessary for a 45 k (310.3 MPa) load on a 4 × 4 × $\frac{1}{2}$ in. (12.70 mm) angle plate? Only side welds are to be used. How should the welds be placed?

> *Solution:* Weld size = 0.500 − 1/16 in. = 7/16 in. $L = 45\,k/5.57\,k/in. = 8.07\,in. =$ total length of weld required.

$$\left(L = \frac{310.3\,\text{MPa}}{38.38} = 8.07\,\text{in.} \qquad \text{or } 20.52\,\text{cm} \right)$$

Weld Placement. By the AISC manual, the center of gravity for this size beam section is given as 1.18 in. from the bottom edge. The moments of *A* and *B* are taken about the load line *P* so that both will be equal:

$$A \times 2.82 = B \times 1.18$$

$$A + B = 8.07$$

Then,

$$A = 8.07 - B$$

And,

$$(8.07 - B) \times 2.82 = B \times 1.18$$

$$22.8 - 2.82B = 1.18B$$

$$B = 5.7$$

$$A = 8.07 - B$$

$$= 2.37\,\text{in.}$$

PROPERTIES OF WELDS TREATED AS LINES

A simple and more direct method of finding the properties of welded connections, proposed by Blodgett, is to treat the welds as lines having neither leg nor throat dimensions. The forces, which can be taken from Table 10.4, are considered

(a)

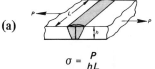 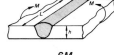 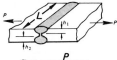

$$\sigma = \frac{P}{hL} \qquad \sigma = \frac{6M}{Lh^2} \qquad \sigma = \frac{P}{(h_1 + h_2)L}$$

(b)

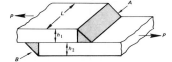 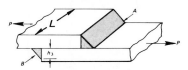

Stress in weld A = stress in weld B

$$\sigma = \frac{1.414P}{(h_1 + h_2)L}$$

Both plates the same thickness

$$\sigma = \frac{.707P}{hL}$$

Stress in weld $A \neq$ stress in weld B

Weld $A = \sigma = \dfrac{1.414P}{(h_1 + h_2)L}$

Weld $B = \sigma = \dfrac{1.414Ph_2}{h_3L(h_1 + h_2)}$

(c)

 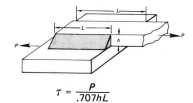 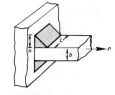

$$\tau = \frac{P}{.707hL}$$

$$\sigma = \frac{.707P}{hL}$$

Note: If subject to bending (M) in place of tension (P), substitute

$$\sigma = \frac{1.414M}{hL(b + h)}$$

(d)

  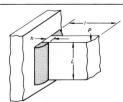

(e)

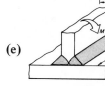

$$\sigma = \frac{4.24M}{hL^2}$$

Average $\tau = \dfrac{.707P}{hL}$

Max $\sigma = \dfrac{4.24Pl}{hL^2}$

$$\sigma = \frac{6M}{Lh^2}$$

(f)

$$\sigma = \frac{5.66M}{hD^2\pi} \qquad \tau = \frac{2.83M}{hD^2\pi}$$

σ	= Normal stress, PSI
τ	= Shear stress, PSI
M	= Bending moment, lbs
P	= External load, lbs
h	= Size of weld
$\left.\begin{array}{c}L\\l\end{array}\right\}$	= Linear distance, inc.

TABLE 10.4. *Stress Formulas for Weld Joints.*

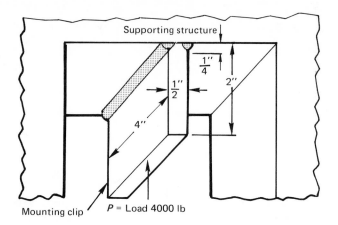

FIGURE 10–42. *A mounting clip is welded to a supporting structure.*

on a unit length basis, thus eliminating knotty problems of combined stresses. This is not to neglect stress distributions within a weld, which can be very complex. However, the actual fillet welds tested, upon which the unit forces are based, also had these same stress conditions.

When the weld is treated as a line, the property of the welded connection is inserted into the standard design formula for that particular type of load and the force on the weld is found in terms of pounds per lineal inch. A sample problem will serve to illustrate the use of this method.

Problem. Shown in Fig. 10–42 is a clip that has been welded to a supporting structure. Will this clip be able to sustain a load of two tons, which at times has some impact force also?

Solution: Choose a matching stress equation from Table 10.4, in this case, part (c). What is the maximum load the clip will hold? The maximum allowable stress for the weld is 3,710 psi with an E70 electrode.

$$\sigma = \frac{0.707P}{hl}$$

Solving for P,

$$P = \frac{\sigma hl}{0.707} = \frac{3710 \times \frac{1}{4} \times 8}{0.707} = 10,495 \, \text{psi}$$

Metric Solution:

$$\frac{3.710 \times 0.70 \, \text{kg/sq m} \times 6.35 \, \text{mm} \times 203.2 \, \text{mm}}{0.707} = 4739 \, \text{kg/sq mm}$$

$$4739 \times 2.2205 \, \text{lb} = 10499 \, \text{psi}$$

TABLE 10.5. *Weld Safety Factors*

Type of Weld	SC Factor (Kf)
Reinforced butt	1.2
Toe of transverse fillet, occasional impact	1.5
End of parallel fillet	2.7
T-butt with sharp corners	2.0

Applying a safety factor for occasional impact (Table 10.5) of 1.5:

$$\sigma = 10495 \times 1.5 = 15.742\,\text{lb}$$

A second problem will serve to further illustrate the use of the line method.

Problem. Shown in Fig. 10–43 is a 1 in. diam gear and motor drive shaft connected by clutch plates. The weld must withstand a shear stress not exceeding 10,000 psi. A moment of 4000 in./lb is developed at the clutch plate. What size weld is required?

Solution: Use the part (b) equation from Table 10.4.

$$\sigma = \frac{2.83M}{hD^2\pi}$$

Solving for h,

$$h = \frac{2.83\,M}{SD^2} = \frac{2.83 \times 4000\,\text{lb/in.}}{10000 \times 1^2 \times 3.14}$$

$$h = 0.360\,\text{in. (use a 3/8 in. fillet weld)}$$

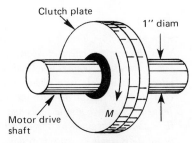

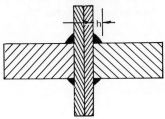

FIGURE 10–43. *The weld as shown must withstand 10,000 psi shear stress. (Courtesy of Welding Design and Fabrication.)*

379

Metric Solution:

$$= \frac{2.83 \times 45300^* \, \text{kg/mm}}{7.03 (\text{kg/mm}^2)^{**} \times 645^{***} (\text{mm}^2) \times \pi}$$

$$= \frac{2.83 \times 45300 \, \text{kg/mm}}{7.03 \times 645 \times 3.14} = \frac{128199.00}{13796.38} = 9.0 \, \text{mm}$$

$$= 9.0 \, \text{mm} \times 0.04 = 0.360 \, \text{in.}$$

DISTORTION

The designer of fabricated structures recognizes the fact that the finished product will not be as shown on the print unless steps are taken to deal with distortion each time a weld is made. The purpose of this discussion is to show the principal causes of distortion and how to control it.

Causes of Distortion. Distortion in weldments is the result of nonuniform heating and cooling. Expansion and contraction of the weld metal and the adjacent base metal may be compared to heating a metal bar between vise jaws, Fig. 10–44. The restricted bar is not able to expand uniformly, therefore, it is upset in the area of greatest heat. Upon cooling, it will contract or shrink in all directions, with the result that the bar is now shorter and thicker than it was previously.

A weld made on restricted members as shown in Fig. 10-45 cannot shrink when cooled; therefore, residual stresses are set up in the weld and in the heat affected zone (HAZ) adjacent to it. In an unrestricted weld, where the members are

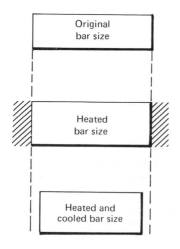

FIGURE 10–44. *Three stages of metal bar sizes showing the effect of restrained heating but unrestrained cooling. The localized heating effect of a weld is very obvious in that the cold adjacent metal acts as the restraining influence. As the hot weld cools and contracts, it is again restrained by the colder adjacent metal. This condition leads to residual stresses, distortion, and cracks.*

* 4000 × 0.453 kg = 1812 kg × 25 mm = 45300 kg/mm
** 10000 psi = 7.03 kg/mm^2
*** 1 in.^{2} = 645 mm^2

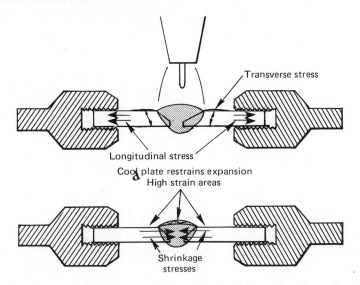

FIGURE 10–45. *Strains develop in the weld and in the immediate heat affected zone (HAZ) upon cooling.*

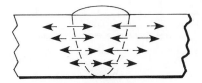

FIGURE 10–46. *Tensile stresses are set up between the weld metal and the cooler base metal as solidification and contraction take place.*

free to expand and contract, the metal next to the HAZ serves somewhat the same function as a vise, as shown in Fig. 10–46. Since the members do move, the degree of stress in the weld and HAZ will not be as great. Some of the locked-in stresses are relieved as the base metal shifts or distorts.

Types of Distortion.   Distortion is evidenced in many different forms, but basically, it is seen in two main types, angular and longitudinal, Fig. 10–47. Transverse shrinkage is the main cause of angular distortion. It is also directly related to the volume of weld deposit as shown in Fig. 10–48.

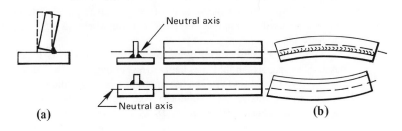

(a)

(b)

FIGURE 10–47. *Distortion types: (a) angular, and (b) longitudinal.*

381

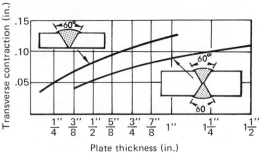

Transverse contraction—single Vee vs. Double Vee

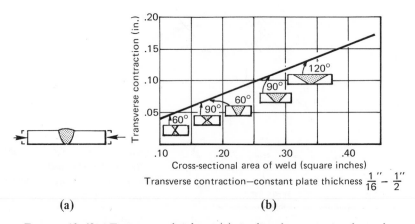

Transverse contraction—constant plate thickness $\frac{1}{16}'' - \frac{1}{2}''$

(a)　　　　　　　　　　　　　　(b)

FIGURE 10–48. *Transverse shrinkage (a) is directly proportional to the amount of weld deposit (b). (Courtesy of The Lincoln Electric Co.)*

Distortion Control. Shrinkage cannot be prevented, but a knowledge of how it operates is useful in preventing or minimizing distortion. The following principles are used in minimizing distortion.

1. *Minimize Shrinkage Forces.* Shrinkage forces can be minimized by using only that amount of weld metal required. Overwelding not only causes shrinkage forces to build up, but it is also uneconomical. Proper edge preparation and fit-up will help minimize the amount of weld metal required. Intermittent welding can also be used to minimize the weld deposit where strength requirements are not critical.

    Where transverse shrinkage forces cause distortion, as in a butt weld, the number of passes should be kept to a minimum. The shrinkage of each pass tends to be cumulative. The use of larger electrodes allows more weld deposit per pass with a smaller total volume of heat input into the base metal.

2. *Help Shrinkage Forces Work in the Desired Direction.* Parts may be positioned or preset out of position before welding so that shrinkage forces will bring them into alignment as illustrated by the tee joint and butt weld, Fig. 10–49.

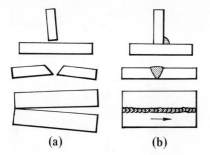

FIGURE 10–49. *Prepositioning (a) allows weld shrinkage forces to bring parts into alignment, (b).*

(a) **(b)**

3. *Balance Shrinkage Forces with Other Forces.* A common example of balancing shrinkage forces is the use of fixtures and clamps. The components are locked in the desired position and held until the weld is finished. Of course, internal stresses are built up, as mentioned previously, until the yield point of the metal is reached. For typical welds on low carbon plate, this would be about 45,000 psi. Perhaps you may visualize distortion taking place as soon as the clamps are removed. This does not happen due to the small amount of unit strain (ε) compared to the amount of movement that would occur if no restraint existed during the welding process. The strain for steel can be calculated as follows:

$$\varepsilon = \frac{\sigma}{E}$$

$$= \frac{45000}{30 \times 10^6}$$

$$= 0.0015 \text{ in./in.}$$

Another common example of balancing shrinkage forces is a butt joint welded alternately on each side.

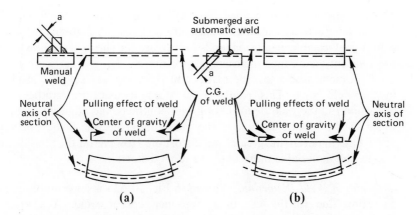

(a) **(b)**

FIGURE 10–50. *Deeper penetration (b) achieved with the automatic submerged arc results in welds that are nearer the center of gravity with less distortion than the manual deposit (a). (Courtesy of The Lincoln Electric Co.)*

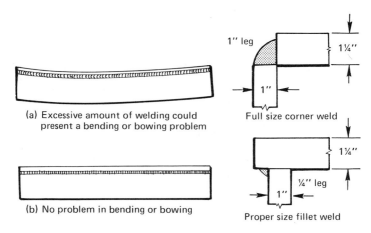

(a) Excessive amount of welding could present a bending or bowing problem

Full size corner weld

1" leg

1¼"

1"

(b) No problem in bending or bowing

Proper size fillet weld

1¼"

¼" leg

1"

FIGURE 10–51. (a) *Longitudinal distortion due to excessively large corner weld and* (b) *elimination of distortion by a change in design and a smaller weld.* (*Courtesy of The Lincoln Electric Company.*)

4. *Remove Shrinkage Forces After Welding.* Peening is often used as a means of stress relieving after welding. In theory the metal is made to expand under the force of the blow, thus relieving the stress. Care must be exercised in peening to avoid concealing a crack or work hardening the metal.

   In more specialized cases such as the alloy steels, preheating and post-heating are used to minimize the residual stress and distortion.

5. *Place Welds Near the Center of Gravity.* It is not always possible to place welds on or near the center of gravity, however, even the amount of penetration as shown in Fig. 10–50 can change the amount of distortion. Shown at Fig. 10–50a is a tee weld made by manual welding, and that shown in Fig. 10–50b is by the submerged arc automatic process.

Metal Properties and Distortion. Distortion is related to the coefficient of expansion, thermal conductivity, modulus of elasticity, and yield strength of the material welded. A high coefficient of expansion tends to increase the shrinkage of the weld metal and the adjacent metal, thus increasing distortion.

A metal with a low thermal conductivity, such as stainless steel, retains heat in the weld area. This results in a steep temperature gradient and greater distortion. As the weld deposit cools and contracts, the adjacent metal "stretches" to satisfy the volume demand of the weld joint. Thus, the higher the yield strength of the base material, the greater the distortion. However, if the modulus of elasticity is high, the material is more likely to resist distortion.

Examples of Weld Distortion. Shown in Fig. 10–51a is an example of longitudinal distortion due to the use of a one inch corner weld. A change in design resulted in the welds as shown by Fig. 10–51b. The amount of weld shrinkage varies with the cross-sectional area of the weld, which is the square of the leg size. Thus, substituting a 1/4 in. weld reduces the shrinkage force to 1/16 of its original value.

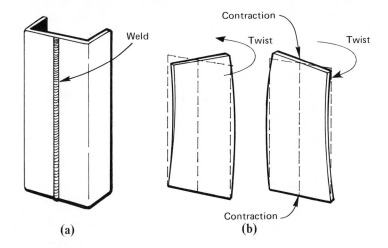

FIGURE 10–52. *Twist results after a weld is made (a) because of the low torsional resistance to the longitudinal contraction (b) of the centerline weld. (Courtesy of The Lincoln Electric Company.)*

Twisting is a problem in welding thin materials because of the low torsional resistance. When a weld is made down the centerline of a member, as shown in Fig. 10–52, the contraction upon cooling will make the outer edges longer than the centerline. The result is the twisting action shown at Fig. 10–52b.

To minimize the twisting action, the heat input may be controlled by increasing the welding speed or making intermittent welds. A small increase in the material thickness will help considerably since torsion is a function of the thickness cubed. Thus, doubling the thickness will reduce the torsional resistance by a factor of 8.

EXERCISES

Problems and Questions

10–1. (a) Make a sketch similar to Fig. P10–1 and indicate the direction of the principal stresses in the welds of each type of joint shown.

(b) Name the principal type stress in each of the joints shown.

10–2. (a) Assume a load of 10,000 lb (453.6 kg) for the center of the beam in problem 10–1. The beam is 20 in. (50.8 cm) long and 2 in. (5.08 cm) thick and 4 in. (10.16 cm) wide. The weld at the top has a 3/8 in. (9.53 mm) leg and the one at the bottom has

a $\frac{1}{4}$ in. (6.35 mm) leg. What is the average shear stress on the welds?

(b) What would the maximum shear stress be if only the top welds were put in?

10–3. State what type of stress is encountered in the welds shown in Fig. P10–3.

10–4. Shown in Fig. P10–4 is a joint requiring two 4 in. (10.16 cm) fillet welds. Assume the tensile load is 50 k. Are the welds adequate if they are made with an E60 electrode?

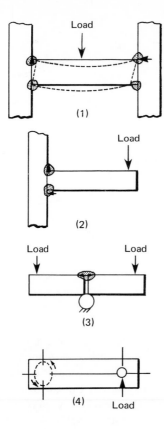

FIGURE P10–1. *Use arrows to indicate the direction of stress in the welds.*

10–5. (a) A butt weld is made in an HSLA steel, A572, that has a maximum carbon content of 0.22 %. The material is 1.5 in. (3.81 cm) thick and 4 in. (10.16 cm) wide. What is the normal stress on the joint if it is placed in tension at 20 ksi (137.9 MPa)?

(b) If the butt weld of Problem 10–5(a) has a $\frac{1}{2}$ in. (12.70 mm) thick by 3 in. (7.62 cm) wide plate placed on each side of it and welded along both ends of the plate only, what will be the normal stress on these plates?

10–6. As shown in Fig. P10–6, plate B is to be welded to plate A. Upon completion the weldment

must resist a 48,000 lb tensile stress as indicated by arrows at axis of plate. The following information is provided:

1. Plate B is low carbon steel, 5/16 in. × 2 in. × 15 in. (7.93 mm × 5.08 cm × 38.1 cm) with a tensile strength of 48,000 psi (331.0 MPa).

2. Plate A is also low carbon steel, 5/16 in. × 4 in. × 12 in.   (7.93 mm × 10.16 cm × 30.48 cm), with a tensile strength of 48,000 psi (331.0 MPa).

3. As specified in the structural code of AWS, the allowable load for a fillet weld in shear for this grade of steel is based on a stress of 21,000 psi (144.8 MPa) in the throat section, which includes a safety factor of about three.

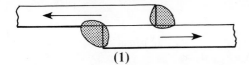

(1)

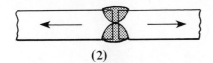

(2)

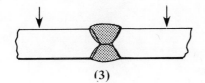

(3)

FIGURE P10–3.   *Name the principal types of stresses encountered by the welds shown.*

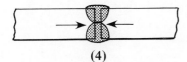

(4)

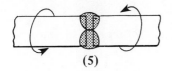

(5)

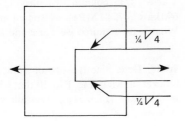

FIGURE P10–4.   *The load applied to the fillet welds is 50 k.*

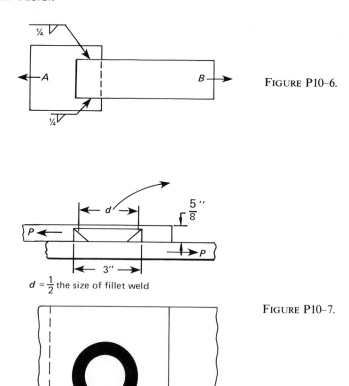

FIGURE P10–6.

FIGURE P10–7.

4. An E7016 electrode is prescribed.

(a) How many inches of fillet weld are required on each side of Plate *B* to guarantee the proper strength?

(b) A fillet weld loaded at right angles is about 30 % stronger than if loaded parallel. Applying the information of Fig. P10–4, what must the length of each fillet weld be? In your opinion, would this be a satisfactory weld? Explain?

10–7. A 3 inch (7.62 cm) diameter hole is made in a 5/8 in. (15.875 mm) mild steel plate that is to be welded to another plate by means of a continuous 1/2 in. (12.70 mm) fillet weld at the edge of the hole, as shown in Fig. P10–7. Compute the load the weld will support if an E6010 electrode is used.

10–8. Find the size of the two welds required to attach a plate to the machine shown in Fig. P10–8, *if* the plate carries an inclined force $P = 10,000\,lb$ (4530 kg). The allowable shearing stress is 18,000 psi (124.1 MPa) for weld metal.

10–9. Design a column base plate attachment for a 14W, 87 lb column to withstand a moment of 690 kips (4754.1 MPa), as shown in Fig. P10–9. What is the required weld size and length?

10–10. Determine the size of a fillet weld for the bracket shown in Fig. P10–10 that will carry a load of 25,000 lb (11 325 kg). The allowable shearing stress is 18,000 psi (12.66 MPa) for the weld material.

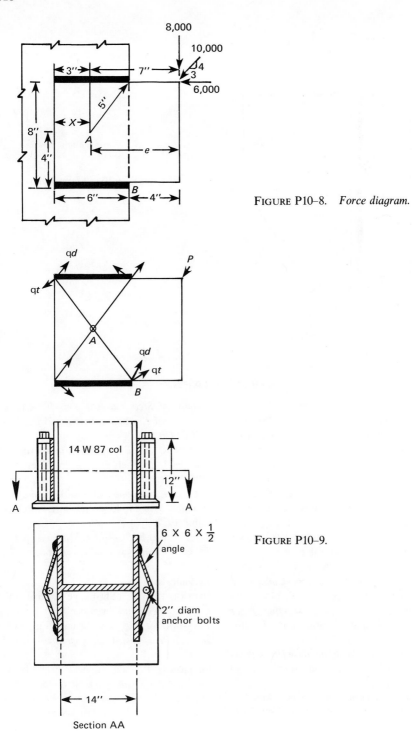

FIGURE P10–8. *Force diagram.*

FIGURE P10–9.

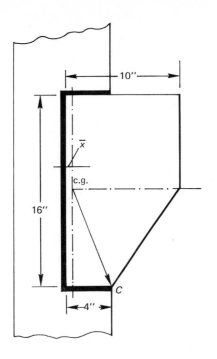

FIGURE P10–10. *Force Diagram.*

BIBLIOGRAPHY

Books

BLODGETT, O. W., *Design of Welded Structures*, The James F. Lincoln Arc Welding Foundation, Cleveland, Ohio, 1966.

————*Design of Weldments*, The James F. Lincoln Arc Welding Foundation, Cleveland, Ohio, 1963.

BROCKENBROUGH, R. L., AND JOHNSTON, B. G., *Steel Design Manual*, United States Steel Corporation, Chicago, Ill., 1968.

Designer's Guide for Welded Construction, Lincoln Electric Company, Cleveland, Ohio, 1962.

GRAHAM, I. D., SHERBOURNE, A. N., AND KHABBAZ, R. N., *Welded Interior Beam-to-Column Connections*, Am. Inst. of Steel Const., New York, N.Y., 1959.

HARMAN, R. C., ed., *Handbook for Welding Design*, vol. 1, Sir Isaac Pitman & Sons Ltd., London, England, 1967.

Joint Committee of the Welding Research Council and ASCE Manual of Engineering Practice, no. 41, 1961.

Manual of Steel Construction, 6th ed., American Institute of Steel Construction, Inc., New York, N.Y., 1963.

Manual of Steel Construction, 7th ed., American Institute of Steel Construction, Inc., New York, N.Y., 1970.

PETERSON, R. E., *Stress Concentration Design Factors*, John Wiely & Sons, Inc., New York, N.Y., 1953.

Procedure Handbook of Arc Welding Design and Practice, 11th ed., Lincoln Electric Co., Cleveland, Ohio, 1957.

TIMOSHENKO, S., *Strength of Materials*, parts I and II, 3rd ed., D. Van Nostrand Co., Princeton, N.J., 1956.

Periodicals

BLODGETT, O. W., "Shrinkage Control in Welding." *Civil Engineering*, November 1960.

FREEMAN, F. R., "The Strength of Arc Welded Joints." *Proc. Inst. Civil Engrs.*, vol. 231, London, 1931.

GILLIGAN, J. A., "Higher Strength Structural Steels." *Civil Engineering*, November 1964.

HENSLEY, R. C., AND J. J. AZAR, "Computer Analysis of Nonlinear Truss-Structures." *J. of the Structural Div. Am. Soc. of Civil Engr.*, June 1968.

HICKEY, R. L., AND SHULTZ, R. W., "Welding, the Way to Hybrid Structures." *Metal Progress*, September 1966.

HINKEL, I. E., "Joint Designs Can Be Both Practical and Economical." *Welding Journal*, June 1970.

JENSEN, C. D., "Welded Structural Brackets." *Welding Journal*, October 1936.

JOHNSTON, B. G., AND DEETS, G. R., "Tests of Miscellaneous Welded Building Connections." *Welding J.*, January 1942.

MEARS, R. B., "The Prevention of Corrosion by Use of Appropriate Design." *Australasian Corrosion Engineering*, vol. 6, no. 12, December 1961.

MUNSE, W. H., AND GROVER, L., *Fatigue of Welded Steel Structures*, Welding Research Council, New York, N.Y., 1964.

PRAY, R. F., AND JENSEN, C. O., "Welded Top Plate Beam Column Connections." *Welding J.*, vol. 35: **7**, July 1956.

PRIEST, H. M., "The Practical Design of Welded Steel Structures." *Welding J.*, August 1933.

ROBERTS, J. A., "The Causes of Stress-Corrosion Cracking." *DuPont Innovation*, vol. 1, no. 3, Spring 1970.

SOLAKIAN, A. G., "Stresses in Transverse Fillet Welds by Photoelastic Methods." *Welding J.*, February 1934, p. 22.

STALLMEYER, J. E., MUNSE, W. H., AND GOODAL, B. J., "Behavior of Welded Built-up Beams Under Repeated Loads." *Welding J.*, January 1957.

———NORDMARK, G. E., MUNSE, W. H., AND NEWMARK, N. M., "Fatigue Strength of Welds in Low Alloy Structural Steels." *Welding J.*, **35,** June 1956, p. 298-s.

THOMPSON, A. G., AND BROOKSBANK, F., "The Design of Welds." *Welding and Metal Fabrication*, April & May, 1951.

VRIEDENBURGH, C. G. J., "New Principles for Calculation of Welding Joints." *Welding J.* August 1954, p. 743.

WILSON, R. A., "The Steels That Mellow with Age." *Iron Age*, July 1969.

TABLE 10.A. *Properties of Metals* (Courtesy of *Welding Engineer*)

Metal or Alloy	Symbol (Main Elements)	Melting Point °F	Melting Point °C	Tensile Strength ksi	Tensile Strength MPa	Yield Strength ksi	Yield Strength MPa
Aluminum 99%	Al	1220	660	13	89.63	5	34.47
Aluminum 2011-T3	Al			55	379.2	43	296.5
Brass, yellow	Cu Zn	2340	900–1100	40–110	275.8	18	331.0
Beryllium	Be		900–1000	70–200	70–200		
Chromium steel (5100)	Cr	3430	1890				
Copper	Cu	1981	1083	32	220.6	10	68.95
Cast iron							
Grey	Fe C	2065–2200	1130–1204	20–60	172.4		
Malleable (pearlitic)	Fe C			60–105	413–724	40–90	275–620
Ductile	Fe C			65–150	448–1034	30–60	206–413
Iron, wrought	Fe			40	275.8	27	186.2
Lead	Pb	621	327	3	20.68	1.9	13.09
Magnesium	Mg	1202	650	25	172.4	13	89.63
Molybdenum (4150)	Fe Mo C	4760	2625	105–124	724–855	70–100	490–690
Silver	Ag	1760	960	23	158.6	8	55.16
Steel	Fe		1510				
Low carbon	Fe C	2798		50–75	413.7	40	275.8
Medium carbon	Fe C	2600–2730		70–120	579.2	52	351.6
High carbon	Fe C	2600		100–200	675.7	72	496.4
Steel, nickel (2515)	Fe Ni C			90–105	724.0	69–75	579.2
Steel, cast	Fe C			72	496.4	40	275.8
Steel, stainless (304)	Fe Ni Cr			85	586.1	35	241.3
Titanium (Ti-55A)	Ti	3300	1820	35	241.3	25	172.4
Tungsten	W	6170	3387	500	3447		
Zinc alloy	Zn	727		41–47	282–324		

Table 10.A. *Properties of Metals* (continued)

Metal or Alloy	Elongation in 2 in. Percent	Modulus of Elasticity 10^6 PSI	Modulus of Elasticity MPa	Thermal Conductivity BTU/Sq Ft Hr/°F/In.	Thermal Conductivity Cal/Sq Cm/Sec/Deg. C/Cm	Coefficient of Expansion (In/°F) $\times 10^{-6}$	Coefficient of Expansion (Cm/°C) $\times 10^{-6}$
Aluminum 99%	35	10.2	70330	1570	0.53	12	25.3
Aluminum 2011-T3	15	10.2	70330	1570	0.53	9.8	25
Brass, yellow		14.0	96530	2680	0.94	11.0–11.6	12.4
Beryllium		7.1	4891			9.4	
Chromium steel (5100)		15.0	103400			9.8	16.7
Copper				2680	0.94	9.2	
Cast iron							
Grey	0–1	13	89630	310	0.106	6.0	10.2
Malleable (pearlitic)	10–25	12.5	86125			6.6	11.88
Ductile	1–26						
Iron, wrought	3–20	29.0	199900	418	0.143	6.7	
Lead		2	1379	240	0.083	16.4	29.1
Magnesium	15	6.5	5128			14.3	12.06
Molybdenum (4150)				1090	0.37		
Silver	21–16	10.5	72345	2900	1.00	10.6	18.8
Steel							
Low carbon	27–38	30.0	206800	460	0.18	6.7	12.06
Medium carbon	12–32	30.0	206800	460	0.18	6.7	12.06
High carbon	2–23	30.0	206800	460	0.18	6.7	12.06
Steel, nickel (2515)	27–24	30.0	206800	400			
Steel, cast		30.0	206800				
Steel, stainless (304)	50	29.0	199900		0.137	9.3	16.74
Titanium (Ti-55A)	24	12.1	83369				
Tungsten		51	351600	783–754	0.34	2.6	4.5
Zinc alloy	10–16	18.5	127460		0.264	15.2	26.3

11

Nondestructive Evaluation of Joining Techniques

EVALUATION OF WELD DEFECTS

Reliability is a vital aspect of materials joining, whether it be in shipbuilding, missiles, bridges, or small electronic components. Many techniques are now available for testing joined materials without harming the structure. These methods are commonly referred to as nondestructive tests (NDT). Even though these tests can often be made quickly and even on a production line basis, the question arises: Is 100 % inspection of critical areas required for full quality assurance (QA)?

Acceptance standards for NDT are often based on requirements other than the fabrication problem at hand. In fact, most of the present day acceptance criteria can be traced to the ASME Boiler Code. Although it is a good code, many of the considerations that go into the design of a boiler are not valid for many other steel structures. Thus, rather than accept previously designed standards, it may be better to develop an inspection philosophy. This entails prevention of defects as well as inspection, and may trace problems all the way back to the design. If the part is designed such that welding or other joining methods become difficult, the chances of defects will be much greater. The materials chosen may be altered at the design stage such that they have better weld or joining characteristics.

Welder qualifications, if strictly maintained, can also be a big factor in QA. Before an individualized testing program can be established, certain data should be made available, such as all the service conditions required during the normal life expectancy of the component or assembly, the material used, fabrication methods, and so forth.

One approach to QA is to make separate broad classifications of the joining processes that will be used and the stresses applied to each joint during the expected service life. The following classifications are given as one example:

Class I. Welds that are highly stressed, the bead is oriented within 45 degrees of normal to the principal stress, and the design factor is 2.0 or less.

Class II. Welds that are moderately stressed, the bead is within 45 degrees of parallel to the principal stress, and the design factor is greater than 2.0.

Class III. Welds that are lightly stressed and have a design factor greater than 3.0.

A further division can be made for cast and wrought steels in Classes I and II as follows:

Highly stressed:

$$\text{Class IA:} > 150 \, \text{ksi} \, (1034 \, \text{MPa})$$

$$\text{Class IB:} > 150 \, \text{ksi} \, (1034 \, \text{MPa})$$

Moderately stressed:

$$\text{Class IIA:} < 150 \, \text{ksi} \, (1034 \, \text{MPa})$$

$$\text{Class IIB:} < 150 \, \text{ksi} \, (1034 \, \text{MPa})$$

In general for Class IA, voids and inclusions with sharp tails are unacceptable, and only two pinhole size porosities per linear inch are allowed. In Class II, the void size may be as large as 0.06 in. (1.52 mm) in its longest dimension. In Class III, weld voids of 0.06 in. (1.52 mm) are still the maximum dimension; however, three pinhole porosities/lineal inch are allowed.

The foregoing plan is presented only for illustrative purposes, to show one systematic approach to quality assurance; it is not intended to represent a standard procedure.

Inspection and testing procedures may be broadly classified as follows:

1. Visual inspection.
2. Nondestructive testing.
3. Destructive tests.

Visual inspection and nondestructive testing will be discussed in this chapter, destructive type tests are the subject of the next chapter.

Visual Inspection. Good joining results can be no better than the preparations. The inspector must satisfy himself that the welders have been properly qualified for the particular type of joint being made in simulated field conditions. He will also be familiar with code requirements, and see that the proper materials are used and that storage facilities are adequate. As an example, low hydrogen electrodes, now extensively used in pipeline welding, should be stored in a humidity controlled cabinet or drying oven.

The on-sight inspection will also verify that the workpieces have been properly aligned for tack welding. The working drawings should indicate the acceptable details and tolerances. The visual examination of both the preparatory steps and the completed welds is an extremely important part of a sound welding program. Weld defects or discontinuities that can be checked by visual examination include: undercutting, overlap and incomplete penetration, porosity, lack of fusion, and cracks.

WELD DISCONTINUITIES

Undercutting. Undercutting usually occurs when an excessive current is used, it may also result if the welder uses an incorrect technique and leaves a narrow groove in the base metal on the edge of the weld deposit. After necessary cleaning has been done, undercutting can be corrected, as required, by adding further metal to fill the groove flush.

Overlap. Overlap consists of unfused weld metal laying on the base metal. The condition may be remedied by removing the unfused metal. If this leaves the weld size too small, additional filler material may be added.

Incomplete Penetration. Incomplete penetration is apparent anytime the weld deposit fails to penetrate fully into the root corner of a fillet weld, into the root of a single groove weld, or when the weld metal of a double groove weld deposited from the two surfaces fails to meet and fuse at the root of the weld. Inspection will have to be made after the first pass to determine the possibility of complete fusion at the root before the second or closing pass is made.

Porosity. Gas pockets and similar type globular voids are generally the result of improper welding procedures or faulty techniques. Porosity may exist as large scattered pores or as groups of pores. When severe, that portion of the weld must be corrected by removal and replacement.

Slag Inclusions. Failure to remove slag between successive layers of multi-pass welds can cause inclusions to occur. The slag inclusions are generally angular in nature and may be elongated. If they are well dispersed, they may not be serious; however, certain tolerance limits will have to be established.

Lack of Fusion. The presence of excessive rust, slag, oxides, paint, or other foreign matter may prevent proper fusion. Low current settings, improper procedures, wrong electrode sizes, and lack of proper preparation will also cause insufficient fusion. To remedy this type of defect, it is usually advisable to remove the defective portion in total, then reweld. The use of backup rings or plates will allow the weldor to obtain better root penetration and fusion into the sides of the joint without fear of burnthrough. A smooth root pass will also make it easier to obtain good build-up passes.

Cracks. Of all welding defects, cracks are the most serious and can not be tolerated. Cracks occur by tearing while the metal is in a hot plastic condition, or by fracture of the metal after it has cooled to a low temperature. They may be caused by temporary shrinkage stresses during cooling, particularly when such stresses occur in combination with conditions that do not allow sufficient moving or deforming of the components. Cracking is most severe in heavy sections that are rigidly restrained. Cracking tendencies increase at low temperatures and are often greater in welds of high tensile steels of greater hardenability. Under almost any condition of loading, the strength of a welded structure will be substantially reduced even with a very small crack. Crack defects can be rectified only by removing the defect completely, and rewelding.

Static Loading. Static loads can accommodate a significant defect without an appreciable effect on the static strength of a joint, since, in a properly made weld, failure will occur in the parent material. Investigations by Green et al. show the effect of porosity on the ultimate strength of submerged arc welds and indicate that the static capacity is not materially affected by porosity in amounts up to about 7 % of the cross-sectional area. Also, porosity is not significant in bending loads up to the same porosity. Greater amounts of porosity near the outside face of the weld tend to be more critical, as may be expected.

NONDESTRUCTIVE TESTING

A greater awareness of NDT and its total capability stems largely from the success of "zero defects" as required by aerospace programs and demands from consumers and government for more consistent quality levels in manufactured goods. Also, recent improvements have provided NDT with instrumentation that is smaller, simpler, more reliable, and more economical than ever before. In addition, greater experience is now available relating a particular physical feature such as a nick, an internal inclusion, bubble, or even a different electrical property, to the working properties of a material. This pertains both to the manufacturing process and to performance of the product in its end use. This experience can be used to set limits for acceptance of parts in progress or to adjust the process to produce a higher yield of acceptable products.

Although NDT as discussed in this text is mainly concerned with the detection of flaws in metals and joints, it has many other applications in industry. Examples are numerous, such as inspecting the hardness of metal feeding into a stamping line, or checking moisture of tobacco going into a cigarette packing line and wetting or drying it as required. Established methods of NDT include: radiography (x rays and gamma rays), magnetic particles, eddy currents, ultrasonics, and dye penetrants. Newer techniques that are being evaluated are: neutron radiography, color radiography, flash radiography (pulsed x rays), holography, liquid or cholesteric crystals, and infrared–thermal techniques.

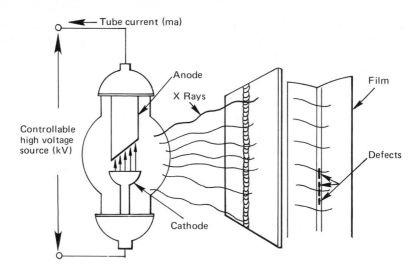

FIGURE 11–1. *X-ray inspection principle.*

Radiography. Radiography, long the best known means of nondestructive flaw detection, embraces a number of different techniques, but all are basically alike. A penetrating beam of radiation passes through an object. As it does, different sections of the object absorb varying amounts of radiation so that the intensity of the beam varies as it emerges from the object. This variation is detected and recorded on film to form an internal picture of the object, Fig. 11–1.

A stream of high speed electrons is emitted from the cathode and directed onto the target or anode. They, in turn, cause the target to give off *x rays*. X rays are rays given off when the electrons inside the atom are temporarily forced out of their normal positions by the energy of the bombarding electrons. The electrons that are temporarily displaced lie very close to the center of the atom and require a high energy source to move them out of their orbit. Thus, the higher the voltage applied to the cathode, the shorter the wavelengths of the x rays and the closer to the wavelength of gamma rays. Two classes of x rays are available depending upon the voltage used. High voltages produce hard x rays used for heavy plate; soft x rays produced by lower voltages are used for thinner materials. As the beam passes through the object, it is recorded in several ways. The most common method is with the use of film, but Geiger counters and fluorescent screens are also used and television screens show considerable promise.

X-ray radiography is used primarily for detecting flaws such as voids, porosity, cracks, and inclusions in both ferrous and nonferrous metals. Properly applied, an x-radiograph can determine with considerable accuracy the size and shape of any flaw, provided the flaw is at least 2 % as large as the thickness of the metal. With high energy sources (about 1 meV), defects as small as $\frac{1}{2}$ % can be detected.

Although x-radiography is one of the older methods used in NDT, it is still very much in demand today. This is due, in part, to its accuracy and the develop-

Basic Fluoroscopic Image Intensifier

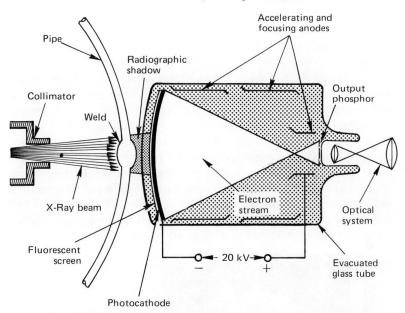

FIGURE 11-2. *An image intensifier, the X rays fall on the phosphor, which fluoresces and excites the photocathode. Electrons emitted by this excitation are accelerated and focused on the output phosphor where an intensified image is viewed by mirrors, binoculars, or television. (Courtesy of American Machinist.)*

ment of newer equipment and techniques. With the recently developed cold cathode (field-emission) tube, it is now possible to generate extremely high energy radiation for short periods of time. Thus, exposure time can be decreased significantly, and in addition, such equipment is smaller and generally costs less than the conventional heated wire (thermal emission) tubes.

Fluoroscopy. As normally used, film radiography is a lengthy and tedious process not compatible with on-line inspection of production parts. Fluoroscopy has been developed as a filmless means of showing the x-ray shadow image. At first, the light output from these screens was very low, but since the image intensifier, Fig. 11-2, the process is practical for industrial applications. The output phosphor, which produces the image, is 3,000 times brighter than the input. The intensified image may be viewed directly through binoculars, projected on a ground glass screen, or picked up by a special TV camera and displayed on a monitor. Production equipment for 100% fluoroscopic inspection with a TV image of the longitudinal welds in submerged arc welded pipe lines is now in use. Weld imperfections such as slag inclusions, gas pockets, incomplete fusion, and weld cracks are easily detectable using the fluoroscope. Although the fluoroscopic image lacks some of the

399

contrast and resolution of a good radiograph, it does show strings of gas pockets less than 1/64 in. in diameter and the fine trailing ends of cracks. At the present time, fluoroscopy is one of the best methods of inspecting pipeline welds as fast as they are produced.

Gamma Radiography. The use of gamma rays is similar to that of x rays, but due to their shorter wavelength gamma rays have greater penetrating power and can be used on thicker materials. The radiation results from atomic disintegration of isotopes. Until 1942 the only source of gamma rays was radium; however, with the advent of nuclear fission, the door was opened to several types of man-made isotopes. Cobalt 60, an isotope produced by neutron irradiation, is now being used more than radium since it is cheaper. The gamma rays have various wavelengths, which may be compared to x rays generated at different voltages. The source of the radioactive rays is, however, constantly decaying. The period in which the energy of the isotope decays to half its original value is called its *half life*. Radium, for example, has a half life of 1,620 years. Cobalt 60 is equivalent to x rays of about 1.3 meV

FIGURE 11–3. *Gamma radiography is intended for portable, on-the-spot testing. The flexible source tube is positioned inside the pipe casting making possible a radiograph of the entire girth in a single exposure. (Courtesy of Automation Industries, Copyright, Penton, Inc., Cleveland, Ohio 44114.)*

and has a half life of 5.3 years. Its capacity range on steel is from 1 to 6 in. (2.54–15.24 cm) in thickness.

Gamma radiography has the added advantage of being completely portable except in its large units, Fig. 11–3. A typical unit weighs 40 lb (18.14 kg) and can be wheel mounted. Personnel must be extremely careful not to be exposed to either x rays or gamma rays.

Neutron Rays.   Neutron rays, also produced by radioactive materials, have the penetrating power of x rays, but are not directly limited by the density of the material. The tendency of rays to be absorbed by a material is called its attenuation coefficient. The higher this coefficient, the poorer will be the results of x rays. The attenuation coefficient of neutrons, on the other hand, has no direct relationship to atomic number. The relationship is entirely random. As an example, hydrogen absorbs neutrons readily whereas lead does not. The mass absorption coefficient for tungsten tube x rays for all elements ranges only from 0.10 to 5, whereas, the mass absorption coefficient for thermal neutrons ranges from 0.03 to 48.5.

This comparison is shown in Fig. 11–4. The absorption rate for x rays is shown almost as a straight-line linear relationship with atomic numbers, but N rays are very scattered. Thus, it is virtually impossible to N-ray through hydrogen, yet lead is almost transparent. For this reason, a neutron shield should be made of concrete, water, or some other hydrogenous material. The attenuation coefficients for a few materials are shown in Table 11.1.

A neutron radiograph is made by exposing the object to neutron or N rays. Some neutrons will be absorbed by the object, while others pass through. The latter must be transferred into some other form of energy, since neutrons are neutral charges and won't expose x-ray film. To accomplish this, a specially treated transfer or

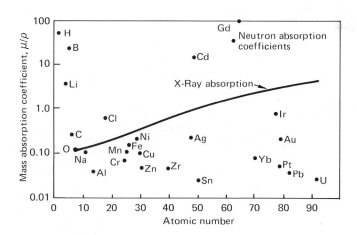

FIGURE 11–4.   *The relationship between atomic number and X-ray/N-ray absorption. Tungsten tube X rays show a linear relationship while that of N-rays are very scattered.*

TABLE 11.1   *Attenuation Coefficients*

Element	X rays	N rays
Hydrogen	0.280	48.5
Lithium	0.125	3.7
Boron	0.138	24.0
Aluminum	0.156	0.036
Titanium	0.217	0.119
Iron	0.265	0.141
Lead	3.5	0.034
U-235	3.9	1.89

conversion screen is used as shown schematically in Fig. 11–5. As the neutrons hit this screen, usually containing indium or gadolinium, it produces ionizing radiation, which in turn exposes ordinary x-ray film.

Advantages of Radiant Energy Methods.   Although x rays, gamma rays, and neutron rays are all used to detect flaws in metals and other materials, each has certain advantages and limitations. X rays are better than gamma rays for the detection of small defects in sections less than 2 in. thick; however, in thicknesses of about 2 in. to 4 in. (5.08–10.16 cm), they are about equal. Gamma rays have less scatter and appear to be better for checking parts of varying thickness.

The x-ray method is faster than the gamma-ray method, requiring only seconds or minutes instead of hours. This time factor for gamma rays is offset when checking cylindrical items, since the film can be placed around the outside of the object with the radiation in the center to expose all surfaces at once.

Neutron rays are a good complement to x rays. They are used to their best advantage where x rays are unsuitable, as in checking for hydrogen embrittlement. They are used to check large castings by filling all voids with water or by seeding all cores with a neutron absorbing material, so that any residual material can be easily detected. Another important application is in checking nuclear fuels. The only other way to check nuclear fuel elements is to cut them apart. Steps required to make a neutron radiograph are shown in Fig. 11–5. Some other potential uses of this NDT method are:

1. Bonding in metal honeycomb structures.
2. Adhesive bonds in "massive metal."
3. Composites, such as boron filaments in metal.

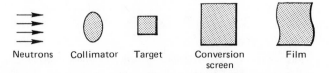

Neutrons	Collimator	Target	Conversion screen	Film

FIGURE 11–5.   *The steps used in making a neutron radiograph.*

4. Lubricant migration in a sealed system.
5. Pressed metal powder parts for uniformity and segregation, as well as component interfaces.
6. Brazed joints.
7. Laminated wood.

One of the principal advantages of radiographic methods of NDT is that a permanent record in the form of a film can be kept for each item checked. This record will show all significant defects, their size and location.

Limitations of Radiant Energy Methods. All radiation testing methods present a health hazard. Ordinarily, lead shielding is used. For x rays, lead shields in front of and behind the film permit an operator to remain inside a vessel being radiographed. Radium elements themselves must be handled with extreme care. When not in use, they are kept in a container having a minimum wall thickness of six inches. The portable container has a two-inch (5.08 cm) wall thickness. When the radium is removed from the lead case, everyone should remain ten feet away unless shielded by heavy metal. Exposure to radiation may cause a decrease in the number of white corpuscles, lower blood pressure, and in severe cases, cause death.

Radiography, when used on parts of variable thickness, requires special attention so as to avoid over- and under-exposure. One approach is to immerse the irregular part in a lead salt solution. Another problem is in interpreting the exposures. In order to interpret a radiograph accurately, it is necessary to know the size of the smallest defect that the radiograph can be depended upon to show. This relies upon the definition or sharpness of the record and the contrast between adjoining areas. The common practice is to employ a thickness gauge or "penetrameter" as an assist in determining the magnitude of the smallest detectable defect. An example of the ASME code for penetrameter use is shown in Fig. 11–6.

One of the major restrictions of neutron radiography is the need for a neutron source. At the present time a nuclear reactor is the most suitable source of neutrons. There is no low cost, compact source of N rays. The most promising source is the use of a radioisotope. This can be a gamma emitter such as antimony, which produces an intense neutron flux when it is put into beryllium.

Eddy Current Testing. Briefly, eddy current testing consists of observing the interaction between electromagnetic fields and metals. It is termed *eddy current* inspection because the basic principal is derived from eddy current phenomenon. Whenever a conducting material is moved into a magnetic field or is subjected to a changing magnetic flux, eddy currents circulate in the mass. The eddy currents are closed loops of induced current circulating in planes perpendicular to the magnetic flux. In electromagnetic testing, a probe, a test sample, and an instrument to check the electromagnetic field are required. The eddy current produced on the alternating electromagnetic field is proportional to the homogeneity and composition of the test material and is opposed to the original electromagnetic field. The vector summation of the two magnetic fields produces a change in the probe signal. As shown

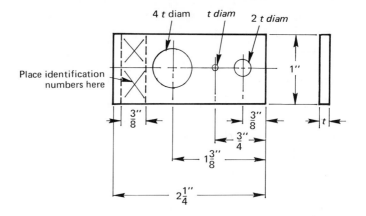

FIGURE 11–6. *An example of a penetrameter used in determining whether or not the radiographic technique used is satisfactory. The penetrameter is placed 1/8 in. (3.18 mm) from the edge of the weld normal to the radiation beam. The various size penetrameters are given in ASME Boiler and Pressure Vessel Code SE-142.*

in Fig. 11–7, the eddy current distribution is altered in the area of the defect. The magnitude of this difference depends on the size of the defect, i.e., its length, width, and depth. Test frequencies are used to probe various designs. Instruments can be selected to measure specific metallurgical properties such as hardness, alloy composition, and case hardening depth.

Two types of eddy current testers are available, absolute and differential. Absolute testers measure permeability against a known standard and are useful in controlling processes that depend on a particular permeability level. An example is the arc adjustment in a welding process, based on the permeability of the arc into the material. Differential testers are used to sense the change in permeability from a standard. Typical uses include detecting surface cracks, determining material hardness, or controlling a correction for hardness during processing. Many newer eddy current systems have both absolute and differential capability in a single unit.

Eddy current techniques can be adapted to high speed automatic production lines. Pieces to be tested may be manually or motor driven through (or under) the test coils, which send into and pick up signals from the part under test, Fig. 11–8. These signals are then electronically analyzed with the instrumentation adjusted to respond to preestablished limits. The data can be presented on an oscilloscope, a recording chart, or simply in terms of ringing a buzzer. The signals caused by faulty parts can be used to actuate equipment, which will automatically discard or mark faulty parts.

Advantages. The eddy current testing method is fast and relatively inexpensive. It can yield information not only on the presence of defects, but also on mechanical, metallurgical, and chemical properties. Through suitable electronic control systems linked to such instrumentation, defective parts or materials can be

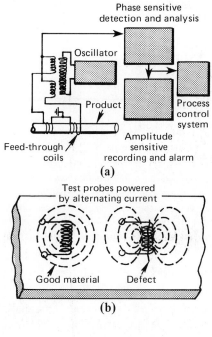

Phase sensitive
detection and analysis

Oscillator

Product

Process
control
system

Feed-through
coils

Amplitude
sensitive
recording and alarm

(a)

Test probes powered
by alternating current

Good material   Defect

(b)

FIGURE 11-7. (a) *The basic system of eddy current testing uses two coils to check the impedance of a long linear product. The difference in impedance between the two coils generates an output voltage if a defect is located. At* (b) *is shown how a structural defect alters the current distribution in the field. The size of the flaw is indicated by the magnitude of the current distribution change. The flow of current around a defect is shown at* (c).

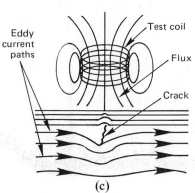

Eddy
current
paths

Test coil

Flux

Crack

(c)

rejected, sorted, and rated. A newer development in eddy current testing is a system for sensing fitup for joining. As in the case of welding tubing on an automated line, the process settings will change to compensate for variations.

Disadvantages.   The chief drawback of this method is its extreme sensitivity. The instrumentation must be carefully designed and adjusted so that the variables affecting test results can be held constant. As with other NDT equipment, the establishment of suitable standards poses the biggest problem, for the detection of flaws that cannot be seen or physical discontinuities that are immeasurable except by destructive tests. Because it deals with electrical properties, eddy current testing

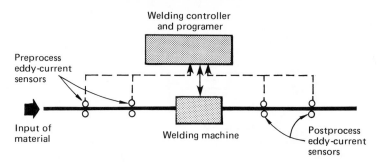

FIGURE 11–8. *Preprocess and postprocess eddy-current sensors can be set up to provide total process control. The sensors ahead of the welding machine detect part mismatch, joint straightness, and changes in magnetic permeability. After welding, a second group of sensors check on the depth of penetration, cracks or inclusions, grain size, and hardness. All information is fed back to a welding controller and programmer. (Reproduced from Automation. Copyright, Penton, Inc., Cleveland, Ohio 44114.)*

can only be used to examine conductive materials, although they can be combined with nonconductors.

Magnetic Particle Testing. Magnetic particle and fluorescent magnetic particle testing are similar techniques developed for detecting surface and subsurface defects in magnetic materials. The use of these methods involves the establishment of a magnetic field in a ferromagnetic part. Fine particles of magnetic iron are either dry-blown or applied in liquid suspension to the surface of the part. In either approach, the magnetic particles are attracted to the cracks and subsurface defects in such a way as to be visible to the naked eye or to examination under ultra-violet (black) light. The dry method permits visual inspection under normal lighting conditions and is generally used to detect defects in welds, large forgings, castings, and other parts having extremely rough surfaces. The wet method provides more emphatic highlighting of surface and subsurface flaws.

Three modes of part magnetization are used: *longitudinal magnetization*, wherein the part is magnetized lengthwise to find transverse cracks; *circular magnetization*, to create a magnetic field around and within the part to detect longitudinal defects; and *multidirectional magnetization*, for simultaneous detection of defects over the entire surface. The first two modes are used where significant cracks are likely to lie in one direction. The third mode is used on large castings, forgings, welded assemblies, and pressure vessels.

Advantages. Magnetic particle testing is able to detect cracks caused by quenching, fatigue, welding, embrittlement, and by seam and subsurface flaws. At production speeds, costs are relatively low, ranging typically from 0.1 to 10 cents per part, depending upon size and nature of the material. This is generally less costly than visual inspection.

Disadvantages. The materials examined must be capable of being magnetized to an appreciable depth. By far, the largest number of damaging defects in welded joints, such as porosity and lack of fusion, are found to lie too far beneath the surface making it difficult to detect by this process. A change in permeability is frequently found at the junction between the weld metal and its parent metal that is sufficiently abrupt to cause a line of powder to adhere, but does not signify a lack of fusion. In some cases it is hard to tell whether this is, in fact, poor fusion. Finally, the tested parts require cleaning after inspection.

Fluorescent and Penetrant Dye Testing. Fluorescent penetrant testing is similar in some respects to fluorescent magnetic particle testing. In this method, a fluorescent penetrant liquid is applied to a part surface by dipping, spraying, or brushing. If defects are present, the liquid is pulled into them by capillary action. The surface is then rinsed with water or solvent and allowed to dry. It is then coated with a dry developer to draw the penetrant out of any surface defects. Under black light, the location, size, and shape of a defect is then defined in bright fluorescence, Fig. 11–9.

A faster, more universal method of penetrant checking makes use of a dye base without the black light. The part is precleaned and the penetrant is applied directly from a spray can. After a short wait, to allow for penetration, the excess penetrant is wiped off and a thin film of developer is applied, also from a spray can. Part defects become clearly visible as colored lines appear, highlighted by the white developer.

Advantages. All types of materials can be processed including ferrous and nonferrous alloys, glasses, ceramics, nonporous powder metal parts, certain plastics,

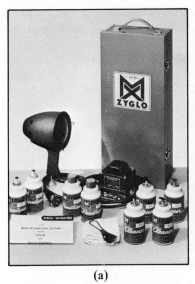

(a)   (b)

FIGURE 11–9.   (a) *The "black light" reveals cracks, and the location and size of surface defects in* (b) *bright fluorescence.* (*Courtesy of Magnaflux Corp.*)

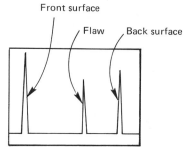

Front surface

Flaw   Back surface

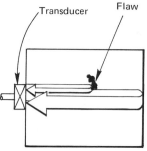

Transducer   Flaw

FIGURE 11–10. *Ultrasonic waves are used to find flaws and measure thickness. (Courtesy of Chemical Engineering Journal.)*

and synthetic materials. The process is easily applied whether for in-process, final, or maintenance inspection. Depth of surface discontinuities may be correlated with the richness of color and speed of the bleedout. Automated penetrant testing systems have been set up that will handle 8,000 jet engine blades in a single day. The only personnel required are an operator to start baskets of parts into the first cleaning station and inspectors to view the processed parts under a black light.

Disadvantages.   The one main limitation of liquid penetrants is that they are not applicable to porous materials, since leakage from the porous surface would act to mask individual defects. Surface finish methods affect both the penetration and removal of the penetrant. Hand ground surfaces, for example, often contain smeared material that fills up the crevices and seals them shut. Defects may also be closed by shot and sand blasting.

Ultrasonic Testing.   One of the most powerful NDT developments to emerge in recent years is that of ultrasonic testing. Commercial equipment in use consist of a transducer capable of producing frequencies between 1 and 25 MHz. At these frequencies, mechanical vibrations can be generated in the form of well-defined beams, which are characteristic of small cross sections. These beams can be directed into products to detect discontinuities such as cracks, voids, inclusions, serrations, and bonding defects. Since the beams are subject to the laws of reflection and refraction, a defect can cause the beam to change its velocity of travel, reflect the energy of the wave, and attenuate the beam. These changes can be detected by various types of instrumentation.

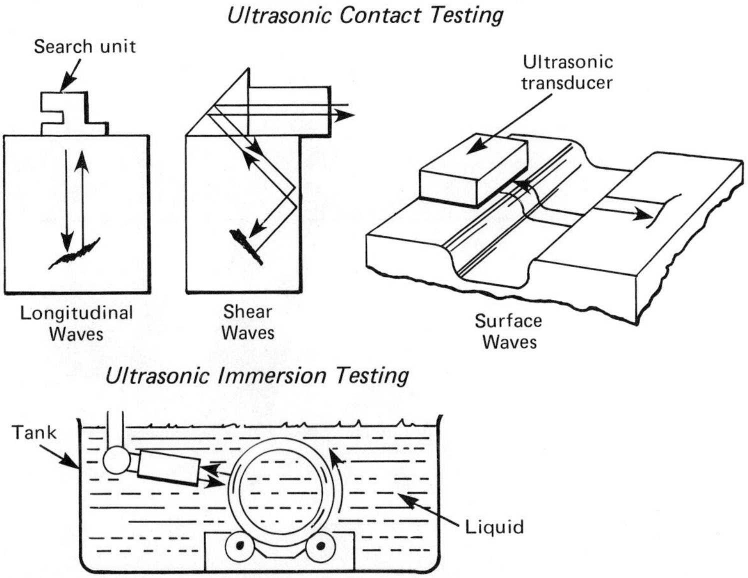

Ultrasonic Contact Testing

Search unit

Ultrasonic
transducer

Longitudinal
Waves

Shear
Waves

Surface
Waves

Ultrasonic Immersion Testing

Tank

Liquid

FIGURE 11–11. *The ultrasonic transducer is designed to produce longitudinal waves that reach a discontinuity away from the surface. The transducer also generates shear waves to detect flaws perpendicular to the surface and surface waves to locate a discontinuity near the surface of the product. In immersion testing, the liquid serves as a coupling for odd-shaped products.*

Pulse Ultrasonics. The most common type of ultrasonic flaw detection is based on a pulse reflection principle. Short pulses are generated by the search unit, which is an electromechanical transducer such as a piezoelectric crystal made of barium titenate. The search unit must be coupled to the product being tested by either a layer of water or oil, or by immersion of the search unit and product in a liquid filled tank. The pulses are reflected back to the transducer by discontinuities within the part as well as from the opposite side of the part. The returned signal is transferred from the transducer to a cathode ray oscilloscope or a printout. A readout display containing only two waveforms indicates that the part under inspection is free of defects: a first spike indicates the transmitted pulse striking the part surface and the second, the opposite side of the workpiece. If a third spike is shown on the display this is the indication of a defect, Fig. 11–10. Ultrasonic signals may also be read out on meters of various kinds. A newer technique called acoustographic imaging combines ultrasonics and liquid crystals (discussed later) to provide a color display on a TV screen.

Three types of waves may be used in the testing operations: longitudinal or straight waves, shear or angle waves, and surface waves, Fig. 11–11.

Resonant Ultrasonics. Resonant ultrasonic detection, as opposed to pulse ultrasonics, is particularly well adapted to measuring thicknesses of complex shapes. It is also applicable to flaw detection for both magnetic and nonmagnetic metal

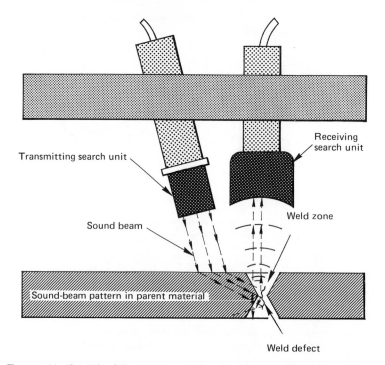

Transmitting search unit

Receiving search unit

Sound beam

Weld zone

Sound-beam pattern in parent material

Weld defect

FIGURE 11–12.   *The delta system gets its name from the triangular relationship between the search unit and the receiver transducer. It has been used in detecting flaws in butt welds at the rate of 50 ft/h (15.24 m/h). (Reproduced from Automation. Copyright, Penton, Inc., Cleveland, Ohio 44114.)*

parts, plastics, or glass that can transmit ultrasonic frequencies. Resonant instruments operate through the steady transmission of changing frequencies, one of which is the natural frequency of the material. The thickness of each material being measured has its own resonant frequency, which results in a standing wave that is detected by the ultrasonic unit and read out directly as a thickness reading. While pulse ultrasonic instruments can range in size from portable units to fully automated test or measurement systems, resonant ultrasonic units are generally portable.

The Delta Technique.   A newer development of ultrasonic testing is the delta technique, which uses redirection of energy for the detection of flaws in butt welds, Fig. 11–12. The sound propagates as a shear wave of energy and is introduced into the parent metal adjacent to the weld. Any feedback differing in acoustic impedance from the parent material provides information about the defect, regardless of which of several paths it followed to get there. Shuttling the transmitter toward and away from the weld permits the sound beam to scan up and down through the weld zone, assuring complete sonic coverage of the area. Weld inspection rates of 50 ft/h have been achieved with a shuttle scan, permanent record system. Excellent correlation has been shown by comparing the NDT results with destructive tests.

Advantages. Ultrasonic testing can be used on a wide variety of materials with only negligible loss in great thicknesses [up to 12 in. (30.48 cm)]. Most parts can be checked in place. Steel billets can be checked for internal defects at the rate of 200 ft/min (60.96 m/min). Pulse ultrasonics can be used both in the contact and noncontact modes. Where contact with the part is difficult, as in the case of a rough surface, quick checks can be made by using the immersion technique. A strip chart will show the location and size of the flaws. Compared to most other methods of NDT, ultrasonic testing is very inexpensive.

Disadvantages. Since the ultrasonic beam is very narrow, it is necessary that every bit of the surface be covered by a progressive movement of the search unit. Also, some discontinuities will be parallel to the waves and give very little reflection. Two tests should be made, one normal to the other. This disadvantage is now being overcome by the delta technique. The performance of the test equipment needs to be checked by means of standard reference blocks having various size holes drilled in them at various distances from the test surface, similar to those shown in Fig. 11–6. Ultrasonics also have difficulty in testing austenitic stainless steels due to the relatively large grain size. An array of signals is produced on the CRT screen; this requires a large attenuation that, in turn, leads to a complete loss of back wall reflections.

Microwave Testing. Microwave testing employs electromagnetic radiation with frequencies between 300 MHz and 300 GHz, falling roughly in the range between radio and infrared waves on the electromagnetic spectrum. Microwaves act like light waves; they travel in a straight line, reflect, refract, and scatter. But unlike light, microwaves move easily through practically all plastic and ceramic materials. When the signal interacts with the surface of an object, it is reflected back in a characteristic "mapping" pattern. In detecting internal flaws, the microwaves scatter as they encounter defects. Microwaves are used to detect cracks, voids, inclusions, and unbounded areas in a wide variety of metallics and nonmetallics. They are, however, unable to penetrate some metallic conductors such as steel.

Recent advances in this area have utilized the microwaves to detect cracks, scratches, or other surface defects as small as 100 μ. Changes in dimension or makeup of material are easily within the sensitivity of phase and amplitude of microwaves. Dielectric constant and loss tangents (microwave absorption) are functions of material composition, structure, homogeneity, orientation, moisture content, density, and similar factors.

Advantages. Microwaves are not only useful in detecting both internal and external defects through the measurement of dielectric properties, but also variations in thickness can be shown. In addition, chemical compositions such as polymerization and degree of cure, oxidation, vulcanization, and glass-to-resin ratio in fiber-reinforced plastic can be detected. Unlike ultrasonic testing, no liquid couplant is required. A permanent film record can be maintained for the items checked.

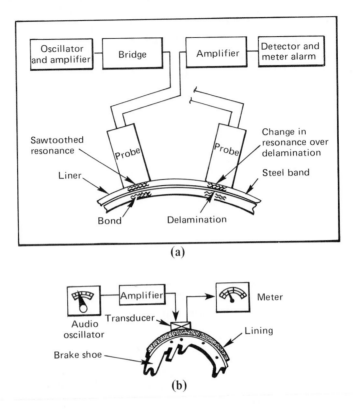

FIGURE 11-13. *Sonic testing consists of a bridge that is preset to the natural frequency of the workpiece.* (a) *As the part is made to vibrate under the probe, any change in resonance will be shown on the meter alarm as a subsurface defect. Shown at* (b) *is a brake shoe and lining bond being checked.* (*Reproduced from Automation. Copyright, Penton, Inc., Cleveland, Ohio 44114.*)

Disadvantages. The initial cost of the equipment is high and there are some radiation hazards in operating it. The depth of the defect is not shown on the film.

Sonic Testing. Sonic testing is relatively new in the NDT field. It is used both as a complement and as a replacement for ultrasonic testing. Unlike ultrasonic testing, sonic inspection does not depend upon the interpretation of returning sound waves. It is rather based on the resonance of the workpiece. Any variation in resonance is taken as the presence of a flaw. The principles of the process are shown schematically in Fig. 11-13a. Figure 11-13b shows sonic testing for the bond between a brake shoe and the lining.

The sonic vibrations, generated by a piezoelectric transducer, are amplified and sent to the probe. The desired frequency is predetermined by testing a piece of known quality. The design of the circuit is such that any significant change in resonant frequency overloads or underloads the piezoelectric crystal in the oscillator.

This, in turn, activates the meter alarm; or, if automation is used, opens or closes a gate.

An example of one problem solved by sonic testing was unexplained explosions in welded clad tubing. At first the problem was thought to be in the welding process, but sonic testing showed incomplete bonds in the cladding. Hence, outgassing from voids caused the explosions during welding.

Advantages. Sonic testing does not require the use of a liquid couplant. A roller is used on the probe that is coated with a proprietary material. It is applicable to a wide variety of materials and speeds. The cost of a basic sonic unit is about one-third that of a basic ultrasonic unit. It can be used in testing adhesive bonded parts and parts in which defects are extremely close (within thousandths of an inch) to the surface. One side inspection makes it possible to check complete assemblies in which interior sides are inaccessible. High speeds can be used, in some cases as much as 190 ipm (482.6 cm).

Disadvantages. Sonic testing may be too sensitive. Hence, an out of specification signal may not always mean a reject but rather grounds for further checking. Sonic testing will not pick up stacked defects, however, this is also true of some other means of NDT.

Acoustic Emission. For many years it has been known that deformed materials make noise due to the generation of stress waves, but only recently has this principle lent itself as a comparatively new NDT technique. Its most successful use has been in finding defects in welds. The amount of acoustic emission is directly related to the size and number of defects in the weld. Flaws such as porosity, inclusions, incomplete fusion, and cracks create emissions due to stresses surrounding them. The flaws can be detected about 20 to 45 seconds after the weld is made. Depending upon circumstances, identifiable sounds will still be produced for about 20 minutes after the weld is made. The system has been successfully applied to resistance, gas tungsten arc, and submerged arc welding. Shown in Fig. 11–14 are three different qualities of welds with the emission levels of each.

The acoustic emission technique is also being used to predict failures in pressure vessels and high pressure piping, to detect stress corrosion cracking, and to provide as an in-flight monitor for structural fatigue in aircraft.

Advantages. Acoustic emission techniques have the advantage of evaluating a structure while it is under load. The only practical limit to the technique's sensitivity is the signal to noise ratio. Flaw growth as small as 0.0001 in.^{2} (0.003 mm^2) can be detected.

Limitations. A time limit exists for acoustic emissions, as in the case of welds. The method is not as applicable to very ductile materials since they have low amplitude emissions. Except in the case of weld structure, parts must be stressed

ACOUSTIC EMISSION RELATED TO WELD QUALITY

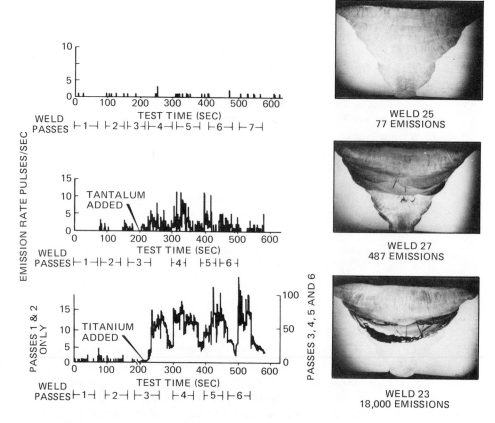

WELD 25
77 EMISSIONS

WELD 27
487 EMISSIONS

WELD 23
18,000 EMISSIONS

FIGURE 11–14. *Acoustic emission is used to monitor weld quality. In a given time span these three welds showed (a) 77 emissions, (b) 487 emissions, and (c) 18,000 emissions. (Courtesy of Battelle Columbus Laboratories, Columbus, Ohio.)*

or operating before they can be checked. It is somewhat of a problem to filter out extraneous noise.

Thermal NDT. There are two approaches to NDT by thermal techniques. The most used method is to apply heat over the specimen, the other is to measure the thermal emissivity of the object.

Applied Heat. Applied heat will diffuse uniformly throughout the part unless a discontinuity is encountered to impede the flow. When the heat flow is interrupted, a "hot-spot" will appear at the surface of the specimen. Heat may be applied in a number of ways: by direct contact with a heat source, by an electrical current, or by the use of an infrared heat source.

Several methods are used to check the differences in surface temperatures. One method is to apply a coating of heat sensitive material that will react predictably when the temperature reaches a certain point. Paints that change color when they reach a given temperature have been widely used; however, a more recent development has been the use of *liquid crystals*.

There are three main branches of the liquid crystal family; namely, the cholesterics, the nematics, and the smectics. Of these three, cholesterics are the only ones used in NDT. The nematics have the property of changing viscosity significantly with a change in electric field density, and the smectics are mainly laboratory curiosities with potential medical and electronic applications.

Operation of Cholesterics. Cholesteric crystals, in the presence of white light, selectively reflect one wavelength of light at each angle of reflection, the light is circularly polarized. An analogy can be made to that of an oil film. Two surfaces of an oil film produce reinforcement and show color reflection when a light wavelength that is a multiple of the film thickness is reflected by each surface of the film. The rod shaped molecules in cholesteric film orient themselves parallel to each other in layers much like cigars in a box. Thus the effect is that of a many layered oil film. At a fixed temperature and a reduction in viewing angle (away from perpendicular), the reflected color is entirely a function of the film's molecular spacing; and therefore, is an accurate index of temperature.

When a calibrated light source such as a mercury lamp is used with its several color peaks, the intensity at various points can be varied sharply enough for a photocell pickup to detect a specific temperature at each peak. The optimum film thickness is 0.001 in. (0.03 mm). Less than 0.001 in. (0.03 mm) does not scatter enough light and more is a waste of crystals.

In NDT the surface to be checked for voids, cracks, or poor bonds is coated with a suitable cholesteric and then heated or cooled through a range of the crystal. Heat flows around the discontinuities creating temperature gradients that show up as irregularities in the color pattern. The pattern is reversible, changing each time the object is recycled through the color-play region. The colors are unique for specific temperatures, and quantitative measurement is possible to $1/10\,°C$. By proper mixing of the liquid crystal systems, the temperature span (violet to red) can be made as narrow as $2\,°C$ or as wide as $150\,°C$, within an overall range of -20 to $250\,°C$ (-4 to $480\,°F$). Sensitivities to energy changes as low as 10^{-6} calories/cc (4×10^{-8} Btu) are possible, and the response time is $1/5$ sec. In dynamic testing, strain heat can be monitored to locate potential fractures.

Advantages. Organic crystals are relatively quick and easy to use. Their biggest advantage over other thermal mapping systems is cost. Liquid crystals are obtained from a wool wax of sheep. The cost of raw liquid crystals may range from \$500 to \$1100/lb, so not many users buy it in this quantity. Costs, as volume increases, are projected as low as \$50/lb. At a 1-mil thickness coating, this would be equal to $5¢/ft^2$. When the test surface is clean and care is used in testing, the crystals can be reclaimed by redissolving in solvent or scraping off with a squeegee.

If contamination is uniform, the liquid can simply be recalibrated. Photographs showing the color variations form a permanent record.

Disadvantages. Like most thermal NDT techniques, liquid organic crystals are useful in detecting surface flaws or those relatively near the surface. Some of the disadvantage of brushing or spraying the liquid has been overcome by an encapsulated sheet form now available. However, the spherical shape of the capsules diffuses the color, thereby sacrificing some of its intensity and photographability. On the other hand, the capsule form eliminates variations in color that occur from different viewing angles.

Infrared NDT. Infrared techniques are often discussed as being a distinct type of NDT technique. However, it is just another method, albeit a very important one, of measuring temperatures. One infrared method, the line scan, is shown in Fig. 11–15a. It consists of focusing a visible radiation beam across a sample and then measuring the surface temperature remotely with an infrared radiometer that is focused a short distance behind the moving heated area. A pen type recording, or oscillogram, can be used to indicate the discontinuities by a vertical deflection on a horizontal trace. A scan begins at room temperature, increases sharply, and stabilizes at about 50 °C (122 °F). Test areas are usually mapped by performing these scan tests at one-inch (2.54 cm) intervals. Another technique, shown in Fig. 11–15b, utilizes an oscillating radiometer. The thermal source applies a band of heat uniformly across the surface area of the structure under test. After an appropriate dwell time, the surface temperature reaches a critical state. Defects in the structure are measured by the oscillating radiometer as the structure is indexed past the scanning area. The scanning motion and the test structure speed are synchronized with the recording mechanism of a facsimile recorder, which produces a permanent record. In most cases, a one-to-one size scale is used so that the recording is a direct image of the test structure.

An oscillograph recorder may also be used to provide a fast and accurate recording method. In this case, light sensitive paper is exposed by the scanning beam of a cathode-ray tube.

Advantages. Thermal NDT methods such as heat sensitive paints and liquid crystals have the advantage of low initial cost. They are relatively simple to apply and do not require a skilled operator to judge the results. Infrared systems are not only useful in detecting surface and subsurface discontinuities, but they can also be used in locating surface corrosion under paint, corrosion between thin skins, and intergranular corrosion. Thermal techniques are also useful in detecting moisture trapped in honeycomb panels and pinpointing overheating in electrical and mechanical systems.

Limitations. Thermal methods of NDT are limited to thin-walled surfaces and relatively shallow penetration of other parts. When thermo-chromic paints or organic liquid crystals are used, there is a critical time–temperature relationship.

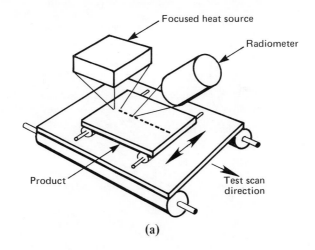

(a)

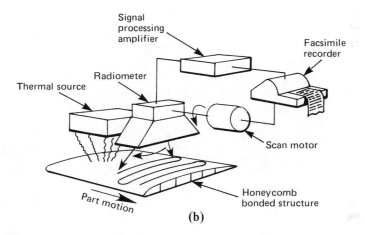

(b)

FIGURE 11-15. *The two methods of thermal scanning are shown schematically. (a) The focused thermal source and radiometer are stationary while the part is moved in two directions for the scanning pattern. (b) The product moves in one direction as the radiometer detector oscillates to provide the scanning motion. (Reproduced from Automation. Copyright, Penton, Inc.)*

Definition suffers when flaws are deep because the hot spot tends to diffuse as heat travels to the surface. Typically, the diameter of the flaw should be about three times its depth to be readily detectable by thermal methods.

Holography.   In conventional holography, a powerful beam of coherent light, usually from a laser, illuminates an object and, through a two-stage process, recreates a three-dimensional image of the object, Fig. 11–16. Recently researchers have been able to combine two techniques, laser photography and laser interferometry, into a new NDT technique known as holographic interferometry. This technique permits

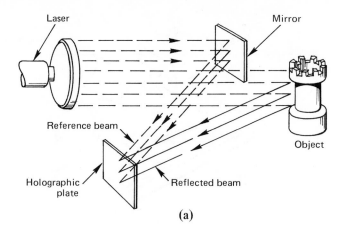

(a)

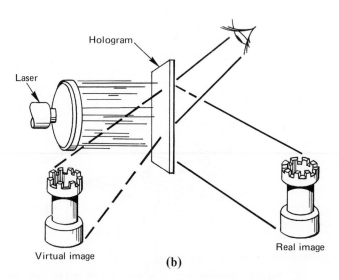

(b)

FIGURE 11–16. *A schematic showing the principles of holography. (a) A film plate is made by the interaction of two coherent beams of light. One is a reference beam, the other is bounced off the object being holographed. (b) Real and virtual images appear when the original wavefront is reconstructed by shining a coherent light on the hologram. (Courtesy of Machine Design.)*

the comparison of an object in more than one state, i.e., at rest and pressurized. Thus it is possible to superimpose a reference hologram of an object over some state of deformation of the object. With this technique, flaws of various kinds, voids, inclusions, nonbonds, and laminations can be detected. Shown in Fig. 11–17 is a hologram of an adhesive void between the skin and the core of an aluminum honeycomb sandwich panel. First, a holographic image is made of the panel "at rest." Then, the panel is bent slightly and a second holographic image is made. The two

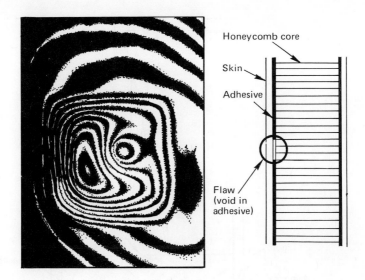

FIGURE 11–17. *A hologram showing the location, size, and shape of an adhesive void between the skin and core of an aluminum honeycomb sandwich panel. (Courtesy of Machine Design.)*

images are then superimposed, and the differences between them show up as fringe patterns at the point of discontinuities.

Ultrasonic Holography. In ultrasonic holography, the first stage image is obtained by ultrasonic "illumination" using a plane-wave ultrasonic source. A variety of detectors can be used to make the interference pattern visible for photographing. As an example, a point receiver can be scanned over the isonified area. If the input of the receiver is amplified and fed to a light that is mechanically linked to the receiver, the intensity of the light at each point in the area scanned will be proportional to the intensity of the sound falling on that area, Fig. 11–18. The result, when photographed, is an optical hologram duplicating the ultrasonic interference pattern. The ultrasound is generated by a transducer located in the lower corner of the liquid filled tank. After passing through the object, the ultrasound beam is focused by an acoustic lens onto the liquid surface. Ultrasound from a reference transmitter is also directed to this surface. The surface detects variations in the acoustical pressure generated by the interference between the reference beams. These variations are displayed as differences in the surface elevation.

Advantages. The holographic technique for NDT is fast (several hundred images per second) and does not require special preparation of the surface, mechanical attachments, or coatings. It can be extremely useful in detecting stress corrosion and thermal fatigue cracking. Complex configurations can be handled easily. The direction of loading can be reversed so that the part can be studied from both aspects. The size of the object is limited only by the area that can be bathed

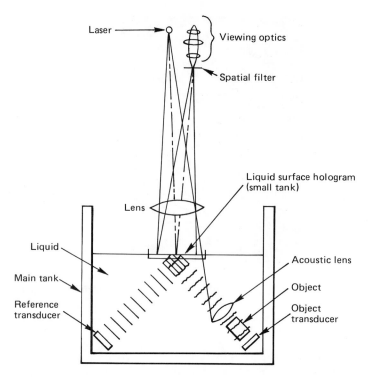

FIGURE 11–18. *A schematic of ultrasonic holography. A typical test configuration is shown whereby the ultrasonic energy is passed through the test object. The pressure waves are focused on the liquid surface to form the hologram. The hologram is then converted, through the use of the laser beam and optics system, into a real-time visual presentation.*

in the coherent light. The greatest advantage of ultrasonic holography is in the visual presentation. The operator is not required to interpret signals on an oscilloscope; he can actually look inside the test piece in real time. The optics can be adjusted to view the top surface, through the object, or the back surface. Adjustments of depth can be made in $\frac{1}{4}$ in. (6.35 mm) increments. The presentation may also be photographed, presented on closed circuit television, or video taped.

Limitations. Holography requires a vibration free environment. Thus, a heavy base is required to dampen the vibrations. The detector must be level, but objects inspected may be at any angle. Although the flaws are readily found, it is often difficult to identify the type.

For ease of study and review, the principle means of NDT have been summarized in Table 11.2.

TABLE 11.2. *Nondestructive Testing Methods* (Courtesy of *Metal Progress*)

Method	Measures or Defects	Applications	Advantages	Limitations
Acoustic emission	Crack initiation and growth rate Internal cracking in welds during cooling Boiling or cavitation Friction or wear	Pressure vessels Stressed structures Turbine or gear boxes Fracture mechanics research Weldments	Remote and continuous surveillance Permanent record Dynamic (rather than static) detection of cracks Portable	Transducers must be placed on part surface Highly ductile materials yield low amplitude emissions Part must be stressed or operating Test system noise needs to be filtered out
Eddy current	Surface and subsurface cracks and seams Alloy content Heat treatment variations Wall thickness, coating thickness Crack depth Metal sorting	Tubing Wire Ball bearings "Spot checks" on all types of surfaces Proximity gauge Metal detector	No special operator skills required High speed, low cost Automation possible for symmetrical parts Permanent record capability for symmetrical parts No couplant or probe contact required	Conductive materials Shallow depth of penetration (thin walls only) Masked or false indications caused by sensitivity to variations, such as part geometry Reference standards required Permeability
Holography (interferometry)	Strain Plastic deformation Cracks Debonded areas Voids and inclusions Vibration	Bonded and composite structures Automotive or aircraft tires Three-dimensional imaging	Surface of test object can be uneven No special surface preparations or coatings required No physical contact with test specimen	Vibrationfree environment is required Heavy base to dampen vibrations Difficult to identify type of flaw detected

TABLE 11.2. *Nondestructive Testing Methods* (Continued)

Method	Measures or Defects	Applications	Advantages	Limitations
Infrared (radiometers)	Lack of bond Hot spots Heat transfer Isotherms Temperature ranges	Brazed joints Adhesive-bonded joints Metallic platings or coatings; debonded areas or thickness Electrical assemblies Temperature monitoring	Sensitive to 1.5 °F temperature variation Permanent record or thermal picture Quantitative Remote sensing; need not contact part Portable	Costly equipment Liquid-nitrogen-cooled detector Critical time–temperature relationship Poor resolution for thick specimens Reference standards required
Magnetic particles	Surface and slightly subsurface defects; cracks, seams, porosity, inclusions Permeability variations Extremely sensitive for locating small tight cracks	Ferromagnetic materials; bar, forgings, weldments, extrusions, etc.	Advantage over penetrant in that it indicates subsurface defects, particularly inclusions Relatively fast and low cost May be portable	Alignment of magnetic field is critical Demagnetization of parts required after tests Parts must be cleaned before and after inspection Masking by surface coatings
Penetrants	Defects open to surface of parts; cracks, porosity, seams, laps, etc. Through-wall leaks	All parts with non-absorbing surfaces (forgings, weldments, castings, etc.). Note: Bleed-out from porous surfaces can mask indications of defects	Low cost Portable Indications may be further examined visually Results easily interpreted	Surface films, such as coatings, scale, and smeared metal, may prevent detection of defects Parts must be cleaned before and after inspection Defects must be open to surface

Method	Measures or Detects	Applications	Advantages	Limitations
Radiography (thermal neutron)	Hydrogen contamination of titanium or zirconium alloys; Defective or improperly loaded pyrotechnic devices; Improper assembly of metal, nonmetal parts	Pyrotechnic devices; Metallic, nonmetallic assemblies; Biological specimens	High neutron absorption by hydrogen, boron, lithium, cadmium, uranium, plutonium; Low neutron absorption by most metals; Complement to x-ray or gamma-ray radiography	Very costly equipment; Nuclear reactor or accelerator required; Trained physicists required; Radiation hazard; Nonportable; Indium or gadolinium screens required
Radiography (gamma rays)	Internal defects and variations, porosity, inclusions, cracks, lack of fusion, geometry variations, corrosion	Usually where x-ray machines are not suitable because source cannot be placed in part with small openings and/or power source not available	Low initial cost; Permanent records; film; Small sources can be placed in parts with small openings; Portable; Low contrast	One energy level per source; Source decay; Radiation hazard; Trained operators needed; Lower image resolution; Cost related to energy range
Radiography (X rays—film and image tubes)	Internal defects and variations; porosity, inclusions, cracks, lack of fusion, geometry variations, corrosion; Density variations	Castings; Electrical assemblies; Weldments; Small, thin, complex wrought products; Nonmetallics; Solid propellant rocket motors	Permanent records; film; Adjustable energy levels; High sensitivity to density changes; No couplant required; Geometry variations do not effect direction of x-ray beam	High initial costs; Orientation of linear defects in part may not be favorable; Radiation hazard; Depth of defect not indicated; Sensitivity decreases with increase in scattered radiation
Radiometry (x-ray, gamma-ray, beta-ray)	Wall thickness; Plating thickness; Variations in density or composition; Fill level in cans or containers; Inclusions or voids	Sheet, plate, strip, tubing; Nuclear reactor fuel rods; Cans or containers; Plated parts	Fully automatic; Fast; Extremely accurate; In-line process control; Portable	Radiation hazard; Beta-ray useful for ultrathin coatings only; Source decay; Reference standards required

TABLE 11.2. *Nondestructive Testing Methods* (Continued)

Method	Measures or Defects	Applications	Advantages	Limitations
Sonic	Debonded areas or delaminations in metal or nonmetal composites or laminates Cohesive bond strength under controlled conditions Crushed or fractured core Bond integrity of metal insert fasteners	Metal or nonmetal composite or laminates brazed or adhesive-bonded Plywood Rocket motor nozzles Honeycomb	Portable Easy to operate Locates far-side debonded areas May be automated Access to only one surface required	Surface geometry influences test results Reference standards required Adhesive or core thickness variations influence results
Thermal (thermochromic paint, liquid crystals)	Lack of bond Hot spots Heat transfer Isotherms Temperature ranges	Brazed joints Adhesive-bonded joints Metallic platings or coatings Electrical assemblies Temperature monitoring	Very low initial cost Can be readily applied to surfaces which may be difficult to inspect by other methods No special operator skills	Thin-walled surfaces only Critical time–temperature relationship Image retentivity affected by humidity Reference standards required
Ultrasonic	Internal defects and variations; cracks, lack of fusion, porosity, inclusions, delaminations, lack of bond, texturing Thickness or velocity Poisson's ratio, elastic modulus	Wrought metals Welds Brazed joints Adhesive-bonded joints Nonmetallics In-service parts	Most sensitive to cracks Test results known immediately Automating and permanent record capability Portable High penetration capability	Couplant required Small, thin, complex parts may be difficult to check Reference standards required Trained operators for manual inspection

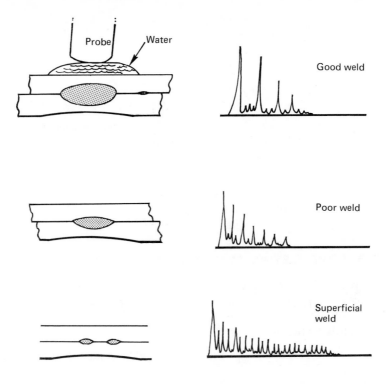

FIGURE 11-19. *Ultrasonic testing of spotwelds.*

NONDESTRUCTIVE TESTING OF SPOT WELDS

Nondestructive tests for spot welds are somewhat similar to nondestructive tests for other welds, such as eddy current, radiographic, and ultrasonic. The ultrasonic test with a special adaptation to the spot weld is described here. The ultrasonic weld checker, the principal described in Figs. 11-10 and 11-11, can also be used to check spot welds. Good, poor, and superficial welds are shown in Fig. 11-19. The good weld has reflections that originate from the far surface but are rapidly absorbed by the coarse grain structure. The small weld reflects partly from the far surface and partly from the faying surface with less absorption. The superficial weld reflects mainly from the faying surface with the least absorption.

As resistance welds are made, the metal between the electrodes, particularly at the faying surface, is raised in temperature. The thermal expansion acts to push the electrodes apart. This action can be observed when welding heavy gauge materials. Measuring the expansion can be used to give a good indication of the weld heat. The expansion can be measured by strain gauges, inductive or capacitive transducers, linear potentiometers, or contacts operated by a beam.

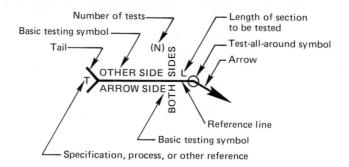

FIGURE 11–20. *The general appearance of an NDT symbol.* (*Courtesy of American Welding Society.*)

NONDESTRUCTIVE TESTING SYMBOLS

The NDT testing symbols were originally developed by the American Welding Society to identify the type, location, and other specifications governing particular types of defects. They have since been approved by the American National Standards Institute.

Basic Test Symbols. Each type of NDT is identified by a two-letter symbol as follows:

RT—Radiographic.
MT—Magnetic Particles.
PT—Penetrant.
UT—Ultrasonic.

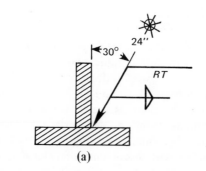

(a)

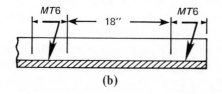

(b)

FIGURE 11–21. (*a*) *NDT symbols may share the same arrow. The joint is to be made with a fillet weld on both sides of the tee. It will also be tested on both sides with radiographic equipment from a distance of 24 inches (60.96 cm) at 30 degrees.* (*b*) *Dimension lines are used to show the exact location, as well as the length, of the section to be tested by magnetic particle testing. When no length dimension is shown, the full length of the part will be tested.* (*Courtesy of American Welding Society.*)

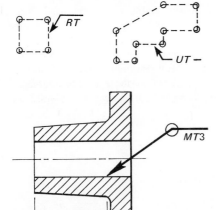

FIGURE 11–22. *The area to be tested may be shown on the drawing by broken lines joined with circles. Dimensions of the enclosures may or may not be indicated. (Courtesy of Manufacturing Engineering and Management.)*

FIGURE 11–23. *The test-all-around symbol used here requires complete magnetic particle testing of the whole inside diameter to within three inches of the face. The lower symbol indicates which portion of the outside diameter will be tested radiographically. (Courtesy of Manufacturing Engineering and Management.)*

A complete testing symbol consists of a reference line, an arrow, the applicable two-letter test symbol, a test-all-around symbol, a number indicating the number of tests to be performed, a tail, another number indicating the length of section to be tested, and a reference indicating the test specification, process, or reference. Only as many elements as are necessary are used and they have standard locations in the testing symbol. The general appearance of the symbol and the location of its elements are shown in Fig. 11–20. Nondestructive testing symbols may share the same arrow with welding symbols, see Fig. 11–21a. Dimension lines are used to show the exact location, as well as length, of the section to be tested, Fig. 11–21b. When a whole area is to be tested, it may be enclosed by broken lines and indicated as a plane on the drawing, Fig. 11–22.

When used in connection with an area of revolution, the test-all-around symbol indicates that the whole area be tested to the limits shown, if any, Fig. 11–23. In this illustration, the testing symbol pointing to the i.d. indicates that the complete bore of the flange is to be tested by magnetic-particle inspection to a distance of three inches from the face. The other symbol indicates what portion of revolution is to be inspected radiographically.

This brief description of NDT symbols is not intended to be complete but merely to make you aware of them and their use. The AWS Handbook, section one, will provide a full discussion.

SUMMARY

The nondestructive tests available today are important tools enabling manufacturers and fabricators to better define the integrity of their products. How and when they are used requires careful consideration. A standard procedure to assure quality control is:

See that all personnel are properly qualified. *Recommended Practices for Personnel Qualifications and Certification* is a publication that can be obtained from ASTM. It will prove to be very useful in training personnel for NDT.

Standard procedures for testing must be set up. Manuals on this may be obtained from ASTM, AWS, and the American National Standards Institute.

As recommended earlier, certain acceptance criteria must be established. These can be made as already described, or recommendations can be obtained from such groups as the AWS and ASTM.

EXERCISES

Questions and problems

11–1. What is the principal difference between pulse ultrasonics and resonant ultrasonics?

11–2. For what type of testing are N rays better than x rays or gamma rays?

11–3. What is meant by a "black" light as used in magnetic particle testing?

11–4. How does sonic testing differ from ultrasonic testing?

11–5. To be certain your plant is turning out high quality welded pipe, you suggest 100% inspection. Management objects stating this will be too expensive because the production rate will go down and the cost of the inspection will also be prohibitive. What type of proposal can you make that will overcome these objections? The pipes are between 2 and 3 ft (60 and 90 cm) in diameter, 40 ft (120 cm) long, and have wall thicknesses ranging from 0.250 to 0.625 in. (6.35 mm to 15.875 mm). Each pipe is made from a 40 ft (120 cm) steel plate that is rolled into a cylindrical shape and then welded by two automatic submerged arc welding machines, one from the outside and one from the inside. The pipe is rotated 180 degrees between welds. After welding,

the pipes are expanded to size against dies. They are then hydrostatically tested, and dried. The problem lies in being sure the welds are 100% sound.

11–6. Nondestructive testing symbols may share the same arrow with welding symbols. Explain each of the following welding and NDT combination symbols.

11–7. What method would be equally suitable for testing welds suspected of hydrogen embrittlement and checking a large casting by first being sure all voids have been filled with water? Why is the method you recommend the most suitable?

11–8. Which of the NDT techniques are dependent on the development of an electrical signal?

11–9. A manufacturing plant has obtained a contract that calls for a severe ring forming operation. The material is heated to facilitate the maximum amount of forming in one operation. The ring is roll formed from titanium bars 1 in. (2.54 cm) thick and 2 in. (5.08 cm) wide. Each forming operation costs $500.00. What NDT testing method could be used to insure flaw free material on a production line basis?

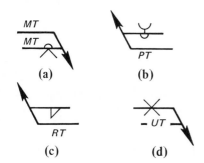

(a) (b)

(c) (d)

FIGURE P11–6.

11–10. What method used in NDT can be adapted for sensing the electrical properties of metal? The information obtained can be fed into the controls of an arc welder to automatically adjust the machine setting for optimum welding conditions.

11–11. Automobile steering knuckles, which are made of ductile iron, are produced in large quantities. It is necessary to have 100% inspection of these items. What would be the preferred method of NDT for: (a) the integrity of the surfaces, (b) the internal soundness, and (c) hardness and matrix structure?

11–12. Give examples of defects that can be detected by NDT methods under the following classification:
(a) Inherent (resulting from melting and solidification).
(b) Processing (resulting from fabrication of the finished article such as rolling, forging, welding, grinding, heat treating, and plating).
(c) Service (resulting from use).

11–13. What are some of the principle benefits of NDT?

11–14. What is the advantage of using the delta technique in checking welds?

11–15. Aircraft structural engineers are increasing the number of applications of adhesive bonding. What single method would you suggest for NDT detection of bondline voids, bondline variations, and porosity?

11–16. Why is lead not a shield for N rays, but water is?

11–17. What is the principle of thermal NDT?

11–18. (a) As an engineer responsible for QA, how would you set up a program to insure proper joining of the frame to the body of an automobile body on an assembly line basis? (b) What NDT means would be used and on what basis would the decision for pass or reject be made?

11–19. High strength steels have made possible the rapid growth in welded continuous frame, high rise buildings. The moment is transmitted from the column to the girder directly through the weld. The weld must be full penetration, free from cracks, unfused areas, or any other defect. This is especially demanding when one considers the higher degree of skill required for welding high strength steels.

A test is needed to give positive assurance of the weld integrity. Criteria that must be met are:

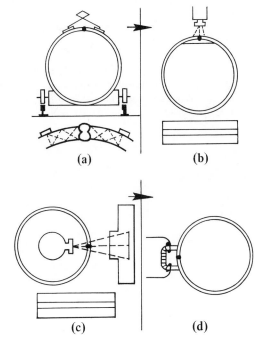

(a)

(b)

(c)

(d)

FIGURE P11–20.

(a) positive detection of surface and subsurface defects, (b) it must be economical, and (c) it must be applicable to field testing. Tell what NDT method you would use and explain why.

11–20. What type of nondestructive tests are symbolized in Fig. P11–20?

BIBLIOGRAPHY

Books

DAVIS, H. E., TROXELL, G. E., AND WISKOCIL, C. F., *The Testing and Inspection of Engineering Materials*, McGraw-Hill Book Company, New York, N.Y., 1964.

FORDHAM, P., *Nondestructive Testing Techniques*, Business Publications, London, England, 1968.

LIBBY, H. L., *Introduction to Electromagnetic Nondestructive Test Methods*, Wiley-Interscience, New York, N.Y., 1971.

MCGONNAGLE, W. J., *Nondestructive Testing*, McGraw-Hill Book Company, New York, N.Y., 1966.

MCMASTER, R. C., *Nondestructive Testing Handbook*, vols. I and II, The Roland Press, New York, N.Y., 1959.

NASA, *Nondestructive Testing; Trends and Techniques*, U.S. Govt. Printing Office, Washington, D.C., 1966.

Welding Inspection, American Welding Society, Miami, Fla., 1968.

Periodicals

ARONSON, R., "What Good is Holography?" *Machine Design*, January 23, 1969.

BAILEY, J. A., "Testing and Inspecting with Magnets." *Automation*, December 1967.

FRIELING, G. H., AND GRIFFITH, E. C., "Consider NDT to Control Processes." *Automation*, April 1972.

GREEN, O. V., AND STOUT, R. D., "A Study of the Influence of Speed on Torsion Impact Test." *ASTM*, vol. 39, 1939.

GRIFFITH, E. G., "The Advent of Sonic Testing." *Manufacturing Engineering and Management*, March 1972.

LAVOIE, F. J., "Neutron Radiography, New Tool for NDT." *Machine Design*, February 1969.

———"Nondestructive Testing." *Design Guide*, September 1969.

MAYO, W. H., AND HENRY, E. B., JR., "Nondestructive Inspection in the Steel Industry." *Science and Technology*, vol. XVI, no. 27, United States Steel Corp., 1969.

NITTINGER, R. H., "Nondestructive Testing." *Chemical Engineering*, February 1966.

PECK, E. S., "The Modern Look of Nondestructive Testing." *Manufacturing Engineering and Management*, March 1971.

SEIDEL, R. A., "Eddy Current Testing." *Metal Progress*, May 1963.

SPROW, E., "Liquid Crystals—A Film in Your Future." *Machine Design*, February 6, 1969.

STOLL, G. J., "Ultrasonic Inspection of Steel Castings. What's in the Future?" *Casteel*, vol. 7, no. 1, Spring, 1972.

12

Destructive Testing, Codes, Certification, and OSHA

In recent years manufacturers have had to place greater emphasis on quality control and testing programs. The general emphasis between the retailer and consumer has changed from "buyer beware" to "seller be wary." US courts are more and more finding the plaintiff need not prove negligence in manufacture. He need only prove that the product had a defect when it left the manufacturer's hands. *It need not be shown that the defect was unknown to the manufacturer or whether it could have been prevented.* Thus, today's manufacturer must place greater emphasis on quality control with both destructive and nondestructive testing programs having an important place.

The traditional approach in manufacturing has been to first make the part and then inspect it. Today, the concept is to merge inspection with manufacturing. The result has been a growing development of automatic in-process testing.

Statistics are used to determine the amount of quality control necessary to obtain an acceptable level at a reasonable cost for a given product. The success of a statistical quality control program is based on accurate observations and recording of the test data. This information can be used by the welding engineer to predict the reliability level of the product.

STANDARDIZED TESTS

Destructive tests are designed to show where failure may occur for various loading and environmental conditions. Tests are conducted and then used to predict a service life with some reasonable safety factor added. Fortunately, a great many tests for materials have been standardized. The most widely used and comprehensive are the tests set up by ASTM. Mainly through the efforts of this society, more than 3,400 standard tests of materials have been established. Each test is complete with material specifications, a detailed description of the specimen, test procedure, and acceptance standards. It is of interest to note that ASTM has now developed a compilation of selected standards for the use of engineering students. Representative illustrations of nationally accepted standard specifications and standard test methods are discussed and illustrated.

The testing standards established are the painstaking work of a committee assigned to the particular problem. On materials having a commercial bearing, the policy is to maintain a balance between representatives of producer interest, consumer interest, and general interest. It is this committee that evolves a standard as to the material and test procedure. Approval is required by the membership at the annual meeting. The committee approved standards are then published in tentative form, for at least one year, during which time criticism is elicited. After corrections, the standard is submitted to the entire membership for a ballot vote. Revisions may be considered at any time by the standing committee concerned. Standards may also be withdrawn at any time by appropriate action.

In addition to the work of ASTM, many specifications are developed by bureaus of the Federal Government and by technical societies. The AWS, for example, has developed standard methods for the mechanical testing of welds as well as other standards.

It is important that the engineer understand the principles of each test he uses or recommends. He must develop the ability to visualize the physical changes that take place during the test: the deformation with accompanying stress patterns, the movement of component parts, material flow, and so forth. He must also be aware that tests are not precise and infallible but that they are always made subject to limiting conditions. And finally, tests are not useful until they have been properly interpreted and reported.

Two problems are involved in the testing of representative specimens. The first is in setting up the procedure to obtain the samples; and the second, is in determining the number and frequency of samples required. Fortunately, standards have now been quite well established as to the type of specimen to use for a given test. As regards the second problem, sampling procedures based on statistical theory have been established and are given in the bibliography at the end of this chapter. Statistical techniques are also being used to establish relationships between variables in testing. Some destructive tests that are now used can be changed to nondestructive types when definite relationships as to defects have been statistically established.

Specimen Selection and Preparation. The data obtained from destructive tests are based on careful selection of representative specimens. As for example, coupons cut from a metal plate may be in the direction of rolling or at right angles to it. The effect of the rolling will be quite pronounced in the case of wrought iron. The cooling rates of a large casting will have considerable variation in hardness and strength according to the structural mass from which it was taken.

The methods of obtaining the specimens must not influence its properties, e.g., a coupon flame cut from a plate must have at least $\frac{1}{4}$ in. machined off for most tests. The machined surface in turn must be smooth so that it will not influence the test. The dimensions and tolerances of the specimens must be closely adhered to. Complete descriptions of materials tests for metals, plastics, and ceramics are given in the ASTM handbook. Tests for welded joints are given in the AWS Handbook, Section 1.

Modes of Failure. Failure may be defined, in a broad sense, as a part not fulfilling its intended function. Some failures are easy to anticipate and prevent, as in the case of knowing a bolt will rust and then supplying a protective coating to prevent oxidation. On the other hand, failure may be complex and catastrophic, as for example, the case of engine separation from an airplane in flight or a ship breaking in half. Facts gathered from an analysis of a failed part can be more valuable than those derived from laboratory or field tests, and are usually much less expensive.

Failures can seldom be assigned to a single cause. Usually they result from the combined effects of two or more factors that are detrimental to the life of the part or structure. From a detailed analysis of failures made by steel companies, vehicle manufacturers, and electrical equipment manufacturers, it has been shown that nearly 50% of all failures can be attributed to faulty design, the rest are distributed between production and service problems.

Classification of Failures. Metal failures may be classified as ductile shear, creep, brittle fracture, and corrosion. The terms are defined here but will be discussed in more detail as they apply to specific destructive tests.

Ductile Shear. Plastic or ductile shear takes place when one crystal plane slides over another and causes a permanent displacement. Ductile shear usually produces a dull fibrous appearance, denoting high energy absorption.

Creep. Creep is the continuing distortion or strain of a material under a steady load. Dislocations in the metal structure are believed to be the cause. The dislocation movements are stopped by grain boundaries or impurity atoms. However, at high temperatures the dislocations are able to "climb" over these barriers. Thus, for common structural metals, creep is an important consideration only at elevated temperatures. The temperature range considered significant is from about .0035 to 0.7 times the melting point of the material.

Brittle Fracture. Unlike plastic deformation or creep, brittle fracture does not involve the slip of an appreciable volume of metal. Fracture takes place before noticeable structural damage is done. It may start at some localized stress concentration and spread across the part to cause a total fracture. This process occurs in brittle materials under a steady load and in ductile materials under repeated stress. In the latter case, failure is due to the exhaustion of the ductility of the material (*fatigue*), and once started, will spread even under relatively low stress.

Corrosion. Corrosion is the deterioration of materials through chemical or electrochemical attack. The resultant decrease in volume of sound material causes an obvious decrease in strength of the corroded part.

Metal corrosion is often accelerated by galvanic action, which always takes place when dissimilar metals are exposed together in a common electrolyte. The severity of the action is dependent upon the difference in potential between the galvanic cells. The metal that serves as the anode (least noble) is the one that corrodes.

The combined action of static tensile stress and corrosion causes a comparatively recent phenomena discovery termed *stress-corrosion cracking*. Generally, localized corrosion of a part forms pits or other surface defects that in turn cause localized stress concentrations. The combined action of the corrosion and stress results in an accelerated propagation of cracks. The time required for corrosion to develop from lines along the surface to cracks may vary from a few minutes to years. The susceptibility of a surface to stress corrosion cracking is largely dependent on the residual stresses of the base metal.

DESTRUCTIVE TESTS

Common destructive tests include tension, hardness, bend, Kinzel, impact, fatigue, explosion bulge, and metallographic examination.

Tension Tests. Tensile tests furnish design information as to the strength and ductility of a material. The designer is able to ascertain the yield point and other characteristics of the material, Fig. 12–1. Tension tests of butt weld joints are of several types, and for all weld metal, longitudinal or transverse, Fig. 12–2. Standard specimens are machined from the cut-out section as shown in Fig. 12–3. The flat longitudinal weld specimen, Fig. 12–4a, is used to test the properties of the weld, the heat affected zone (HAZ), and the base metal. The weld must elongate with the base metal until failure occurs. Poor welds, or those that have a lower HAZ ductility, often force fracture initiation to occur at a strength level considerably below that of the base plate. Because of these effects, weld evaluation should always include longitudinal weld specimens as well as transverse. The transverse specimens, Fig. 12–4b, point up the relative flow strengths of each area of the joint. In general, breaks will occur in the base metal or HAZ due to the higher flow strength of the weld deposit.

Although tensile tests come closer to evaluating fundamental mechanical properties for use in design than any other test, they are not necessarily sufficient for the prediction of the material performance under all loading conditions. In a tensile test of ductile materials, flow or slip takes place until fracture. The condition

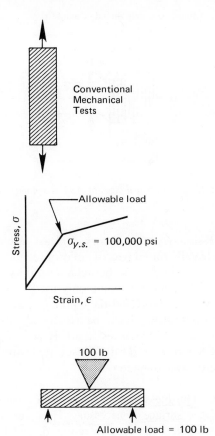

Conventional
Mechanical
Tests

Allowable load

$\sigma_{y.s.}$ = 100,000 psi

Stress, σ

Strain, ϵ

100 lb

Allowable load = 100 lb

FIGURE 12–1. *Establishing the allowable load by tests.*

of fracture is dependent on a host of conditions, e.g., stress, strain, strain rate, environment, and the metallurgical state of the materials and their histories.

Observations of the fractured area of a tensile specimen can reveal information as to the type of material and cause of failure. If the break is unsymmetrical, it may indicate a defect or flaw such as segregation, blowhole, slag inclusion, or other foreign matter. Streaks or ridges radiating outward from some point, or "star," may indicate stress conditions due to cold working or heat treatment. Straight or flat breaks with little or no deformation represent a very brittle material.

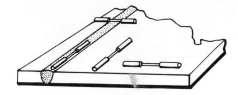

FIGURE 12–2. *Tensile test specimens may be all weld metal, either longitudinal or transverse. Base metal specimens are taken for reference.*

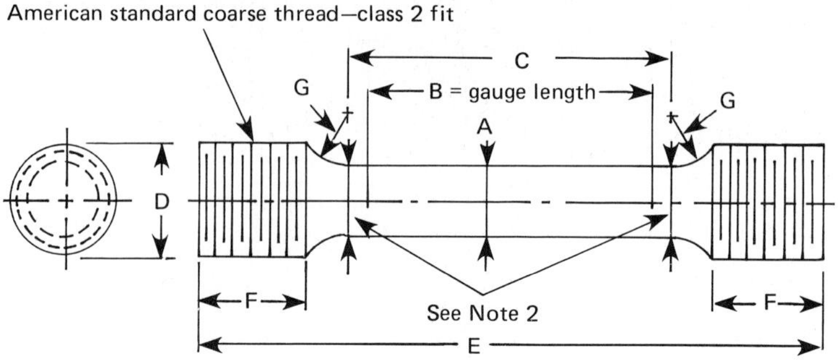

American standard coarse thread—class 2 fit

FIGURE 12–3. *Standard sizes of ASTM cylindrical test specimens.*

Compression tests, in theory at least, are the opposite of tension tests. Specimens are usually round bars of uniform section. The ratio of length to diameter is more or less of a compromise between several undesirable conditions. A length to dimater ratio of 2 or more is commonly used. The actual size will be determined by the material to be tested and the equipment available. For brittle materials, the ultimate compressive strength can be associated with the gauge reading at the time of fracture. For ductile materials, there is no unique phenomenon to mark the ultimate strength. However, it can be associated with the instant a longitudinal crack appears on the outside diameter.

Hardness Test. Hardness tests of materials are classified as destructive since some indentation of the surface is necessary. However, it will be dependent upon the type of material tested, the surface required, and its end use. Hardness testing is often used to determine material characteristics in preference to the more expensive and time consuming tensile test. The most common hardness tests are Brinell,

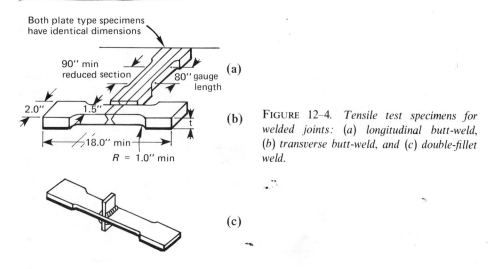

FIGURE 12–4. *Tensile test specimens for welded joints:* (*a*) *longitudinal butt-weld,* (*b*) *transverse butt-weld, and* (*c*) *double-fillet weld.*

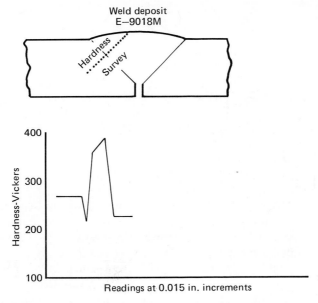

FIGURE 12–5. *Example of a weld hardness test and corresponding graph of the results.*

Rockwell, and Vickers, as discussed in Chapter 10. Shown in Fig. 12–5 is a weld that has been subjected to hardness testing and its properties determined as shown on the accompanying graph.

Bend Tests. Bend tests for weld specimens are of two main types, smooth and notched, or nicked.

Smooth Bend Tests. Smooth bend tests are frequently used to estimate the ductility of weldments. They are also used in qualification tests for welding operators, because they often reveal the presence of defects that do not show up in tension tests. The smooth bend test may be guided or free. In preparation for the test, the weld reinforcement is machined off flush with the surface of the plate. The guided bend consists of bending a joint 180 degrees around a pin that is $1\frac{1}{2}$ in. (2.31 cm) in diam, see Fig. 12–6. In the case of steel, the test specimen is $\frac{3}{8}$ in. (9.53 mm) thick by $1\frac{1}{2}$ in. (2.31 cm) wide (the length is not important, usually 4 to 8 in. (10.16–20.32 cm). Root face and side bend tests are made.

Free Bend Tests. The free bend test specimen is prepared as for the guided bend test with the reinforcement machined or ground off. The width of the specimen is approximately one-and-one-half times the thickness. Any machine marks should be parallel to the bend. The length is not important but varies with the thickness of the specimen. The weld should be in the center of the test length.

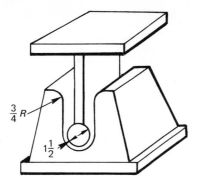

FIGURE 12–6. *A guided bend testing jig.*

The gauge lines, which are approximately 1/8 in. (3.18 mm) less than the width of the face of the weld, are lightly scribed on the weld face surface.

The initial bend may be made in several ways. One method is to employ the fixture shown in Fig. 12–7a. The loading block is centered on the weld and depressed until a bend of 30 to 45 degrees is made. Another method is to clamp the specimen in a vise about one-third of the way from the end and produce the initial bend with a hammer. To insure the bend is made in the middle of the specimen, the operation is repeated on the other end.

The final bend is made as shown in Fig. 12–7b. If a crack or other defect occurs that exceeds the specified size in any direction on the convex face of the specimen, the load is released. The elongation may be evaluated from a measurement taken with a flexible scale between the gauge marks.

Nick Break Tests. Nick break test specimens may be flame cut from the welded plate. The reinforcement need not be machined off. A hacksaw or

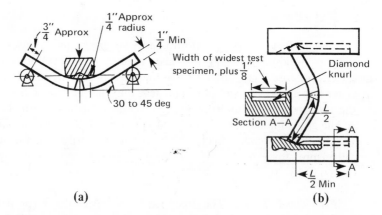

FIGURE 12–7. *The initial bend for free bend testing may be started as shown (a). In place of the rollers, hardened and greased shoulders may be used. (b) The final bend is made as shown. (Reprinted by permission, copyright American Society for Testing and Materials.)*

438

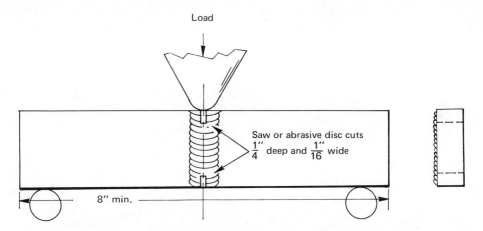

Load

Saw or abrasive disc cuts
$\frac{1''}{4}$ deep and $\frac{1''}{16}$ wide

8" min.

FIGURE 12–8. *A nick-break test specimen.*

abrasive wheel cut, as shown in Fig. 12–8, is made to insure a break in the weld. The maximum load and the bend angle are observed. If instrumentation is available, the results may be plotted.

Kinzel Test. The Kinzel notched bend test is designed to measure the effect of welding on the base metal. The base metal adjacent to the weld may have grain coarsening, partial hardening, and perhaps some flow. The HAZ is usually more notch sensitive than the unaffected base metal. Control over this area is affected by weld current, weld speed, preheat, and postheat. The actual weld metal seldom exerts influence on the bend test. The specimen and load placements are shown in Fig. 12–9.

The angle of bend is measured at maximum load and is considered a toughness indicator. Two transition zones occur as the specimen is bent. The first, a ductile transition zone, is a measure of the ease of fracture. The second zone indicates the resistance of the weld to fracture propagation. This test is proposed as being especially sensitive to the microstructure that is formed by heating into the austenitic range, above A_1 but below A_2, and cooling rapidly. The ferrite is allowed to precipitate on certain crystallographic planes in the austenite. The resulting structure is known as "Widmannstatten," as discussed in Chapter 10.

Impact Tests. Impact tests are designed to study the reaction of materials to a suddenly applied load. Common tests of this type are the notched bar impact or Charpy "Vee", explosion bulge test, drop weight tests, and fracture mechanics testing. Before each of these tests is discussed, some general considerations regarding impact loads will be considered.

Impact Loads. The energy of an impact load may be received by a structure in a number of different ways. When designing for impact loads in machines and structures, the aim is to provide for as much energy absorption as possible through

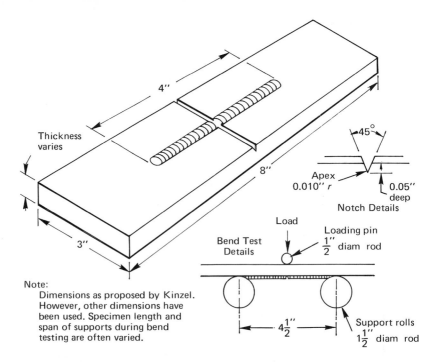

FIGURE 12–9. *A longitudinal bead-on-plate weld converted to a notched-bend specimen. The notch depth is limited by weld metal penetration if there is to be weld metal as well as heat-affected base metal at the root of the notch. In some cases it is necessary to groove the test plate before welding in order to leave some weld metal under the machined vee-notch.*

elastic action, then relying on damping to dissipate it. This elastic energy capacity is often referred to as *resilience*.

Brittle Fracture and Impact Loads. Although many tests have been designed to test impact properties of materials, it is difficult to make tests that will contribute directly to basic design data. In the so-called notched bar test (described briefly in Chapter 10), only the energy to cause rupture is considered. A relative measure of the tendency toward brittleness as affected by the temperature change and chemical composition was shown earlier (see Fig. 10–7). Thus, the engineer designing for impact will base his choice of materials and fabrication procedures on the performance criteria as shown by transition temperature curves.

Brittle Fracture and the HAZ. Most brittle fractures in mild steel weldments have been observed to initiate at flaws and to propogate independently of weld seams. Now however, with a greater use of high strength low alloy (HSLA) steels, fractures have been observed to initiate at flaws and to propagate along weld seams. This is attributed to the fact that under certain conditions the HAZ structure in the HSLA steels is more brittle than in the low carbon steels. Also,

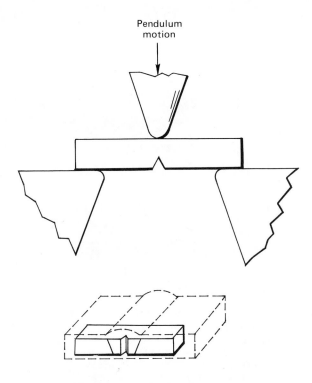

FIGURE 12–10.   *A Charpy vee-notch impact test.*

high strength matching filler metals often exhibit poorer toughness than corresponding high strength base metals.

Charpy Vee Test.   The Charpy Vee test has been very popular for testing the impact properties of structural steels. The test consists of supporting a notched bar horizontally between two anvils and allowing it to be struck by a pendulum as shown in Fig. 12–10. The test is repeated at different temperature levels so that a curve of absorbed energy vs. temperature may be plotted. The three regions of the curve (as were shown in Fig. 10–7) are: a region of brittle fracture at low energy and low temperatures, a transition zone where ductile shear and brittle fracture occur, and a region of ductile shear failure occurring at higher temperatures where high shock loads would be absorbed by the welded specimen. Two significant reference points can be taken from the test curves. One is the midpoint transition temperature and the other is the transition temperature at a specific energy level. The latter is most popular and is usually designated as the transition temperature at the 15 ft/lb (2.07 kg-m) level. The level is chosen because the American Standards Department found a statistically significant relationship between brittle fracture in ship plate and transition at the 15 ft/lb (2.07 kg-m) level.

Drop Weight Tests.   The drop weight test was designed to study brittle

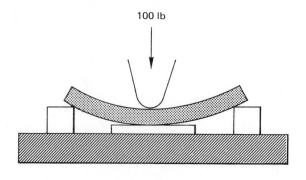

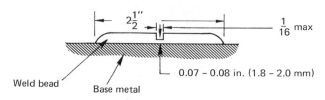

Weld bead   Base metal

0.07 – 0.08 in. (1.8 – 2.0 mm)

FIGURE 12–11. *Yield point loading in the presence of a small crack is terminated by contact with a stop block on the anvil. (ASTM Standard E 208-66T.)*

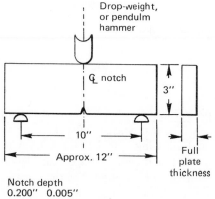

FIGURE 12–12. *The drop-weight tear test as developed at Battelle Institute. (Courtesy of Battelle Columbus Laboratories, Columbus, Ohio).*

fracture at low energy input levels. This level has been known to cause failure in weldments that have residual stresses. A stop block is used as shown in Fig. 12–11 to limit the amount of strain induced by the dropped block. The weight is usually 100 lb (45 kg), dropped from a height of 6 to 12 ft (18 to 36 m). The test specimen has a notch machined in the weld bead as shown at Fig. 12–11b. The test data is used to show fracture strength vs. temperature and flaw size. The test is simple to execute and simulates impact loading. The disadvantage is an assumed flaw, which could be conservative.

Drop Weight Tear Test. The drop weight tear test was developed by Battelle Institute for brittle fracture analysis of steel pipe. The mechanics of the test are shown in Fig. 12–12. A hammer or beam is made to strike the notched specimen on the compression side. Tests are conducted over a range of temperatures and the percent of brittle fracture to surface area is measured. The results for ASTM A36 steel are shown in Fig. 12–13. Studies have shown that velocity of impact, excess energy needed to break the specimen, and notch depth have no

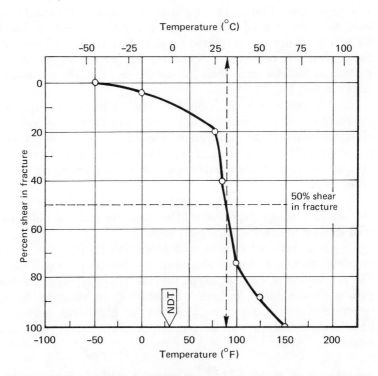

FIGURE 12–13. *A curve plotted from drop-weight tear test data. It depicts the transition from tough (shear) fracture to brittle (cleavage) fracture. The temperature at which 50% shear takes place is used as a significant value to mark the fracture-toughness of ASTM A36 steel. (Reprinted by permission, copyright American Society for Testing and Materials.)*

effect on the transition region location. Increasing the root radius causes more scatter of data and a decrease in the slope of this region.

Explosion Bulge Test. The explosion bulge test, as shown in Fig. 12–14, was developed by the US Navy for high energy fracture analysis of ship hulls. Plates are butt welded and a small groove is machined into the weld bead to simulate a notch. The welded plate is then placed on top of a die and explosively deformed to a specified depth. The fractures are classified as:

1. Nil-ductility, flat break.
2. Elastic fractures.
3. Plastic fractures.

Fracture energy vs. temperature can be plotted and a transition zone between brittle and elastic, and/or plastic fracture, will exist. An advantage of this test is that energy for crack propagation is continuously supplied at a high rate (by gas expansion) even after initial failure.

Testing of Spot Welds. The appearance of spot welds can be very deceiving. A weld may appear good even though, in essence, no weld exists. Destructive tests for spot welds are numerous and are described in the American Welding Society's publication, "Recommended Practices for Resistance Welds." Three common destructive tests for spot welds include the peel test, the tensile test, and the cross-sectional test, Fig. 12–15.

The peel test is probably the simplest to execute and yet gives effective results. All that is required is a sample strip containing welds and a means of pulling it apart. If the penetration is good, the weld nugget will pull a hole in either piece. This is particularly true in materials up to 0.094 in. (2.39 mm) thick. Greater thicknesses may only pull out a slug of metal, leaving a crater in either piece.

Portable tensile testing machines are available for testing spot welds. These machines are of 10,000 psi (68.95 MPa) capacity and will pull single spot welds in mild steel up to 0.094 in. (2.39 mm) thick.

Cross-sectional tests consist of cutting through the middle of the weld, polishing, etching it in a suitable acid, and then inspecting it under a microscope. The penetration of the weld nugget into the base sheet should be between 40 and 70 percent of the sheet thickness. In addition to penetration, the metallurgical structure of the weld can be identified for grain size, microstructure, and any harmful effects such as carbide precipitation, which may occur in certain stainless steels.

Testing of Seam Welds. The most common method of testing seam welds is to cut a sample of the area and pull to destruction. A satisfactory weld will generally show a failure in the metal rather than in the seam.

Seam welds can also be tested by "pillow tests." Two rectangular or square sheets are joined together by welding a seam all around the edges. A pipe

For Details See
NAVSHIPS 0900-500-5000

Test conditions altered with
changes in plate thickness

Crack Starter Test Plate

Bead of brittle weld metal atop welded
joint is notched to provide small flaw.
Weld reinforcement on face of welded
joint is ground flush at ends to permit
intimate contact with supporting die.
Two shots are applied to plate.

Explosion Bulge Test Plate

Welded (and unwelded) test plates are
subjected to repeated shots until a certain minimum bluge depth is developed
without visible failure occurring. Reduction of plate thickness in apex of bulge
also is measured.

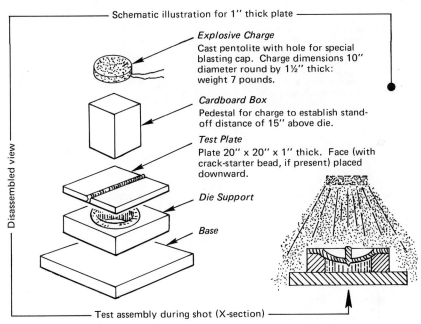

Schematic illustration for 1″ thick plate

Explosive Charge

Cast pentolite with hole for special
blasting cap. Charge dimensions 10″
diameter round by 1½″ thick:
weight 7 pounds.

Cardboard Box

Pedestal for charge to establish stand-
off distance of 15″ above die.

Test Plate

Plate 20″ x 20″ x 1″ thick. Face (with
crack-starter bead, if present) placed
downward.

Die Support

Base

Disassembled view

Test assembly during shot (X-section)

Determination of three critical fracture transition temperatures
from failure mode in crack-starter test plates

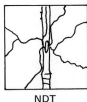 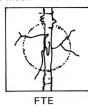 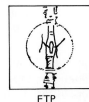

NDT	FTE	FTP
Nil ductility tem-perature. Flat break. Brittle frac-tures extend to edge of plate	Fracture transition elastic. Fractures are arrested in elastically loaded die-supported region	Fracture transition plastic. Fractures are arrested in plastically loaded (bulged) region

FIGURE 12–14. *The explosion-bulge test as designed by the U.S. Navy
to test the results of high energy fracture.*

445

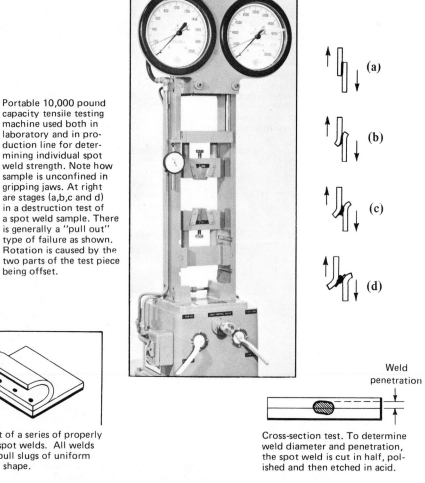

Portable 10,000 pound capacity tensile testing machine used both in laboratory and in production line for determining individual spot weld strength. Note how sample is unconfined in gripping jaws. At right are stages (a,b,c and d) in a destruction test of a spot weld sample. There is generally a "pull out" type of failure as shown. Rotation is caused by the two parts of the test piece being offset.

Peel test of a series of properly spaced spot welds. All welds should pull slugs of uniform size and shape.

Cross-section test. To determine weld diameter and penetration, the spot weld is cut in half, polished and then etched in acid.

FIGURE 12-15. *Three common destructive tests for spot welds.* (*Courtesy of ACCO, Wilson Instrument Div.*)

connection is welded to the center of one sheet. Hydraulic pressure is applied through the pipe until the sheets expand like pillows. The bursting pressure is recorded, as is the length of the weld line. Failure should occur in the parent metal rather than in the weld.

FATIGUE STRENGTH

Most structures are subject to variations in applied loads, which cause stress fluctuations in the various parts. If the stresses are repeated a sufficient number of times, failure occurs. An important factor is that the load is comparatively small

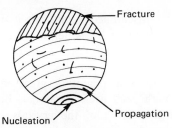

FIGURE 12-16. *A schematic showing the three stages of fatigue.*

and would be insufficient to cause failure if applied slowly and repeated less frequently. Failure is progressive and usually originates in a highly localized region of structural weakness.

Fatigue Theory. The mechanisms of fatigue are not thoroughly understood at present. Numerous theories have been proposed since it was first recognized as one of the causes of material failure about the middle of the 19th century. It is now generally agreed that the process is associated with three sequential stages: nucleation, crack growth, and fracture, see Fig. 12-16. The process is always associated with repeated or cyclic loading and always involves plastic deformations even if in localized microscopic areas.

One theory advanced as to why cracks develop in a material where there are no observable stress raisers is based on the crystal orientation of the metallic structure. Shown at *A*, Fig. 12-17, is an unfavorably oriented crystal. As the shear load is increased to a maximum positive value, the stress produced in crystal *A* lags that in the surrounding crystals because of its lower resistance to deformation. As the load is reversed and approaches zero, the surrounding crystals return to equilibrium, forcing crystal *A* to return the same distance. Because the stress in *A* was lower originally, upon returning it passes through its equilibrium point sooner and is negatively strained. As the shear load reaches the maximum negative value, the stress in *A* is greater than the surrounding crystals by the amount of lag involved. Crystal *A* continues to yield with each cycle, and each time it yields, its strain hardens a little more until a submicroscopic fissure forms at the slip planes.

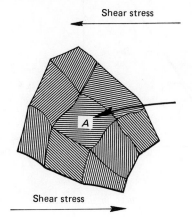

FIGURE 12-17. *The unfavorably oriented crystal (A) is surrounded by others more favorably oriented in relation to the shear stress.*

447

This simplified explanation lacks disciplined detail, but it does propose a mechanism by which repeated loading, at stress levels below the elastic limit, can result in ultimate failure of a structural member. Failures from this type of action are termed "high-cycle" fatigue failures.

The effect of induced cyclic strains of a magnitude great enough to cause immediate gross plastic deformation with each cycle is similar, but the slipping action takes place in a relatively large portion of the crystals composing the structure. The term "low-cycle" is used to distinguish conditions under which most of the total deformation is plastic and results in the appearance of macroscopic cracks, or complete fracture, in less than 10,000 loading cycles.

Fatigue occurs only where the applied load is cyclic. The load may be unidirectional (tensile, compressive, or shear), completely positive, completely negative, or unequal in magnitude.

A clear distinction must be made between induced strain and induced stress. The relationship between the two is not only nonlinear, but it is variable depending on the previous history of the material. A typical stress–strain curve, Fig. 12–18, shows both uniaxial elastic and plastic action as the strain varies both positively and negatively to beyond the proportional limit. In the range of strain between points A and B, approximately linear proportionality exists between stress and strain (elastic range). The deviation from linearity shown within this range is the result of recoverable time dependent deformation. This deviation is ignored in elastic calculations because it is generally insignificant.

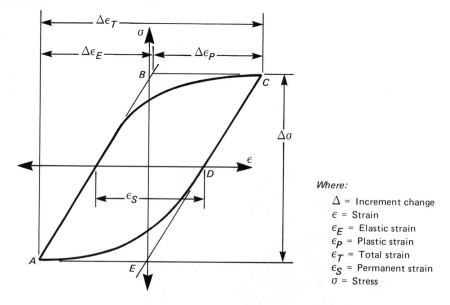

Where:

Δ = Increment change
ϵ = Strain
ϵ_E = Elastic strain
ϵ_P = Plastic strain
ϵ_T = Total strain
ϵ_S = Permanent strain
σ = Stress

FIGURE 12–18. *A typical stress–strain curve in the plastic range (constant strain amplitude) of a metal.*

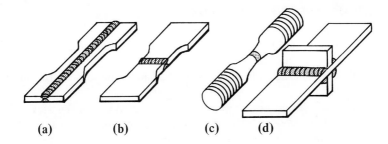

(a)   (b)   (c)   (d)

FIGURE 12–19.   *Common fatigue weld specimens: (a) longitudinal butt weld for axial or flexural loading, (b) transverse butt weld for axial or flexural loading, (c) transverse butt weld for axial or rotary loading, and (d) tee-type, load carrying transverse fillet weld.*

Beyond point *B* is the plastic range wherein strain is due mainly to permanent deformation in the structure of the material. The negative strain from points *C* to *D* takes place as the positive load is decreased to zero. The permanent set, or strain ($\Delta\varepsilon_S$), at this point is approximately equal to the plastic (nonrecoverable) strain ($\Delta\varepsilon_P$). From points *D* to *E*, the load is negative and the strain remains elastic. Plastic strain takes place from points *E* to *A*, and the cycle is complete.

Weld Fatigue Testing.   The methods and equipment used for fatigue testing weldments are essentially the same as those used for determining the fatigue strength of the base metal. The type of specimen is determined by the geometry of the weldment. Examples of some commonly used fatigue specimens are shown in Fig. 12–19. Irrespective of the weld geometry tested, the specimen should include a full cross section of the weld. Round specimens, Fig. 12–19c, are only satisfactory for comparing fatigue strengths of different weld metals.

Specimen size is determined by the capacity of the machine available. It is reasonable to assume that the larger the specimen, the greater is the probability of a defect being present to reduce fatigue life. Tests have shown, however, that the frequency of defects in welds is sufficiently high that the smallest specimen size commonly used covers a representative length of weld. The fatigue load is usually applied axially, ranging from full compression to full tension, and is expressed as an algebraic ratio:

$$R = \frac{\sigma_{\min}}{\sigma_{\max}}$$

The results have already been shown on *S–N* plots (Fig. 10–10), depicting the number of cycles to failure attained at various stress levels. The tests are also used to show the fatigue strength for a certain life, usually 10^5, or 2×10^6 cycles. When testing is performed at various values of *R*, the results are usually presented in the form of a modified Goodman diagram (Fig. 10–13).

FATIGUE AND WELDED JOINTS

Sound welding procedures and good weld design are the main deterrents for preventing fatigue in fabricated structures. Smooth bead contours free of defects such as cracks, lack of fusion, and inadequate penetration are needed.

Butt Joints.   Sound butt joints have been tested above the yield stress for as many as two million cycles. Unless there is an appreciable reversal of stress or secondary bending, conventional butt joints in compression are not subject to fatigue failure. Butt joint fatigue strength is not improved by adding reinforcing plates unless the weld itself is grossly deficient. The reinforcing plates actually introduce stress concentrations that reduce the fatigue strength of the joint.

For a simple butt weld with the reinforcement left intact, fracture occurs at the edge of the weld reinforcement due to the change of cross section, Fig. 12–20. The fatigue strength of transverse butt welds has been shown to increase in proportion to the included angle between the weld reinforcement and the base plate, with a maximum approaching 180 degrees.

Single-vee and single-U welded joints have rather higher fatigue strengths than double-vee joints due, presumably, to the stress concentration at the toes of the weld, Fig. 12–21. Partial penetration longitudinal butt welds have been found to have fatigue strengths as high as full penetration longitudinal butt welds, thereby justifying the common practice of using them when their axes lie in the direction of the major applied stress. However, when a partial penetration butt weld is used such that the stress is in the transverse direction, the weld fatigue strength is severely reduced.

Fillet Welds.   Fillet welds often have incomplete penetration producing "cracks" or "notches" at the vertex of the weld. Also, abrupt changes in cross section at the toe, Fig. 12–22, produce stress concentration points. Excessive weld reinforcement increases the included angle, thereby increasing the stress concentration.

Most fatigue tests of structural details involving fillet welded lap joints have shown better fatigue resistance for transverse welds than longitudinal, Fig. 12–23.

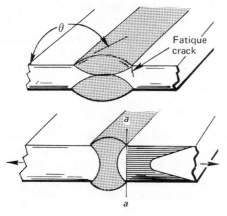

FIGURE 12–20.   *The fatigue strength of butt welded joints is related to the weld reinforcement angle and the base plate. It approaches maximum when the included angle is 180 degrees.*

FIGURE 12–21.   *The axial fatigue stress distribution on a double-vee butt weld.*

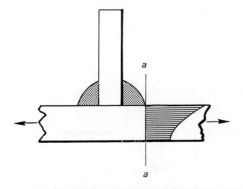

FIGURE 12–22. *The axial fatigue stress distribution in a transverse fillet weld.*

Other examples of the proper placement of fillet welds to avoid stress concentration are shown in Fig. 12–24. Partial cover plates are often fillet welded to beams or girders for added strength in a given section. Studies have shown that anything that can be done to make the cross-sectional change from the plate to beam more gradual is beneficial in increasing the fatigue strength, Fig. 12–25. The placement of the welds is also very important. As shown in Fig. 12–25, the welds are to be placed along the sides only. Even if no special preparation is made and the cover plate is made with square ends, the fatigue life will be increased by welding along the sides only.

Weld Joint Geometry. Weld joint geometry is the single most important factor in determining fatigue strength. The following joints are listed in order of their ability to withstand fatigue, from the least susceptible to the most susceptible*:

Longitudinal full penetration butt welds.
Continuous longitudinal fillet welds.
Transverse manually welded butt welds.
Transverse nonload carrying fillet welds.
Longitudinal nonload carrying fillet welds.
Transverse load carrying fillet welds.
Longitudinal load carrying fillet welds.
Plate with longitudinal attachment on its edge.

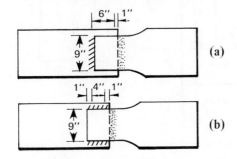

FIGURE 12–23. *The transverse fillet welds (a) distribute the stress in a more uniform manner to the plate material. Severe stress concentrations are encountered at the ends of the longitudinal fillet welds shown at (b).*

* *Welding Handbook*, 6th ed., section 1, "Fundamentals of Welding," American Welding Society, Miami, Fla., 1968.

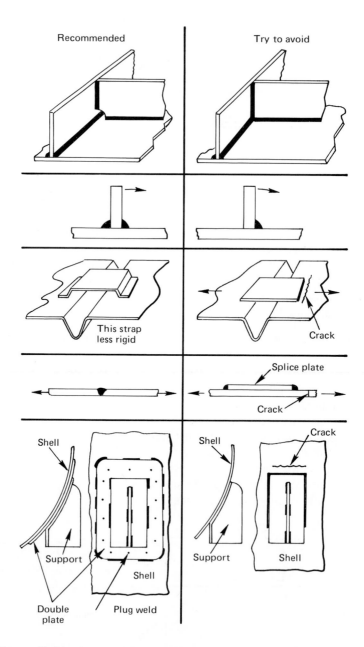

FIGURE 12–24. *Recommended practices in the placement of fillet welds that avoid stress concentrations and fatigue failure.* (*Courtesy of The Lincoln Electric Co.*)

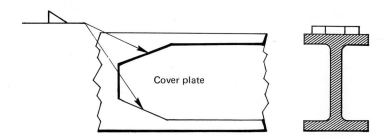

FIGURE 12–25. *A method of reducing fatigue failure in welded cover plates.*

Practical design tips to minimize geometrical fatigue stresses are listed below:

1. Use butt joints rather than lap joints wherever possible.
2. Streamline fillet weld joints.
3. Change sections gradually to avoid re-entrant angles.
4. Align parts to avoid eccentricity.
5. Locate welds where fatigue stresses are not likely to be severe whenever possible.
6. Give preference to designs with multiple load paths, so that fatigue cracks may follow any one of several members and are not likely to precipitate a collapse of the whole structure.

Weld Defects.   Fatigue stresses are initiated at weld defects such as porosity, slag inclusions, undercutting, and lack of penetration. The sensitivity to weld defects increases with the strength of material. A small amount of slag inclusion or porosity, well dispersed, may not reduce fatigue strength by more than acceptable amounts. Slag inclusions near the surface of the weld can reduce fatigue by 40% relative to a specimen with inclusions in the center of the weld.

Residual Stresses.   Welded structures usually contain residual stresses due to the differential contraction between the weld deposit and the base metal. This results in

TABLE 12.1.   *Residual Stresses Caused by Manufacturing Operations* (Courtesy of *Machine Design*)

Tensile Stresses	Compressive Stresses	Either Tensile or Compressive
Welding	Nitriding Shot peening Flame and induction hardening Heat and quenching Single-phase materials	Carburizing Rolling Casting Abrasive metal cutting (tensile stresses most common) Nonabrasive metal cutting Heat and quenching materials that undergo phase transformation (tensile stresses most common)

the creation of tensile residual stresses in the weld material and balancing compressive stresses in the base plate. These tensile residual stresses reduce fatigue strength, particularly where notches exist. Common manufacturing processes that lead to residual stresses are listed in Table 12.1.

POSTWELD TREATMENT

A number of postweld treatments have been developed to improve the fatigue strength of welds. These may be placed into three broad categories: (1) reduction in the stress concentration at the toe of the weld by changing the geometry, (2) modification of the residual stress system in the vicinity of the weld, or (3) protection of the weld toe from the environment.

Reduction in Stress Concentrations by Changing Geometry. A substantial reduction in the stress concentration at the weld toe can be obtained by grinding off the weld reinforcement. The amount of improvement will be dependent upon the initial reinforcement angle ϕ, the soundness of the weld deposit, and the type of joint. Improvement may range from none to 100%. A significant improvement in the fatigue strength of fillet welds can also be obtained by grinding the toes to obtain a smooth junction with the base plate.

Modification of Residual Stresses—Thermal Stress Relief. A weldment heated to a temperature at which the yield strength is low (usually 1200 °F, 649 °C, for steel) will relieve the residual stresses and increase fatigue life. The weldment is cooled slowly to prevent further residual stresses from forming.

A technique that is sometimes used for increasing the life of fillet welds involves slowly heating the end of the weld to a temperature just below the lower critical and then quenching the "notch" with a jet of water. The notch cools much faster than the surrounding metal without appreciable restraint. By the time the surrounding mass cools, the material at the notch is strong and resists the contraction of the material around it.

Peening. Compressive residual stresses at the toe of a weld may be produced by peening the surface with a pneumatic hammer. Peening on nonload carrying transverse fillet welds has shown fatigue life improvements of 75 to 90%, and longitudinal weld increases of 40 to 80%. The higher value for longitudinal fillet welds was obtained when the weld was continued around the corner.

Protection in the Vicinity of the Weld—Plastic Coatings. Plastic coatings applied at the toe of the weld have shown increases of 75% in fatigue life. Presumably, the coating reduces corrosion by the atmosphere. Only certain coatings result in an improvement, so as yet the mechanism is not well understood.

Application of Postweld Treatments. Only a few methods are widely used to increase the fatigue strength of welds. Grinding of butt welds is most frequently used because of its simplicity. Peening is widely used for increasing the fatigue life of rotating machine parts, but it has not been used much on welded structures, although applicable to all weld geometries. Thermal stress relief is only beneficial if the weldment is subject to alternating loading, and even then, only a moderate increase in fatigue strength is possible. Also, heat treatment of a complete welded structure is generally not possible. Large weldments subjected to localized stress relief can produce unfavorable stress distributions if incorrectly applied. Local heating and quenching can produce substantial increases in fatigue strength on discontinuous longitudinal fillet welds but are not applicable to continuous welds.

Although plastic coatings are quite easily applied, they have the disadvantage of requiring frequent inspections to maintain the coating integrity in service.

CODES, QUALIFICATION, AND CERTIFICATION

Introduction. An essential factor in the development and use of materials and processes is a concern for those people who will be exposed to the possible hazards that may be produced. Therefore, it is essential that the material used and the capabilities of the process be at a prescribed level. This implies that detailed laws, or codes, will be set up to govern various materials and processes. Also implied is the presence of a prescribed means of inspection.

Codes. A code may be described as a systematic collection of existing laws relating to a particular subject. It is considered to be a very forceful document containing positive statements as to what "shall" or "must" be done for given operations under certain conditions. By general agreement, for maximum effectiveness a set of minimum standards should also be established for any given area. It should, ideally, contain but one standard, and thus, the manufacturer need only reference the applicable existing standard. At the present time, however, there are several codes in the same area. Fortunately, the codes originated by such nationally known organizations as ASTM, ASCE, and AWS do not conflict.

In 1966 the American Standards Association (ASA) adopted a new name, American National Standards Institute (ANSI). This organization provides, through the working of its sectional committees, a means of approving virtually all standards of national importance. Standards approved by ANSI are known as USA standards.

Organizations providing standards do not have a means of enforcing codes they have approved. They do not become law until adopted by a municipality, county, or federal governmental agency. Adoption may be made by reference or citation of the particular code number and date, requiring that all operations in that area be performed accordingly. Another alternative is to copy the code verbatim and write it into the law.

Welding Codes. When welding was in its infancy, failures in welded equipment structures prompted engineering socities and manufacturers to develop proper codes. Some well-known codes and specifications in this area cover the following: pressure vessels and boilers; buildings, bridges, and other structural work; and pressure piping, cross-country pipe lines, ships, military equipment, and shipping containers.

AWS Structural Welding Code. Codes, as stated previously, are developed by national societies. As an example, AWS has developed a code for the design and construction of structures, which is identified as AWS D1.1–72. This code, adopted by many municipalities, provides among other things for: the design of welded connections, filler metal and flux, workmanship and techniques, qualification of welding operators and procedures, and the strengthening and repair of buildings.

Design considerations such as the computation of stresses and the proportioning of load carrying members is assumed to have been covered in the applicable general building code. Provision is made, however, for the qualification of processes, procedures, and joints not covered by prequalification.

The ASME Pressure Vessel Code. Almost every pressure vessel purchaser today wants a "coded" vessel. The reason: any vessel bearing the ASME cloverleaf emblem can be considered 100 % safe. To be entitled to stamp the ASME insignia on a vessel, the fabricator must first hire an inspection agency (possibly an insurance company) and must then have his shop inspected and approved by ASME. He must also have his welding procedure qualified; but this can be done by anybody having inspectors commissioned by the National Board of Boiler and Pressure Vessel Inspectors, such as independent testing laboratories, insurance companies, or state inspectors.

National Board of Boiler and Pressure Vessel inspectors must have a certain minimum of experience, either in operation of high pressure boilers as supervisors or by being actually involved in the design and fabrication of pressure vessels. They must also pass a rigid two-day examination on the code as prescribed by the National Board.

In recent years, independent testing laboratories have become increasingly important. Quite a large number of such companies belong to the American Council of Independent Laboratories (ACIL). Although the cost of using an independent laboratory is sometimes a consideration, it does have the advantage of providing trained experts to perform the mechanical tests, eliminating the possibility of bias.

The ASME Boiler and Pressure Vessel Code is published in several parts:

Section I.   Power Boilers.
Section II.   Material Specifications—Part A–Ferrous.
Section II.   Material Specifications—Part B–Nonferrous.
Section II.   Material Specifications—Part C–Welding Rods, Electrodes, and Filler Metals.
Section III.   Nuclear Power Plant Components.

Section IV. Heating Boilers.

Section V. Nondestructive Examination.

Section VI. Recommended Rules for Care and Operation of Heating Boilers.

Section VII. Recommended Rules for Care of Power Boilers.

Section VIII. Pressure Vessels—Division 1.

Section VIII. Pressure Vessels—Division 2–Alternative Rules.

Section IX. Welding Qualifications.

Section X. Fiberglass–Reinforced Plastic Vessels.

Section XI. Rules for Inservice Inspection of Nuclear Reactor Coolant Systems.

Many subcommittees are formed to work on such topics as steel plate, steel castings, strength of nonferrous castings, and so forth.

Copies of the ASME Boiler and Pressure Vessel Code, in its several sections, and the interpretations and cases issued by the Code committee are obtainable from ASME.

Standards and Specifications. In addition to codes, standards and specifications are available as part of a system needed for consumer and public protection. Standardization has an important influence on the terminology, procedures, and processes. Standards, once agreed upon, have the advantage of promoting uniformity throughout the industry.

AWS Standards. The AWS A3.0 standard has been largely responsible for providing the uniform nomenclature and symbols used in welding. The master chart of welding processes and terminology was provided through this standard. It also defines and illustrates what is meant by various positions of welds as well as basic types of joints. Other significant standards developed by the AWS are: Welding Symbols, A2.0; Nondestructive Testing Symbols, A2.2; and Methods for Mechanical Testing of Welds, A4.0.

ANSI Standards. Safety in welding and cutting, ANSI standard Z49.1, is organized under the rules of ANSI and sponsored by the AWS. It is the industries' safety standard and covers such phases as:

1. Installation and operation of gas welding and cutting equipment.
2. Application, installation, and operation of arc welding and cutting equipment.
3. Fire prevention and protection.
4. Health protection and ventilation.

Examples of other USA standards are: Safety Code for Eye Protection, ANSI87; Safety Code for Respiratory Protection, ANSI788; and Project for Industrial Head Protection, ANSI89.

Underwriters' Laboratories, Inc., Standards. The Underwriters, Laboratories, Inc. (UL), is a nonprofit organization sponsored by the American Insurance Association. It is organized to perform scientific investigations and tests so that it can determine the relation of various materials and construction methods of life, fire, and casualty hazards. The results of UL tests are published for the benefit of insurance organizations and other interested parties. Provision is also made for periodic checking of the tested items to assure continued acceptance. The UL label placed on the item under their supervision indicates an approved product. The list of UL standards is long. Copies of their standards may be obtained from Underwriters' Laboratories, Inc.

Specifications. Specifications such as the AWS, "Specifications for Welded Highway and Railway Bridges," or the American Institute of Steel Construction (AISC), "Specifications for the Design, Fabrication, and Erection of Structural Steel for Buildings," contain detailed information as to materials and processes. Specifications are revised and updated to take advantage of new technology as it becomes available. As an example, in 1972 the AISC specifications were rewritten to provide more leeway in the use of GMA, flux cored wire, electrogas, and electroslag welding. At the same time, higher strength levels were accepted for welding. The previous AWS specifications, A5.17 (SAW), A5.18 (GMA), and A5.20 (flux-cored), were used for weld metal deposit up to 70 ksi (482.6 MPa).

Twenty-one specifications have been set up by AWS to cover filler materials. Each of the twenty-one filler metal specifications is made in two sections. The first section contains the mandatory requirements, such as mechanical properties, usability, and chemical analyses, together with the system for identifying the classification covered. This system was explained in Chapter 8. The second section, which is not mandatory, provides a guide to the method of classification and explains the reasons for various features in the mandatory section.

ASTM Ferrous Metals Specifications. The ASTM ferrous metals specifications are prefixed with the letter "A." Since there are several hundred in this category, only a few will be presented to show how they are designated.

A–7–61   Structural Steel for Bridges and Buildings.
A–36–63   Structural Steel.
A–53–64   Welded and Seamless Steel Pipe.
A–105–64   Forged or Rolled Steel Pipe, Flanges, Forged Fittings and Valves, Valves and Parts for High Temperature Service.

ASTM Nonferrous Metal Specifications. The ASTM nonferrous metal specifications carry the prefix letter "B." There are also a few hundred in this category. A few examples follow:

B–43–62   Seamless Red Brass Pipe—Standard Sizes.
B–75–62   Seamless Copper Tube.
B–96–61   Copper Silicon Alloy Plate and Sheet for Pressure Vessels.

Society of Automotive Engineers Specifications. Perhaps the best known of the SAE specifications are those for the analysis of steels. These specifications were originated by the SAE Iron and Steel Division. The same numbering system is used by the American Iron and Steel Institute, AISI. This numbering system, used throughout industry to identify alloy steels, was discussed in Chapter 9.

Aerospace Specifications. The Aerospace Materials Division of SAE publishes specifications that are popularly referred to as "AMS Specifications." Commodity Committees prepare specifications in the following areas: nonmetallics, refractory materials, electronic materials and processes, finishes, processes and fluids, corrosion and heat resistant alloys, carbon and low alloy steels, and nonferrous alloys. Currently there are 10,000 "AMS Specifications" listed with appropriate indexes.

Federal Specifications. Federal specifications are issued under the supervision of the US Federal Supply Service, General Services Administration. The titles given the specifications are so arranged that the most significant word appears first followed by modifiers and descriptive words, e.g., "Rods, Welding, Copper and Nickel Alloys." The initial letter appears in the identification symbol of the specification. Thus in the specification QQ–R–571, "R" represents rods, "QQ" covers metals, and "571" is the serial number.

Military Specifications. Military specifications are issued under the Department of Defense. The Defense Supply Agency issues a Defense Standardization Manual (M 200), which provides an explanation of policies and procedures.

Military specifications are identified by means of a title and a letter-number designation, followed by the date of issuance. The following may be taken as a typical example: MIL-T-18068(NAVY), titled Torches, Welding; and Torches, Cutting-Hand, Commercial, Oxyacetylene and Oxy-hydrogen Gases. The first three letters identify this as a military specification. The letter "T" is taken from "torches," the significant word in the title. Following the second dash is the serial number of the specification. The enclosure, "(NAVY)", means it is limited to that branch of the service. If no branch of service is designated, it means it has been coordinated and used in all branches. The example given is as it appears in a first issue. If it is revised, the serial number will be followed by a capital letter. Immediately following, and usually printed beneath the designation, is the date of issuance. Amendments, if involved, are numbered serially and printed below the specification identification.

Other Specifications. In addition to the bodies mentioned that provide extensive specifications for materials and procedures, there are others of a more specialized nature. Many large corporations, for example, prepare their own specifications where applicable and make additions when necessary.

Most public utility companies also operate under established specifications, such as the AWS Filler Metal Specifications and the ASME Boiler and Pressure Vessel Code. Power plant piping comes under the ANSI standard Code for Pressure Piping, B31.1, Power Piping.

Certification for Welding Operators. Due to the large number and variety of codes, it is sometimes confusing to determine what the necessary qualifications are for a welding operator. However, one section of the code is usually involved in presenting qualifications. As an example, Section IX of the ASME Boiler Code serves as the basis for all welding qualification specifications involved with pressure vessels and pressure piping.

Welding operators cannot be certified on their own. Only manufacturers, contractors, or other authorized testing agencies can certify that an operator is able to do a specific job, based on his ability to pass certain tests. Many codes do not allow operators to be certified until after the procedure has been qualified. Other codes allow established processes combined with specified joints to constitute a qualification procedure. No deviation is allowed from the applicable code or specifications for the given work. An operator who is certified under a code for one employer is not certified, even under the same code, with another employer. There is however one exception, operators certified by the National Certified Pipe Welding Bureau may transfer from one member company to another without recertification.

Qualification for Specific Work. Most welding codes establish tests that will qualify the operator to do specific work. This work may be limited by position, thickness, or base metal analysis. In some codes, changing the welding technique, use or absence of a backing strip, and changing the welding process would require requalification.

The AWS standardized welding position terminology and welding joint types are uniformly accepted by all specifications; however, the joint design details and the specimen details have not been so well standardized. A typical fillet weld used in qualifying operators is shown in Fig. 12–26. Although this fillet weld is very similar in specifications for both AWS and ASME, it is not tested in the same way.

Qualification for Structural Welding. The most widely used test is the groove weld made in either plate or pipe. If plates are used, tests may be given for welding in each of the positions. If pipe is used, the axis will be for both vertical and horizontal. The positions used for both plate and pipe tests are shown in Fig. 12–27. Welding may be uphill or downhand as the test requires. The thickness of the material also becomes a factor. Usually, specimens can be made in 3/8 in. (9.53 mm) material, which will qualify the operator to weld up to 3/4 in. thick. Material over 3/4 in. (19.05 mm) must be used to qualify the operator for welding material of unlimited thickness. In the thicker specimens, the side bend test is used instead of the root or face bend test. The specimen is machined to a 3/8 in. (9.53 mm) thickness.

Structural tests often include the double fillet weld specimen with a 15/16 in. (23.81 mm) vee opening as shown in Fig. 12–28. Specimens from this weld as well as most groove welds are tested by use of the guided bend test, discussed previously in this chapter.

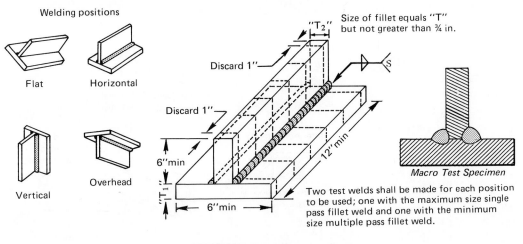

Welding positions

Flat Horizontal

Vertical Overhead

Discard 1"

Discard 1"

6" min

6" min

"T₁"

"T₂"

Size of fillet equals "T" but not greater than ¾ in.

12" min

S

Macro Test Specimen

Two test welds shall be made for each position to be used; one with the maximum size single pass fillet weld and one with the minimum size multiple pass fillet weld.

Weld Size (In.)	T_1 (In.)	T_2 (In./Min.)
3/16	2/1	3/16
1/4	3/4	1/4
5/16	1-1/2	5/16
3/8	2-1/4	3/8
1/2	3	1/2
5/8	3	5/8
3/4	3	3/4
3/4	3	1

FIGURE 12–26. *The fillet weld soundness test for procedure qualification. The macroetch test is used to examine the weld for defects. It must show fusion to the root but not necessarily beyond the root, and both legs must be equal within 1/8 in. (3.17 mm). (Courtesy of American Welding Society.)*

Local Qualifications. Individual states operating through their Industrial Commission determine the necessary qualifying tests to enforce the state building code. The tests must be performed under the supervision of an approved testing laboratory or commerical testing engineer. Operators who pass the required qualification tests are issued a certificate bearing the operator's name, address, signature, and the extent of his successful qualification testing. The certificate remains in force one year, provided the operator is engaged in welding without interruption of more than three consecutive months' duration, in which latter case the ceritficate is automatically void. The certificate is granted only after the qualifications are again passed.

After a weldor becomes certified, he is given an identification number. When welding in critical areas, as on pressure vessels, his number is stamped on the weldment at certain specified places. In case of failure, this information becomes very useful. The results of qualification tests made for each individual become a

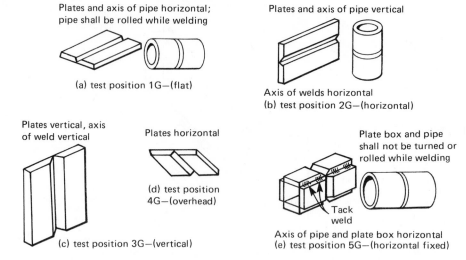

Plates and axis of pipe horizontal; pipe shall be rolled while welding

(a) test position 1G—(flat)

Plates and axis of pipe vertical

Axis of welds horizontal
(b) test position 2G—(horizontal)

Plates vertical, axis of weld vertical

Plates horizontal

(d) test position 4G—(overhead)

(c) test position 3G—(vertical)

Plate box and pipe shall not be turned or rolled while welding

Tack weld

Axis of pipe and plate box horizontal
(e) test position 5G—(horizontal fixed)

FIGURE 12-27. *The groove weld qualification test is made in both plate and pipe and may be given in each of the various positions shown.*

matter of public record and can be used as reference by inspectors and others for specific jobs.

Once a specific welding *procedure* has been qualified, it accompanies the job to the welding floor and is available at any time to inspectors, weldors, foremen, and so forth. By procedure is meant all the necessary steps involved in making the weld such as: base metal specifications; filler metals; joint preparation; welding position; welding process, techniques and characteristics (current setting, electrode manipulation); preheat, interpass, and postheat temperature.

Although the term *certified weldor* is often used, it is somewhat of a misnomer. As stated previously, it is not the weldor who is certified; but it is, rather, the

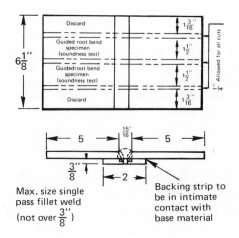

FIGURE 12-28. *The double fillet weld with a 15/16 inch (23.812 mm) vee opening is used as a qualifying soundness test. The specimens are cut as shown in the plan view and tested by means of the guided-bend test.*

manufacturers and contractors who certify that a given operator is capable of performing a specific job according to prescribed procedures. As stated in Section 9 of the ASME Boiler and Pressure Vessel Code, welded products manufacturers and contractors shall establish a welding method (qualified procedure) and prove that a sound weld can be produced by using it.

Paragraph P112 of the same code stipulates that if a weldor leaves the employ of one manufacturer for another, he must qualify under the latter's procedures before he is permitted to work. However, interchange of weldors is permitted among manufacturers and contractors who use the same qualifying procedures, provided they are acceptable to the purchaser. Thus, the term certified weldor refers to a very limited set of conditions.

Code Responsibility. Each manufacturer or contractor is responsible for the welding done in his organization. Each must conduct the tests required to qualify the welding procedure he uses in fabricating weldments under the code and the performance of his welding operators. Since the contractor or the manufacturer is responsible for all welding incorporated into his product, he must replace at his own expense any work that does not meet specifications if it can be shown that there was any attempt to defraud, or for gross nonconformance to the code. It is of course impossible for the manufacturer or contractor to personally observe each weld, therefore, he must rely upon the establishment of sound procedures and practices.

Contractors must also maintain complete records of procedures and operator qualification. They must also, under certain codes, give requalification tests periodically. Up until January 1971, manufacturers, contractors, and assemblers were certified on a corporate basis for a three year period. Now, however, the rules have been changed so that individual plants (of manufacturers) and individual control offices (of engineering contractors and assemblers) are certified for the specific discipline for which they were qualified.

The Occupational Safety and Health Act (OSHA). In April of 1971, the Occupational Safety and Health Act became a federal law. Under the provisions of this law, the federal government is empowered to take a closer look into the safety practices of virtually every business in the United States. Two agencies of the federal government are involved: (1) the National Institute of Occupational Safety and Health (NIOSH) of the US Department of Health, Education and Welfare (HEW) and (2) the Occupational Safety and Health Administration (OSHA) of the US Department of Labor. NIOSH develops the criteria that are to be used by OSHA in developing standards. Thus, NIOSH serves in an advisory capacity through HEW to the Labor Department's OSHA. Documents published by NIOSH are therefore highly important since they indicate the trend of future standards.

Under the provisions of this law, the Secretary of Labor's office is empowered to promulgate and enforce mandatory safety and health standards, establish occupational health research, and assist in safety education and training programs. The law may be divided into three parts, investigation, enforcement, and penalization.

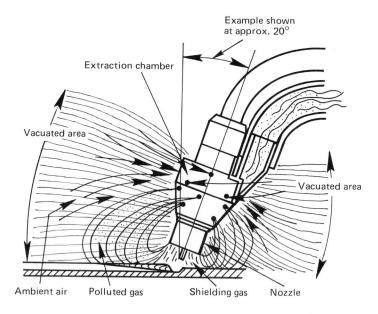

FIGURE 12–29. *A welding tip designed to remove airborne pollutants at the point where they are formed uses an extraction chamber around the nozzle. Suction is provided by a 60 ft³/min (54.86 m³/min) centrifugal fan driven by a 1.5 hp motor mounted remote from the gun. (Courtesy of Engineering Materials and Design.)*

Investigation. Investigation by OSHA may take place as an unannounced visit, or it may come at the request of one of the employees. Any employee who believes a violation of health or safety exists that may cause him serious physical harm can request an investigation by a federal representative of OSHA. As an example, the possible injury to welding operators by air pollutants will be watched more closely. Manufacturers have anticipated this and some redesigned torches have appeared on the market, as shown in Fig. 12–29.

The federal representative can take one of two courses of action: (1) He may proceed with the investigation accompanied by an authorized representative of the employee. (2) Or, he may feel the complaint does not merit investigation and will give a written explanation of his reasons to the employee. If an investigation is made, it may involve private audiences with the person or persons involved. The private examination of witnesses is intended to maintain the employee's anonymity from his employer, and thus, to guarantee his protection under the act.

Enforcement. The Secretary of Labor or any one of his representatives may issue a citation if the results of his investigation show noncompliance with the standards established under OSHA, or if there are any proven "recognized hazards" even though no existing standards cover them. Citations will contain a written explanation of the violation or violations and set a realistic deadline for their elimination. The employer will have fifteen days to respond to the Secretary of Labor's Office. Failure to do so eliminates any future review by the court.

Variances. Should the employer be unable to correct the violation by the date set because of extenuating circumstances, an extension may be granted. This is called a *variance*.

Penalization. In order to make the law effective, a system of penalties was set up as follows:

1. Serious violation, fine of up to $1,000.00.
2. Failure to correct violations, fine of up to $1,000.00 per day until corrected.
3. Willful or repeated violations, fine of up to $10,000.00 for each violation.
4. Willful violation resulting in death, fine up to $10,000.00 and/or imprisonment up to 6 months. Second conviction $20,000.00 and/or one year imprisonment.
5. Failure to display requirements, fine up to $1,000.00.
6. Falsification in any application or report, fine up to $10,000.00 and/or up to 6 months imprisonment.

In addition to the stated penalties, the US District Court has the right to shut down any operation or place of business found to be under imminent danger until that danger has been eliminated.

Additional Legal Powers. Under the act several clerical duties must be performed:

1. All work related deaths as well as injuries and illnesses requiring medical care must be reported.
2. The employer must monitor employee exposure to health hazards.
3. The employer must provide for physical examinations at regular intervals if environmental conditions warrant it.
4. The employer must disclose to all employees all hazards encountered in his work and provide educational programs and safety equipment to help guard against them.

EXERCISES

Problems and Questions

12–1. What principal types of destructive tests are classified as to the load application?

12–2. As an engineer, you would like to substitute steel X in a particular application for steel A. Steel A, which has been used successfully, has a transition temperature of $-20\,°F$ ($-29\,°C$) at the 15 ft/lb (20.3 J) energy level. Could steel X be substituted if it had a similar transition temperature? Explain.

12–3. Make a sketch of the following types of welds:

(a) Longitudinal full penetration butt weld.
(b) A load carrying longitudinal fillet weld.
(c) A nonload carrying transverse fillet weld.

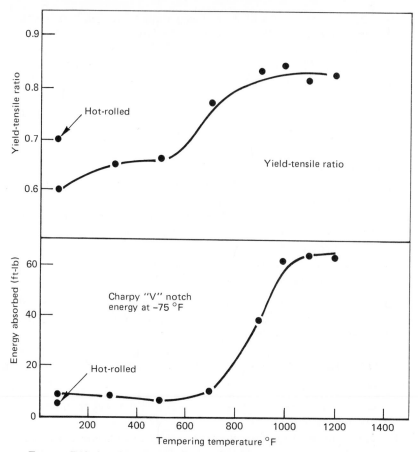

FIGURE P12–4. *Comparison of the effect of quenching and tempering on yield-tensile ratio and toughness. (Courtesy of U.S. Steel.)*

(d) A tee type load carrying transverse fillet weld.

(e) A cover plate type load carrying transverse fillet weld.

12–4. Shown in Fig. P12–4 is a yield-tensile ratio curve in comparison with an *S–N* curve for a high strength steel. Answer the following:

(a) What is the transition temperature at 15 ft/lb (20.3 J) of energy?

(b) What is the effect on toughness and yield strength of increasing the tempering temperature to 1000 F (538 C)?

(c) Does it appear that a high yield tensile ratio also denotes brittleness?

(d) What is the transition temperature for 40 ft/lb (54.2 J) of energy?

(e) Would the 40 ft/lb (54.2 J) energy range be con-

sidered a safe area of operation? Why, or why not?

12–5. (a) What are three different types of impact tests?

(b) What are they designed to measure?

12–6. Is the depth of impression or the diameter of impression measured in the Brinell hardness test?

12–7. Would you expect the Brinell hardness numbers for loads of 500, 1500, and 3000 kg to vary or remain constant for any given specimen? Explain.

12–8. (a) If the Brinell number obtained from measuring the hardness of a mild steel weld was

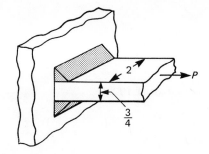

FIGURE P12–13.

108 to 150, what would be the approximate tensile strength?

12–9. Rate the following three welding designs for cover plates on beams in order of their ability to resist fatigue failure.

(a) Partial length cover plate, square ends with continuous weld all around.
(b) Partial length cover plates, square ends with continuous weld along the edges only.
(c) Partial length cover plate, tapered ends with continuous welds along the edges only.

12–10. Make a sketch showing a partial penetration weld that is not adversely affected by fatigue.

12–11. (a) Obtain a copy of the AWS Structural Welding Code, D1.1–72. Using the information given in 4.12, could a butt weld made by the submerged arc process on ASTM A36 steel be specified for tensile strengths up to 100 ksi (689.5 MPa)?

12–12. A two inch (5.08 cm) thick steel plate is to be butt welded to a one inch (2.54 cm) thick steel plate. Make a scale drawing to show two ways this can be done to meet the AWS D1.1–72 Structural Welding Code.

12–13. (a) A double fillet weld is made as shown in Fig. P12–13. The load placed at P is 40 ksi (18,000 kg). The welds are made equal to t of the horizontal member. An E6010 electrode is used. What is the shear stress encountered at the welds?

(b) Will this come within the allowable load? (See Table 11.7.) If any adjustment is needed, what do you recommend?

12–14. Two sheets of 2014–T6 aluminum, 1/8 in. (3.175 mm) thick and 4 in. (10.16 cm) wide, are held together by five spot welds. The shear strength for this material is 56,000 psi (386.1 MPa). According to MIL–W–6858B specifications, the minimum shear load for each weld should be 2120 lb (960.4 kg). What diameter must each weld have to be able to resist this load?

12–15. Shown in Fig. P12–15 is a lap joint with a cover plate. The joint will be subject to a shear load of 50 k (22,500 kg). An E7010 electrode is used and the base material is mild steel. Will the joint as made come within the allowable limits of the AISC and AWS codes for structures?

12–16. The 1972 AISC Building Code and Bridge Specifications allows limited credit for penetration beyond the root of a fillet weld made by the submerged arc process as shown in Fig. P12–16. For fillet welds 3/8 in. (9.53 mm) and smaller, the effective throat (t_e) is now equal to the leg size of the weld (h). For fillet welds larger than 3/8 in. (9.53 mm), the effective throat is obtained by adding 0.11 to .707h, or:

When $h > 3/8$ in., then $t_e = 0.707h + 0.11$

This change represents a 41 % increase in effective weld throat of fillets up to and including 3/8 in. (9.53 mm). Combined with the increase in allowable

FIGURE P12–15.

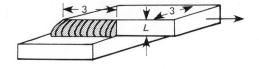

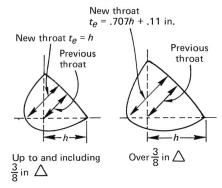

New throat
$t_e = .707h + .11$ in.

New throat $t_e = h$

Previous throat

Previous throat

$\leftarrow h \rightarrow$

$\leftarrow h \rightarrow$

Up to and including $\frac{3}{8}$ in $\triangle$

Over $\frac{3}{8}$ in $\triangle$

FIGURE P12–16.

strength, welds of this type can almost be cut in half and still have the same allowable unit force per inch.

(a) Calculate the difference in allowable unit force for a $\frac{1}{2}$ in. (12.70 mm) E6010 weld, 10 in. (25.4 cm) long, made with a stick electrode under the old AISC specifications and the one made in 1972. The allowable unit load before 1972 was 15.8 ksi (108.86 MPa). (See Table 10.3 for present allowable strengths.)

(b) What would the allowable unit load be for a $\frac{1}{4}$ in. (6.35 mm) fillet weld made under the 1972 code by the submerged arc process?

12–17. Who is it that is able to enforce a specific code such as the one used for bridges? Explain what makes it enforceable.

12–18. Why does a saying state, "there is no such thing as a certified weldor," when there are many testing laboratories that are able to perform certified testing services?

12–19. Is it true that the certification laboratory is responsible for a weldor's performance on the job? Explain.

12–20. Why are there differing bodies making codes in the same area, such as AWS and ASCE?

12–21. How can you tell at a glance whether you are looking at ASTM metal specifications for ferrous or nonferrous metals?

12–22. If you are responsible for the installation and operation of flame cutting equipment, what standard should be used to insure everything is done according to code?

12–23. Who sponsors the Underwriters' Laboratories and why?

12–24. What were some significant changes in specifications made in 1970 by AISC and AWS as regards welding?

12–25. What does "QQ" stand for in the federal specifications?

12–26. If a certified weldor quits his job and goes to work for another employer, is he still certified? Explain.

12–27. Who develops the criteria for the standards used by OHSA?

12–28. If an employer gets a citation from OHSA, what must he do?

12–29. (a) Who can request an investigation by OSHA?

(b) Does OSHA investigate all complaints? Explain.

12–30. What government body in the US provides a means of checking all standards used in industry?

BIBLIOGRAPHY

Books

ASME Boiler and Pressure Vessel Code, secs. VIII and IX, American Society of Mechanical Engineers, New York, N.Y., 1968 and 1971.

AWS Structural Welding Code, D1.1–72, American Welding Society, Miami, Fla., 1972.

DAVIS, H. E., TROXWELL, G. E., AND WISKOCIL, C. T., *The Testing and Inspection of Engineering Materials*, McGraw-Hill Book Company, New York, N.Y., 1964.

JUVINALL, R. C., *Stress, Strain and Strength.* McGraw-Hill Book Company, New York, N.Y., 1967.

Manual of Steel Construction, 7th ed., American Institute of Steel Construction, New York, N.Y., 1970.

Selected ASTM Standards for Mechanical Engineering Students, American Society for Testing and Materials, Philadelphia, Pa., 1965.

Welding Handbook, 6th ed., sec. I, American Welding Society, New York, N.Y., 1968.

Periodicals

AHLBECK, D. R., "A Dictionary of Statistical Jargon." *Machine Design*, March 1966.

BRIGGS, C. W., "The Effects of Surface Discontinuities on the Fatigue Properties of Cast Steel Sections." *Research Report*, Steel Founders' Society of America, August 1966.

CARY, H. B., "What Do You Really Know About Welding Specs and Codes?" *Welding Design & Fabrication*, March 1961.

———, "What Should You Know About Qualifying Operators?" *Welding Design & Fabrication*, June 1961.

DIVERS, K. C., "Statistics Help Solve Problems in Heat Treating and Forming." *Metal Progress*, December 1965.

GURNEY, T. R., "Fatigue of Welded Military Structures." *British Welding Journal*, vol. 15, no. 6, 1968, pp. 276–282.

INGLE, S. R., "Regression Analysis Made Easy." *Modern Machine Shop*, August 1971.

KLING, R. E., "Understanding Fatigue in Metals." *Machine Design*, October 1965.

KOZIARSKI, J., "Fatigue Aspects in Aircraft Welding Design." *The Welding Journal*, May 1955.

LIPSON, C., "Basic Course in Failure Analysis." *Machine Design*, October 16, 1969 through January 28, 1970.

MINDLIN, H., "Influence of Details on Fatigue Behavior of Structures." *Journal of the Structural Division*, ASCE, vol. 94, no. ST12, December 1968.

PETERS, R. L., "How To Estimate Mechanical Safety Factors." *Tooling and Production*, September 1969.

POLLARD, B., AND COVER, R. J., "Fatigue of Steel Weldments." *Welding Journal Research Supplement*, vol. 51, no. 11, November 1972, pp. 544-s–554-s.

SHANNON, J. L., JR., "Fracture Mechanics, Reducing Theory to Practice." *Machine Design*, October 12, 1967.

"So You Want To Be A Certified Weldor." Editor, *Welding Engineer*, March 1972.

THOMPSON, N. C., "Decision Modeling: The Art of Scientific Guessing." *Machine Design*, November 12, 1970.

WANG, K. K., AND DEVRIES, M. F., "Investigation of Manufacturing Processes by Statistical Experimental Design Techniques." *SME, Report No. 263*, Dearborn, Michigan, 1969.

"You and The Codes." *Welding Engineer Annual Fact File*, 1965.

13

Materials Joining Economics

Today's materials joining engineer realizes the necessity to adapt variations in fabricating techniques whenever possible. He must utilize any opportunity to reduce manufacturing costs or produce a better product at competitive costs. Newer technologies may be readily available, but are too often thwarted by a reluctance of management to change a process that is now performing well. As an example, one manufacturer who produces a large volume of cabinets, chassis, switch boxes, and other enclosures found, after some research, that a savings of 30 to 40% could be effected if the assembly process was changed from spot welding to adhesives. He, like many others, refused the change even though tests revealed that the change would also provide a stronger product.

A study of this text has shown many of the factors to be considered in selecting a joining process. Each one has advantages and disadvantages that must be weighed. The proper selection can only be resolved by those who are familiar with the design requirements and the process capabilities.

The economics of joining begins with the design and follows through to the final inspection of the finished product. Joining costs may be controlled through each of the following steps:

1. Design.
2. Process selection.
3. Cost factors.
4. The use of fixtures and positioners.
5. Preparation, finishing, and inspection costs.

DESIGN

The initial design of a unit to be fabricated will play an important role in the types of joints that will be used and their cost. Some parts can be fabricated more economically by forming and the integrated use of forgings and castings than the use of any one process alone. Shown in Fig. 13-1a is a stanchion socket that was originally designed as a casting that required machining. A web plate was fitted to the upright and fastened with three rivets. Fifteen operations were required for the completed assembly. Shown at Fig. 13-1b is the redesign, as a cored forging that requires no machining and no assembly. Numerous examples of this nature can be cited, whereby a designer has taken a second look at an assembly and asked, "How can this part be simplified and yet perform the same function?" In the foregoing example, other methods of producing the same part are readily apparent—such as brazing a disc on the tube or the use of a flanged disc fastened to the tube with an adhesive.

Redesign. A unit to be redesigned and fabricated by another process is usually considered for redesign on the basis of cost, quality, and function. As a simple example, the lever shown in Fig. 13-2 is reproduced as a weldment. It may not be economical or necessary to round the corners and taper the plate. A comparison of the cost of producing square and rounded end plates is shown in Fig. 13-3.

Standard rolled shapes available from suppliers and bent shapes made in the shop can be incorporated into a pleasing functional design as shown in Fig. 13-4.

Joint Design. The economics of joint design grows in significance as the material thickness increases, which is particularly applicable in welding. However, the width and profile of a weld joint can be designed to control the quantity of metal deposited and hence, to reduce the cost. Obviously, the narrower the groove is, the smaller the quantity of weld metal required. There is, however, a minimum width of

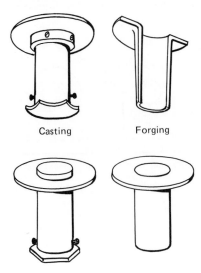

Casting   Forging

FIGURE 13-1. *The redesign from a casting to a cored forging can eliminate fifteen steps in manufacturing.*

471

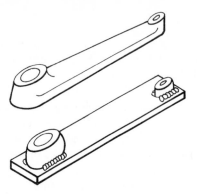

FIGURE 13–2. *It is not economical to round all the corners in changing a design from a cast or forged process to that of welding, brazing, or adhesives.*

seam that can be effectively utilized and still obtain adequate quality and joint efficiency. You may recall that electron beam welding has a particularly high depth to width capability. However, most other processes are based on the electrode diameter. Large diameter electrodes have high deposition rates, however, wider seams are required. Depending on comparative costs (electrodes or filler wire) and labor costs, it may be more economical to deposit a larger quantity of weld metal per minute with a wider seam than that afforded by a small electrode with a narrow seam. There usually is, of course, an optimum condition as to width of seam and diameter of electrode used.

The profile and joint spacing will determine the quantity of weld deposit. Some basic profiles for butt joints are: (1) square butt, closed, open, and open with a backing bar; (2) single vee; (3) double-vee; (4) single "U"; and (5) double "U" (as shown in Fig. 10–22). The square butt joint is most economical since it requires the minimum of weld deposit. However, it is limited to relatively thin plates unless an RP electrode is used, in which case it may go as high as 1/8 in. (3.17 mm).

After the thin, square butt joint, the single vee joint is the most economical for thicknesses up to 5/8 in. (15.87 mm), and the double vee, for thicknesses up to

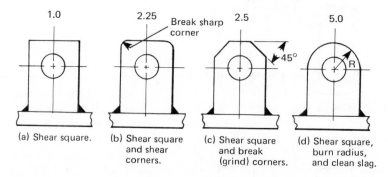

FIGURE 13–3. *A comparison of costs in finishing the corners of a steel plate by methods shown. (a) Shear square. (b) Shear square and shear corners. (c) Shear square and break (grind) corners. (d) Shear square, burn radius, and clean slag.*

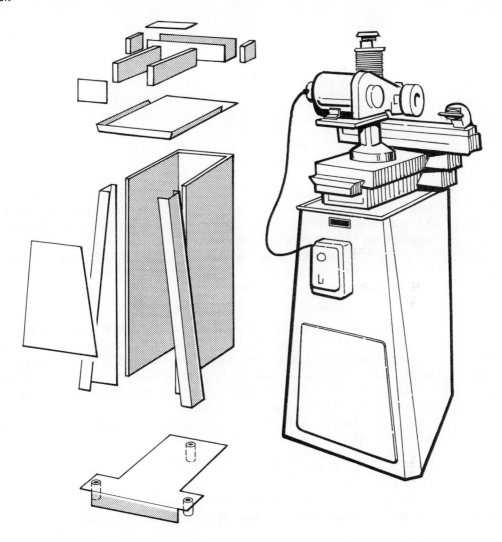

FIGURE 13–4. *The use of standard shapes and formed members provides an effective method to cut fabricating costs and to achieve a pleasing functional design. (Courtesy of Lincoln Arc Welding Foundation.)*

$1\frac{1}{4}$ in. (31.75 mm). The double vee machining cost is higher, but it requires only half the amount of weld deposit. For greater plate thicknesses, the width of the top of the seam becomes too wide for practical purposes. Thus, for heavier thicknesses, the single "U" or the double "U" become most economical. Although the "U" groove requires less metal, the preparation costs are generally higher.

Overwelding and Stub Loss. The practice of overwelding is often the result of two mistaken concepts, that strength is added to the joint and weld deposit is cheap.

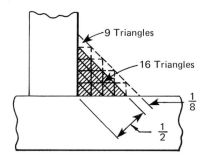

FIGURE 13-5. *The overweld of 1/8 in. (3.175 mm) is represented by nine triangles— 9 : 16 = 51 % more weld deposit than required.*

Neither of these is true. Shown in Fig. 13-5 is a fillet weld with a $\frac{1}{2}$ in. (12.70 mm) throat. The weld cross section is divided into smaller triangles each with a $\frac{1}{8}$ in. (3.17 mm) throat. When this fillet is overwelded $\frac{1}{8}$ in. (3.17 mm), the triangles will increase from 16 to 25 or a 51 % overweld.

Electrode stub losses can account for a 30 to 40 % increase in electrode costs. Shown in Table 13.1 is the cost of electrodes per 100 ft (30 m) of welding, with various stub losses and overweld.

Joint Fitup. The largest weld deposit losses are made from poor joint fitups. Shown in Fig. 13-6 are three fillet welds. The first is shown with a tight fitup; the second, with a 1/16 in. (1.58 mm) gap, and the third, with a 1/8 in. (3.17 mm) gap. In order to obtain the same size fillet, approximately 20 % more weld deposit is required with the 1/16 in. (1.58 mm) gap, and approximately 45 % more weld deposit is required for the 1/8 in. (3.17 mm) gap.

The cost of poor fitup plus overwelding and stub losses can become excessive, particularly, when using alloy electrodes. Based on a normal cost of $18.00 per 100 ft of weld in carbon steel, the cost of a 1/16 in. (1.58 mm) gap and a 2 in. (5.08 cm) stub loss would be $29.55; for a 1/8 in. (3.17 mm) gap, it would be $34.20. For stainless steel, and the same conditions, the costs would soar from $108.90 per 100 ft to $126.20. If larger stub losses are taken into consideration, the cost picture could change from a basic $108.90 to $183.50 for stainless steel, as shown in Table 13.2.

TABLE 13.1. *The Effect of Stub Loss and Overweld on Electrode Costs*

	Electrode Costs (per 100 ft (30 m) of Weld)			
	Carbon Steel		Stainless Steel	
Stub Length in. (cm)	1/32 in. (0.79 mm) Overweld	1/16 in. (1.58 mm) Overweld	1/32 in. (0.79 mm) Overweld	1/16 in. (1.58 mm) Overweld
2 (5.08)	$21.60	$25.95	$79.60	$95.60
3 (7.62)	23.55	28.20	87.00	104.50
4 (10.16)	25.50	30.60	94.30	113.40
5 (12.7)	27.60	33.15	101.70	122.00
6 (15.24)	29.55	35.40	109.00	131.00
7 (17.78)	31.50	37.80	116.30	139.50

FIGURE 13–6. *Poor fitup requires increased weld deposits.*

TABLE 13.2. *The Effect of Stub Loss Plus Poor Fitup on Electrode Cost/100 ft of Weld*

| | Electrode Costs (per 100 ft of Weld) | | | |
| | Carbon Steel | | Stainless Steel | |
Stub Length in. (cm)	1/16″ Gap	1/8″ Gap	1/16″ Gap	1/8″ Gap
2 (5.08)	$29.55	$34.20	$108.90	$126.20
3 (7.62)	32.25	37.50	119.00	138.00
4 (10.16)	39.95	40.50	129.00	149.00
5 (12.7)	37.65	43.50	139.00	160.50
6 (15.24)	40.35	46.65	149.00	172.00
7 (17.78)	43.05	49.68	159.00	183.50

Gaps of 1/16 in. (1.58 mm) and 1/8 in. (3.17 mm) are not uncommon in most fabricating shops. Even gaps of 1/4 in. (6.35 mm) are not too unusual, particularly in field welding of large diameter, heavy walled vessels. Usually, the time spent in obtaining the proper type of joint preparation and fitup will be readily offset by reduced weld deposit and the accompanying reduction in labor, power, and overhead.

PROCESS SELECTION

In Chapter 1, various joining processes were compared. In this chapter, process selection will be confined to that of welding.

Manual Vs. Automatic Welding. Before making a comparison of the various modes of welding, it may be well to define the terms. *Manual welding* refers to hand manipulation of the electrode as in shielded metal arc and GTA welding. *Semiautomatic welding* is defined as those processes where the filler wire is automatically fed through the welding torch, which is hand manipulated by the weldor. *Fully automatic welding* provides machine control over both electrode feed and travel. With these definitions in mind, attention can be focused on when to select each one. A simple answer may be to select semiautomatic or fully automatic welding whenever an overall cost reduction might be achieved; however, to be reasonably sure of the answer, a systematic approach is required.

The cost of manual welding is quite well established and can be based on the following: labor and overhead, $10/hr; cost/lb of deposited electrode, $6.67; and 8 hr work day and 5/32 in. electrodes. The cost/lb of deposited metal by the semiautomatic method, such as in-hand manipulation of GMA processes or flux-cored wires, drops to about $1.90 for labor and overhead. If fully automatic welding can be used, the cost is further reduced to about $1.00/lb.

The basis of welding process selection should also be made on its application. Shown in Fig. 13–7 is a comparison of costs for welding a seam in a $\frac{1}{4}$ in. thick plate used to fabricate a cylinder. The semiautomatic GTA process proved to be the most expensive method. In this case, the capital and weld material cost with the semiautomatic inert arc, using a small solid wire and argon, is 50% more costly than the manual arc. However, several other semiautomatic and automatic welding processes might be used for this application that will produce substantial savings over manual arc welding, even though they have a higher initial capital cost.

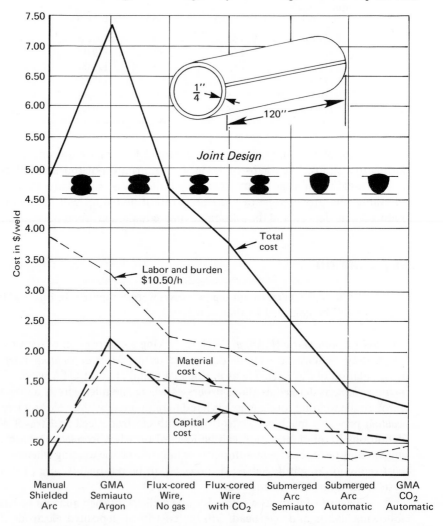

FIGURE 13–7. *A cost comparison of manual, semi-automatic, and automatic welding processes.*

The higher initial capital cost of automatic equipment can only result in greater savings if the operating time is favorable. Usually this is a problem of production scheduling and the availability of versatile tooling to keep the set-up time down and to minimize downtime during loading and unloading.

Arc Welding Process Comparisons. A brief comparison of the principal arc welding processes will serve to point up selection factors.

The GTA process is used almost entirely for more specialized applications such as welding light gauge materials, particularly high alloy and nonferrous types. GTA is frequently used where it is the only process that provides satisfactory results due to material thickness and/or type of material. Since this process is somewhat limited in application, it will not be considered in an economic comparison.

Shielded metal arc, GMA, submerged arc, and flux-cored welding processes are compared as to operating costs for a particular application in Table 13.3. As shown, the stick electrode cannot compete with the automatic processes. It would, of course, have a much more favorable place with semiautomatic versions, but it would still require a lower operating factor.

Some factors that are not so readily apparent in changing from manual to semiautomatic or fully automatic welding are: impact on shop personnel, tuition fees for learning the new process, changes in joint preparation, and welding procedure. A useful check sheet in helping to decide what the economic difference is between two methods includes a comparison of preparation required, lbs of weld metal per foot, labor cost per foot, and return on investment as shown by the example in Fig. 13.8.

Joint Design and Process Selection. A joint design must be suited to the process selected. Shown in Fig. 13–9, weld joints are classified as to suitability for welding by manual, semiautomatic, and automatic methods. Generally, the root opening and

TABLE 13.3. *Cost Comparisons of Shielded Arc, GMA, and Submerged Arc*

Cost Data for $\frac{1}{4}$ in.² Steel Butt Joint	Shielded Electrode	GMA CO_2	Submerged Arc
Operating factor	40%	50%	35%
Welding speed, ipm	8	15	24
Welding current	300	420	800
Arc time, h/100 ft of weld	2.5	1.33	.84
Electrode, lbs/100 ft	30	15	17
Welding flux, lb/100 ft	—	.92	2.10
Labor cost/100 ft at 3.50/h	18.75	7.60	8.40
Electrode cost/100 ft	4.50	6.40	2.21
Power cost/100 ft	.50	.50	.50
Direct costs	$23.75	$15.42	$13.21
Overhead	18.75	7.60	8.40
Total/100 ft of weld	$42.50	$23.02	$21.60

Present Method (Manual)

Type 6027 Manual Electrodes, 7/32 in. × 18 in.
    Joint Design: 45 degree bevel in $1\frac{3}{4}$ in. material.
Lbs. of weld per foot   5.16
Man-hours per foot   1.96

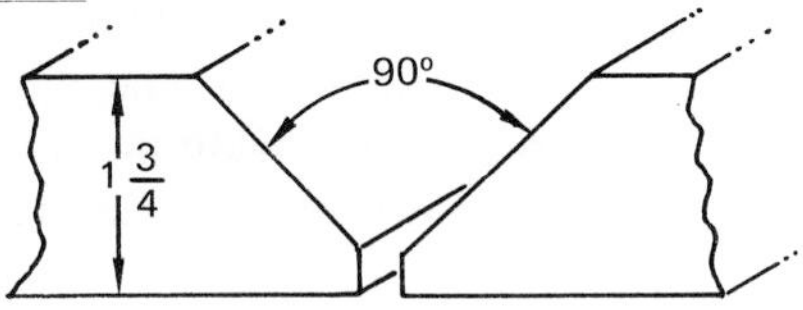

Joint preparation
for stick electrode

Proposed Method (Semi-Automatic)

1/8 in. Flux-Cored Wire, Semi-Automatic Full Weave
    Joint Design: 22 degree bevel in $1\frac{3}{4}$ in. material.
Lbs. of weld per foot   2.5
Man-hours per foot   .5

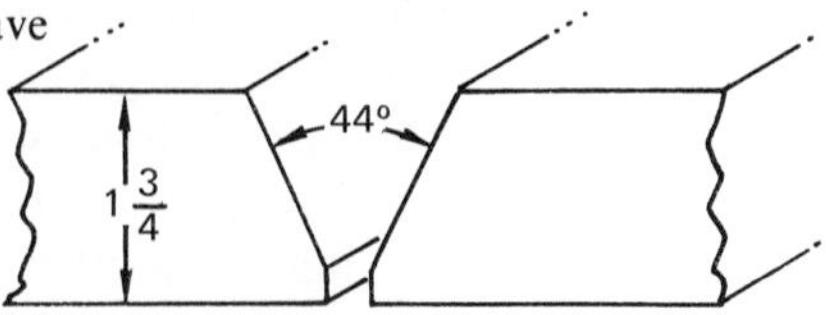

Return on Investment resulting from change

Equipment and learning, cost per unit $1,800.00
Material—Savings per ft of weld 5.16–2.5

= 2.66 lb (cored wire more costly per lb
than coated electrodes). Dollar savings per ft
even with the difference in cost = $0.28
Labor & Overhead savings per ft of weld

1.96-.5 = 1.46 h. Dollar savings per foot
= 1.46 × 12.50 (Labor and Overhead) = $18.30

        Total savings per foot $18.58

$$\frac{\$1800}{\$18.58} = 97 \text{ ft of welding required to amortize each unit}$$

One foot of weld could be done in .5 h by the new
method. Therefore, each unit could be amortized in
4.85 weeks, a 100 % or more return on investment.

Joint preparation for
semi-automatic welding

FIGURE 13–8. *A cost analysis procedure for analyzing the difference from
substituting flux-cored wire for manual electrode welding. (Courtesy of Welding
Engineer.)*

the total included angle are reduced when changing from manual to semiautomatic
and automatic welding. The depth of the root face is also increased. Both of these
differences decrease the amount of weld deposit required to complete the joint.
Thus, productivity is increased and air pollution is decreased.

COST FACTORS

Joining costs may be classified as fixed and variable. Fixed costs include
depreciation and maintenance of plant machinery and equipment; administrative,
finance, marketing, research and development expenses; interest, taxes, insurance,
etc. Variable costs include: direct labor; raw materials; variable factory overhead;
and outside services such as subcontract work.

FIGURE 13-9. *Joint designs for various welding processes made in 3/4 in. (19.05 mm) thick steel in the flat position.*

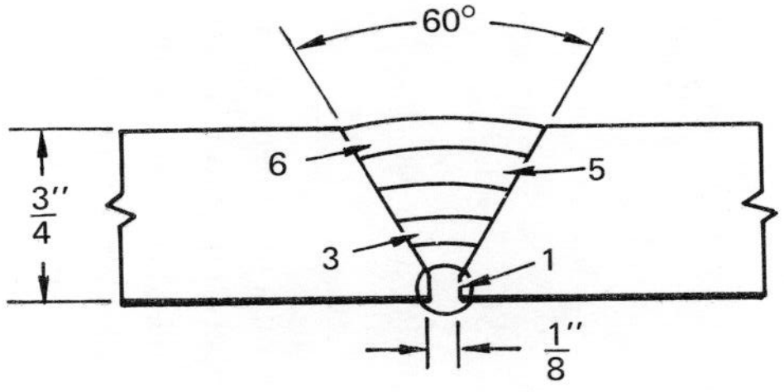

Shielded Arc (Manual)

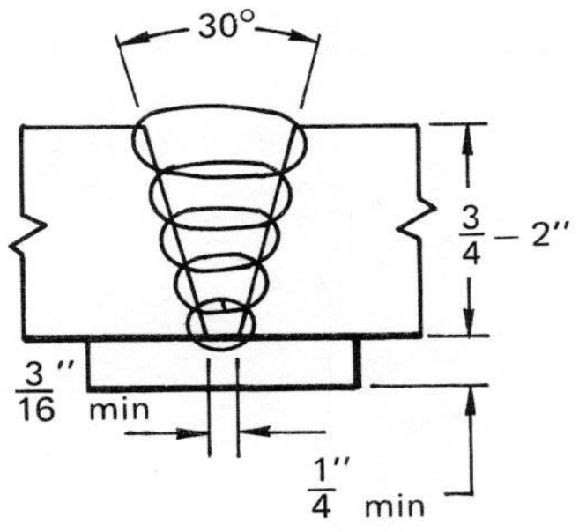

Flux Cored

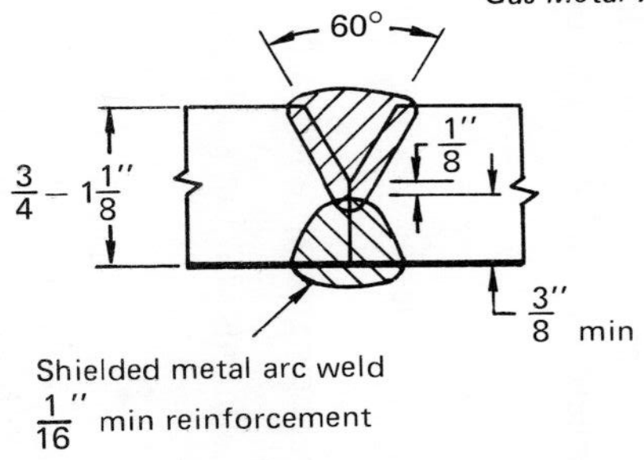

Submerged Arc

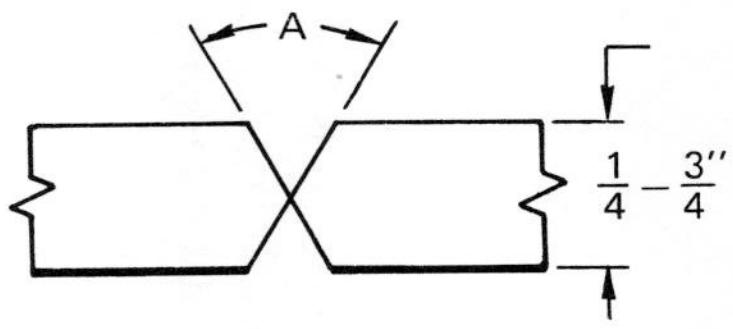

Gas Metal Arc (Spray Transfer)

	Shielded Arc			Flux-Cored	Submerged Arc	Gas Metal Arc
Plate thickness	3/4			3/4	3/4	3/4
(mm)	(19.05)			(19.05)	(19.05)	(19.05)
Pass	1–2	3–6		1–3	1	1–4
Electrode size	5/32	1/4	1/4	7/64	7/32	3/32
(mm)	(3.96)	(6.35)	(6.35)	(2.77)	(5.54)	(2.38)
Current	135	275	400	500 dc(−)	900 dc(+)	450 DCRP
Arc speed (in./min)	6	9	12	10–11	18	12
(cm/min)	(180)	(270)	(360)	(300–330)	(540)	(360)
Electrode req'd (19/ft)	0.168	0.228	1.47	1.43	0.29	
(480 cm)	(0.076)	(0.103)	(0.666)	(0.648)	(0.1315)	
Total time (h/ft weld)	0.122			0.0570	0.0111*	0.0667
(h/30 cm)						

* Does not include stick electrode weld made first.

** A = (Gas Metal Arc) angle in degrees, $\frac{1}{4}$ to $\frac{1}{2}$ in. (6.35–12.70 mm) thick—60 deg is used; on $\frac{3}{4}$ in. (19.05 mm), 90 deg is used.

Profit is generally defined as the net amount after deducting all costs and expenses from sales and other operating revenues. It is usually measured as a percent of total sales and revenues, or a percent return on capital invested.

Everything done in the plant will have some effect on the overall operating cost. Although it is desirable to control all costs, the opportunity to do so is less in the area of labor rates, inventory standards, pricing or administrative, sales, R & D, interest, taxes, etc. The areas where special skills can be used to improve the overall cost and profit picture are:

1. Reducing the cost per pound of weld deposit.
2. Increasing work flow through the plant by better welding procedures, planning, and material handling.
3. Reducing quality control costs by reducing scrap and rework.

The main factors used in determining materials joining costs are: labor, overhead, power, materials, equipment, and profit.

Labor. Although the cost of labor on an hourly basis is considered to be a variable of limited control, it is the largest single cost item. Labor cost is based on the amount of weld produced per hour. This in turn is based on the size of the electrode used and the position. Higher deposition rates can be achieved in the flat or nearly flat position. Travel speeds for manual and semiautomatic welding can be obtained from trial runs or from past records. Travel speeds for full automatic welding can be obtained from manufacturers' tables or from past records.

Since the arc will not be operating 100% of the time, an operating factor, or *duty cycle*, will have to be used. It is based on the time required to set up, change electrodes, chip slag, and so forth. Arc time recorders can be connected into the weldor to record the actual time the arc is being used. It is, of course, desirable to have as high an operating factor as possible, since this means that a high percentage of the weldor's time is being used in joining metal. The labor cost formula is given as:

$$\text{Labor Cost/ft } (L) = \frac{(W)\,(\text{Hourly Rate})}{(S)\,\text{Travel Speed in./min} \times (5)\,\dfrac{(60\text{ min/h})}{(12\text{ in./ft})} \times (C)\,(\text{Duty Cycle})}$$

Or,

$$L = \frac{W}{5SC}$$

A nomograph as shown in Fig. 13–10 may also be used in computing welding labor costs.

Example. What is the labor cost/weld foot on a job where the hourly rate is $3.50, the welding speed is 10 ipm, and the operating factor is 0.5?

Solution: Enter the hourly labor rate ($3.50) and draw a straight line through welding speed (10 ipm). From the intersection of this line with the pivot, draw a straight line through the operating factor (0.5). At the intersection with the second scale from the left, read the labor cost as $0.14/ft of weld.

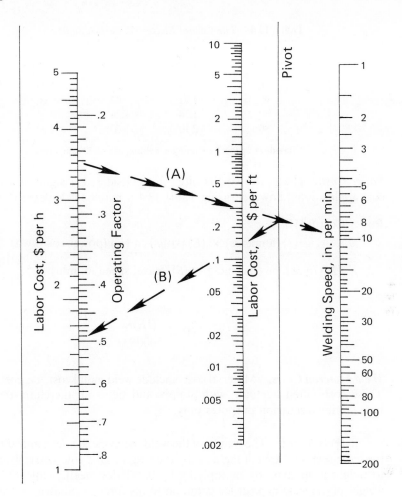

FIGURE 13–10. *A nomograph used to compute welding or flame cutting labor costs on a per-foot basis. (Courtesy of Welding Engineer).*

Overhead. The overhead rate is pre-established and is usually two times the direct labor cost. If lumped together with labor, it will not give a true picture of the costs involved.

Overhead Cost/ft (*O*)

$$= \frac{(W)\, 2 \cdot (\text{Hourly Rate})}{(S)\,(\text{Travel Speed in./min}) \times (5)\,\dfrac{(60\,\text{min/h})}{(12\,\text{in./ft})} \times (C)\,(\text{Duty Cycle})}$$

Or,

$$O = \frac{2W}{5SC}$$

TABLE 13.4. *The Cost of Electric Power/8 h Shift**

% Duty Cycle	dc Motor Generator	dc Rectifier	ac Transformer
20	.77	.45	.37
40	1.21	.84	.68
60	1.74	1.24	1.01
80	2.10	1.70	1.33

* Based on 2¢/kwh, assuming a welding arc of 300 amperes.

Power Cost. The power cost is often not considered a significant factor in weld cost applications; however, using Table 13.4 it may be calculated as follows:

Power Cost/ft (V)

$$= \frac{(I)\,(\text{amperes}) \times (E)\,(\text{volts}) \times (K)\,(\text{power cost/kwh})}{(M)\,(\text{machine efficiency}) \times (S)\,(\text{travel speed in./min}) \times 5\,\dfrac{(60\,\text{min/h})}{(12\,\text{in./ft})}} \times 1000$$

Or,

$$V = \frac{(IE)K}{5000SM}$$

Weld Material Costs. Material costs include weld metal cost, gas cost, and flux cost. These costs often fluctuate with supply and demand, therefore, the figures used here are for discussion purposes only.

Weld Metal. The weight of the weld metal can be determined by multiplying the cross-sectional area of the weld by the length. This volumetric value is converted to weight in lb/cu in. by multiplying by 0.283 (for steel). Table 13.5 provides the weight of the weld deposit for common joints directly. Shown in Table 13.6 is the average time required for fillet and 45 degree bevel welds.

Some loss usually occurs in depositing the weld metal, hence it is divided by a factor termed *deposition efficiency*. For manual welding, a certain amount of stub loss and spatter is always present and accountable. Thus, 60% efficiency is considered normal for shielded arc welding, whereas in the GMA process (using CO_2), it will be about 92% efficient—and submerged arc is about 100% efficient. Most gas shielded processes other than CO_2 are considered about 100% efficient. The formula to determine the weld metal cost/ft is:

$$\text{Weld Metal Cost/ft } (T) = \frac{(D)\,(\text{lb/ft of weld}) \times (B)\,[\text{cost/lb (dollars)}]}{(A)\,\text{deposition efficiency}}$$

Or,

$$T = \frac{DB}{A}$$

TABLE 13.5. *Inches in One Pound of Wire* (Courtesy of *Welding Engineer*)

Wire Diam		Material								
Decimal Inches	Fraction Inches	Mg	Al	Al Bronze (10) per cent	Mild Steel	Stainless Steel 300 series	Si Bronze	Copper Nickel	Nickel	De-ox. Copper
.020		50500	32400	11600	11100	10950	10300	9950	9900	9800
.030		22400	14420	5150	4960	4880	4600	4430	4400	4360
.035		16500	10600	3780	3650	3590	3380	3260	3240	3200
.040		12600	8120	2900	2790	2750	2580	2490	2480	2450
.045		9990	6410	2290	2210	2170	2040	1970	1960	1940
.062	1/16	5270	1510	12200	1160	1140	1070	1040	1030	1020
.093	3/32	2350	3382	538	519	510	480	462	460	455

TABLE 13.6. *Pounds of Electrodes (Steel Deposit)/Linear Foot of Weld*

Horizontal Fillet Weld

Size of Fillet L (in inches)	Pounds of Electrodes Required per Linear Foot of Weld* (approx.)	Steel Deposited per Linear Foot of Weld—Pounds
1/8	0.048	0.027
3/16	0.113	0.063
1/4	0.189	0.106
5/16	0.296	0.166
3/8	0.427	0.239
1/2	0.760	0.425
5/8	1.185	0.663
3/4	1.705	0.955
1	3.030	1.698

* Includes scrap end and spatter loss.

Square Groove Butt Joints

Joint Dimensions (in inches)			Pounds of Electrodes Required per Linear Foot of Weld* (approx.)		Steel Deposited per Linear Foot of Weld—Pounds	
T	B	G	Without reinforcement	With reinforcement**	Without reinforcement (lbs)	With reinforcement** (lbs)
3/16	3/8	0	—	0.16	—	0.088
		1/16	0.04	0.20	0.020	0.109
1/4	7/16	1/16	0.05	0.23	0.027	0.129
		3/32	0.07	0.26	0.039	0.143
5/16	1/2	1/16	0.06	0.27	0.033	0.153
		3/32	0.09	0.30	0.050	0.170
1/8	1/4	0	—	0.21	—	0.119
		1/32	0.03	0.24	0.013	0.132
3/16	3/8	1/32	0.04	0.36	0.020	0.199
		1/16	0.07	0.39	0.040	0.218
1/4	7/16	1/16	0.10	0.47	0.053	0.261
		3/32	0.14	0.53	0.080	0.288

* Includes scrap end and spatter loss.
** r = (height of reinforcement).

Square Groove Butt Joints

. . . welded one side

R = 0.07

If root of top weld is chipped or flame gouged and welded, add 0.07 lb. to steel deposited (equivalent to approx. 0.13 lb. of electrodes).

. . . welded two sides

R = 0.07

R = 0.07

Vee Groove Butt Joint

Joint Dimensions (in inches)			Pounds of Electrodes Required per Linear Foot of Weld* (approx.)		Steel Deposited per Linear Foot of Weld—Pounds	
T	B	G	Without reinforcement	With reinforcement**	Without reinforcement (lbs)	With reinforcement** (lbs)
1/4	0.207	1/16	0.15	0.25	0.085	0.143
5/16	0.311	3/32	0.31	0.46	0.173	0.258
3/8	0.414	1/8	0.50	0.70	0.282	0.394
1/2	0.558	1/8	0.87	1.15	0.489	0.641
5/8	0.702	1/8	1.35	1.68	0.753	0.942
3/4	0.847	1/8	1.94	2.35	1.088	1.320
1	1.138	1/8	3.45	4.00	1.930	2.240

* Includes scrap end and spatter loss.
** r = (height of reinforcement).

NOTE: Variations in joint configuration not shown may substitute appropriate figures in the formula $W_e = M_d/(1 - L)$. *Where,* W_e = (weight of electrode required), M_d = (weight of deposited metal), and L = (total electrode losses). The weight of the steel deposited is based on the volume of the joint (area of the groove × length) which is converted to weight of steel by the factor 0.283 lb/cu in.

NOTE: Before the weld condition figures can be applied to this formula, a value for D must be determined:

$$D = \frac{60X}{5NS}$$

where
X = wire speed in ipm (Table 13.6).
N = number of in./lb of wire.

Gas Cost. For ease of calculation, gas cost is determined on a per foot of weld basis. The gas flow in cu ft/h is multiplied by the cost of the gas/cu ft divided by the welding speed in inches/min:

$$\text{Gas Cost/ft } (G) = \frac{(R) \text{ (gas flow, cu ft/h)} \times (P) \text{ (gas cost/cu ft)}}{(S) \text{ (welding speed in./min)} \cdot (5) \cdot \dfrac{(60 \text{ min/h})}{(12 \text{ in./ft})}}$$

Or,

$$G = \frac{RP}{5S}$$

Flux Cost. The flux cost is primarily considered as a material cost in submerged arc welding. A rule of thumb can be used to estimate the amount consumed based on the electrode weight ($0.86 \times$ electrode weight in lb).

Total Cost/Ft of Weld.

Total Welding Cost = *(material cost)* + *(operational cost)* × *(length of the weld)*

Example. Determine the cost of welding two plates of 1/4 in. thick material with a square groove butt joint. The weld is made on one side with reinforcement. The space between the plates is 3/32 in. The following data applies:

Process—semiautomatic GMA (CO_2).
Wire, 0.045 in., price—$1.50/lb.
Operating factor   —90%.
Total weld length   —30 ft.
Gas flow   —25 cu ft/hr.
Gas cost   —$0.05/cu ft.
Labor   —$5.00/hr.
Overhead   —$5.00/h.
Travel speed   —40 in./min.
Duty cycle   —65%.

Solution:

MATERIAL COST:

$$\text{Steel Deposited/ft} = 0.143 \text{ lb (Table 13.6)}$$

$$\text{Electrode Cost/ft} = \frac{.143 \times 1.50}{90} = \$0.023$$

$$\text{Gas Cost/ft} = \frac{25 \times .05}{40 \times 5} = \frac{1.25}{200} = \$0.062$$

OPERATIONAL COST:

$$\text{Labor Cost/ft} = \frac{5.00}{40 \times 5 \times 65} = \$0.038$$

$$\text{Overhead Cost} = \frac{2(5.00)}{40 \times 5 \times 65} = \frac{10}{13000} = \$0.076$$

$$\text{Power Cost} = .0024 + 0038 \times .10 = \underline{\underline{\$0.0006}}$$

TOTAL COST/FT: $\$0.1996/\text{ft.}$

Therefore, the total cost of semiautomatically welding 30 ft of $\frac{1}{4}$ in. plate under these conditions is $5.99.

Estimating Welding Costs. Welding costs can often be quickly estimated without the more detailed cost formulas. The approximate cost will be based on: (*labor and overhead*) + (*welding consumables*) + (*profit*).

Table 13.6 can be used to determine the pounds of electrodes required per linear foot of weld or the weight of the deposited metal per foot for various types and sizes of joints. In Table 13.7 is a tabulation of the cost of deposited weld metal in $/lb, which includes electrode cost—assuming labor and overhead at $6.00/h.

TABLE 13.7. *Approximate Cost of Deposited Weld Metal $/lb (Includes Electrode Cost and $6.00/h for Labor and Overhead)*

Type	Size	Current	Operating Factor				
			60%	50%	40%	30%	20%
E6010	1/8	100 dc(+)	4.38	5.22	6.47	8.55	12.72
	5/32	140 dc(+)	3.20	3.80	4.70	6.19	9.17
	3/16	180 dc(+)	2.67	3.16	3.90	5.12	7.57
E7018	1/8	130 dc(+)	4.05	4.82	5.96	7.85	11.63
	5/32	180 dc(+)	2.89	3.43	4.23	5.55	8.20
	3/16	260 dc(+)	2.21	2.61	3.20	4.18	6.15
	1/4	325 dc(+)	1.89	2.23	2.73	3.55	5.20
E7028	5/32	230 ac	1.94	2.28	2.78	4.63	5.32
	3/16	280 ac	1.51	1.76	2.14	2.77	4.04
	7/32	330 ac	1.22	1.42	1.71	2.20	3.18
	1/4	400 ac	0.97	1.12	1.33	1.70	2.43
E6012	3/16	245 ac	2.44	2.88	3.55	4.66	6.87
	7/32	285 ac	1.85	2.18	2.67	3.49	5.12
	5/16	405 ac	1.50	1.76	2.19	2.81	4.07
E7024	1/4	375 ac	1.16	1.35	1.62	2.07	2.97
Automatic submerged arc	Full	1000 dc(+)	0.64	0.70	0.80	0.96	1.29
	Semi	500 dc(+)	1.00	1.15	1.57	1.71	2.41

TABLE 13.8.   I. *Average Cost Data for Manual Oxyacetylene Welding of Iron and Steel, and* II. *Approximate Weight of Weld Metal in 60° and 90° Single Vee Joints* (Courtesy of *Welding Engineer*)

I.

Thickness of Steel Inches	Joint Preparation No Spacing	Diameter of Rod Inches	Tip Drill Size	Oxygen—Cubic Feet		Acetylene—Cubic Feet		Pounds of Rods		Speed Foot per Hour
				per Hour	per Linear Foot Welded	per Hour	per Linear Foot Welded	per Hour	per Foot	
1/64	Square Butt	1/32	75	0.7	0.03	0.7	0.03			26.0–30.0
1/32	Square Butt	1/32	75–60	1.0	0.05– 0.04	1.0	0.05– 0.04			22.0–25.0
1/16	Square Butt	1/16	60–56	2.4	0.13– 0.11	2.3	0.13– 0.11	0.23–0.27	0.013	18.0–21.0
3/32	Square Butt	3/32	60–54	5.1	0.36– 0.30	4.9	0.36– 0.29	0.42–0.51	0.030	14.0–17.0
1/8	Square Butt	1/8	56–53	8.8	0.80– 0.68	8.5	0.77– 0.65	0.58–0.69	0.053	11.0–13.0
3/16	90° Single V	3/16	53–49	17.7	2.36– 2.08	17.0	2.27– 2.00	1.13–1.28	0.150	7.5–8.5
1/4	90° Single V	3/16	49–44	27.0	4.50– 3.86	26.0	4.33– 3.72	1.59–1.86	0.265	6.0–7.0
5/16	90° Single V	1/4	44–40	33.0	7.40– 6.05	32.0	7.11– 5.82	1.87–2.28	0.414	4.5–5.5
3/8	90° Single V	1/4	43–36	45.7	11.42– 9.13	44.0	11.0 – 8.80	2.39–2.98	0.597	4.0–5.0
1/2	60° Single V	1/4	40–36	58.2	11.65– 9.70	56.0	11.2 – 9.33	2.90–3.48	0.637	5.0–6.0
5/8	60° Single V	5/16	36–32	73.8	21.10–16.42	71.0	20.30–15.79	3.06–3.92	0.872	3.5–4.5
3/4	60° Single V	5/16	32–30	91.5	36.60–26.16	88.0	35.20–25.17	3.27–4.57	1.307	2.5–3.5

II.

Thickness of Metal Inches	Weld Metal in 1" Length of 60° Vee cu. in.	Weight of Weld Metal in 1-inch Length of 60° Vee Joint in Pounds					Weld Metal in 1" Length of 90° Vee cu. in.	Weight of Weld Metal in 1-inch Length of 90° Vee Joint in Pounds				
		Steel	Armco Iron	Stainless Steel	Nickel	Page Bronze		Steel	Armco Iron	Stainless Steel	Nickel	Page Bronze
1/4	0.035	0.0098	0.0099	0.0101	0.0112	0.0105	0.062	0.0174	0.0176	0.0179	0.0198	0.0187
3/8	0.080	0.0224	0.0227	0.0232	0.0255	0.0240	0.140	0.0392	0.0397	0.0405	0.0446	0.0421
1/2	0.144	0.0403	0.0408	0.0417	0.0459	0.0432	0.250	0.0700	0.0709	0.0723	0.0796	0.0751
5/8	0.225	0.0630	0.0638	0.0651	0.0716	0.0676	0.390	0.1092	0.1105	0.1128	0.1241	0.1172
3/4	0.324	0.0907	0.0918	0.0937	0.1031	0.0973	0.562	0.1574	0.1593	0.1625	0.1789	0.1689
7/8	0.441	0.1235	0.1250	0.1275	0.1404	0.1325	0.765	0.2142	0.2168	0.2211	0.2435	0.2298
1	0.577	0.1616	0.1635	0.1668	0.1837	0.1734	1.000	0.2800	0.2833	0.2890	0.3182	0.3004
1 1/8	0.729	0.2041	0.2066	0.2107	0.2320	0.2190	1.265	0.3542	0.3584	0.3656	0.4026	0.3801
1 1/4	0.901	0.2523	0.2553	0.2604	0.2867	0.2707	1.562	0.4371	0.4425	0.4515	0.4971	0.4690
1 3/8	1.090	0.3052	0.3088	0.3151	0.3469	0.3275	1.890	0.5292	0.5355	0.5463	0.6014	0.5678
1 1/2	1.298	0.3634	0.3678	0.3752	0.4131	0.3899	2.250	0.6300	0.6375	0.6503	0.7160	0.6760
1 5/8	1.523	0.4265	0.4315	0.4402	0.4847	0.4576	2.640	0.7392	0.7480	0.7630	0.8401	0.7932
1 3/4	1.766	0.4945	0.5003	0.5094	0.5620	0.5306	3.062	0.8574	0.8675	0.8850	0.9744	0.9200
1 7/8	2.028	0.5679	0.5746	0.5861	0.6454	0.6094	3.515	0.9842	0.9958	1.0159	1.1185	1.0560
2	2.308	0.6462	0.6539	0.6673	0.7345	0.6934	4.000	1.1200	1.1332	1.1560	1.2728	1.2018
2 1/4	2.920	0.8176	0.8273	0.8439	0.9292	0.8773	5.062	1.4174	1.4341	1.4630	1.6108	1.5209
2 1/2	3.606	1.0097	1.0216	1.0422	1.1475	1.0834	6.250	1.7500	1.7707	1.8063	1.9888	1.8778
2 3/4	4.363	1.2217	1.2361	1.2610	1.3884	1.3109	7.562	2.1174	2.1423	2.1854	2.4063	2.2720
3	5.196	1.4549	1.4721	1.5017	1.6534	1.5611	9.000	2.5200	2.5497	2.6010	2.8638	2.7040

NOTE: The data in the above table considers the metal veed from only one side of the joint; if the metal is veed from both sides of the joint, the volume and weight of the weld metal required are one-half of the above figures.

489

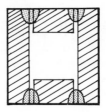

FIGURE 13–11. *One method of fabricating a square beam using 1 in. (25.40 mm) plate and $\frac{1}{2}$ in. (12.70 mm) weld penetration. A root gap of 1/8 in. (3.175 mm) was provided.*

Estimating Example. Four 20 ft long, $\frac{1}{2}$ in. horizontal vee groove butt joints are required to make a square beam, Fig. 13–11. The electrodes used are $\frac{1}{4}$ in. E7024. The operating factor is 40%. What is the cost of the deposited metal?

Total weight of weld deposit $= 20 \times 4 \times 0.489$ (Table 13.6) $= 39.12\,\text{lb}$

The cost of the deposited metal $= 39.2 \times 1.62$ (Table 13.7) $= \$63.50$

If full automatic submerged arc can be used, the operating factor can be changed to 60%, and the cost would then be $39.12 \times 0.64 = \$25.03$.

Estimating Manual Oxyacetylene Welding Costs. In determining manual welding costs, it is necessary to establish average conditions with the full realization that many variations will be encountered under actual working conditions. If a weldor works solely on one welding operation, a high degree of skill is usually developed with the result that production speeds increase to far above the average. The figures given in Table 13.8 are based on average working conditions with allowances made for lost time and other such factors. The size of the welding rods and tip will be dependent, to a great extent, on the skill and the speed of the individual weldor; and for this reason, ranges have been given for these factors.

Example. Estimate the cost of welding a $\frac{1}{4}$ in. 90° single vee steel butt joint 3 ft long. Assume the cost of oxygen/cu ft is 0.02¢, acetylene/cu ft is 0.03¢, and the rod cost is $1.70/lb. Labor is $3.50/h and overhead is 100% of the base labor rate. Assume a 70% operating factor.

Solution:

MATERIAL COST:

Oxygen required $= 4.18$ cu ft/ft of weld (*average*)
Acetylene required $= 4.00$ cu ft/ft of weld
Pounds of rod $= 0.26$/ft of weld
Time $= 6.5$ ft/h
$= [(4.18 \times 3) \cdot 0.02(O_2)] + [(4.00 \times 3) \cdot 0.03(C_2H_2)] + (0.26 \times 3)(rod)$
$= 0.25 + 0.36 + 0.78$
$= \$1.39$

LABOR COST:

$= 0.35$/ft (*use nomograph*)
$= 0.35 \times 3 = 1.05$

OVERHEAD COST: $= \$1.05$

Therefore, the cost of welding two $\frac{1}{4}$ in. plates, 3 ft long, and by the oxyacetylene method would be $3.49.

TABLE 13.9. *Cost of Gases Purchased in Single Cylinder Quantities/Cu Ft*

Natural Gas	$0.0014
Propane	0.017
Propane base	0.06
Acetylene	0.065
Oxygen	0.03

The cost of various gases used in welding and cutting, if purchased in single cylinder or small quantity orders, is given in Table 13.9. Prices vary considerably.

Estimating Oxygen and Plasma Arc Cutting Costs. The oxygen cutting speeds shown in Table 13.10 reveal quite a wide range for each thickness and tip used. The gas flow and cutting speed values shown are only intended to act as a guide for determining more precise settings for a particular job. When completely new material is cut, a few trial cuts will be necessary to obtain the most efficient cutting conditions.

TABLE 13.10. *Data for Manual and Machine Cutting of Clean Mild Steel* (*Not Preheated*) (Courtesy of American Welding Society)

Thickness of Steel, in.	Diameter of Cutting Orifice, in.	Cutting Speed, ipm	Gas Consumptions, Cu Ft per Hour			
			Cutting Oxygen	Acetylene	Natural Gas	Propane
1/8	0.020–0.040	16–32	15–45	3–9	9–25	3–10
1/4	0.030–0.060	16–26	30–55	3–9	9–25	5–12
3/8	0.030–0.060	15–24	40–70	6–12	10–25	5–15
1/2	0.040–0.060	12–23	55–85	6–12	15–30	5–15
3/4	0.045–0.060	12–21	100–150	7–14	15–30	6–18
1	0.045–0.060	9–18	110–160	7–14	18–35	6–18
1 1/2	0.060–0.080	6–14	110–175	8–16	18–35	8–20
2	0.060–0.080	6–13	130–190	8–16	20–40	8–20
3	0.065–0.085	4–11	190–300	9–20	20–40	9–22
4	0.080–0.090	4–10	240–360	9–20	20–40	9–24
5	0.080–0.095	4–8	270–360	10–24	25–50	10–25
6	0.095–0.105	3–7	260–500	10–24	25–50	10–30
8	0.095–0.110	3–5	460–620	15–30	30–55	15–32
10	0.095–0.110	2–4	580–700	15–35	35–70	15–35
12	0.110–0.130	2–4	720–850	20–40	45–95	20–45

Preheat oxygen consumptions. Preheat oxygen for acetylene = 1.1 to 1.25 × acetylene flow (cu ft per hr); preheat oxygen for natural gas = 1.5 to 2.5 × natural gas flow (cu ft per hr); preheat oxygen for propane = 3.5 to 5 × propane flow (cu ft per hr).

Operating notes. Higher gas flows and lower speeds are generally associated with manual cutting, whereas lower gas flows and higher speeds apply to machine cutting. When cutting heavily scaled or rusted plate, use high gas flows and low speeds. Maximum indicated speeds apply to straight line cutting, for intricate shape cutting and best quality, lower speeds will be required.

If the same figures as given for oxyacetylene welding are used, the cost of cutting steel plate by the oxygen method can be easily determined.

Example. Find the cost of cutting a $\frac{1}{2}$ in. × 4 × 8 ft sheet in four equal 2 × 4 ft sections. Use an operating factor of 40% and an overhead of 100% on the base labor rate.

Solution:

MATERIAL COST:

= *oxygen,* 65 cu ft/h *(average),* *acetylene,* 8 cu ft/h *(average),* and *time,* 18 ipm *(average)*

3 cuts = 144 in.

$$\text{Time} = \frac{144}{18} = 8 \text{ min}$$

Oxygen = 1.8 cu ft/min

Acetylene = 0.13 cu ft/min

GAS COST: = [(1.8 × 8) · 0.02] + [(0.13 × 8) · 0.03] = 32 ¢

LABOR COST: = 0.10/ft *(use nomograph)* 0.10 × 12 = $1.20

OVERHEAD: = $1.20

The total cost of cutting a $\frac{1}{2}$ in. thick sheet into four 2 × 4 ft sections is $2.72.

Estimating Plasma Arc Cutting. Plasma arc cutting, also referred to as constricted tungsten arc cutting, employs an extremely high temperature, high velocity constricted arc, Fig. 13–12. When cutting mild steels and cast iron, increased cutting speeds can be achieved by using oxygen bearing cutting gases. Typical cutting conditions for plasma arc cutting of mild steel are shown in Table 13.11. Mixtures of argon and hydrogen or nitrogen and hydrogen are generally used for cutting stainless steel, aluminum, and other ferrous metals. Nitrogen–hydrogen mixtures and compressed air exhibit desirable properties for cutting carbon steel, cast iron, and alloy steels.

Example. Compare the cost of cutting a $\frac{1}{2}$ in. × 4 × 8 sheet, as given in the previous problem, between oxyacetylene cutting and plasma arc cutting. Add 10% to the

TABLE 13.11. *Data for Plasma Arc Cutting of Mild Steel* (Courtesy of American Welding Society)

Plate Thickness in.	Speed Range, ipm	Current Range, amperes	Gas Flow Rate, scfh**			Dual Flow Gas Flow Rate, scfh**	
			Air*	$N_2 + H_2$		Shield N_2	Sheath Air or O_2
1/2	35–110	150–250	200	100		60	400
1	30–50	250–350	300	110		60	400
2	15–25	250–500	350	180	10	100	400
3	7–10	500–750	...	200	20	100	400
5	3–6	1000	...	200	20	...	...

* Multiport nozzles used when cutting with compressed air.
** Standard cu ft of gas/h.

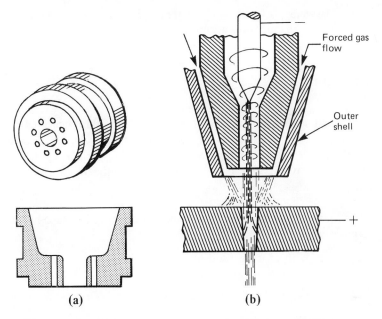

FIGURE 13–12. (a) *A multiport plasma-arc cutting nozzle facilitates cutting in any direction on the plate.* (b) *A dual flow plasma-arc cutting nozzle has a nonoxidizing gas around the inner electrode, and air or oxygen forms a sheath around the plasma.*

total cost for power and compressed air. The cost of $N_2 = 4\,¢/\text{cu ft}$. Use a combination of O_2 and N_2 for shielding.

Solution:

MATERIAL COST:

$$= N_2 = \frac{100}{60}\,\text{cu ft/min } (\textit{inner electrode}) = 1.68$$

$$N_2 = \frac{60}{60}\,\text{cu ft/min } (\textit{sheath}) = 1.0$$

$$O_2 = \frac{400}{60}\,\text{cu ft/min } (\textit{sheath}) = 6.8$$

$$\text{Time} = \frac{144}{72}\,\text{ipm} = 2.00\,\text{min}$$

$$= (1.68 \times .04)2 + (6.8 \times .02)2$$

$$= \$0.406$$

LABOR COST: $= \$.023/\text{ft } (\textit{nomograph}) \times 6 = \$.138$
OVERHEAD: $= .138$
TOTAL COST: $= (.406 + .138 + .138) + 10\%$

$$= \$0.75$$

Therefore, the total cost of cutting a $\frac{1}{2}$ in. thick plate, as designated, will be $0.75 with the plasma arc torch as compared to $2.72 by the oxyacetylene torch.

WELDING FIXTURES

Most semiautomatic and automatic welding applications require some type of fixturing in order to achieve a reasonable duty cycle. It can be extremely simple as in the case of a supporting device to hold the work beneath a traveling welding head, or it may be very complex as in hydraulically or numerically controlled tooling.

Fixture Classification. Welding fixtures may be placed in one of three broad categories:

1. Fixtures with moving welding heads.
2. Fixtures for holding and moving the work.
3. Special application fixtures.

Fixtures with Moving Welding Heads. Several methods of fixturing and moving the welding head include: ram type, beam type, straddle type, portable carriages, lathe type, and spud type, Fig. 13–13.

Ram Type. The ram type manipulator provides a mount for the welding head that allows it to be raised, lowered, extended, retracted, or swung in an arc of 360°. Thus, the head can be quickly placed in the desired position within reach of the boom. This allows the ram type manipulator to be used in conjunction with welding positioners, head and tailstock positioners, turntables, and other devices that permit high speed positioning of the work.

Beam Type. In this class of fixture, the welding head is mounted on a carriage that travels the length of the beam. The beam is usually supported on both ends. Beam fixtures are most often used in connection with turning rolls to weld tanks or other cylindrical work where both longitudinal and circumferential welds are needed.

Straddle Type. As the name implies, the fixture is made to straddle the workpiece. It may be fixed on tracks so that the entire fixture can be made to travel in relation to the workpiece. Large fixtures of this type can accommodate several welding heads.

Portable Carriages. For flat sheet, pipe, or vessel welding, portable carriages or tractors often can replace expensive tooling. Two types are in general use, track and trackless. Trackless types employ a template or a sensitive guidance system. With either type, power, gas, and water cables are supported by

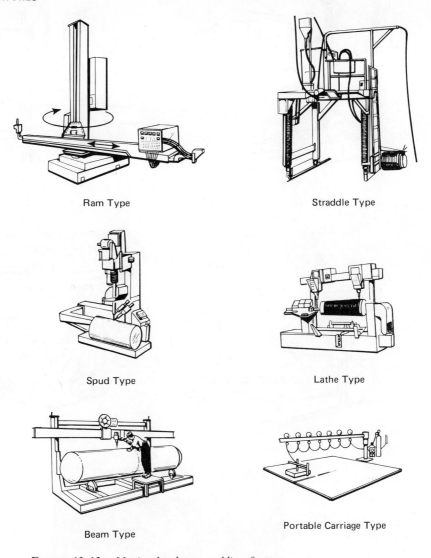

Ram Type

Straddle Type

Spud Type

Lathe Type

Beam Type

Portable Carriage Type

FIGURE 13–13. *Moving head type welding fixtures.*

means of a boom or a similar arrangement equipped with roller hooks for the cables (as shown schematically in Fig. 13–13).

Lathe Type. The lathe type fixture (also shown in Fig. 13–13) is a combination of a work positioner and a beam type fixture. It is especially useful in tank manufacture and for hard surfacing rolls.

Spud Type. The spud type fixture is especially adapted to welding small

Power Rolls

Idler Rolls

Headstock and Tailstock

Positioner

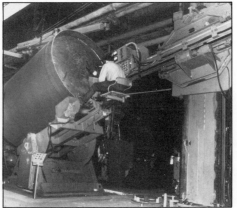

Manipulator and Tilting Turning Roll

FIGURE 13–14. *Fixtures and positioners for moving the work.* (*Courtesy of Ransome Company.*)

pipe connections onto tanks and cylinders. It provides a centering clamp and a rotating head.

Fixtures and Positioners for Moving the Work. A welding head that is stationary calls for work movement and invariably leads to more complex work fixturing. Sometimes the weld head mounting for a stationary type fixture will allow movement up and down or to one side for passage of the workpiece. Fixtures and positioners used for moving the work are: power rolls, headstocks and tailstocks, positioners, and turntables, Fig. 13–14.

Power Rolls and Idlers. Rolls are a convenient means of moving the work when doing circumferential welding. Power and idler rolls are also used to position the work for longitudinal seam welding.

Headstocks and Tailstocks. When circumferential work is to be done in quantity, set-up time can often be reduced by the use of headstocks and tailstocks. The headstocks are power driven and can be rotated at a variable speed by remote pushbutton control. For smaller work, welding positioners can be used as lathe headstocks.

Positioners. The most versatile of the tools for holding the welding workpieces is the welding positioner. It can be rotated within a variable speed range and can be tilted through a wide angle. As stated previously, the main advantage of being able to position the workpiece is to take advantage of the ease and speed of flat or downhand welding, Fig. 13–15. It is also an accepted fact that consistently higher quality is obtained in the downhand position. Positioners may range in size from small table models with a capacity of 100 lb (45.35 kg) to very large sizes rated at 100,000 lb (45 000 kg) capacity.

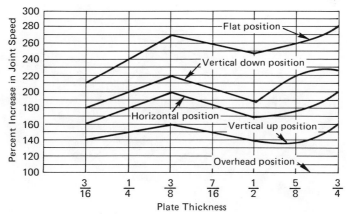

FIGURE 13–15. *Comparison of welding speeds for various welding positions. Joint speed for downhand welding can approach 280% faster than other positions.*

Turntables. Circular parts can be rotated for welding by means of a turntable, which is usually equipped with a variable speed drive. Turntables do not have the versatility of a positioner, but they are also less expensive.

Special Application Fixtures. Large and complex assemblies that are of a low production type often require special fixturing in order to maintain accuracy. Large quantity production also makes it feasible to build fixtures that are especially adapted to the particular design. Many such fixtures may be seen in automobile body assembly plants. Specialized fixturing may be very complete and in addition to incorporating welding and positioning devices already mentioned, they may include:

Machines capable of cutting the edges of plates in preparation for welding (saws, mills, routers, flame "U"-grooving torches, etc.).
Forming machines for rolling or bending sections to the desired shape.
Rotational devices.
Loading and unloading devices.
Timing and automatic sequencing, and remote controls.

Fixtures for Welding Beams. One example of a special fixture will serve to show how it is designed and some of the problems associated with it. Shown in Fig. 13–16 is a specialized fixture used in fabricating "I" beams. The designer considered the following requirements:

1. Preferably, one man can operate it.
2. Extensions should be possible.
3. It should be adaptable to nonparallel flanges, as in roof trusses.

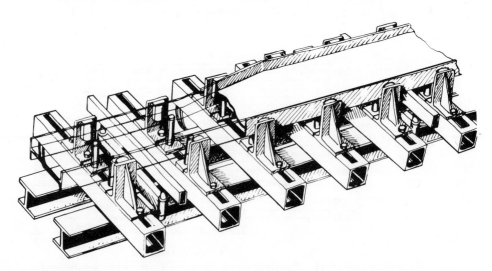

FIGURE 13–16. *A specialized welding fixture used to fabricate special "I" beams.* (*Courtesy of ESAB.*)

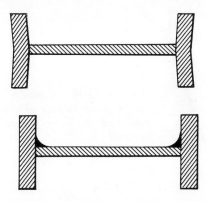

FIGURE 13–17. *Prebending of the "I" beam flanges will counteract small shrink forces produced in welding.*

4. Tack welding should be eliminated as far as possible.
5. Dimensional accuracy must be maintained and distortion minimized.

If both sides are welded at the same time, the shrinkage forces are comparatively small due to the fact the moment arm is equal to only half of the web thickness. If the flanges must be absolutely straight, the simplest and safest way is to pre-bend them as shown in Fig. 13–17. The degree of accuracy required should be determined on the basis of function so that production costs will not be excessive.

The initial cost of this type of fixture is very high, but it is a necessity for all workshops who do this type of work on a more or less regular basis. To determine the breakeven cost of when the fixture will pay for itself, the following formula can be used:

$$Break \; Even \; Point = \frac{A}{B - C}$$

where A = cost of tooling.
   B = cost/part without tooling.
   C = cost/part with tooling.

Fixture Selection.   Elaborate welding equipment or work fixturing usually requires a high production volume of a single item or of reasonably similar parts. If volume does not exist, the next consideration is fixture adaptability. This refers to combining a variety of holding devices with a standard fixture, such as a beam with a carriage fixture.

Clamping and Fitup.   A good clamping system has many small clamps or points of pressure application. Box beams, for example, are difficult to hold. They should be clamped at closely spaced intervals along their entire length, or gaps will develop during welding (which will permit burn through) and cause distortion of the finished part. A six or eight inch (15.24–20.32 cm) channel box section as shown in Fig. 13–18 should be clamped every six inches (15.24 cm) to avoid joint openings as the weld progresses.

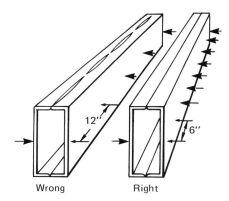

FIGURE 13–18. *Numerous pressure points are required to keep a seam from opening ahead of the weld. (Courtesy of Welding Journal.)*

Wrong   **Right**

Preparation, Finishing, and Inspection Costs.   Each of these factors will be treated in regard to the basic joining processes presented in Chapter 1; namely, welding, brazing, soldering adhesives, and mechanical fasteners.

Welding.   The preparation costs for welding are largely involved with edge preparation and fixturing. The edge preparation will differ with the various thicknesses of materials and the welding process chosen. Whenever possible, edges for heavier plates are prepared in the flat by means of side planers or flame machining. Thinner materials are usually cut on a power shear. Straight oxyacetylene and plasma cutting costs may be estimated, from Tables 13–10 and 13–11, when material and labor costs are known. Special preparation times and gas consumption for "J" grooving are shown in Table 13.12.

Preheating cost calculations are necessary in gas welding all heavy ($\frac{1}{2}$ in. and above) sections for most metals, to assure fusion. In welding with the electron beam process, preheating can often be minimized. The same is true for several of the solid state processes. In metal arc welding of steel, preheating above 400 °F (205 °C) is effective in reducing the rate of cooling, with resulting lower hardness, a wider HAZ, and less possibility of forming microcracks. Actual tests have shown that

TABLE 13.12.   *Gas Consumptions and Weld Areas for Single Pass, Oxy-Fuel Gas "J" Grooving* (Courtesy of American Welding Society)

Plate Thickness, in.	Groove Cross Section Area, sq in.	Total Gas Consumption, Cu Ft per Linear Ft of "J" Groove					
		Oxygen	Acetylene	Oxygen	Propane	Oxygen	Natural Gas
1-1/2	0.375	7.9	0.35	8.2	0.22	7.9	0.27
2	0.650	13.4	1.26	14.0	0.66	13.2	1.00
2-1/2	0.900	22.1	1.26	22.9	0.66	22.6	1.00
3	1.35	38.6	1.56	40.5	1.13	38.9	2.13
3-1/2	1.65	38.6	1.56	40.5	1.13	38.9	2.13
4	2.20	53.5	1.56	55.3	1.13	54.7	2.13

* Weld area—2 X groove area.

in steel structures having plate thicknesses up to 3/4 in. (19.05 mm), preheating at 400 °F may produce 9% greater strength than is obtained by 1000 °F (540 °C) postheat treatment, even though preheating has very little effect on the residual stress.

Plate fabricating shops often use natural or manufactured gas as a fuel for local preheating of weldments, since it is relatively clean and convenient. This gas is used in torch type burners, premixing the gas with air or using compressed air with an oxygen torch. Induction heating is also used, especially for preheating pipe.

Brazing. Brazing filler metal is relatively costly, and therefore, careful preparation of the joint such that it will require the minimum of filler metal is essential. Heating is another major economic factor in joint design. The time and material necessary to raise a large mass to a brazing temperature can be substantially reduced if joints are designed to require only a small mass be heated. Clean oxide free surfaces are imperative to insure uniform quality. This must be considered in the joint preparation cost. Cleaning is done either chemically or mechanically, with trichlorethylene and trisodium phosphate as the usual cleaning agents. The cleaning agent residue must be removed by careful rinsing. Scale and other objectionable surface conditions may be removed by mechanical means such as filing and wire brushing.

Brazing preparation costs also include jigs, fixtures, and clamps to hold the members in alignment during the heating and cooling cycle. Brazing fixtures differ from welding fixtures in that they are usually more intimately involved in the heating and cooling cycle. The materials used in making the fixture may be metal or ceramic. Each must be considered from the standpoint of overall cost. Metal fixtures have the advantage of ease of fabrication and moderate cost; however, they are subject to oxidation, corrosion, and chemical attack. In addition, metal fixtures are subject to change of hardness particularly at elevated temperatures, high thermal expansion and electrical conductivity, and have a tendency to absorb process heat. They are also susceptible to wetting by brazing alloys. Ceramic fixtures are not subject to many of the disadvantages of metal fixtures, however, they are difficult to machine (with few exceptions) and have lower impact strength. They also have a somewhat higher initial cost (except for some castable types). In some cases, it is possible to eliminate the cost of a brazing fixture by the use of tack welding.

Soldering. Preparation costs for soldering are similar to those for brazing.

Adhesives. The preparation cost for adhesives is primarily that of cleaning and fixturing. An example of the preparation costs for steel would be:

1. Sandblast with fine grit abrasive.
   (a) Pickle for two or four minutes at 60–70 °F (15–21 °C) in a solution of: 10% by volume concentrated sulfuric acid, 10% by volume concentrated nitric acid, and 80% by volume tap water.

2. Rinse.
3. Bright dip for one-half to one minute at 60–70 °F (15–21 °C) in a solution of: 50–60% by volume concentrated hydrochloric acid, 2% by volume (30% hydrogen peroxide), and 38–48% by volume tap water.
4. Hot water rinse and force air dry.
5. Bond or prime coat as soon as possible to prevent rust.
   (a) Or, treat with one of the proprietary systems that form a complex phosphate coating on the surface. The system should be controlled to accomplish a thin, highly adherent, nonpowdery deposition.

Fixturing equipment ranges from the more simple clamping arrangements to the more complex types actuated by pneumatic and hydraulic forces. In the latter case, automatic followup pressure may be programmed in to compensate for adhesive flow-out.

Vacuum equipment provide an inexpensive method of applying pressure during the bonding operation. Assembled components are enclosed in a rubber blanket and a vacuum is drawn. A cure is accomplished by placing the assembly in an oven. Vacuum techniques can be used with a heated platen to provide both heat and pressure simultaneously.

Mechanical Fasteners. The preparation costs for mechanical fasteners is usually associated with providing a drilled or punched hole in the parts to be joined. The present trend is to provide the holes by punching rather than drilling.

Further cost reductions are being accomplished through self-drilling and self-tapping screws. They are now available in a hardened throughout condition rather than case hardened, mainly for use in noncritical semistructural applications. Self-drilling screws also have the advantage of eliminating hole alignment problems, clearance holes, and separate washers and nuts. As shown in Fig. 13–19, self-drilling fasteners are capable of joining mild steel sections totalling 5/16 in. (7.93 mm) in thickness in one operation.

Metal-piercing rivets also eliminate separate punching operations. They are capable of penetrating and joining sandwiched thicknesses of mild steel ranging up to 0.150 in. (3.81 mm) or more.

Fasteners are often selected on the basis of appearance or convenience without due regard to cost, since this is usually quite cumbersome to calculate. It can, however, be very worthwhile to evaluate competitive fastening methods. A simple formula is used to calculate the in-place cost of fasteners:

$$C_a = \frac{C_l \left(1 + \dfrac{R_o}{100}\right) + C_f}{R_a}$$

where C_a = in-place cost of fasteners/assembly.
C_l = labor cost/h.
R_o = overhead or burden rate as a percent of the base labor rate.
C_f = the fastener cost/assembly.
R_a = assemblies made/h.

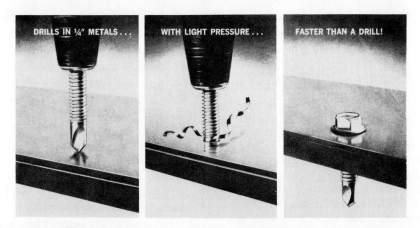

FIGURE 13–19. *Self-drilling and tapping screws eliminate the need for separate operations. Shown here is a total thickness of 5/16 in. (7.9375 mm) mild steel being joined in one operation. (Courtesy of Shakeproof Division of Illinois Tool Works Inc.)*

Example. Assume a single set screw is required per each assembly. It is put in place and tightened by one operator using a pneumatic torque wrench at the rate of 50 assemblies/h. The operator's wage is $2.65/h. The overhead rate is considered to be 100% of the base labor rate. The setscrews cost $40.00 per thousand.

Solution:

$$C_a = \frac{2.65\left(1 + \dfrac{100}{100}\right) + .04}{50}$$

$$= 0.1068$$

The in-place fastener cost/assembly is 10.68 ¢.

Inspection Costs. Inspection procedures will usually be dictated by code or by the customer's specifications, or both. Defects found during NDT procedures can be very costly to repair. In the case of adhesive bonding, the part will usually have to be scrapped. The repair of welding defects can more than double the initial joining cost.

As an example: A 60 in. (152.4 cm) long seam is welded by the submerged arc process in 4 in. (10.16 cm) thick low alloy, high tensile strength steel. A material of this type will require a relatively high preheat that must be maintained after welding until the seam is stress relieved. If by radiographic inspection 10% defects are revealed, this will necessitate removal and repair welding. Six inches of defect will have to be removed, but in the process, 8 to 15 in. (20.32–38.1 cm) of seam will be removed. Grinding or flame gouging will be used to remove the defective weld material. The seam will have to be preheated prior to the welding repair, which in all probability will have to be done manually. After welding, the seam will have to be again stress relieved and radiographed.

The designer, welding engineer, manufacturing engineer, and others who work with materials joining problems must be constantly kept informed as to the wide variety of joining processes available and the merits of each. Once a product is in the field the percentage of failures or excessive assembly costs can easily spell the difference between operating at a profit or at a loss.

EXERCISE

Problems and Questions

13–1. What would it cost to cut 25 pieces of mild steel 2 in. thick and 60 in./cut with the oxyacetylene torch? Use an operating factor of 40%. Labor cost is $3.50/h. Overhead is taken at 100% of the base labor rate. Gas costs are as given in the text.

13–2. (a) Shown in Fig. P13–2 is an outlet cover box. Use the summary sheet as presented here to help you estimate the cost of welding this cover. Labor and overhead values are from problem 13–1. Note that the operating factor has already been taken into account if Table 13.7 is used. No welds are calculated for the major component (in this case the cover). Welds are calculated as they are added. For the second part, only the amount of weld required to join it to the first part is counted on the summary sheet. The calculated actual weight of the parts proves useful in solving material handling and positioning problems.

The estimated weight of deposited metal proves useful in determining the weight of electrodes for manual welding. The deposited weld metal weight divided by the factor .55, is a value based on the fact that the actual deposited weld metal is 55% of the electrode's original weight. Estimate the time required for each of the welds listed.

(b) How many pounds of shielded type electrodes would be required?

(c) What is the total cost of welding the outlet cover box? Assume labor at $3.50/h and overhead at 100% of the base labor rate. Electrode cost = 45 ¢/lb.

(d) What are two factors that could throw the material estimate way off?

(e) How can the material estimate be made more accurate?

(f) Assume a welding positioner was used and it was able to put all welds in a downhand position, whereas before, approximately half had to be welded in a vertical position. How much time could be saved?

13–3. In your estimation, would it be more or less economical to weld part No. 8 in Fig. P13–2 by intermittent or skip welding? State reasons for your choice.

13–4. Could the left lugs, part no. 10 in Fig. P13–4, be attached with adhesives and be expected to perform satisfactorily? Explain.

13–5. (a) Tanks are being manufactured by rolling 2 in. × 6 ft × 12 ft plates. A butt weld is made in the 2 in. × 6 ft joint with the single wire submerged arc process. The tank will be 4 ft in diameter when finished. Sketch the joint preparation recommended.

(b) Domes will be welded on each end and pipe fitting connections installed on the circumference of the tank. What type of welding fixtures and positioners would you recommend?

(c) The present deposition rate of weld metal for the job is 40,000 lb/year. Labor and overhead total $12.50/h/man. Each man averages 4 lb/h. The total time to weld one cylinder is 15 h. It is proposed that the process be changed to one wire electroslag welding. Assume the deposition rate/h is 8 lb/man-hour due to the new process and a 20% duty cycle. Determine the total cost/year of each process.

(d) What is the estimated savings per year by the new process if the equipment and training cost equal $2,000?

(e) What is the percent return on the investment?

13–6. Estimate the cost of cutting the mild steel plate as given in problem 13–1 with the plasma arc torch. Use O_2 in the dual flow gas shield. The cost of H_2/cu ft is $.05.

13–7. Inspection revealed that in five cover boxes the end flange, part no. 4 in Fig. P13–2, was made

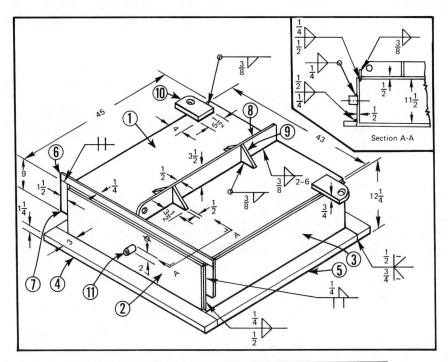

Part No.	Part Name	No. Req	Estimated Weight of Parts	Measured Joining Welds	Estimated Weight of Deposited Weld Metal	Estimated Time (from tables or calculated)
1	Cover top	1	261 lb			
2	Cover end	2	136	7.16 $\frac{1/4}{1/2}$ @.106# 1 ft @ 425# 1 ft	0.76 lb 3.04	
3	Cover side	2	144	11.33 $\frac{1/4}{1/2}$	1.21 1.83	
4	End flange	2	104	7.16 $\frac{1/4}{1/2}$	0.76 3.04	
5	Side flange	2	96	7.5 $\frac{1/4}{1/2}$		
				1.0 $\frac{1/2}{3/4}$ @.425# 1 ft @.957# 1 ft	0.43 0.96	
6	Top bar	1	5	3.6 $\frac{}{1/4}$ @.106# ft	0.38 0.38	
7	Side bar	2	2	20 15 $\frac{}{1/4}$	0.21 0.16	
8	Rib	1	21	7.5 (or 2.66′ 3/8 $_{2-6}$ × 1.5 factor)	0.65	
				.25 3/8 @.239# 1 ft	0.06	
9	Gusset	4	4	5.0 3/8	1.20	
10	Lift lug	2	18	3.2 3/8	0.77	
11	1–1/2″ Pipe nip.	1	2	10 1/4	0.106	
Total Weight			793 lb		22.94 lb	
Total Weight - Assembly			815.94 lb		Estimated Welding Time	

FIGURE P13–2. *An outlet cover box showing 11 components used in fabrication.*

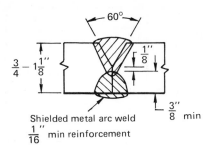

FIGURE P13–4

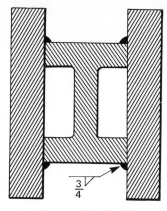

FIGURE P13–9. *A welded building column is made in 20 ft lengths.*

with a 5/8 in. fillet rather than 1/2 in., as called for in the print. What is the cost of this mistake if labor costs and the operating factor is as given in the example on p. 504.

13–8. An assembly requires 24 semi-tubular rivets that are clinched at the rate of 250 assemblies/h by a group of 8 operators. The wage scale is $3.00/h and the overhead burden rate is assessed at 100% of the base labor rate. Rivets cost $1.63/hundred. What is the in-place cost of these fasteners/assembly?

13–9 (a) Estimate the cost of making 10 columns as shown in Fig. P13–9 with a semiautomatic flux-cored wire, a semiautomatic submerged arc, and a manual stick electrode. Assume a 5/32 in. electrode and the standard labor and overhead costs suggested by the text.
(b) If the fixture used to build this beam costs $45,000, how many will have to be ordered before it will pay to build it? Assume it will be welded by the semiautomatic process with fixture. Without a fixture it costs $1,500/beam for welding.

13–10. Find the cost of making 500 feet of 1/8 in. mild steel fillet welds by the Mig process under the following conditions:

Wire speed   = 280 ipm
Electrode cost/lb   = 0.30
Gas   = CO_2
Gas flow   = 25 cfh
Gas cost/cu ft   = 0.01
Weld progress speed   = 15 ipm
Duty cycle   = 0.60 (60%)
Deposition efficiency   = 0.95 (95%)

Weldor rate/h   = 4.00
Overhead rate/h   = 8.00
Power cost/kwh   = 0.01
Power supply mach. efficiency   = 0.50 (50%)
Amps (DCRP)   = 140
Volts   = 20
Wire diameter   = 0.035 in. (E60S3)

where L = labor cost/ft of weld.
     W = weldor rate in dollars/h.
     S = weld travel speed in ipm.
     C = duty cycle.
     O = overhead cost/ft of weld.

The factor 5 (used in the formula) is derived from the ratio of 60/min/h, 12 in./ft.

13–11. Estimate the approximate difference in labor cost of making 30 ft of butt weld in 3/4 in. thick steel plate by the following methods: manually with a stick electrode, semiautomatically with GMA, and semiautomatically with flux-cored electrode. Labor and overhead are given as $8.00/h. (Use slowest travel rate in each case.)

13–12. Compare the labor cost of making 30 ft of flat position butt weld in 3/4 in. thick steel plate between the submerged arc process and the GMA process. Labor and overhead are given as $8.00/h. Assume the submerged arc weld is made with a 60 degree vee down to 3/8 in. of the bottom of the plate. The bottom side weld is put in with a stick electrode at the rate of 0.663 h/ft.

13–13. Compare the cost of making a butt weld in 1/8 in. thick mild steel sheet stock/ft by gas welding and by stick electrode on a small quantity basis.

No edge preparation and no spacing is used. Labor and overhead is given at $6.00/h. An ac transformer is used for arc welding with a 60% duty cycle. An E6010 electrode is used. NOTE: The filler rod for gas welding is not considered.

13–14. Compare the basic cost of cutting ten 12 in. diameter circles in 1/4 in. thick mild steel plate by oxyacetylene and by oxypropane. Since labor and overhead will be approximately the same for each method, only the difference in the cost of the gases used need be considered.

BIBLIOGRAPHY

BENES, J. T., "The Cost of Fastening." *Machine Design*, June 10, 1971.

BLODGETT, O. W., "The Design, Fabrication and Erection of Steel Structures with Mechanized Arc Welding." *Welding Journal*, August 1968.

BROWN, C. F., "ABC's of Weld Estimating." *Welding Engineer*, October 1958.

HAYLING, L., *Welding of Girders*. Svetsaren, English ed., **3**, Göteborg, Sweden, 1969.

HUMM, R., "Figuring the Real Cost of Rivets." *American Machinist*, October 1962.

MIKULAK, J., "Economy of Welding." Paper presented at the 15th annual Midwest Welding Conference held at the University of Wisconsin, Madison, Wis., 1970.

NORCROSS, J. E., "Cost Considerations for Maximum Profit Improvement." *Welding Engineer*, February 1969.

PAYNE, S., "The Engineering of Arc Welding Fixtures." *Manufacturing Engineering and Management*, January 1970.

——, "Fixtures and Manipulators for Mechanized Arc Welding." *Welding Journal*, December 1969.

Appendix A

CHAPTER 1 ANSWERS TO PROBLEMS AND QUESTIONS

1–1. (a) Three methods that can be used in making this joint are:

1. Arc welding.
2. Brazing.
3. Adhesives.

NOTE: Other welding methods can be used but they are not discussed until later in the text.

(b) Since the joint is not highly stressed and the principal stress will be in longitudinal shear, adhesives will be a good choice. Although brazing and welding can be used, the big discrepancy in the thickness of the materials to be joined makes adhesives much easier and faster to use.

1–2. The large tubing i.d. = 1.875 − 0.090 (two wall thicknesses) = 1.785 in. The next smaller size standard tubing has an o.d. of 1.750 in. This leaves a clearance of approximately .035 in., or about 0.016 in./side between the two tubes, which is excessive for brazing or adhesives. Thus, the best method of joining the two pieces of tubing will be welding by the GMA process.

1–3. (a) Originally the hat sections were attached to the boat by riveting with a sealing compound placed in the joint. Although this method was

satisfactory, the joining method was changed to a two part epoxy type adhesive.

(b) The adhesive method is preferred because there is less danger of developing leaks, also, the actual placement requires less labor. The cure time (in this case 12 hours at 180 °F using an electric heater under cover) is longer. Of course, the student will not have this type of information, but a brief study of Table 1.2 would indicate that for most adhesives some elevated temperature cure would be necessary.

1–4. (a) Submerged arc. (b) Stick electrode. (c) GMA welding. NOTE: Other methods are possible, these answers are within the presentation of this chapter.

1–5. There are several methods of fastening the fan on the shaft.

(a) A conventional method would be a keyway, press fit with a retaining ring to form a shoulder and another to hold it on.

(b) Another conventional method is a press fit and a setscrew.

(c) Make a clearance fit of 0.004 to 0.005 in. and use an anaerobic adhesive. It will develop handling strength at room temperature in 10 to 20 minutes and full strength in 3 to 4 hours. The parts may be disassembled by overcoming the bond strength or by heating the joint to 450 °F to weaken the sealant. Disassemble while hot.

1–6. Methods of improving the strength of the adhesives bonded corner joint are shown below.

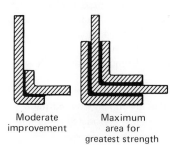

Moderate
improvement

Maximum
area for
greatest strength

FIGURE A1–6.

1–7. Three methods that would improve the joint are shown in Fig. A1–7 for adhesive joining:

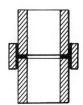 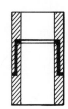 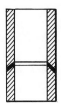

FIGURE A1–7.

1–8. Soft soldering.

1–9. Any method of welding can be used to join aluminized steel, provided the techniques are adapted to the properties of the material. Since welding heat damages some of the protective aluminum coating in the weld area, resistance welding is recommended. In this case, seam welding would be used on the cylinder and spot welding on the end where the muffler joins with the tube. Spot welding does not materially affect the corrosion resistance of the material.

If fusion welding is used, the corrosion resistance would have to be restored by metalizing. If fusion welding is chosen, the preferred method would be the GMA process using either helium or argon as a shielding gas.

Another approach is to make a lock seam which may be brazed, spot welded, or left as folded.

Brazing

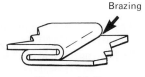

FIGURE A1–9.

1–10. (a) The area in double shear = $3 \times 2 = (6 \text{ in.}^2)/(2) = 3 \text{ in.}^2$. Epoxy phenolic has a temperature range of -423 to $500 \,°F$, with a shear strength of 1500 psi at 500 °F.

$$1500 \times 3 = 4{,}500 \text{ psi shear strength}$$

(b) Advantages:

1. No metallurgical effect on the metal.
2. More flexibility and less danger of fatigue.
3. More contact or bonding area.

(c) Improvement of joint for adhesive bonding.

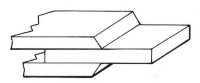

FIGURE A1–10.

1–11. The preferred method of fastening the electrical terminal would be by having it made with a small tip protruding from the end for capacitor discharge stud welding. Stud welding would be very fast and would have a complete bond over its entire surface. An added advantage is that there would be a minimum of distortion and heat marking on the opposite side.

An alternative method would be to braze the connection to the sheet. If strength requirements are not high, it can be soldered in place. Both methods would have minimum heat requirements.

Another slight modification would be to make the stud for projection welding; however, since this would provide less contact area, it would not be as good for an electrical connector.

CHAPTER 2 ANSWERS TO PROBLEMS

2–1. $\dfrac{40 \times 600 \times 60}{21}$ or 68,571 joules.

2–2. (a) The differences in the melting temperatures of the two metals (A1/1218 °F and carbon steel/2750 °F) is so great that the A1 will have vaporized long before the steel melts.
(b) The atomic structure of the metals is not compatible to each other.

2–3. $C = 5/9(°F - 32)$

$5/9(6300 - 32)$

$5/9 \times 6268$ or 3462 °C

2–4. $F = 9/5(C) + 32$

$F = 9/5 \times 889 + 32$ or 1632 °F

2–5. The more efficient the system is, the less distortion one can expect in the finished product. For example, the oxyacetylene torch is said to be only 2 % efficient because any melting must come about from conduction, therefore, a great deal of distortion. Compare this with electron beam welding where the metal being joined is completely pierced by the electrons and the molten metal resolidifies around and behind the beam of energy, the distortion is considered minimum.

CHAPTER 2 ANSWERS TO QUESTIONS

2–1. Unless the metal is melted or under great pressures, no known way exists to get metal surfaces smooth enough for atomic joining. Likewise, except for the electron beam process where joining is done in high vacuums, cleanliness and purity is hard to control.

2–2. (a) Remove all possible contamination from the surfaces of the metal to be joined. (b) Prevent further contamination from taking place before, during, and immediately following the joining process. (c) Produce an adequately smooth surface so that atomic bonding can take place.

2–3. Mechanical:

Forge welding.
Explosive welding.
Friction welding.
Ultrasonic joining.
Diffusion joining.

Chemical:

Combustible gases.
Atomic hydrogen.
Thermit.

Electrical:

Gas resistance

Liquid resistance
Solid resistance.
Special group.

2–4. $H = \dfrac{E \times I \times 60}{S}$

2–5. $H = I^2RT$

2–6. Induction heating.

2–7. Electrons completely penetrate the thickness of material being joined, therefore, EB welding does not depend upon thermal conduction for penetration.

2–8. The green light from the xenon flash tube excites the Cr atoms to a higher energy level, eventually triggering a pulse of monochromatic unidirectional light from the partially reflective end of the crystal.

2–9. The advent of lunar travel and oceanography exploration.

2–10. He must be well versed in metallurgy, chemistry, mechanics, electronics, and mechanical applications.

2–11. The atmosphere.

2–12. Mechanical, chemical, and electrical.

CHAPTER 3 ANSWERS TO PROBLEMS

3–1. The friction welding process would be best for joining this number of idlers. There are several reasons for this selection. However, the three most obvious are: (1) this process can be easily automated, (2) the process creates a narrow heat affected zone, and (3) it is a very rapid joining process.

3–2. The explosive process. This process gives good bonding across wide surfaces and is readily adaptable to the joining of heavy plates of dissimilar metals.

3–3. The diffusion bonding process. If 100 % bonding is necessary, then an intermediate layer of bonding material would be required.

3–4. There is no such thing as the ideal fuel gas; however, each has its desirable qualities and each, its undesirable qualities. Acetylene, for example, gives a clean burning neutral flame with a propagation rate of 17.7 ft/sec. However, it is atomically unstable under pressure and can be dangerous. This compared to natural gas, which is a stable gas, has a propagation rate of only 8.2 ft/sec, and does not give a clean burning neutral flame.

3–5. The two pieces being joined by the thermit process are preheated prior to welding. This along with the heat stored in the mold from the preheating and that from the action of the thermit provides for a slow rate of cooling. Also with thermit welding, the pieces being joined are well shielded from atmospheric contamination.

3–6. The oxyacetylene torch takes equal parts of oxygen and acetylene from the cylinders. Through the design of the torch the other $1\frac{1}{2}$ parts of oxygen is taken from the air.

3–7. Tolerance allowance on 1 in. thick plate is 3/32 in. Therefore, the diameter may vary 6/32 or $\frac{3}{16}$ in., considering both sides of the diam., $10 \pm 3/16 = 9\,7/16$ in. to $10\frac{3}{16}$ in. (size range).

3–8. No, this amount of drag would not be excessive.

$$Drag \text{ (percentage)} = \frac{d}{t}(100)$$

$$\frac{0.125}{2\text{ in.}}(100) = 6.25\%$$

CHAPTER 3 ANSWERS TO QUESTIONS

3–1. The six principles for mechanical joining are forge, cold, explosive, friction, ultrasonic, and diffusion joining.

3–2. The metal at the faying surfaces is generally stretched until the oxides are broken and dispersed, except for the diffusion processes in which the faying surfaces must be very clean from the start.

3–3. The faying surfaces are adequately smoothed through pressure and heat except for ultrasonic, where an excitation of atoms takes place.

3–4. The oldest known welding process is "forge welding"; it was generally used to weld low carbon steels and wrought iron.

3–5. At elevated temperatures, metals oxidize more readily and make it increasingly difficult to get metal to metal bonding.

3–6. Fluxes are used in several welding and most brazing operations, for and in, lowering the melting temperature and dissolving the oxides.

3–7. Forge welding produces better stress relieved joints over most other processes because the entire piece, or at least large sections, is at elevated temperatures when being worked, which contributes to better stress relieved weldments.

3–8. Often the "cold welding" process is used to seal off the ends of copper tubes of the compressor on a refrigeration system.

3–9. Explosive welding is a directional detonation process. As the two pieces of metals collide, most of the air between the two pieces is expelled. The collision plastically deforms the metal surfaces and bonding takes place.

3–10. Pressure, timing, and the braking device influence the quality of weldment with a friction welder.

3–11. The difference between the conventional friction welding process and Caterpillar's inertia welding process is that the conventional process uses a braking device whereas Caterpillar uses the flywheel principle of the proper mass to complete the weld.

3–12. Elastic hysteresis is internal friction of the atoms caused by the cycles characteristic to the ultrasonic process.

3–13. The four variables necessary for making quality joints by diffusion welding are: surface cleanliness, smoothness, time, and pressure.

3–14. The most commonly used applications for the oxy-fuel flame are soldering, brazing, braze welding, fusion welding, metal spraying, hardsurfacing, cambering, heat treatment, and ferrous metal cutting.

3–15. The oxy-hydrogen flame is used only for underwater welding and cutting where the depth of water creates too much pressure for safety with acetylene.

3–16. The seven most commonly used fuel gases for the oxy-fuel process are: natural gas, butane, propane, acetylene, Flamex, Mapp, and hydrogen.

3–17. Oxygen does not burn, it only supports combustion, therefore, is a gas other than a fuel gas.

3–18. The only fuel gas used successfully to fusion weld carbon steels is acetylene. The reason for this is that only the oxyacetylene flame gives a completely neutral flame.

3–19. An oxidizing flame burns with an excess of oxygen, preventing a neutral flame or complete combustion. It is used primarily for removing oils, grease, and paints from metal surfaces. The oxyacetylene cutting flame would be considered as highly oxidizing.

3–20. A carburizing flame burns with an excess of fuel gas preventing a neutral flame. It is used primarily for applications requiring a soft flame such as soldering and brazing.

3–21. The chief function of a gas regulator is to automatically reduce the gas pressure in the tank to a working pressure and to control its flow.

3–22. The atomic hydrogen process is used most often in the resurfacing of worn parts.

3–23. Thermit welding gives a joint free of contamination; the slow cooling gives good metallurgical properties. It requires little initial investment in equipment and has good control over distortion and residual stresses.

3–24. Drag in flame cutting refers to the difference in horizontal distance from where the oxygen enters the top of the metal and the point where the slag emerges at the bottom of the cut.

3-25. It is advisable to preheat medium and high tensile strength steels, thus avoiding hard edges and possible cracking under load.

3–26. Cast iron has a higher melting point oxide than the base metal and contains considerable graphite with more impurities that break up the cut. To make a successful cut, the metal must be worked (poked with a rod as it is brought to the molten state).

3-27. Nonferrous metals and stainless steel can be successfully flame cut using the powder cutting process.

3-28. (a) It is particularly advantageous to use the oxygen lance for piercing holes in steel. It is also used to advantage in very heavy plate cutting.
(b) The powder lance is used to advantage in cutting concrete.

CHAPTER 4 ANSWERS TO PROBLEMS

4-1. $(I \text{ Load})^2 = \dfrac{(200)^2}{100} \times 60$

$\dfrac{40000 \times 60}{100} = \dfrac{2400000}{100}$ or 24,000

$(I \text{ Load})^2 = \sqrt{24{,}000}$ or 141 amperes at

100 % duty cycle

4-2. % duty cycle $= \dfrac{I \, (\text{Rated})^2}{I \, (\text{Load})} \times$ rated duty cycle

% duty cycle $= \dfrac{(200)^2}{(400)^2} \times 60$ or $\dfrac{2{,}400{,}000}{160{,}000}$

% duty cycle $= 15\%$

4-3. The EB process is best for welding reactive metals such as zirconium because the system is free from oxygen and nitrogen atmosphere.

4-4. The current density for a 1/8 in. electrode using 124 amperes is:

$125 \div 0.01227 = 10{,}106 \text{ amps/in}^2$

4-5. The energy in both the EB welder and the microwave oven completely penetrates the thickness of the material, whereas traditional processes heat only the surfaces and penetration must be achieved through conduction.

4-6. The circle to be cut has a 5 ft. diameter. Plasma arc cutting is done at 70 ipm. Flame cutting is done at 20 ipm.

$Distance = \pi D = 3.1416 \times 5 \times 12 = 188.5 \text{ in.}$

1.88.5/70 = 2.6 min. for plasma arc cutting.
188.5/20 = 9.42 min for flame cutting.

CHAPTER 4 ANSWERS TO QUESTIONS

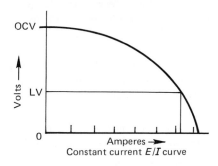

Constant current E/I curve

Constant potential E/I curve
OCV—Open Circuit Voltage
LV— Load Voltage

FIGURE A4-2.

4-1. The three natural divisions for the electrical energy joining group are: gas resistance, solid resistance, and liquid resistance.

4-2. The schematic for the CP (constant potential) and CC (constant current) · E/I curves are shown opposite.

4-3. The penetration patterns for each of the three currents when using a stick electrode with CC current is:

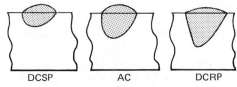

DCSP   AC   DCRP
DCSP—Direct Current Straight Polarity
DCRP—Direct Current Reverse Polarity
AC—Alternating Current

FIGURE A4-3.

4-4. For overhead welding with the stick electrode, DCRP current is recommended.

4–5. The only control the operator has over the voltage of his power source when welding with a CC machine is by the length of the arc. As the arc becomes longer the voltage rises.

4–6. Every welding power source has two voltages, namely, open circuit voltage (OCV) and short circuiting voltage (SCV). The OCV is with machine running and no amps flowing. The SCV is when metal is being deposited from the electrode to the base metal.

4–7. Arc blow is the result of the magnetic field deflecting the arc. It is encountered only with dc current.

4–8. Pinch is the result of the magnetic field forming perpendicular to the wire and acts to overcome the surface tension of the molten metal, pinching it off.

4–9. The steeper the slope, the more often the short circuits of the arc take place.

4–10. The modes of metal transfer for the GMA welding process are short circuit transfer, globular transfer, and spray transfer.

4–11. Duty cycle for a welding power source is the percentage of the time that the power supply must deliver its rated output in each of a number of successive ten minute intervals.

4–12. The rectifier is a device used to change ac current to dc current by preventing the current from flowing in more than one direction. The two commonly used rectifiers in a welding power source are the selenium stack and the silicon diode.

4–13. The only control the operator has over the amperage with a CP power source is through the wire feed speed.

4–14. The HF permits arc starting without bringing the electrode in contact with the base metal. The reverse cycle of the ac current cleans the aluminum oxide from the base metal while the straight cycle cools the tungsten electrode.

4–15. The wire feed drive motor is controlled by the arc voltage; therefore, as the arc becomes long, the voltage rises and the motor speeds up feeding the wire fastener.

4–16. The electro-slag and the electro-gas processes weld only in the vertical position and complete heavy plates in one pass.

4–17. The critical points of resistance for the resistance spot welder are each of the two electrodes, each of the points of contact of electrode to base metal, the two pieces of metal being joined, and the faying surfaces where the nugget is to be found.

4–18. The two kinds of guns used in plasma arc processing are transferred and nontransferred. The transferred gun uses the base metal as the anode whereas the anode is a part of the gun for the nontransferred.

4–19. To change high voltage low amperage to low voltage high amperage, a transformer would be used. This principle is used in the resistance spot weld.

4–20. The thermal efficiency of the EB welder is 95 to 98 % compared to a 10 % thermal efficiency of the shielded metal arc process.

4–21. The letters in the word LASER stand for: *L*ight *A*mplification by *S*timulated *E*mission of *R*adiation.

4–22. The two kinds of welding lasers in use today are: the pulsed ruby and the CO_2.

4–23. The arc-air torch can be automated for cutting and grooving by attaching it to a radiograph. (Not covered in the text, but the student should be able to think this one through from previous information.)

4–24. Plasma arc cutting is limited to conductive material because it employs a transferral type of torch, where the workpiece is the anode.

CHAPTER 5 ANSWERS TO PROBLEMS AND QUESTIONS

5-1. (a) Butt joint $= \frac{1}{4} \times 2 \times 30{,}000 = 15{,}000$ lb. Scarf joint $= 0.750 \times 2 \times 30{,}000 = 40{,}500$ lb.
(b) Yes ($4t$ is sometimes used).
(c) It is time consuming to prepare.
(d) It provides a smooth surface with more strength than a butt joint.

5-2. (a) Butt joint $= \frac{1}{4} \times 2 \times 45000 = 22{,}500$ lb. Lap joint $= 3(\frac{1}{4}) \times 2 \times 45000 = 67{,}500$ lb.
(b) It does not present a smooth surface. It has a tendency to peel.
(c) It can be made with an offset.

5-3. (a) Machine the 1.750 tube end.
(b) A clearance of 0.003 in./side would require the inside tube to be machined to $1.750 - 0.006 = 1.744$ in. This is optimum. The range would be 1.746 to 1.740 in.

(c) $L_1 = \dfrac{TtF}{Ss}$.

Small tube o.d. $= 1.750 |$ weakest
 i.d. $- 1.500 |$ member

$T = \dfrac{\pi d^2}{4}$ of o.d. $- \dfrac{\pi d^2}{4}$ of i.d. $\times 64000$ psi

$\quad = 2.4 - 1.76 \times 64000$

$\quad = 40{,}960$ lb

$\quad t - 1/8$ in. $\quad F = 3$

$\quad = \dfrac{40{,}960 \times 0.125 \times 3}{20{,}000}$

$\quad = 0.77$ in.

The required overlap will be 0.77 in., or approximately 3/4 in.

5-4. Basic clearance desired per side $- 0.004$ in., or 0.008 in. on o.d. Coefficient of expansion of stainless pipe $= 0.0000085$. $(2100 - 70) = 0.017$ in. Coefficient of expansion of copper pipe $= 0.0000098$ $(2100 - 70) = 0.020$. Reduced clearance due to differential in coefficient of expansion $= 0.020 - 0.017$ in. $= 0.003$ in. Therefore, the copper pipe o.d. should be $1.750 - (0.008 + 0.003) = 1.739$ in.

5-5. Pressure on end of the tank $= \dfrac{\pi d^2}{4} \times 8000$
$= 402{,}123$ lb. Area in shear $= \pi d \times 3/8 = 9.4$ in.^{2}. $402{,}123 \div 9.42 = 42{,}688$ psi. The $3t$ brazed lap joint will not be adequate since 42,688 psi is more than twice the required 20,000 psi, the rated tensile shear value of the braze joint.

5-6. Stress $= L/A$

$\quad = \dfrac{5000}{\dfrac{\pi d^2}{4}} = \dfrac{5000}{0.196}$

$\quad = 25{,}510$ psi.

The yield strength of mild steel is 30,000 lb. Therefore, the coating will not be strained.

5-7. (a) Cut out sections of the elbow, apply hardsurfacing, and reweld as shown.
(b) Martensitic alloy iron is chosen because of its excellent abrasion resistance. It can be applied by the manual stick electrode process.

Figure A5-7.

5-8. The statement is quite inaccurate since the brazing process is well suited to mass production techniques, particularly by the use of jigs, furnace brazing, or index tables similar to that shown in Fig. 5-15.

5-9. (a) A eutectic solder has one temperature point at which it is completely liquid rather than a range.
(b) Two eutectic solders are: tin-lead, 63% lead and 37% tin, liquidus $= 361\,°F$; and zinc-aluminum, 95 Zn and 5 % Al, liquidus $= 720\,°F$, used for soldering aluminum.

5–10.

Qualities				
Joining Method	Strength	Appearance	Visual Inspecta-bility	Ease of Field Repair
Brazing	best	good	fair	good
Solder-ing	fair	good	fair	good

5–11. (a) Since the wear is uneven, it will be best to machine it down on a lathe. At that time it could be grooved with a threading or sharp pointed tool and knurled for better bonding. However, it could also be machined and blasted with aluminum oxide abrasive to provide a good surface for bonding.

(b) The recommended bond coat is 0.010 in. per side plus 0.005 in. for each additional inch up to a total of 0.040 inches. For this shaft it would be $2(0.010) + 2(0.005) = 0.030$ in.

(c) 9,500 psi.

CHAPTER 6 ANSWERS TO PROBLEMS

6–1. (a) An epoxy adhesive will be satisfactory.

(b) 1. It has a long term service temperature rating of 400 °F.
    2. Resists brine and moisture.
    3. Cure time may be as low as 10 min at 350 °F.
    4. No clamping pressure is required during cure.
    5. The lap–shear strength is more than adequate, 2,500 psi, and all that is required is 500 psi.

6–2. (a) Recommended surface preparation for the aluminum coupler consists of a vapor degrease in trichloroethylene followed by a rinse, and a sulfuric acid chromic etch followed by forced air drying. The pipe ends will be cleaned by abrading with a tungsten carbide abrasive disc, followed by degreasing with chloroethylene.

(b) Epoxy or fluorosilicone.

(c) It will be subject to moisture and gas. Either epoxy or fluorosilicone do not require clamping pressure during cure.

(d) Make up a test pipeline of a hundred feet or more with several of the intended type adhesive joints. Bury it in the ground where it will be subject to as much of the environmental conditions as can be expected and run the gas through it at the intended pressures for a period of a year. Take it out after the test period and subject it to pressures two times higher than expected operating pressure.

6–3. (a) Adhesive chosen, Table 1.2, highest shear strength, epoxy nylon is 6,000 psi at room temperature (RT).

$$\text{Lap area} = 10t \times 2 = 10(0.060)2 = 1.2 \text{ in.}^2$$
$$= 1.2 \times 6000 = 7,200 \text{ lb}$$

(b) Four rivets 0.175 in. diam.

$$A = .0232 \times 4 = 0.0928$$
$$\text{Rivet } S_s = 26,000 \text{ psi}$$
$$= 26,000 \times 0.0928 = 2,412 \text{ lb}$$

Basic tensile strength for 2024–T3 = 64,000 psi. Tensile strength for one thickness = $0.060 \times 2 \times 64000 = 7,680$ lb.

(c) No, the joint strength will not exceed the strength of the material.

(d) Epoxy at RT = $2,500 S_s = 1.2$ in.$^2 \times 2500 = 3,000$ lb.

6–4. (a) *Static load:*

$$S = S_s \times A$$

$$= 30000 \times \frac{Td^2}{4} = 30000 \sqrt{\frac{5 \times 4}{3.14}}$$

$$= 30000 + \sqrt{.636} = 75,693 \text{ lb}$$

Dynamic load:

$$S = S_s \times \text{joint length in inches}$$

$$= \frac{5000 \times 2}{2} = 5,000 \text{ lb}$$

The adhesive joint 2 in. wide would be able to withstand a dynamic load of 5,000 lb with a safety factor of 2.

(b) They reduce the tendency to peel.

(c) No, not if you want the joint to have some flexibility. (Note the bolt as drawn would not be directly involved in stress transmission because of the clearance.)

6–5. (a) The rivets being hollow can allow some flexibility.

(b) $S = S_s \times A$

$$= 41000 \times \frac{\pi d^2}{4} \text{ of o.d.} - \frac{\pi d^2}{4} \text{ of i.d.}$$

$$= 41000 \sqrt{\frac{.468 \times 4}{3.14}} - \sqrt{\frac{.218 \times 4}{3.14}}$$

$$= 41000 \quad .771 - .526 = .245$$

$$= \frac{41000 \times .245}{3} = 3,350 \text{ lb}$$

6–6.

Adhesive	Type TS	Type TP	Advantages	Limitations
Straight epoxy		x	High shear strength Contact bonding	Low peel strength Poor impact strength Temp. up to 300 °F
Neoprene phenolics	x		Excellent peel strength Good flexibility Absorb vibration	Must be cured with heat and pressure
Modified epoxy	x		Either heat cure or room temp. by chemical activator Excellent, metal to metal/concrete/glass	Lack flexibility in bond line Temp sensitive
Cyanoacrylate (Eastman 910)		x	Rapid room temp. cure High strength	High cost Poor heat resistance Poor shock resistance Thin glueline required

6–7. (a) See Fig. 6–20. Creep = 200 h at 270 °F = 0.0035 in. 100 h at 75 °F = 0.0024 in. Therefore, there would be 0.0035 − 0.0024 = 0.0011 in. more creep at 270 °F for 200 h than there would be for 100 h at 75 °F.

(b) Creep is stabilized after about 50 hours. Therefore, it should stay at about 0.0025 in. and never reach 1/8 in.

6–8. See Fig. 6–23. Lap shear strength of aluminum at 300 °F = 3,800 psi. 300 °F = 2,500 psi. Therefore, aluminum = 2 × 2 × 3800 = 15,200 lb. Stainless steel = 2 × 2 × 2,500 = 10,200 lb.

CHAPTER 7 ANSWERS TO PROBLEMS

7–1. $S_s - \dfrac{P_s}{An}$,

$$A = \frac{P}{S_s} = \frac{100000}{50000} = 2.00 \text{ in.}^2 = \frac{\pi d^2}{4},$$

$$d = \sqrt{\frac{2.0 \times 4}{3.14}} = \frac{1.59}{4} = .398$$

Yes, four $\frac{1}{2}$–13 UNC bolts will be large enough.

7–2. (a) M9.525 (Table 7.0).
(b) No.
(c) M6 refers to 6 mm for the o.d. (Table 7.0).
$6 \times .03937 = 0.2362$ in.

7–3. (a) $S_y - \dfrac{P}{A}$,

$$A = \frac{P}{S_y} = \frac{2000}{37500} \quad \text{(Assume } S_y = 75\ \% \text{ of tensile strength)}$$

$$= 0.0534 \text{ in.}^2 = \frac{\pi d^2}{4},$$

$$d = \sqrt{\frac{.0534 \times 4}{3.14}} = 0.260$$

The nearest bolt size is $\frac{1}{4}$–20 UNC thread.
(b) Depth of hole to be tapped = 1.5×0.250 (Table 7.5) = 0.375 in.
(c) No. 7 (Table 7.5).
(d) Taper, plug, and bottoming.
(e) 0.3 in. + thread run out (2 threads) + clearance (2 threads) $\approx$ 0.3 in. + 2(1)/(20) + 2(1)/(20) = 0.500 in.

7–4. Shear stresses will be encountered.

$$G = \frac{S_s A N}{2W} \quad A = \frac{2GW}{S_s N} = \frac{2(25)(1900)}{30000(4)}$$

$$= 0.792 \text{ in.}^2$$

$$= 0.792/4 = 0.198 \text{ in.}^2$$

The nearest standard bolt will be 9/16–18. A safety factor of 2 (noncritical) will make the bolt size required $2 \times 0.198 = 0.396.^2 = \frac{7}{8}$–9 (see Table 7.4).

7–5. (a) $0.750 + 3\% = 0.772$
(b) Nearest standard drill sizes 0.7656 (49/64) or 0.7812 (25/32).
(c) Yes. A drill will not be within the required size.

7–6. (a) $G = [268(8)]/[2(40)] = 26.8$ lb.
(b) $G = [(60000) \times (0.0085) \times (8)]/[2(40)] = 51$ lb.

7–7. (a) $T = CDP = 0.12 \times 0.750 \times 10000 = 900$ lb, $= 900/10 = 90$ lb reading on a 10 in. torque wrench.
(b) 1,530 lb (Table 7.3).
(c) $P = 11200 \times 4 = 44{,}800$ lb.

7–8. (a) The drill size should be the pitch diameter of the thread.

$$PD = \text{o.d.} - \frac{1}{n} = 0.250 - \frac{1}{20} = 0.200 \text{ in.}$$

The nearest standard drill size, 0.201 in., is no. 7.
(b) BT or BF (thread-cutting) types.

7–9. (a) $P_s = S_s(\pi d^2)/(4) = 10 \times 14000 \times (\frac{1}{2})^2 (\pi)/(4) = 27{,}500$ lb.
(b) $P_b = S_{std} = 10 \times 26000 \times \frac{1}{2} \times \frac{1}{2} = 65{,}000$ lb.

7–10. (a) Tension load support = $0.0503 \times 4 \times 37500 = 7{,}545$ lb. The four bolts selected could actually support 7,500 lb.
(b) $G = (50{,}000 \times 0.0503 + 4)/(2 \times 50) = 100.6$ lb.

The four bolts could support this motor with a shock load of up to 100 lb.

7–11. (a) Since there are two planes of shear, the load $P = F/2 = 20000/2 = 10{,}000$.

$$A = \frac{P}{S_s} = \frac{10000}{30000} = 0.333 \text{ in.}^2 = \frac{\pi d^2}{4}$$

$$d = \sqrt{\frac{0.333 \times 4}{3.14}} = 0.651$$

Use a 3/4 in. diam bolt (0.750). (Friction between the plates has not been considered.)
(b) The root diam = $D_r = D - 2 \cdot (1/n) = 0.750 - 0.222 = 0.528$ in. No, this bolt will not be large enough.

7–12. (a) One quick and easy way is to measure the elongation of the bolt or screw. Measure the bolt length before tightening it. A micrometer can be used to get the length accurately. The elongation should be about $+3\%$.

　　Another method will be by the turn-of-the nut. When the nut is firmly seated on the surface, the elongation can be determined by the pitch of the thread. As an example, a 16 pitch thread will advance 1/16 in./rev. or 0.0625 in. This method is simple but requires more care. The accuracy is considered to be within $\pm 15\%$.

7–13. The base plate thickness should be a minimum of 1/3 the weld base diam, $1/3 \times 1/2 = 1/6$ or 0.166. The nearest standard gauge would be 8 or 0.164 in.

7–14. (a) The approximate tensile strength of $\frac{1}{2}$ in. diam stud is: $S_t = (\pi d^2)/(4) \times 50{,}000 = 9{,}800$ lb.

(b) $S_y = (\pi d^2)/(4) \times 40000 = 7{,}800$ lb. Please note these are only approximate answers. For exact answers use threaded area as given in Table 7.4.

$$S_t = .1376 \times 5000 = 6{,}800 \text{ lb.}$$

$$S_y = .1376 \times 40{,}000 = 5{,}500 \text{ lb.}$$

7–15. The tap drill size for $\frac{1}{4}$–20 $= D - 1/N = 0.250 - 1/20 = 0.200$. Table 7.5 $= 0.201$.

7–16. (a) Tap drill size of 75 % thread (Table 7.5) $= 0.6562$ in., or 21/32 in.

(b) Start with a plug tap and finish with a bottoming tap.

(c) Drill for hole in top plate will have to make clearance for the bolt, o.d. $= 0.750 + 1/16$ in. $= 13/16$ in.

(d) (Table 7.3) $= 1259$ in./lbs.

CHAPTER 8 ANSWERS TO PROBLEMS

8–1. (a) Point of eutectic—the tin and lead is alloyed as to meet its lowest possible melting temperature.

(b) 30/70 tin lead solders have a reasonable range of plastic properties.

(c) A 75/25 costs approximately 5 times that of 25/75.

(d) 50/50–general purpose–reasonable cost.
60/40–electrical–good strength.
30/70–auto body solder–low in cost and long plastic range.

8–2. (a) 361 °F = 183 °C [Equation: $C° = (F - 32) \times 5/9$].

(b) 1510 °C = 2750 °F [Equation: $F° = 9/5C + 32$].

8–3. To gain the deepest penetration with the GTA process helium gas will be used. Helium has an ionization potential of 24 electron volts, which provides a hotter arc than that of argon.

8–4. (a) 95 % Ar and 5 % O_2.

(b) 100 % helium.

(c) 75 % H–25 % Ar.

8–5. The end product would be acetylene (C_2H_2) gas.

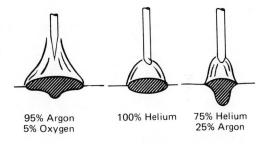

| 95% Argon 5% Oxygen | 100% Helium | 75% Helium 25% Argon |

FIGURE A8–4.

8–6. Which fuel gas to choose. This may depend upon the following:

1. Is natural gas abundantly available in the area?
2. Is natural gas already available within the shop?
3. What fuel gases are available through the local distributor?
4. Costs of each per lb of metal cut.
5. Quality of cut desired.
6. Desired time for preheat.

CHAPTER 8 ANSWERS TO QUESTIONS

8–1. American Welding Society and American Society for Testing Materials.

8–2. A.5.XX–XX—American Welding Society.

8–3. American Welding Society specification for surfacing welding rods and electrodes.

8–4. On the coating near the brushed end.

8–5. An electrode, iron powder coating, low hydrogen, a tensile strength of 70,000 psi.

8–6. A steel which has not been deoxidized.

8–7. A filler rod giving 60,000 psi–Ts (tensile strength) for carbon steel to be used with a gas flame.

8–8. A low hydrogen electrode giving 100,000 psi–Ts in the as welded condition with a specific chemical composition.

8–9. A copper aluminum bronze coated electrode.

8–10. A deoxidized, oxygen free, electrolytic tough pitch.

8–11. (a) Silver. (b) Copper phosphorus. (c) Nickel. (d) Copper and copper zinc. (e) Magnesium (f) Precious metals. (g) Aluminum silicon.

8–12. (a) Solidus—the fine line temperature between a complete solid and where material begins to melt.
(b) Liquidus—the fine line temperature between complete liquid and beginning to solidify.

8–13. Source of contamination.

8–14. The chemistry of a bare electrode and rod filler metal for welding titanium and titanium alloy.

8–15. F = flux.
7 = Minimum Ts in 10,000 psi.

1 = Impact properties with 20 ft lbs at 0 °F.
E = Electrode.
M = Medium manganese content (1.25% max).
12 = Nominal carbon content in electrode.
K = Made from steel that has been silicon killed.

8–16. (a) Shielding. (b) Fuel. (c) Ionized to generate heat within an arc.

8–17. A gas that is chemically inactive. They are composed of atoms whereby their outermost shells of electrons are complete.

8–18. Argon, helium, and xenon.

8–19. A fuel gas when combined with some other gas such as oxygen produces a chemical reaction, which in turn generates heat.

8–20. The seven fuel gases are: acetylene, Mapp, Flamex, natural gas, propane, butane, and hydrogen.

8–21. Influences the penetration patterns.

8–22. To preserve the tungsten electrode, inert gas must always be used with the GTA process.

8–23. The high ratio, O to C_3H_8 produces a highly oxidizing flame, which when used for fusion welding of steel oxidizes it rather than melts it.

8–24. Hydrogen is generally used as a fuel gas for underwater welding and underwater cutting.

8–25. Xenon gas is used in the pulsed laser welders. Its purpose is to excite the ruby atoms to a higher energy level.

8–26. In case of gas seepage, it will rise and disperse. If it were heavier than air, it would have the tendency to collect in cavities.

CHAPTER 9 ANSWERS TO PROBLEMS AND QUESTIONS

9–1. Oxyacetylene welding, stick electrode, GMA, submerged arc, and electron beam.

9–2. 1018, 5100, 4140, and E51100. See Tables 9.6, 9.7, and 9.8.

9–3. They both have precipitation hardening types.

9–4. The nickel makes the transformation very sluggish, hence a quench does not affect it.

9–5. (a) $CE = \%C + \dfrac{\%Mn}{4} + \dfrac{\%Ni}{20}$

$\qquad + \dfrac{\%Cr}{10} - \dfrac{\%Mo}{50} - \dfrac{\%V}{10} + \dfrac{Cu}{40}$

$\qquad = .40 + \dfrac{1.80}{20} + \dfrac{.45}{10} - \dfrac{.25}{50}$

$\qquad = .40 + .09 + .045 - .005$

$\qquad = 0.53$

A 4340 steel would be equivalent to a 0.53% carbon steel.
(b) Use low hydrogen electrodes along with pre-heating.

9–6. (a)

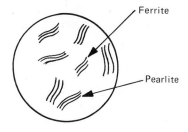

FIGURE A9–6.

It would be a pearlite structure with excess ferrite. (b) Yes, if the steel has more than 0.35 carbon, therefore, it should be welded with low hydrogen electrodes along with preheating.

9–7. Variations in cross section will change the cooling rate. If the steel is hardenable, heavy sections will have martensite in the HAZ. The same would be true for very light sections. Also, the grain size would be smaller where there is any quenching action due to heavy cross sections or very thin sections.

9–8. Intermetallic compounds such as $CuAl_2$ that form during precipitation block the slip planes and cause dislocations to pile up, thus increasing the strength.

9–9. The weld metal becomes molten but is contained by the HAZ that acts as a mold while it solidifies.

9–10. Preheating decreases the rate at which a welded structure will cool. If the metal is hardenable, it will not be as hard or have hard spots, and cracking will be eliminated.

9–11. Malleable irons go through a prolonged heat treatment to obtain the proper structure. Welding with its relatively short heat and cool cycle will produce some martensite, which will be hard and brittle.

9–12. Use a lot of heat so that the structure will not have localized stresses, or use very limited heat and do not allow the casting to get any hotter than can be touched with the bare hand.

9–13. When a single pass weld is made, there is no opportunity for grain refinement. The multipass weld provides the necessary heat for recrystallization of the grain structure from the previous pass.

9–14. A time-temperature-transformation diagram such as the one shown in Fig. 9–8 allows you to determine what type of structure to expect for a given type of steel if the cooling rate is known.

9–15. (a) A metal that has been heated and cooled quickly will not allow the metal to relax but will have stresses locked in it. Also, metals that have been rigidly restrained during a heating cycle will have locked-in stresses.
(b) Locked-in stresses may be relieved by reheating the metal as shown by Fig. 9–4. A process anneal for stresses induced by cold working can be relieved at temperatures below the lower critical. The greater the amount of cold work, the lower the temperature required. A process anneal is usually all that is required for locked-in stresses.

9–16. (a) The structure will be martensitic with hardness of about R_c 30.
(b) The normal properties may be restored by reheating the structure to 900 °F from 3 to 6 hours and air cooling.

CHAPTER 10 ANSWERS TO PROBLEMS AND QUESTIONS

10–1. (a) The stresses in the welds will be as shown below:

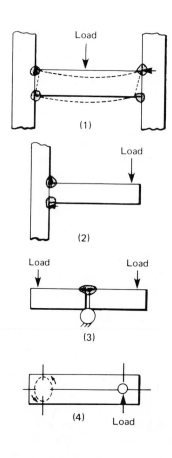

(1)

(2)

(3)

(4)

FIGURE A10–1.

(b) 1. Tension in top welds and compression in the bottom two. 2. Tension and compression. 3. Tension. 4. Shear.

10–2. (a) Formula from Table 10.4.

$$\text{Ave } S_s = \frac{.707P}{hl} = \frac{.707 \times P/2}{hl}$$

$$= \frac{.707 \times 5000}{3/8 \times 4 + 1/4 \times 4} = \frac{3535}{2.5}$$

$$= 1.414 \, \text{psi}$$

(b)

$$S = \frac{P}{hl(b + h)} \sqrt{2L^2 + \frac{(b + h)^2}{2}}$$

$$= \frac{5000}{3/8 \times 4[2 + (3/8)]}$$

$$\times \sqrt{2 \times 10^2 + \frac{[2 + (3/8)]^2}{2}}$$

$$= \frac{5000}{3.56} \sqrt{200 + 2.82}$$

$$= \frac{5000}{3.56} \times 14.24$$

$$= 20,000 \, \text{psi}$$

10–3.

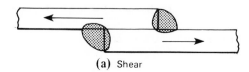

(a) Shear

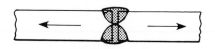

(b) Tension

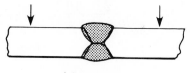

(c) Bending

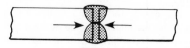

(d) Compression

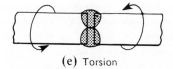

(e) Torsion

FIGURE A10–3.

10–4. $L = \dfrac{50\,k\,(\text{load})}{3.98\,(\text{allowable load/in.})}$

Based on $(1/4\,\text{in}{-}1/16\,\text{in})$

$= 12.5\,\text{in}.$

No, the two 4 in. fillet welds will not be adequate.

10–5. (a)

$$S = \frac{20000}{1.5 \times 4}$$

$$= \frac{20000}{6}$$

$$= 3{,}333\,\text{lbs}.$$

(b) $\quad S = \dfrac{1.414 \times 20000}{(9/16 \times 3)4}$

$$= \frac{28101}{6.75}$$

$$= 4{,}163\,\text{lbs}.$$

10–6. (a) The strength of the fillet weld, regardless of the direction of the applied force, is based on the cross-sectional area of the throat multiplied by the allowable shearing stress for the weld metal.

$$X = 0.25 \cos 45° = 0.25\,(0.707)$$

$$X = 0.177\,\text{in}.$$

Allowable force per inch of weld = 21,000 × 0.177 = 3,700 lb/in. A 48,000 lb load is to be supported. If Y is the total length of the fillet weld: $Y = 48000/3700 = 13\,\text{in}.$, $L = 6\frac{1}{2}\,\text{in}.$ of fillet on each side.

(b) 1. A fillet weld loaded at right angles is about 30% stronger than if loaded in parallel. Allowable for per inch of weld:

$$= 3700 + 0.3 \times 3700 = 3700 + 1110$$

$$= 4{,}810\,\text{lb/in}.$$

Total fillet length = $Y = 48000/4810 = 10\,\text{in}.$ For fillet on each side: $L = 5\,\text{in}.$ If the force of 48,000 lb is resolved in the direction of the throat section, then for the throat area:

Force = 48000×0.707

$$= 34{,}000\,\text{lb in the throat direction}$$

$$Y = \frac{34000}{4810} = 7.05\,\text{in}.$$

$L = $ for fillet on each side $= 3.52\,\text{in}.$

Either of these two lengths can be chosen, but for safety it is advisable to select the longest one.
2. This weld would not be satisfactory because we require 5 or 3.52 in. (whichever is chosen) length of fillet for enough strength, but we have only 2 in. width of plate "B". Therefore, this is not possible in our weld design.

10–7. The length of the weld, as shown in Fig. P10–5, is the perimeter of a circle whose diameter is $D - \frac{1}{2}$, the size of the fillet weld. The length is $(3.0 - 0.5) \times 3.1416 = 7.85\,\text{in}.$ Load the weld will support = $7.85 \times 6370 = 50{,}000\,\text{lb}.$

10–8. The applied force is resolved into horizontal and vertical components. The centroid of the two welds will be at A with:

$$\bar{x} = \frac{2 \times 6 \times 3}{6 + 6} = \frac{36}{12} = 3\,\text{in}.$$

$$e = 7\,\text{in}.$$

$$Px = (3/5) \times 10000 = 6{,}000\,\text{lb}.$$

$$Py = (4/5) \times 10000 = 8{,}000\,\text{lb}.$$

Twisting moment = $T = 8000 \times 7$
$$- 6000 \times 4 = 32{,}000\,\text{lb/in}.$$

The polar moment of inertia can be calculated by using the parallel-axis theorem: $J = 2[(6)^3/(12) + 6(4)^2] = 228\,\text{in}.^3$ The components of the direct force per inch of weld are:

$$(qd)x = \frac{Px}{total\ length\ of\ the\ weld}$$

$$= \frac{6000}{6 + 6}$$

$$= 500\,\text{lb/in}.$$

$$(qd)y = \frac{Py}{total\ length\ of\ the\ weld}$$

$$= \frac{8000}{12}$$

$$= 667\,\text{lb/in}.$$

The horizontal and vertical components of the torsional force per inch of weld are:

$$(qt)x = \frac{Ty}{J} = \frac{32000(4)}{228} = 562 \, lb/in.$$

$$(qt)y = \frac{Tx}{J} = \frac{32000(3)}{228} = 421 \, lb/in.$$

$$q_{max} = \sqrt{(500 + 562)^2 + (667 + 421)^2}$$

$$= 1,520 \, lb/in.$$

Inspection of the direct and torsional forces on welds shows that point B has the highest stress. The allowable force per inch of weld, regardless of the direction of the applied force, is: $q = 18000(0.707)W = 12,750W \, lb/in.$, $12,750W = 1520$, and $W = 1520/12.750 = .1192$. Hence, a uniform size of 3/16 in. or 1/8 in. fillet weld should be used throughout.

10–9. Treating the weld as a line: Section modulus of the weld $Sw = (2L^2)/(6) = L^3/3$ (where L is the total length of the weld). Load on the bolts $= F = 690/18 = 38.4$ kips. Bending force on the weld/linear inch $= Fb = (Fe)/(Sw)$, $(Fe)/(L^2/3)$. Vertical shear on the weld/linear inch $= Fv = F/2L$. Resultant force on the weld

$$Fr/inch = \sqrt{Fv^2 + Fb^2}$$

$$= \sqrt{\frac{F^2}{4L^2} \times \frac{9F^2e^2}{L^4}} \frac{F}{2L^2}\sqrt{L^2 + 36e^2} \quad (1)$$

Let h be the fillet leg size.

Allowable load/linear inch $= 18.0 \times .707 \times h$

$$= 12.75h \quad (2)$$

Length of the weld can be calculated by equating (1) and (2), and assuming a value for h:

$$12.75h = \sqrt{\frac{F}{2L^2} \times L^2 + 36e^2}$$

Solving this for L^2:

$$12.75h \times 2L^2 = F\sqrt{L^2 + 36e^2}$$

$$(25.5)^2h^2L^4 - F^2L^2 - 36F^2e^2 = 0$$

$$L^2 = \frac{1}{2}\left[\frac{F}{(25.5)h}\left|\frac{F}{(25.5)h} + \sqrt{\left(\frac{F}{(25.5)h}\right)^2 + 144e^2}\right|\right]$$

Assume $h = \frac{1}{4}$ in. i.e., $\frac{1}{4}$ in. fillet is used.

$$\frac{F}{25.5h} = \frac{38.4}{(25.5)(\frac{1}{4})} = 6.023$$

$e = 2$,

$$L^2 = \frac{1}{2}[6.023\{6.023 + \sqrt{(6.023)^2 + 576}\}]$$

$$= \frac{1}{2}[6.023\{6.023 = 24.744\}]$$

$$= 92.655$$

$$L = 9.625 \, in.$$

Hence use $L = 10$ in.

10–10. The eccentric loading here produces moment in addition to shear. The resisting shear is assumed to act through the center of gravity of the weld group. First, find out the C.G. of the weld group. By symmetry, as shown in the figure,

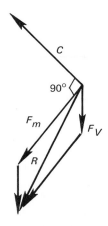

FIGURE A10–10. *Force diagram.*

the C.G. is located halfway up, or 8 in. from the bottom of the bracket. To find the distance over, take moments of weld "areas" (assuming unit width as size of the weld) about the left side and divide by the total "area."

$$X = \frac{2 \times 4 \times \frac{4}{2}}{4 + 4 + 16} = 0.67 \, in.$$

Since the moment of inertia is the sum of the I's about each axis, we have: $I = I_0 + Ay^2$, and $I_0 = (1/12)(bd^3)$ from which $I_x = (0 + 4 \times 8^2) + (0 + 4 \times 8^2) + (1/12)(1 \times 16^3 + 0)$, $I_x = 853$ in^4. and $I_y = 16 \times 0.67^2 + 2 \times 4(2 - 0.67)^2 + (1/12)(2 \times 4^3) = 32.0$ in^4.

The polar moment of inertia is then: $I_p = I_x + I_y = 853 + 32 = 885$ in.^{4} Also, $C^2 = 8^2 + (4 - 0.67)^2$, and $C = 8.67$ in.

The moment arm is $10 - 0.67 = 9.33$ in., and the moment is $25 \times 9.33 = 233$ in./kips. The moment force is:

$$F_m = \frac{M_c}{I_p}$$

$$= \frac{233 \times 8.67}{885}$$

$$= 2.28 \text{ kips}$$

and the shear force is:

$$F_v = \frac{P}{L}$$

$$= \frac{25}{4 + 4 + 16}$$

$$= 1.04 \text{ kips/in.}$$

Plotting them to some convenient scale, the resultant is found to be:

$$R = 2.85 \text{ kips}$$

Find the size of the weld:

$$18000 \times 0.707 \times W = 2,850$$

$$W = \frac{2850}{12750} = 0.218$$

3/16 in. weld can be used.

CHAPTER 11 ANSWERS TO PROBLEMS AND QUESTIONS

11–1. Pulse ultrasonics utilize short pulses as generated by the search unit. These pulses are reflected back by any discontinuities within the part or from the opposite side. Resonant ultrasonics utilize a steady transmission of changing frequencies. One of the frequencies is the resonant frequency of the material whose thickness is being measured. This will produce a standing wave that is detected by the ultrasonic unit and read out directly as a reading of thickness.

11–2. N-rays are better than X rays or gamma rays for testing welds for hydrogen embrittlement, adhesive bonds in honeycomb structure, and in "massive metal" composites. Also in checking the following: lubricant migration in a sealed system; powdered metal compact uniformity; and segregation, brazed joints, and laminated wood.

11–3. Black light is another name for ultraviolet light.

11–4. Sonic testing does not rely on the interpretation of the returning sound waves. It is based on the resonance of the workpiece. Changes in resonance are interpreted as flaws in the workpiece.

11–5. This question is designed to generate considerable discussion on the merits of various NDT methods. More than one answer can be right, however, it will depend upon the criteria set up. Here we are considering 100% inspection of welds at production speeds. The process must be economical.

The two NDT most likely to be chosen for this application are: fluoroscopic inspection (with image intensifiers) and ultrasonic equipment. Both of these methods can be used while the pipe is in motion, and imperfections can be detected immediately.

Fluoroscopy is first choice for the following reasons:
(1) The defects; gas pockets, cracks, and lack of penetration, are all pinpointed as to exact location and can be identified.
(2) Ultrasonics will also locate the defects but is less sensitive to the grosser volume type defects such as slag and gas pockets.

11–6. Explanation of NDT and welding symbols:
(a) A backing weld of a single vee groove type, checked by magnetic particle testing. Weld and testing to be done on the side opposite the arrow.

(b) A backing type single "U" groove type weld to be tested with penetrant on the arrow side of the joint.

(c) A fillet weld to be made opposite the arrow side. It is to be tested radiographically.

(d) A double vee groove weld to be tested ultrasonically.

11–7. N-rays are extremely good at detecting even very minute concentrations of hydrogen, therefore, they can be used to check for suspected hydrogen embrittlement. Since water is an ideal hydrogenous material, any casting voids would be easily detected by first soaking it in water and then subjecting it to N-rays.

11–8. NDT techniques that generate an electrical signal are: ultrasonics, sonic testing, eddy current testing, infrared, N-rays, magnetic reluctance, and microwave testing.

11–9. Several methods can be used. The criteria in establishing which one will be based on cost, production requirements, permanent records required, and if any surface defects are expected. From the description it may be assumed that any difficulties experienced in the forming operation will be caused by internal defects. These can be detected by radiography, particularly X rays. Also applicable is ultrasonic testing. This process lends itself more readily to higher production.

11–10. The electrical properties of a metal can be sensed by eddy current testing. This information can, in turn, be fed into the welder for automatic adjustments.

11–11. (a) The integrity of the surfaces can be checked by wet fluorescent magnetic particle inspection. This process involves magnetizing the part, spraying with a water base solution containing magnetic particles, and making visual examination under an ultraviolet light.

(b) Internal soundness can be checked on a production line basis by submerging the casting in water and checking it ultrasonically. The equipment will be calibrated against known masters.

(c) The hardness and matrix structures can be evaluated by means of eddy current techniques. Again the parts tested are compared to known masters.

11–12. Examples of NDT methods that can be used as called for by this classification are:

Inherent—(cold shuts in castings, hot tears, skin laminations, etc.).
  Testing—eddy current, magnetic particles, and penetrants.
Processing—(fabricating defects, rolling, forging, welding, heat treating, etc.).
  Testing—acoustic emission, eddy current, magnetic particle, microwave, penetrants, and radiography.
Service—(cracks resulting from use).
  Testing—As in processing but deleting acoustic emission.

11–13. Some benefits of NDT. Nondestructive testing allows us to evaluate product quality and reliability at any stage of the manufacturing process without harm. This can result in a considerable savings of time and material during manufacture and insure a reliable product for the customer. Specifically, there will be an increase in the number of acceptable products, a savings in raw materials, less wasted production, more efficient production, and better information for management.

11–14. The delta technique allows the ultrasonic beam to approach the weld from the side, thus giving it a better scanning coverage.

11–15. Bondline voids, bondline variations, and porosity in adhesive joints can be checked by ultrasonics, sonics, thermal scanning, and holography.

11–16. Neutron rays have no direct relationship to atomic numbers and are not affected by density, therefore, lead is no barrier. Water, on the other hand, contains hydrogen, which absorbs neutrons, making it an effective shield.

11–17. The principle of thermal NDT is based on the fact that there will be a difference in the heat buildup in an area of discontinuity when a whole surface is subjected to thermal control. Different methods, such as a thermal coating, are used to detect the points of difference in temperature.

11–18. (a) A production line means of joining the frame body would be by spot welding. (b) An NDT method for checking the joints would be ultrasonically or by acoustic emission. The basis for accept or reject can be made in comparison to a master.

11–19. The question is similar to problem 11–5 in that it is designed to stimulate discussion. Possible considerations are: (a) X-ray or gamma ray, however, the cost would run high ($4.00 per foot of weld inspected). Also, the area would have to be cleared of all personnel due to radiation hazards. (This means building delays.) (b) Penetrant inspection would only be suitable for surface defects. (c) Magnetic particle testing could be used, but it would have to be used first on the root pass and then every 1/4 to 3/8 in. as the weld deposit is built up. (d) The best method is by ultrasonics.

It will be sensitive to the subsurface defects most apt to cause trouble (slag inclusions, underbead cracking, lack of fusion, or incomplete penetration). It also has a fair sensitivity to surface cracks. It also provides defect location with respect to the front and back surfaces. It does not provide a permanent film record, however, the weld can be sketched and the size and location of the defects noted. The cost per foot of weld can be counted in cents rather than dollars, and there is no hazard to the personnel in the area.

11–20. (a) Ultrasonic. (b) X rays. (c) Fluoroscopy. (d) Magnetic particle.

CHAPTER 12 ANSWERS TO PROBLEMS AND QUESTIONS

12–1. The principal types of destructive tests classified as to load application are: (1) tension and compression, (2) bend, (3) impact, and (4) fatigue.

12–2. Yes. It would indicate the materials have about the same toughness.

12–3. The welds called for are as shown.

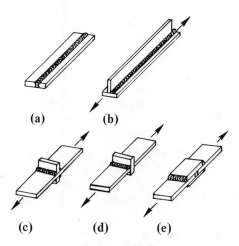

(a)   (b)

(c)   (d)   (e)

FIGURE A12–3.

12–4. (a) About 750 °F. (b) The impact strength or toughness increases. (c) No. (d) About 900 °F. (e) Yes, it is still in the middle of the range.

12–5. (a) Three different types of impact tests are: (1) Charpy or Izod, (2) drop weight, and (3) explosion bulge.
(b) (1) Charpy test measures impact properties. (2) Drop weight test measures brittle fracture. (3) Explosion bulge tests measure high energy rate fracture.

12–6. The diameter of the Brinell impression is measured.

12–7. The hardness numbers will vary according to the load used. Brinell hardness numbers are calculated by the area of indentation in relation to the load.

$$BHN = \frac{\text{load on ball}}{\text{indented area}}$$

12–8. The tensile strength is approximately 500 times the Brinell number: 108 to 150 × 500 = 54 to 75 ksi.

12–9. The fatigue strength rating starting with the best type: (a) Cover plate with tapered ends welded along the edges only. (b) Cover plate, square end welded along the edges only. (c) Square end cover plate welded all around.

12–10. A partial penetration butt weld not adversely affected by fatigue is as shown.

527

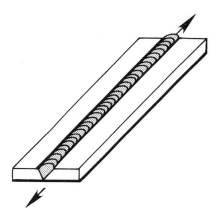

FIGURE A12–10.

12–11. (a) No, only steel Grades F100 and F110 would qualify. (b) TS = 0.500 × 3 × 90 ksi = 135 ksi (minimum).

12–12. Two methods of preparing and welding a 2 in. and a 1 in. plate according to AWS Structural Code D1.1–72. Scale = $\frac{1}{2}$.

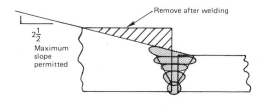

Remove after welding

$2\frac{1}{2}$
Maximum slope permitted

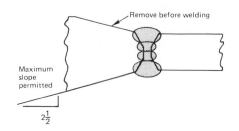

Remove before welding

Maximum slope permitted

$2\frac{1}{2}$

FIGURE A12–12. *Two methods of preparing and welding a 2 in. and a 1 in. plate according to AWS Building Code D1.0–69. Scale = $\frac{1}{2}$.*

12–13. (a) See Table 10.4. $S = (.707P/hL) = (.707 × 40,000/.750 × 4) = 9,426$ psi. (b) Yes, it comes within the 18,000 psi allowable stress.

12–14. (a) Weld area $= P/S_s = 10,600/56,000 = .1895$ in.$^2 = (\pi d^2/4)$.

$$d = \sqrt{\frac{.1895 × 4}{3.14}} = \frac{.491}{5}$$

$$= 0.098$$

The diameter of each weld should be a minimum of 0.098 in., if it is to meet MIL–W–685B specifications.

12–15. Lap joint in shear. $\tau = (P/.707hL) = (50000/.707 × 1 × 6) = 11,800$ psi. Yes. This joint is well within the AISC and AWS codes, which allow 14,850 psi for this joint.

12–16. (a) $F = \frac{1}{2}\Delta = \frac{1}{2} × 0.707 × 15.8 × 10 = 55.8$ kips, $F = \frac{1}{2}\Delta = \frac{1}{2} × 0.707 × 18 × 10 = 63.6$ kips.
(b) $F - \frac{1}{4}\Delta = \frac{1}{4} × 18 × 10 = 45$ kips.

12–17. The state or the local governing unit is able to enforce codes and specifications only after they have enacted it into law.

12–18. A "Certified Weldor" implies that a person is able to do any type of weld anywhere. This is not true. He has been certified only for certain specified welds under certain procedures and conditions.

12–19. No, the certification laboratory only vouches for the fact that a given individual is capable of making a certified weld under certain conditions. It is the responsibility of the contractor to see that the welds are made up to standard.

12–20. Ideally there would be only one set of codes. However, various bodies have seen the need for codes and have gone ahead and developed them. There is now greater cooperation between societies in setting up codes.

12–21. The prefix letter for ferrous metals is A, and for nonferrous, is B.

12–22. USA Standard Z49.1.

12–23. The underwriters laboratories are sponsored by the American Insurance Association. It provides the insurance companies with information on which to base their estimates for life, fire, and casualty hazards.

12–24. More leeway was given to the use of GMA, flux-cored wire, and electro-gas and electro-slag welding.

12–25. (a)

ASTM: A27, Class 60–30.
SAE: Automotive Grade 0030.
Federal: QQ–S–681d, Class 65–35.
Military: Mil–S–15083B, Class B.

(NOTE: There may be other answers—this is only an example of different types of specifications listed.)
(b) Metals.

12–26. No. He must become certified again with the new employer unless there is an agreed to inter-change between the contractors who use the same qualifying procedure.

12–27. The National Institute of Occupational Safety and Health (NIOSH).

12–28. He must reply to the Secretary of Labor's office within 15 days or lose any chance of a future review by the court.

12–29. (a) Any employee who believes he is working under conditions in violation of health and safety that may cause him serious physical harm can request an investigation.
(b) No, the complaint may not merit an investigation. An explanation of the reasons for not making an investigation will be written out and given to the employee.

12–30. The American National Standards Institute (ANSI).

CHAPTER 13 ANSWERS TO PROBLEMS AND QUESTIONS

13–1. Given: No. of pieces = 25. Thickness = 2 in. and 60 in./cut. Operating factor = 40%. Labor cost = $3.50/h. Overhead = 100% of the base labor. Oxygen cost = $0.02/cu ft. Acetylene cost = $0.03/cu ft.

Using Table 13.10, for 2 in. thickness, we have: Cutting speed = 10 ipm (average). Oxygen = 160 cu ft/hr (average). Acetylene = 12 cu ft/hr (average). Time/cost = (60 in./10 ipm) = 6 minutes. Oxygen required/min = 160/60 = 2.66 cu ft/min. Acetylene required/min = 12/60 = 0.20 cu ft/min.

$$\therefore Gas\ cost/cut = (2.66 \times 6) \times 0.02$$
$$+ (0.2 \times 6 \times 0.03)$$
$$= 0.32 + 0.036$$
$$= 0.356 = 36¢$$

Using nomograph of Fig. 13–10, labor cost = $0.18/ft, and labor cost/cut = 0.18 × 60/12 = $0.90/cut.

$\therefore$ *Overhead cost/cut* = $0.90

$\therefore$ *Total cost/cut* = 0.90 + 0.90 + 0.36 = $2.16

$\therefore$ *Total cost of cutting 25 pieces* = 25 × 2.16
$$= \$54.00$$

13–2. (a)

(b) 22.94 lb = wt of weld deposit. 22.94/55 = 41.75 lb or 42 lb = electrode weight.
(c) Total cost = material cost + labor cost + overhead. Material cost = 42 lb × 0.45 = $18.90. Labor cost = $3.50/h × 527.8 h = $30.80. Overhead cost = $30.80.

$\therefore$ *Total cost*
$$= \$30.80 + \$30.80 + \$18.90 = \$80.50.$$

(d) Overwelding and poor fitup.
(e) By inspection prior to welding to see if the fitup is normal or if additional metal will have to be added.
(f) Assume about half of the welds made in the downhand position. Thus a savings of about 100% could be made on the remaining time. (8.8/2) = (4.4/2) = 2.2 h saved.

13–3. This question will provide a basis for discussing the merits of skip welding. In actual practice this part was made both ways, first by intermittent welding, then by solid welding. By welding straight through, the time increased 50%, but in skip welding the operating factor was much

Part No.	Part Name	No. Req	Estimated Weight of Parts	Measured Joining Welds	Estimated Weight of Deposited Weld Metal	Estimated Time (from tables or calculated)
1	Cover top	1	261 lb			
2	Cover end	2	136	7.16 ¼ ▷ @.106# 1 ft ½ @ 425# 1 ft	0.76 lb 3.04	19.2 min 65.7
3	Cover side	2	144	11.33 ¼ ▷ ½	1.21 1.83	30.5 104.2
4	End flange	2	104	7.16 ¼ ▷ ½	0.76 3.04	19.2 65.7
5	Side flange	2	96	7.5 ¼ ▷ ½		20.0 67.9
				1.0 ½ @.425# 1 ft ¾ @.957# 1 ft	0.43 0.96	9.1 18.0
6	Top bar	1	5	3.6 ¼ @.106# ft	0.38 0.38	10.0 10.0
7	Side bar	2	2	20 15 ¼	0.21 0.16	5.3 4.0
8	Rib	1	21	7.5 (or 2.66′ ⅜▽ ⅜▽₂₋₆ X 1.5 factor)	0.65	25.5
				.25 ⅜▽ @.239# 1 ft	0.06	1.5
9	Gusset	4	4	5.0 ⅜▽	1.20	29.9
10	Lift lug	2	18	3.2 ⅜▽	0.77	19.4
11	1-1/2″ Pipe nip.	1	2	10 ¼▽	0.106	2.7
Total Weight			793 lb		22.94 lb	527.8 min
Total Weight - Assembly		815.94 lb		Estimated Welding Time		8.8 hr

Figure A13–2.

lower due to the delay caused by stops and starts and by marking or judging weld locations.

13–4. The total weight of the assembly is 815.94 lb (problem 13–2). The lift lug lap area = $4 \times 5\frac{1}{2}$ in. = 22 in.^{2} × 2 = 44 in.^{2}.

$$S = \frac{L}{A} = \frac{816}{44} = 18.4 \text{ psi}$$

The lap shear strength of epoxy adhesive = 2,500 psi, which is much more than required. However, if lifted straight up the joint will be subjected to peel.

13–5. (a) Referring to Fig. 13–9 (joint preparation) for automatic and semiautomatic submerged arc welding for more than 1 in. thickness, the following illustrated joint is recommended:

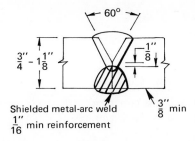

Shielded metal-arc weld
$\frac{1}{16}''$ min reinforcement
$\frac{3}{8}''$ min

FIGURE A13–5.

(b) Power rolls, beam type fixture, and spud type fixture (for pipe fitting).

13–6. Given: No. of pcs = 25. Thickness = 2 in. and 60 in./cut. Operating factor = 40%. Labor cost = $3.50/h. Overhead cost = 100% of base labor rate. Oxygen cost = $0.02/cu ft. Hydrogen cost = $0.05/cu ft.

From Table 13.11 for 2 in. thickness: Speed = 20 ipm (average). N_2 = 180 cu ft/h = 3 cu ft/min. H_2 = 10 cu ft/h = 1/6 cu ft/min. Shield N_2 = 100 cu ft/h = 10/6 cu ft/min. Shield O_2 = 400 cu ft/h = 40/6 cu ft/min.

$$\therefore Time/cut = 60/20 = 3\,min$$

$$\therefore Gas\ cost/cut = \left[\left(3 + \frac{10}{6}\right)3 \times 0.04\right]$$
$$+ \left[\frac{40}{6} \times 3 \times 0.02\right]$$
$$+ [1/6 \times 3 \times 0.05]$$
$$= 0.56 + 0.40 + 0.025$$
$$= \$0.985/cut$$

Using nomograph of Fig. 13–10, labor cost = 0.09 ft.

$$\therefore labor\ cost/cut = \$0.09 \times 5 = \$0.45/cut$$

Overhead cost/cut = $0.45.

$$\therefore Total\ cost/cut = 0.985 + 0.45 + 0.45 = \$1.885$$
$$\therefore Total\ cost\ for\ 25\ pieces = 25 \times \$1.885$$
$$= \$47.12.$$

13–7. For part no. 4, end flange: Length of weld = 7.16 ft. size of weld = $\frac{1}{2}$ in., and overwelded size =

5/8 in. Using Table 13.6, $\frac{1}{2}$ in. fillet weld needs = 0.760 lb/ft (electrode).

$$\therefore Total\ weld\ deposit = 0.760 \times 7.16$$
$$= 5.45\,lb\ of\ electrode$$

5/8 in. fillet weld needs = 1.185 lb.

$$\therefore Total\ weld\ deposit = 1.185 \times 7.16$$
$$= 8.50\,lb$$

$$\therefore Extra\ weld\ deposit = 8.50 - 5.45$$
$$= 3.05\,lb\ of\ electrode$$

Cost of electrode 1 lb = $0.45.

$$\therefore Extra\ cost = 0.45 \times 3.05 = \$1.36$$

For time calculation, extra time required = $(95 - 66)/60 = 29/60$ h. Extra labor and overhead = $7.00 \times 29/60 = $3.38. Total cost incurred for overweld = $3.38 + $1.36 = $4.74.

13–8.
$$C_a = \frac{C_1\left(1 + \dfrac{R_o}{100}\right) + C_f}{P_a}$$

C_1 = labor cost $3.00 × 8 = $24.00. R_o = overhead rate 100% = $48.00. C_f = fastener cost/assembly = .0163 × 24 = 0.39. P_a = assemblies/hr = 250

$$= \frac{24\left(1 + \dfrac{100}{100}\right) + 0.39}{250}$$

$$= \$.193$$

Therefore, the cost/assembly for 24 rivets in place is ≈ 20¢.

13–9. (a) *No. of columns* = 10.
Length of column = 20 ft.
Total length of welding = 4 × 20 = 80 ft.
Weld size = 3/4 in. fillet weld.
From Table 13.6, for 3/4 in. fillet weld (pounds of weld deposit/ft = 0.955 lb/ft).

1. Semiautomatic submerged arc process: ∴ Total amount of weld deposit required—0.955 × 80 = 76.40 lb. From Table 13.7, assuming 60% operating factor for semiautomatic submerged arc: cost/lb = $1.00. Total cost of welding one beam = 76.40 × 1.00 = $76.40/beam. For 10 beams the cost would be $76.40 × 10 = $764.00.

2. Manual arc welding: From Table 13.7, assuming E6010 type 5/32 in. electrode, 40 % operating factor, cost of weld deposit/lb = \$4.70.

$$\therefore \; Total\; cost\; of\; welding = 76.40\,lb \times \$4.70$$

$$= \$358\; for\; one\; beam$$

$$\therefore \; Total\; cost\; of\; welding\; 10\; beams = \$3,580.00$$

(b) Cost of fixture = \$4,500.00. Cost of welding beam with fixture (semiautomatic submerged arc) = \$76.40. Cost of welding beam without fixture = \$358.00.

$$\therefore \; No.\; of\; beams = \frac{\$4,500}{\$358.00 - \$76.40} = \frac{\$4,500}{\$281.60}$$

$$= 15.98 \approx 16$$

Therefore, about 16 beams have to be welded before the welding fixture is amortized. This, of course, is a simple calculation and does not take into account interest on the investment or depreciation on the equipment.

13–10. *Labor*:

$$L = \frac{W}{55C} = \frac{4}{5(15 \times .60)} = \$0.088/ft\; of\; weld$$

Overhead:

$$O = \frac{2W}{S5C} = \frac{8}{5(15 \times .60)} = \$0.177/ft\; of\; weld$$

Gas:

$$G = \frac{RP}{S5} = \frac{25 \times .01}{5 \times 15} = \$0.003/ft\; of\; weld$$

Power:

$$V = \frac{(EI)K}{5000\,SM} = \frac{0.2 \times 20 \times 140}{5000 \times 15 \times .50}$$

$$= \$0.002/ft\; of\; weld$$

Electrode:

$$T = \frac{DB}{A} = \frac{0.30 \times 0.061}{.95} = \$0.019/ft\; of\; weld$$

Total = \$0.289/ft of weld.

To solve for D:

$$D = \frac{280 \times 60}{5 \times 3650 \times 15} = 0.061$$

NOTE: Value 3650 is taken from Table 13.5.

Therefore, the cost of a 500 ft, 1/8 in. mild steel fillet weld by the Mig process is \$114.50.

13–11. Stick Electrode Cost (Fig. 13–9): Time = time/ft of weld = 6 in./min = 0.5 ft/min, 30/0.5 = 60 min total time or 1.0 h. Cost = 1.0 × \$8.00 = $\underline{\$8.00}$.

GMA Cost: (Table 13.9): Time = 1 × 30 − 30 min total time, or 0.5 h. Cost = 0.5 × \$8.00 = $\underline{\$4.00}$.

Flux-Cored Cost: (10 in./min = 0.83 ft.). Time = 30/0.83 = 36 min, or 0.60 h. Cost = 0.60 × \$8.00 = $\underline{\$4.81}$.

13–12. Submerged Arc Cost (Fig. 13–9): Time = 18 in./min + 0.663 in./min = 2.16 ft/min, 30/2.16 = 13.88 min total time, or 13.88/60 = 0.23 h. Cost = 0.23 × \$8.00 = $\underline{\$1.83}$.

GMA cost: (12 in./min = 1 ft). Time = 0.01 × 30 = 0.30 min total time, or $\frac{1}{2}$ h. Cost = 0.5 × \$8.00 = $\underline{\$4.00}$.

13–13. Gas Weld (Table 13.8): 11 to 13 ft/h. Acetylene/h = 8.5 cu ft. Oxygen/h = 8.8 cu ft.

Gas time = 1/12 h, or 5 min. Gas used/ft = (8.5/12) − 0.7 cu ft acetylene, (8.8/12) − 0.73 cu ft oxygen.

Gas cost/ft = acetylene = 0.065 × 0.7 = 0.045 oxygen = 0.03 × 0.73 = $\underline{0.021}$ Total gas cost/ft = 0.066.

Labor cost/ft = \$6.00/12 = \$0.50. Total cost = 0.066 + 0.50 = \$0.566.

Arc Welding: Electrode lb/ft = 0.21 (Table 13.6). Approximate cost/lb deposited for 1/8 in. electrode (Table 13.7) = \$4.38. Cost of deposited electrode/ft = \$4.38/21 = \$0.21.

The cost of making a butt weld with no spacing by gas welding is \$0.566/ft as compared to \$0.21/ft by arc welding.

13–14. Length of cut required = $\pi D \times 10 = 3.14 \times 12 \times 10 = 377$ in. The average cutting speed = 21 in./min. Cutting time = 377/21 = 18.8 min.

Oxygen = 42 cu ft/h (av.),

$$42 \times \frac{18.8}{60} = 13 \, \text{cu ft oxygen}$$

Acetylene consumption = 6 cu ft/hr (av.),

$$6 \times \frac{18.8}{60} = 1.8 \, \text{cu ft acetylene}$$

Propane consumption = 9 cu ft (av.).

$$9 \times (18.8/60) = 2.82 \, \text{cu ft propane}$$

Oxygen cost = 13 × 0.03 = 0.39   }
Acetylene cost = 1.8 × 0.06 = 0.11 } = \$0.50

Propane cost = 2.82 × 0.017 = 0.047 }
Oxygen cost = 13 × 0.03 = 0.39   } = \$0.437

Index